The Habit of Turning the World Upside Down

Books by Howard Mansfield

Cosmopolis

In the Memory House

Skylark

The Same Ax, Twice

The Bones of the Earth

Turn & Jump

Dwelling in Possibility

Sheds

Summer Over Autumn

Editor

Where the Mountain Stands Alone

For Children

Hogwood Steps Out

THE
HABIT
OF
TURNING
THE
WORLD
UPSIDE
DOWN

HOWARD MANSFIELD

Bauhan Publishing · Peterborough · New Hampshire
2018

Library of Congress Cataloging-in-Publication Data
Names: Mansfield, Howard, author.Title: The habit of turning the world
upside down / Howard Mansfield.
Description: Peterborough, NH : Bauhan Publishing, [2018] | Includes
bibliographical references.
Identifiers: LCCN 2018027609 (print) | LCCN 2018033553 (ebook) | ISBN
9780872332713 (ebook) | ISBN 9780872332706 (softcover : alk. paper)
Subjects: LCSH: Land use—United States—Planning—Citizen participation.
| Community life—United States. | Community development—United
States.
Classification: LCC HD205 (ebook) | LCC HD205 .M3535 2018 (print) |
DDC 333.330973—dc23
LC record available at https://lccn.loc.gov/2018027609

Book design by Sarah Bauhan.
Cover design by Henry James.
Author photograph ©Annie Card Creative Services

BAUHAN
PUBLISHING LLC
PO BOX 117 PETERBOROUGH NEW HAMPSHIRE 03458
603-567-4430
WWW.BAUHANPUBLISHING.COM

Printed in the United States of America

Once again, and always, for Dr. B. A. Millmoss

CONTENTS

It would seem that the habit of changing place,
of turning things upside down, of cutting, of destroying,
has become a necessity of [the American's] existence.

—Alexis de Tocqueville, noting his conversation with Joel Poinsett, 1831

If summer were spring and the other way round,
Then all the world would be upside down.

—*The World Turned Upside Down*, an old English ballad

INTRODUCTION

I N A HIGH PASTURE IN NORTHERNMOST NEW HAMPSHIRE I stood next to a farmer looking out on mountains for as far as we could see, mountains that were blue shadows in the distance, and beyond that, grey shadows. This was an exalted view. Seeing it in this moment made me happy; but seeing this every day from the time you could first see would shape you in ways you might not be able to express. This farmer loved this land. A big hydroelectric power project was pushing down from Canada headed south to busy towns and cities and they wanted to go right across his land and his neighbors' with high-voltage transmission towers as tall as a fourteen-story building. "All I ask for," he told me, is "just to get by and hold on to this property that's been in the family. . . . There's a lot of history here. You start wrecking it and the history goes with the wreckage." And when the big company kept coming back and back offering him more and more money—money that he, a dairy farmer, could have used—he told them, "My roots are deeper than your pockets." He said no to four million dollars. Some of his neighbors were less sure; some, like his cousins next door, sold out for much less. Old friendships were destroyed.

This book began in that pasture and it began at a kitchen table in a small Massachusetts town as a man, an engineer who studies ocean acoustics, read to me from his daughter's ninth grade civics notebook all about how in a democracy "the people hold sovereign power" and how "the will of the people translates directly into public policy." That is "the definition of a democracy." His daughter had brought her notes to him and asked: *Dad, is this true? Is this the way America works?* Because it wasn't working that way around her house. One morning her father had walked out the door on his way to work and found a man standing in his driveway who had big news: a major interstate pipeline carrying natural gas more than a hundred miles from Pennsylvania is going to cross your land and tear up that place where you play soccer with your kids and fly kites. It will sit just two hundred feet from your house. He had a big map and legal forms to be signed. His message was blunt: this is going to happen. We are coming for your land and by law we can take it from you. This wasn't what they were teaching in civics class at Bishop Guertin High School.

And this book began in a large meeting hall with rows of tables covered in big floppy property maps, and company men and women standing behind the tables with a well-rehearsed neutral stare and a practiced friendliness. They were representing the company that wanted to build the natural gas pipeline. They wore bright blue shirts, crisp with the folds from the box they had arrived in, and their name tags hung down on red lanyards. The civilians entered the room and tried to find out if their land was somewhere on those maps. The company had all the information, and again the message was blunt: this is going to happen. The route may change here or there, but we are your new neighbors. The deal is done, they implied. We've sold you this car. We can negotiate about the options and the down payment, but the deal is done. This, too, wasn't the way things were supposed to work if you followed the ninth grader's civics notes.

I sought out these confrontations. They were painful to see and to report. Two different lines had been drawn hundreds of miles across the land by two different companies, and what followed was the usual collision of technical information and claims that they were acting in the public interest. The debate moved over many different circuits simultaneously: Was the gas or electricity really needed? What assumptions lay behind the projected need? And above all: who gets to decide? I was especially interested in the emotional toll these projects took. These projects got under the skin. This was about more than kilowatts, power lines, and pipelines. Something in this upheaval felt familiar. I began to realize that I was witnessing an essential American experience: the world turned upside down. And it all turned on one word: property.

~

American property, as we shall see, is always in motion. The world spins; the land under our feet is set spinning by what the law has decided, case by case, over hundreds of years, to be the "highest and best use" of property. The law favors active use and big plans. That deed you may hold is subject to further negotiation by parties unknown to you; it's subject to being called in, even nullified. The "quiet citizen must keep out of the way of the exuberantly active one," is how it was defined in the nineteenth century. Development has the right-of-way.

And yet property is our anchor and our North Star. We take it as a certainty. It's our liberty, our pursuit of happiness, as the nation's founders said. The battles going on around the country today about where to build pipelines, transmission towers, and windmill farms show that our rock-solid certainty is uncertain. That's part of the anguish in each conflict.

The stories in this book are about our belief in property and the cost of that belief. That New Hampshire farmer with his deep roots is part of the story, as is that one Massachusetts family defending their two acres. In other chapters we'll walk the borderlands of the Sonoran Desert, see New Hampshire's White Mountains when they were a clear-cut and smoldering ruin, uncover the buried history of an interstate highway that crushed a Vermont farmer, and visit a very small island off the coast of Maine to try to face the rising seas brought on by the changing climate. Our haste to unsettle the world has led to a warming planet. This is the cost of our property faith, one which we have yet to tally.

Our heritage, as Tocqueville observed, is one of "turning things upside down." This book presents just a few stories of Americans living in a time in which everything is in motion, in which the world will be turned upside down, again and again.

I

Sacred Property

After a bloody war in defense of property,
they forgot that property is sacred.

—John Adams, 1789

DESERT TEACHINGS

I N THE SPRING OF 2000 I WAS WALKING in the Sonoran Desert
with the Tohono O'odham. In school, when we were taught about
the Indians in the third grade, they were one of those peripheral names
spilled out over the blank spaces of the wall map: the Papago Indians.
Papago means "bean eaters." While this does allude to a key characteristic of their desert agriculture, it would be like calling all white people
"the Doritos eaters." We were taught about their charismatic neighbors
to the north, the Pueblo Indians. We built pueblos for a class project
(brown grocery-bag paper over shoe boxes, wire for ladders). As for
the Papago's land, that was vague. There were no boundaries or landmarks. All those tribal names floated above the continent like passing
clouds.

I met the O'odham at the Mexican border and walked with them
from village to village on their large reservation. The Arizona-Sonora
Desert Museum had organized the walk to encourage the revival of
the O'odham's traditional diet. The O'odham had started walking 145
miles south of the border at the Sea of Cortez, the home of their ancient trading partners, the Comcaac, who had joined them. They had
ninety miles more to walk. I was one of four journalists. There was
a gardening columnist from the *New York Times*, a reporter from a
newspaper covering the West, and a freelance Public Radio reporter.
The O'odham were wary of us. I'd catch up with a few, walk awhile,
talk, and in the manner of cross-country walking, we'd go our separate
ways. Miles later I'd join strides with another small group and we'd
talk, or not. I had long hours walking alone sometimes to consider
what they had told me. The flatness of the desert was punishing. You
could see for miles, mocking any notion of forward movement. To
walk fifteen, twenty, twenty-five miles in a day was like standing still.

I'd close my eyes before sleep and see the day's walk—the unmoving desert—blazing before me.

The Comcaac were always far ahead, sometimes miles ahead, eight young men, ages thirteen to twenty-seven, walking at a clip in a tight pack. They were thin, seasoned walkers, who back home had walked forty-one miles in a day carrying their own provisions (including a gallon of water). The Comcaac, also known as the Seri, were among the world's last hunter-gatherers. They jack-rabbited out of the village before us and finished the day's walk hours before we finally arrived. Usually they'd be dancing.

Occasionally, I walked with some of the museum staff, learning the names for different cactuses and the meaning of the dry washes. This dry land is shaped by water. The rain, in some years just three inches, in others a still-slim sixteen inches, arrives in a flood and moves fast. The dry gulches fill with racing, turbulent water. For 10,000 years the O'odham—"the Desert People"—had fed themselves from these sudden rains. Coming from New Hampshire's folded hills, twisting roads, and bounty of water—*Long Pond, First Pond, New Pond . . .*—everything about the desert was astonishing. I didn't know if some twisted brown plant was dead, or just waiting for water. I didn't know if it was a year old or hundreds of years old. Distances were hard to gauge. Without the maples, beeches, and white pines of home, it was as if someone had thrown away the rules of perspective.

Walking along on the first day, trying to understand just where I was, it dawned on me that I was walking in the Gadsden Purchase. *People lived in the Gadsden Purchase.* That was another wall map, fourth grade, I think. Rolled down in the front of the class, west of the big lamb chop of the Louisiana Purchase, west of Texas, was the relatively small Gadsden Purchase—an empty space. No native names on that map as I recall.

A map of Arizona isn't Arizona; a map isn't the place. Property isn't land. Property is the legal right to draw a boundary and exclude others. The most important part of those maps that piece together the march of Manifest Destiny is the boundaries of each added territory. The different tribes, as presented on the maps, have no defined boundaries, and therefore no property. They lived off the land; but they had no right to it. They could be removed.

Our nation is founded on property rights. Our nation is founded, in part, on the distinction between property and land, and this is a source of much sorrow. Property—claiming, owning—comes first. All the "amber waves of grain" talk comes later. We are founded on treating one group of people as slaves, as property, and denying property to another group, the native peoples, because, as John Winthrop wrote in 1628, "they enclose no land, neither have any settled habitation, nor any tame cattle to improve the land by." Throughout the nineteenth century the law redefined Indians until they were "occupants" but not owners of their own lands.

On the desert walk, there was a young Comcaac, Francisco Molina, a twenty-seven-year-old long-distance runner with waist-length, shiny black hair. He was a leader of walks back home. He went by the grand name of El Indio, but he was wise. Speaking through a translator, he said, "For us the desert is a sacred person and that's why we don't want to damage it. We draw upon its resources like we draw upon the resources of a friend." The ironwood that the Comcaac depended upon for food, medicine, carving wood, and slow-burning fuel had been plundered to make fake "mesquite" charcoals and souvenir knock-offs of their own carvings. As we neared sprawling Tucson, walking close by roads busy with long motorhomes towing cars—so close we'd be knocked aside in their wake—he asked that other people in the region, we non-Indians, "think of the desert itself as a sacred person."

But that kind of thinking is not on the maps. Securing property is the usual business of maps. Property was crucial to our nation's founders. It was the guardian of all the other rights, they said. Property was liberty; property was the pursuit of happiness. Property was the making of citizens—without property a white male couldn't vote. The income from property would free them to serve in government; property, widely owned, would free the young republic from the dismal feudal past. Property had to be protected. John Adams complained that the newly freed colonies had impeded commerce. "After a bloody war in defense of property," Adams said, "they forgot that property was sacred."

~

I walked alone for miles with my ignorance, thinking about the blank map of the Gadsden Purchase. A fool's moment, but a representative moment of blindness. We are as ignorant about the land we live on. We have the same blindness about New England, Florida, the Mississippi Valley, California . . . about the land we love, and the land we pave for commerce. We don't see the native history around us, and that absence shapes the stories we tell ourselves. The untold stories about native cultures are like a shadow star. Astronomers sometimes observe visible light that is deflected by the gravity of an unseen object—a planet, a dark star. The native presence is the unseen object that shapes our entire history in America.

This desert was someone's holy land, and someone else's land deal, their property. In an empire built by war and chance, the Gadsden Purchase was a small land deal, a footnote to Manifest Destiny. The treaty of Guadalupe Hidalgo, which ended the Mexican War in 1848, gave the United States about half of Mexico's land. The new border was to be drawn following, in part, the Rio Grande, Gila, and Colorado Rivers. When surveying parties from the two countries met to mark the border, they discovered that the treaty had relied on an inaccurate map. (It was plagiarized from a twenty-year-old map, which itself was a stolen version of an earlier map.) One key place on the Rio Grande, El Paso, was thirty-four miles south and over one hundred miles west from where the map imagined. The two surveying parties proposed different boundaries, with each claiming the 6,000-square-mile Mesilla Valley. Mexico sent troops to Mesilla. The governor of New Mexico led a volunteer force to a village on the contested border where he issued a proclamation claiming the valley. A second war was a possibility.

The United States government thought the valley was a worthless desert full of marauding "savages," but coveted it as the best route for a southern transcontinental railroad. The Mesilla Valley was "the great gateway to the Pacific." To prevent a war, President Franklin Pierce appointed James Gadsden to settle the disputed border by treaty. Gadsden was a southern railroad promoter who dreamt of uniting the southern railroads with the Pacific. At the time the transcontinental railroad project was deadlocked in Congress as each section of the country touted itself as the best route. Gadsden's hunger for "an air line railroad," a straight and flat route, to San Diego was the "secret history" of the treaty, said the *New York Herald*. The Gadsden Purchase in 1854, the last addition to the borders of the continental United States, was shaped by the desire for a railroad empire.

The Gadsden Purchase was hated in both countries. President Pierce had sent Gadsden instructions to bargain for five different borders, with the preferred border claiming the most Mexican land. Gadsden was rude and bullying. If the Mexican dictator Antonio López de Santa Anna refused to sell the land, "imperious necessity would compel [the United States] to occupy it one way or another," he said. Santa Anna was desperate; he needed money to stay in power. The Mexican treasury was in debt, and there was a revolution in one Mexican state. He saw the treaty as the best alternative to a war and the loss of even more territory. Gadsden returned with the least amount of land, disappointing the President. The treaty was negotiated in secret, and ratified in secret by the US Senate, though the entire text and much of the debate leaked out. The Senate defeated the treaty on the first vote, and then rewrote key provisions, reducing the territory and purchase price. Santa Anna was again insulted, but signed the treaty. Mexicans were enraged; Santa Anna was overthrown a year later.

The Gadsden Purchase added 30,000 square miles to the United States—at the "southern end of the liver," as one newspaper said. On the maps it is the flap of land that completes the border of Arizona and New Mexico. (The first US settlers proposed a separate territory, Gadsdonia.) As presented in history books, the Gadsden Purchase is American space—an emptiness awaiting our ingenuity, an obstacle on the way to somewhere else. It was nothing, "so utterly desolate, desert and God-forsaken that . . . a wolf could not make his living upon it," in Kit Carson's frequently quoted judgment. Gadsdonia was an empty, hostile place.

The new border cut the O'odham's world, the Sonoran Desert, in two. There are O'odham living in Mexico. After the border was set, the O'odham continued their travels, some not knowing that this same desert was now in different countries. They continued with their ancient pilgrimage to the ocean, which they considered to be a sacred person. They would trade with the Comcaac, sometimes intermarrying, as they had done for thousands of years. Then right after World War II, with an outbreak of hoof-and-mouth disease, the government killed thousands of head of livestock and prohibited the O'odham from making their ocean pilgrimage on horseback. They lost contact with the ocean, the Comcaac, and part of their aboriginal home. The once permeable border is today an armed and dangerous place. The white helicopters of the Border Patrol chasing "illegals" and drug smugglers are a constant presence.

The O'odham knew the route by a sequence of songs that told of a series of mountain peaks marking the way, including the sacred Baboquivari Peak where the creator, *I'itoi*, dwelled in a cave. The cone-shaped Baboquivari, a signpost on the landscape visible for miles, is the O'odham's spiritual center. This desert walk was a way to return to the old routes and the old songs. Only two of the O'odham walking knew the songs: Danny Lopez and his twenty-seven-year-old student Tony Johnson, who had turned down Harvard Divinity School to return home and learn the old ways. Lopez, at sixty-three the oldest walker, had been waiting a long time for a student with Johnson's promise.

"We were at a point one day when we were walking through the desert and it was just hot and we were tired—tired of walking, and wanted to get home. We were both trying to keep each other's morale up. It was just too much," said Johnson. "There was a small mountain, and when we came around it, we could see the top of Baboquivari Peak. It was the best feeling just seeing it to know that we were home." He couldn't explain that feeling, he told me. "But we explained in the song that *I'itoi* gave us about our home." They sang their way home. There was, of course, no sacred peak on those plagiarized maps.

For 10,000 years the O'odham had practiced a flash-flood agriculture, cultivating a desert that the newest landlords said couldn't support a wolf. They planted white tepary beans, Papago peas, Spanish watermelons, squash, corn, and muskmelons. They gathered wild plants: fruit from saguaro and organ pipe cactus, prickly pear, mesquite pods, ironwood and plantain seeds, broomrape stalks, amaranth, and wild chilies. For meat they hunted desert tortoises, packrats, bighorn sheep, mule deer, doves, quail, and jackrabbits. They wove watertight baskets from bear grass, cattail, willow, devil's claw, and yucca.

In modern times the old ways lost out to the "mini-mart diet" of white bread, soda, and sweet snacks, and this has led to serious health problems for the O'odham, and increasingly the Comcaac. The O'odham have the highest rate of Type 2 or adult-onset diabetes in the world. (The disease is epidemic among Native Americans.) Diabetes was virtually unknown until the native diet was abandoned between 1890 and 1940. The native diet was high in soluble fibers and pectins with only 15 percent fat. Today, fat makes up about 40 percent of the O'odham's diet. Half of all O'odham adults have diabetes.

White bread was destroying the culture, the O'odham's *Himdag*, the ancient wisdom of the "desert peoples' way." In the last few years, there had been a small revival as fifty families, at first, planted traditional crops and picked the fruit of the saguaro cactus. The families gathered to "sing down the rain," reviving the traditional rain ceremony for the first time in twenty-five years. A few songs had lingered, but without the crops, the songs had lost their force and were on their way to becoming artifact. "Most of the planting songs are gone," said the teacher Danny Lopez. Few of the young speak O'odham. The language, traditions, and ceremonies face extinction. Their map, too, was increasingly blank. To plan the desert walk they had to ask the elders because no one else knew where the old dirt tracks ran.

The O'odham, having lived successfully in the Sonoran Desert for thousands of years, now find themselves at this border: They have the lowest per capita income of all US reservations ($3,113), 70 percent unemployment, and the highest school drop-out rate of all US Native American tribes—fewer than half of the adults have completed high school. The homicide rate is three times the national average and twice that of all Native American communities.

The Comcaac are facing similar challenges. There are only 650 Comcaac, and though 90 percent speak their language, it too is faced with extinction. El Indio, the long-distance runner, was working to use modern ways to help his people. "I would like to see a future," he said, "in which everything—traditional and outside knowledge—fits together perfectly."

I was visiting two deserts: the Sonoran Desert of the O'odham, and the freeway, go-go, cable-view desert. I was in the airlock of the airport, car rental, motel, franchise world. The kerosened air, the motel-perfumed air, the rental car smell—a plasticky mix of a flower that never was, a pine that never was. The dense block of warnings you sign to rent a car, a room, an airplane seat. You waive liability, you waive your existence. It all comes at you like a murmur, and you take this instead of breezes, home life, and the softer world. You accept, then welcome, surveillance, warnings, and the murmuring of voices telling you, selling you, warning you.

Phoenix and Tucson are raceways. Each driver knows the finish line—the race is to the wealthy. Money is all—it is sex, beauty, time, things and more things. So drink down the Colorado River. Fill the desert with green lawns and a thick brown haze. Never leave the bubble. Turn up the AC, HBO, MP3, and go, go, go. Let sprawling Phoenix light up the night, all of it like shards of broken glass, a smashed view of the world whose glitter we mistake for something of value.

The O'odham live on the collision line of these two deserts, at the storm line. Chant and story, elder and kinship, hitting up against alcohol and drugs, TV and the beautiful people, money and having and getting. And when the storm breaks there's alcoholism, diabetes, poverty. The white cross markers along the reservation's laser-straight roads, the litter of quart bottles of Budweiser, and the men wandering around doing nothing, drunk or coming off a drunk, or getting drunk.

After a few days of walking and talking with the O'odham, I knew just the barest outlines of their suffering. Along with the Desert Museum and the tribe's community groups, I was rooting for the strength of the old ways, the other side of the storm front. The side that says "in spite of," "and yet," "and still." The language remains; the speakers dwindle. I wanted to believe that even if it came down to one O'odham speaking the old stories, that would be force enough to stand against the storm. One witness. And one night I did get to hear that witness.

Near the walk's end, Danny Lopez's student, Tony Johnson, spoke powerfully after dinner. When I had talked with him as we walked, he was thoughtful and reserved, but on this night he spoke as a leader. He was a commanding figure. He filled the village's feast hall with his story. He could be anything, I thought. He spoke to make up for all the sorrow and anger. He cleared that out of the hall, out of the village, for that moment with his testimony.

"We started this walk ten days ago down at the ocean. Started out, walked all the way. I was afraid that we weren't going to make it. But we did. It wasn't because of any strength that I had or it wasn't because of anything I did or said, but it was because of the people here with me. The people here. The O'odham people. That's why I walked, because I believed that we're not dead," he said to applause.

"We're not dead. People say we're dead because we're diabetic. They say we're dead because we're alcoholics. Because we don't do this or we don't do that. But I haven't seen a dead O'odham yet walking from Mexico to here." Johnson had transformed the hall; there was relief, a lightness that comes from being freed from daily worries. "They're alive. They may look dead, but they're alive," he said to laughter and more applause.

"And all you O'odham kids out there, don't ever let them tell you that you can't make it. All these people that have walked, all these people that have helped, all these people that are here to support you: They're the ones that are going to bring you to places that you never thought.

"When we were so far away, and we missed home, and we didn't know where we were going, and we didn't know where the next road was going to lead to, the only thing that was familiar to us, the only thing that made us feel at home was our songs, was our *Himdag*. That no matter wherever we were, wherever you go in this world—you can go to the other side of the world—but if you have that in your heart, you know your songs and you know your language, no matter where you go, you'll always be home, no matter where you are. . . . That's where your strength is. And that's what's going to keep you going to make it over those humps in those hills, those dirt roads, those cliffs, those valleys. Most of all I thank our creator, because if it wasn't for him, we wouldn't be here today."

We had walked for days on the reservation, which taken together is almost as big as Connecticut. (Back home I'd measure out Connecticut on my wall map with the first joint of my thumb and then move my thumb to the bottom of Arizona. It made no sense.) Each day another community hosted, and joined the walkers.

At day's end the O'odham stood in front of that community's feast house in a receiving line. Each footsore walker was welcomed with songs, blessings, holy water, confetti, a hug, and a thank you from each person in the line. Perfect hospitality. The O'odham's welcome could melt an iceberg.

The last sun slanted down into the dirt parking lot of Our Lady of Lourdes, Little Tucson, as a line of thirty or so welcomed the latest walkers. When I had arrived earlier, four O'odham came over in a small receiving line and shook my hand, saying "Thank you."

The receiving line continued to grow. This was one of the most gracious welcoming ceremonies I had ever seen. Each night there were feasts of traditional desert food. Little Tucson (Ali Cukson) laid out a festive meal. The O'odham filled up the long tables in the ramada's outdoor kitchen. Tables were added in the kitchen inside as more residents arrived. There were tables of little children, and tables of the elderly. All of Ali Cukson—population about 130—may have been here.

The O'odham served us, ladling out heaps of white beans and venison, red sauce and rabbit, red beans, squash, cholla cactus buds, a mashed blend of local corn, and the softest tortillas, which they cook over an open fire. They don't ask; they put everything on your plate—of course you'll take everything. When a second bowl is added, someone appears to carry it to your table. They bring over juice and mesquite cookies. On this evening in the adobe feast house, Little Tucson seemed serene.

After dinner each night, short speeches followed in English and O'odham, with Spanish translations for the Comcaac, and there was dancing. The Comcaac usually dance first. With a walking staff to steady themselves, they perform an elegant tap dance as an elder sings and shakes tin-can rattles with trailing ribbons. The feet tap out an answer to the song, the feet and the voices a duet. The Comcaac used to hunt sea turtles, and they used the great shells as their dance floor. On this trip they dance on an old piece of particle board. Later the museum will present them with a wooden replica of a sea turtle shell to dance upon.

The O'odham followed with an entirely different dance: a circle of the young around the elders who chant in the middle. As the circle grows larger, and moves to the chanting, the effect is almost hypnotic. On some nights the Comcaac join in, or share their own dance with an O'odham to great cheering. On this evening, the O'odham and Comcaac jested each other. An O'odham man with bells on his legs, much like a Morris dancer, began by dancing with an older woman as a fiddler played. The Comcaac answered him with their tap dance, and the man in bells stepped back in and gently mimicked their steps to much laughter. Soon they were mixing it up, a young Comcaac dancing in the O'odham manner with the elder woman. A circle dance began with a few, and hours later, swirled through most of the red dirt plaza in front of the church.

At the beginning of that evening's dancing, the O'odham and Comcaac had formed a prayer circle. At the center the elders were arguing about something. I didn't know what language they were speaking, but the tension in their expressions was universal. Our little press corps, now just two, stood off to the side, where we should be, as witnesses. The native peoples I have known, Onondagans and Abenaki among them, were plagued by the good intentions of white folks who thought that they, too, were some part native, some part Cherokee, or who hung up Dream Catchers around their house and believed that they could add the ways of the Lakota Sioux to their New Age collection. Or they arrived at the reservation on a mission to undo the centuries of missionary work, convinced that they could solve all the problems as if this tribe were another disease-of-the-week, a few reforms away from being cured. The natives I knew were also plagued by disbelief, by whites who thought that all the *real* Indians were dead, gone with the buffalo. How could a Wampanoag still be a Wampanoag—the natives who had helped the Pilgrims at Plymouth—and yet also be another guy stuck in traffic, worried about his retirement account, his mortgage, and his suddenly monosyllabic teenage son? This was some kind of trick to gain advantage—fishing rights, or a casino, or who knew? So the O'odham had good reasons to be wary of their visitors.

The argument concluded. The elders called for us non-Indians to come and stand in the center of the circle. At the still center as the dancers moved around us, we stood, as instructed, facing two elders, one from each tribe. They chanted, each in their own language, with their outstretched palms to Baboquivari Peak. It felt ancient; it felt familiar, the rabbi and the cantor facing the open tabernacle. Turn around, we were told. The full moon was just rising over the sacred mountains.

It was magic. I could have stood there forever, humbled by their generosity. We stepped aside; we had no equal gift. The chanting, drumming, and shaking rattles defined the coming night. Some lean reservation dogs threaded through the circle dancers. Dusty little boys kicked a soccer ball, stirring up more red dust in the twilight. The tall saguaro cactuses stood as sentinels against the distant, dark, shadowy, holy mountains. We were just fifty miles from Tucson and a million miles from sacred property.

LAND RUSH: A BRIEF HISTORY OF TURNING THE WORLD UPSIDE DOWN

1. Adventurers for Draining the Dismal Swamp

GEORGE WASHINGTON BOUGHT 40,000 ACRES of swamp in 1763. Washington, along with eleven others, formed a company, Adventurers for Draining the Dismal Swamp. They paid $20,000 and began to clear the cypress and cedar, the bears and snakes, to make the land they wanted to see. The Adventurers planned to farm the swamp. It was, said Washington, "a glorious paradise." That was an uncommon opinion.

The Great Dismal Swamp was a vast cypress swamp of more than 2,000 square miles crossed someplace by the border of Virginia and North Carolina. Thirty-five years before Washington and his Adventurers, William Byrd II, an eminent Virginian, led a survey crew into the swamp to set the state boundaries. To Byrd the swamp was a "miserable morass," inflicting nearby inhabitants with "agues and other distempers occasioned by the noxious vapors [that] rise from that vast extent of mire and nastiness." The survey work was slow, each footstep pooling with water, "the ground moist and trembling under our feet like a quagmire." They fought their way over windblown cypresses that were heaped upon each other with "sharp snags" growing on them "pointing every way like so many pikes" and pushed their way through reeds twelve feet tall. Sometimes they could only advance one mile, from morning 'til night. "Never was rum that cordial of life, found more necessary than it was in this dirty place," Byrd wrote.

But Byrd thought he could make something worthwhile from this "nastiness." He petitioned the king to drain the swamp. He wanted to grow hemp and dig canals. He planned to sell shares in his enterprise and put 300 slaves to work making "the Dismal good as any surrounding land."

His unfulfilled designs for the Dismal were only a small part of his ambition. Byrd added thousands of acres at a time to his inherited land, including a 105,000-acre acquisition. "As the years passed a veritable hunger for land possessed him," says one historian. "All told he increased the Byrd possessions from 26,231 acres to 179,440." In his land acquisition and public service, "Byrd was typical of the ruling class of Virginia."

Byrd's contemporary, Robert "King" Carter, is a representative man of the Virginia aristocracy. He piled up 300,000 acres, much of it by granting it to himself and his relations, and by fees and trades. He enslaved more than 750 people. Carter led his generation in acquisition. The lesson wasn't lost on young George Washington. "Men in these times, talk with as much facility of fifty, a hundred and even 500,000 Acres as a Gentleman formerly would do of 1000 acres," Washington later wrote. "In defiance of the proclamation of Congress, they roam over the Country on the Indian side of the Ohio, mark out lands, survey, and even settle them." This is our heritage of takings.

Property—land—is what will free the colonists from the old hierarchy, and the hereditary ways of England. City air may make free, as the medieval saying has it, but in America, it's land. What George Washington and his Adventurers were trying to do was make wild lands into property. This was the endeavor of the age.

Property is the air we breathe. We are immersed in it. We scarcely notice because we can't imagine an airless, property-less world. Our lungs would cease; we would cease. Property is how we have ordered the world. Property lines mark the limits of what we think is possible—it marks what we think about, and even how we think. "How we imagine property is how we imagine ourselves," writes the poet and critic Lewis Hyde. Property is one of the major works of our civilization; it's our pyramids and our Parthenon.

The first Europeans in America defined the Indians by what they didn't own—the Indians didn't own land they fenced off or cattle or church pews. Property rights propelled the Revolution; what right did the king have to deny his subjects in the colonies a say in how they were taxed? And who was he to say that they couldn't settle beyond the Appalachian Mountains? The Declaration of Independence is a list of a tenant's grievances delivered to his landlord.

Property was liberty; the pursuit of happiness was the pursuit of property. "Liberty and property are not only joined in common discourse," said *The Boston Gazette*, "but are in their own nature so nearly ally'd that we cannot be said to possess one without the other." Property is present even when it's not named. When you're reading about liberty in the early republic, you're reading about property. When you're reading about individual freedoms, you're reading about property. And when you're reading about the framers of the Constitution creating a government of divided sovereignty, you're reading about institutions that are designed to protect property. They sought stability; they sought protection against tyranny. Property would help shelter their rights.

The founders were land hungry; they consumed vast draughts of land, speculated, traded, and legislated to open new tracts. Washington, Madison, Morris, Hamilton, and others bought and sold hundreds of thousands of acres. They were busy surveying, forming land companies, looking for European investors, angling for an advantage.

Most of the land was taken by the wealthy. (About 70 percent of the population owned land after the Revolution; less than a hundred years later this had fallen to 45 percent.) In Virginia the elite granted themselves sprawling tracts of land and then shipped in slaves to work that land. Slavery was essential to create their wealth. These leading men might hold patents to 100,000 or 200,000 acres, and they were on the hunt for more, appealing to England to grant them 300,000 or 500,000 acres in Ohio. They held a paper empire of deeds and a cultivated empire in river houses and plantations, as well as a rougher frontier empire of blazed and felled trees and muddy roads. They coveted land, and leveraged debt and petitioned the crown to get as much as they could without toppling their estates. The acreage from the land grants, land deals, and patents spins by like a drunken odometer: 5,000, 15,000, 100,000, 200,000 acres, hundreds of thousands of acres, millions. "Hoarding cheap land was a universal madness in Virginia and the other colonies," says historian Ron Chernow.

The Federalist ideal put money-getting at a far remove. Officeholders were to be free of any direct interest in the market. Rent and interest was the proper genteel income. A gentleman needed property, enough land so he didn't have to work. Washington strove to uphold this ideal of the disinterested public servant—he refused a salary as commander-in-chief during the Revolution, and only reluctantly accepted a salary as president—but he was always in debt. (He had to borrow money to get to his own inaugural. After two terms as President, he had to sell off land to finance his homecoming and to "lay in a few necessaries for my family.")

At age sixteen, Washington had his first job away from home, surveying in the Blue Ridge Mountains. He was a good surveyor and observer. He understood that the leading Virginia estates were formed "by taking up and purchasing at very low rates the rich back lands, which were thought nothing of in those days, but are now the most valuable lands we possess." He soon made his first significant investment, buying 1,500 acres in the Shenandoah Valley. By age twenty, he owned 2,315 acres in that valley. A year earlier he could not afford corn for his horse, says Chernow. Buying land was the ideal way to "amass riches, a way to invest in his own future and that of the country, mingling idealism with profit," says Chernow.

Washington looked west for bigger land deals. He worked in secret to buy land in Ohio beyond a line that forbade purchase. "Any person therefore who neglects the present opportunity of hunting out good lands and in some measure marking and distinguishing them for their own (in order to keep others from settling them) will never regain it," he wrote to his boyhood friend William Crawford in 1767, directing to find him good land. The British had set that line in the Proclamation of 1763. It was a failed attempt to contain the colonists and stop them from driving the Indians off their land. Washington and many others claimed land knowing that the Proclamation line would be overrun. Settlers grabbed land without a clear title. "I can never look upon that Proclamation in any other light (but this I say between ourselves) than as a temporary expedient to quiet the minds of the Indians," he wrote to Crawford.

"I would recommend it to you to keep this whole matter a profound secret, or trust it only with those in whom you can confide and who can assist you in bringing it to bear by their discoveries of land . . . because I might be censured for the opinion I have given in respect to the Kings Proclamation and then if the scheme I am now proposing to you was known it might give the alarm to others, and, by putting them upon a plan of the same nature (before we could lay a proper foundation for success ourselves)." He asked Crawford to notify him once he had located the land. "I will have the lands immediately surveyed to keep others off, and leave the rest to time and my own assiduity to accomplish."

As a veteran of the French and Indian Wars, Washington was also promised "bounty lands" in the Ohio country. He secretly bought more land rights from cash-strapped veterans by having his brother purchase the claims. He also used a distant cousin as a front. Washington ignored a law restricting the extent of riverfront holdings. To avoid riverfront monopolies, a landowner was forbidden to own long strips of land that would block access for other landholders. Most veterans had narrow properties with a mile and a half of riverfront; Washington had one property hugging the river for more than forty miles. In the end, Washington owned 30 percent of the Ohio bounty lands by the Kanawha River. He owned the best of the bounty lands, more than 30,000 acres with survey rights to another 10,000 acres.

More land came to him when he married Martha Custis. With the addition of Martha's land and slaves, Washington was one of the richest men in Virginia. When his sister-in-law died, Washington inherited Mount Vernon. He kept adding to Mount Vernon until it was 8,000 acres. Two years after his marriage he was in debt. He owed 2,000 pounds. George Washington had started with nothing—lacking even corn to feed his horse—and now he owned ninety-six square miles of land and he was still in debt. He was cash poor and land rich. After the Revolutionary War, rents were his chief income. He leased farms but retained timber and mineral rights. He chased squatters off his land and took them to court. The great general was a bill collector. "No theme appears more frequently in the writings of Washington than his love for the land—more precisely, his own land," says Dorothy Twohig, editor of George Washington's papers.

The Adventurers—as speculators in a commercial "adventure" or venture were sometimes called—were each required to cover expenses and to send five of their slaves to work in the Great Dismal Swamp. Since the crown had started to limit land grants to 1,000 acres per person, Washington's company added 138 fake names to their petition. Virginia's House of Burgesses and the Governor's Council gave the company the power of eminent domain to cut ditches and causeways "through the lands of any person whatsoever adjacent to the said Dismal Swamp." If a property owner thought himself "injured," the company had to agree to arbitration to determine compensation. However, a property owner could not sue the company because the Adventurers' work benefited everyone. Cultivating the swamp "will be attended with public utility," said the law.

At Dismal Plantation, which the company had rented, Washington judged the market value of each of the fifty-four slaves his partners had sent (forty-three men, nine women, a boy, and a girl). He set them to work digging a ditch that would be five miles long, three feet deep, and ten feet wide, digging this through the mud, working all day in standing water and 100-degree heat in a swamp clouded with mosquitoes. This ditch was supposed to drain the land to make it ready to farm. Washington Ditch, as it is known today, is the oldest canal in the United States (and oddly, the nation's earliest monument to Washington). Locals thought that the Dismal was "a low sunken Morass, not fit for the purposes of agriculture." But Washington was sure it would prove to be "excessive rich."

Washington's company had underestimated their task. Five months after this start, each shareholder voted to send five additional slaves and commit more money to expenses, setting a pattern for years to come. A few years later, finding the number of slaves "considerably lessened by deaths," the company bought more slaves.

After ten years the Dismal Swamp Company, as it was now known, had fifty slaves, thirty cattle, and a few hogs and sheep "not worth mentioning." They grew some rice, seven tons in a good year, but could not find a buyer. They operated Dismal Plantation and had completed Washington Ditch. The Dismal was still a vast swamp. The locals were right: peat made for poor farming. The company gave up growing rice and corn. They cut the cedar and the cypress trees, hundreds of years old, into shingles, which were shipped to Philadelphia, New York, and Boston. By 1795 the company was cutting 1.5 million shingles a year. Most of this work was done by slaves.

The Revolution was another setback. General Cornwallis destroyed Dismal Plantation, taking food, tools, and slaves. Washington, away at war, heard nothing about the company. He still believed that these "sunken lands . . . will in time become the most valuable property in this country." After the war, at the first meeting of the Dismal Swamp Company in ten years, Washington found the business in a "deranged state." There were only a few "worn out" slaves, four of whom were retired; the drainage ditches were a mess. As they met, heavy rains flooded the plantation. They decided to take out a loan to hire as many as 300 indentured servants, "laborers acquainted with draining and other branches of agriculture" from Holland, Germany, or other European countries. The company would stop using slave labor. Washington and his partners tried to borrow the money in London but were refused. Four years later, Washington, at last discouraged, stepped aside from managing the company.

Twenty-eight years after the Adventurers had gone into the Dismal to improve and save the land, they invested in a new venture, a canal to be dug through the swamp to the busy port of Norfolk, Virginia. The Dismal Swamp Company was the largest shareholder. The prospects were bright for shipping out shingles and for all the traffic that would use the canal.

This project followed a familiar pattern. The Dismal Swamp Canal Company expected that it would take three or four years to dig a canal, by hand, twenty-two miles, with one crew working from the north and another from the south. They were still digging six years later. A hard day of work might advance the canal only ten yards. Shares that had sold for $250 were only worth $100. The canal company repeatedly asked shareholders for more money. Each year the company's president promised that the north and south trenches were about to meet. "I think the boats may pass through by next summer," he would say. At the outset, Washington had warned that the canal venture was "better stocked with good wishes than money."

It took slaves twelve years to dig the long canal. The work was brutal. Moses Grandy was a slave working in the swamp around 1800. He was a skilled waterman guiding canal boats loaded with shingles and boards. He witnessed the harsh treatment of other slaves in the swamp. "The labor there is very severe," Grandy said. "The ground is often very boggy: the negroes are up to the middle or much deeper in mud and water, cutting away roots and baling out mud: if they can keep their heads above water, they work on. They lodge in huts, or as they are called camps, made of shingles or boards. They lie down in the mud which has adhered to them, making a great fire to dry themselves, and keep off the cold. No bedding whatever is allowed them; it is only by work done over his task, that any of them can get a blanket."

He saw cruel overseers who "tie up persons and flog them in the morning, only because they were unable to get the previous day's task done: after they were flogged, pork or beef brine was put on their bleeding backs, to increase the pain . . . the sufferers often remained tied up all day, the feet just touching the ground, the legs tied, and pieces of wood put between the legs. All the motion allowed was a slight turn of the neck. Thus exposed and helpless, the yellow flies and mosquitoes in great numbers would settle on the bleeding." He went on to describe even more extreme torture.

Grandy was able to buy his freedom. He was allowed to moonlight sometimes to earn money. With "my free papers, so that my freedom was quite secure, my feelings were greatly excited," Grandy said. "I felt to myself so light, that I almost thought I could fly, and in my sleep I was always dreaming of flying over woods and rivers. My gait was so altered by my gladness, that people often stopped me, saying, 'Grandy, what is the matter?' I excused myself as well as I could; but many perceived the reason, and said, 'Oh! he is so pleased with having got his freedom.' Slavery will teach any man to be glad when he gets freedom." Moses Grandy sailed to England in 1842 and told his story, addressing the World's Anti-Slavery Convention in London. His life story was published; the proceeds were used to free his relatives.

⁓

Toward the end of his life Washington tired of land speculation. He wanted to sell his land and invest in safe securities, using the interest to support a "more tranquil" life. With a secure income, he wanted to free his slaves, which he only achieved at his death. (His 123 slaves were freed, but not Martha's 153 slaves. She had inherited them from her first husband. By law she had a widow's "dower share," a lifetime use, which required her to pass the slaves to her heirs. In any case, Martha had no interest in freeing slaves.)

His investment in the Dismal Swamp Company, now thirty-three years old, had yet to produce a dividend, only calls for Washington to send more money to finance the latest scheme to achieve a profit. Three years before his death, Washington had a deal to sell his share in the Dismal Swamp Company to Henry Lee, Robert E. Lee's father, for $20,000. Lee would pay Washington once other land speculators had paid him. But the land market collapsed, taking with it men like Lee who had bought and sold land with now worthless promissory notes from other debtors. The sheriff took sixty-seven debtors to jail in six weeks. Lee scrambled to buy the share, offering Washington stock, house lots, and even corn, offers that arrived years late and proved to be too little. None of it was worth cash. Lee couldn't afford the share; he, too, was jailed for his debts. Washington's share passed on to his heirs upon his death in 1799. The company finally turned a profit eleven years after his death.

The Great Dismal Swamp would be logged and diminished for 150 years. By 1830 the swamp was cut up with nearly 200 miles of ditches, towpaths, and railroads. The Dismal Swamp Company's successor sold out to a lumber baron in 1899. The swamp was logged until it could be logged no more. The last of the old growth went in the 1950s. On Washington's birthday in 1973, the Union Camp Corporation donated 49,100 acres for what would become the Great Dismal Swamp National Wildlife Refuge. It's home to the largest black bear habitat on the East Coast and is an important sanctuary for songbirds. The once-dominant cypress is now found in only 12 percent of the swamp. In places the swamp is so dry that lightning strikes have started big fires, with the peat burning for more than three months.

Today, the government is spending millions to "rewet" the swamp and plant cedars. The refuge manager wants to "let the wetland be a wetland." He points out that the Dismal is a great asset to fight climate change: Peat land can sequester thirty-three times more carbon than any other soil or forest. Restoring a swamp will be difficult; it's never been done. "It took 250 years to get to this point," says the refuge manager. "Maybe it's going to take another fifty to understand the swamp." The goal is to restore the remaining swamp to how it was on the day before Washington and his Adventurers arrived. The Great Dismal Swamp has been reduced to just 15 percent of its original size.

2. Further Adventures in Turning the World Upside Down

Alexis de Tocqueville was a young man of twenty-five when he visited America in 1831, eager to get to the edge of his known world, "the utmost limits of European civilization," and confront the primitive wilds. Haste was the hallmark of the cities and towns he visited. The Americans' restlessness was extraordinary, thought Tocqueville. "He grasps at everything but embraces nothing and soon lets things slip from his grasp so that he may go chasing after new pleasures," he writes in *Democracy in America*. "A man carefully builds a home to live in when he is old and sells it before the roof is laid. He plants a garden and rents it out just as he's about to savor its fruits. He clears a field and leaves it to others to reap the harvest." The nation he saw was in a feverish chase after land. "It would be difficult to describe the avidity with which the American hurls himself upon the immense prey that fortune offers him. In pursuit of it he fearlessly braves the Indian's arrows and the maladies of the wilderness. He is undaunted by the silence of the woods and unmoved by the approach of wild animals. A passion stronger than the love of life constantly spurs him on. Ahead of him lies a continent virtually without limit, yet he seems already afraid that room may run out, and makes haste lest he arrive too late."

This fever of speculating on land, of settling and selling, is the crucible that formed our notions of property. "There is no country in the world where the feeling for property is keener or more anxious than in the United States," writes Tocqueville. Land rush, land grab, land grant—a hunger for land is at the heart of it all, driving the migrations to America and across America, driving every rotten deal with the natives and giving us slogans like Manifest Destiny and American Exceptionalism. It's a banquet and the tables are quickly claimed and cleared. The appetite is rapacious. From the start of the republic, America was an ownership society, with newly minted dispossessed and owned peoples.

Even if the earliest European arrivals had come in search of religious freedom, they, too, caught the fever. The Puritans had started out working on communal farms, but the harvest was poor, so the leaders relented, setting up individual farms. William Bradford regretted how it weakened his community of believers. Increase Mather scolded his flock for coveting what lay just outside their bounds—"*Land! Land!* hath been the Idol of many," he preached, but to no avail. The first Puritans had been satisfied with an acre to farm, but now his flock wanted "many hundreds, nay thousands of Acres."

"Land Rush," writes historian Daniel Boorstin, is "only another name for much of American history." The settlers occupied land in advance of completed surveys or laws. They would band together to protect their claims. They looked to the federal government to dole out land, survey it, and protect their claim. They claimed land by "preemption" and "priority." In a word, squatting. Settlement before the law arrived is the pattern for the continent. The land is divided up and treated as if each parcel were the same as the next. "The disorder of the Land Rush covered the continent," writes Boorstin.

The push and pull over land is largely disguised or forgotten in our history. There were serious but sporadic back-country rebellions and riots in about twenty areas after the Revolution. These Liberty Men, Wild Yankees, Anti-Renters, and Whiskey Rebels were contesting property rights, questioning who really owned the land, what they should be paid, and the fairness of the surveyors who were sent out to mark the official boundaries. They didn't want to pay the proprietors, the wealthy absentee landlords, so they squatted on the land. They had fought the Revolution; this land was theirs, they believed and it should be free for the taking. Many of them couldn't pay the proprietors, and why should they? Their labor cleared the land. They barely had food to get through the year and sometimes not that. The settlers saw the proprietors as the new imperial order. They burned barns and hay-stacks, destroyed fences, wounded and chased off surveyors, dressed as "white Indians" and surrounded houses at night throwing stones, shooting guns, and shrieking like a war party, and, in Pennsylvania, massacred a government-protected tribe. "Hang the proprietors," they said: It's despotism for one person to own thousands of acres. But this wasn't an all-out shooting war. The settlers didn't want to lose the more moderate among them and they didn't want the militia sent out. The back-country rebellions were a violent negotiation. The right price quieted the protestors. If the proprietors backed off their top price and accepted smaller payments, the squatters got their title. This diffused the resistance and opened the way for the surveyors to enter the rest of the land.

Dividing up all this land was once the business of America. The Land Ordinance of 1785 diced the country into a neat grid. The Home-stead Act of 1862 followed the appeal of the National Reform Associa-tion to "vote yourself a farm" and handed out up to 160 acres to a pio-neer who stayed five years and cultivated the land. Congress awarded vast grants of land to private companies to encourage the building of a transcontinental railroad.

Behind all this is the press for the Indians' land, which is relent-less. It exists in every breath, every transaction. It doesn't matter the era, or the change in policy, it's an equation that runs one way only, in the white man's favor. It's more than broken treaties, warfare, and removal. The white "settlers" would provoke the Indians into a fight, knowing that the army would join and eventually remove the Indians by way of a forced treaty or a forced land sale. They had already bet on the outcome, buying and selling "preemption rights"—the right to buy the land once the Indians were removed. (Washington was one active buyer.) Attempts to limit the market for Indian land failed.

Essentially, the Indians sold their land in a "broken marketplace," says historian Stuart Banner. Legally, they could only sell land to the Federal government or its licensed agent. (This is what happens when a house is seized under eminent domain. The party taking the house also determines the price it will pay.) No competing bidders are allowed. The government licensed bidders or bought land itself. Even though the government claims exclusive right to buy land, private dealing goes on. There's almost no redress for fraud, constant pressure to sell by the growing population of whites and their cattle and dams, and confusion about who had the right to sell Indian lands, which depressed prices. In many tribes no one has the authority to sell land. There are drunken sales, fraudulent deeds that claimed more land than was sold, and on and on. The Indians had little recourse to courts or to government. When they won a hearing, the rulings were not enforced on the frontier. It's a barrel with a hundred holes. By the 1700s, the Indians had almost no land to speak of in the East. Indian ownership of their land, which had been recognized from first contact, was reduced in a precedent-setting Supreme Court case of 1823 to a "right of occupancy." They were reduced to being "tenants at will" on their own land, tenants awaiting removal.

One Creek Chief, Speckled Snake, summed up the experience of most tribes: "I have listened to a great many talks from our Great Father. But they always began and ended in this: Get a little farther; you are too near me."

Throughout the nineteenth century the definition of property expanded and expanded like a balloon that might burst. Land, water, mineral rights, and other people (slaves) were the first property. (The Constitution devotes more space to discussing slavery than to any other type of property.) Women were invisible, isolated in their own property sphere, part of a husband's "chattel"—the moveable property of an estate: the furniture, tools, and animals. While one set of people were enslaved, treated as property, another set—the natives— were denied property rights in the land they had lived on for over a millennium.

Many things were considered property by the century's end, including patents, copyright, and brand names. In the twentieth century you could own the news (Can the news be owned? That question went all the way to the Supreme Court in 1918), parts of the broadcast spectrum, the "air rights" above buildings, a celebrity's fame, a set amount of pollution (which could be bought, sold, or traded), and life itself. In 1988 Harvard patented a genetically altered mouse. One-fifth of all human genes had been patented before the Supreme Court ruled unanimously in 2013 that genes are not property.

The way we think about land also changed. By the end of the nineteenth century, land, as it lived in the law, was no longer just a physical holding; it was a "bundle of rights" that owners had *in* that property. The owner didn't have a property right, he had many rights. America grew a super property. For us property is like that bit of (wrong) received wisdom that went around years ago saying that the Eskimos had hundreds of words for snow. (They don't.) But we have a dizzying number of types of property. It's challenging to find that one-tenth of the law that doesn't deal with possession.

There are many clever definitions of property but they all share this characteristic: exclusion. Property is about drawing lines. Each boundary severs what was once whole. Each line drawn across the land reorders the world and buries history. Each line is an erasure. Property is made by boundaries, but the earth is a totality. It's whole, complete. It just is, and we can't own that "just is."

~

Property rights should anchor us in certainty, but they only leave us unstable, left to protect our claim to exclusive dominion against others who may claim the minerals and oil below our feet, or assert a public interest that takes our property. In the last few years I have met with homeowners and farmers defending their land against power lines, pipelines, and windmills. Their distress went beyond the aspects of the particular project. They were now the Indians facing a land grab. Their treaty with the bank, the taxes they paid, the work they'd put in, was nullified and they stood before the law, invisible. They could accept their loss or join a back-country rebellion against the absentee corporations.

Every claim has the possibility of a claim that can be set against it. Turf wars—boundary fights, challenged claims, eminent domain—are daily news. American law favors "property in motion, or at risk, rather than property secure and at rest," says historian J. Willard Hurst. If you are happy planting dahlias and mowing your lawn, you may find yourself having to justify why your pursuit of happiness should stay a big public or private project. It's a straight (surveyed) line from John Winthrop's complaints in 1628 about Indian indolence—"they enclose no land"—to a pipeline or road taking aim at a nature refuge. That land is just sitting there; we can use that.

We like to think of the property we own as a sure thing. We call it *real* estate, after all. But just because we have a deed doesn't mean that it's locked up. The community or a corporation may assert a claim on our domain. Throughout American history, the law's attitude has been that the "quiet citizen must keep out of the way of the exuberantly active one." This is how many dams, mills, railroads, and highways were built—by taking someone else's land. Economic development—"property in motion"—has the right of way over "property at rest."

This is our dual inheritance: land is our security; it is never secure.

Property is our right; it can be taken away. Our autonomy is on call to be surrendered or taken. Property—land—was never just its market value to the country's founders, says historian Gregory S. Alexander. Land was the anchor of the social order. If "the right of property is the guardian of every other right," as Arthur Lee said in 1775, then eminent domain is like an icebreaker plowing through the social order. It takes what was once perceived as solid—my land, my house—and cuts right through it, opening it to traffic and commerce. Our home place, where generations may have lived, is always "on the market" whether we've put it there or not.

Our faith in property betrays us. We're wed to property, but property is in motion. We believe in a fluid, tradable property, not the frozen property of the Old World. From the first, Americans were determined to move beyond European feudalism. In letters home, America was defined as the anti-Europe: rough, lacking history, with land without end and wood to build bigger fires than even the old nobility could command, with the chance for riches or the chance to walk away from your failures, reinvent yourself, change your name, lie, start over on the frontier, move on. Feudalism could be called property at rest. All social relations are locked down. You do not move from your "station" in life. Removing the remaining restrictions on property was seen as a necessary break from the encumbered feudal world. The American system created property in motion.

This can be seen starkly in the laws governing mineral rights. A landowner was entitled to extract as much oil and gas as he could acquire, even if this diminished his neighbor's holdings. He could even explode nitroglycerine to increase the gas flow. What could his neighbor do to protect his interests? "Nothing, only go and do likewise," said one court. The law encouraged landowners to exhaust resources as quickly as possible. The country was hungry and in a hurry. The energy supply per capita leapt an astonishing forty-fold from 1820 to 1930.

Colonial laws promoted the active use of the land. Laws required landowners to develop their land or forfeit it, to sell to a mining interest or mine it themselves, to cooperate with neighbors in draining wetlands, to build with proper materials and, in some cities, at the required distance from the street. Passive ownership was penalized. Colonial land was sometimes granted with require- ments to clear an acre or build a house. A Massachusetts ordinance from 1634 said that land must be improved within three years or the court may "dispose of it to whom they please." In New York if land was not inhabited within three years it could be forfeited. Threatening the forfeiture of unimproved land was a typical Co- lonial-era regulation, says historian John Hart. Under Virginia's Forfeiture Act of 1661–'62 (still enforced 110 years later) land could be assigned to those who would actually use it. There were similar laws in Delaware, Georgia, North and South Carolina, and among the Dutch in New Netherland. Land was not to be left idle; that was how the Indians lived. The earth belonged to men to labor and to live on, as Thomas Jefferson had written.

Utility ruled in ways that we would find surprising. Several colo- nies aimed to increase the pace of mining by decreeing that the govern- ment could appoint someone to work a mine that had been dormant for a year. Under Connecticut's copper mine laws, the assembly could seize private mines they deemed to be lagging, without compensating the owner. The Maryland Ironworks Act of 1719 authorized citizens to condemn one hundred acres of uncultivated land next to "any run of water" to develop a water-powered iron furnace or forge. The target- ed property owner could only avoid having his land condemned by agreeing to erect an ironworks.

This part of our history hides like invisible ink in our property deeds and mortgages, until someone comes along with plans for a highway or a power line or a pipeline. The quiet citizen is left to face the exuberant citizen.

America is still not "settled." It may never be. We live in an era of competing property rights. Property is always on the move, redefined and fought over each generation. New property rights are invented and they frequently shoulder aside old rights.

Today's land rush is about power lines, pipelines, oil and gas drilling, cell towers, and windmills. We tend to see them as isolated intrusions in a settled landscape—a community fights a new pipeline there, another fights fracking for natural gas. But each project is part of a new order that puts industrial uses right next to someone's home. The extent of developed land is growing faster than the population. From 1982 to 2012 the amount of land we built on has outpaced population growth by 22 pecent in America, according to the Environmental Protection Agency. We have an ever-larger footprint.

"All locations have been converted into inter-changeable parts and parcels that are traded in international markets," writes the architecture critic Grady Clay. "Nature is now abstracted for world data-banking and trading. All places are now designated, and we are going through the painful process of deciding which shall be saved, which used, which exhausted, and to whose benefits or loss."

LAND OF MANY USES

WE HAD A BORDER COLLIE WHO WAS crazy happy in the presence of waterfalls. She'd bark and dash about looking at the rushing water. She'd have her distinctive, puffy-lipped smile. Sally was a dog we'd adopted from a shelter. We didn't know how old she was. We did know her past and how her previous owners had let her puppies freeze to death. She had been abused; she had a Dickensian past. Sally was a complicated dog whose past shadowed her daily, but she was a joyous, playful spirit, and if waterfalls made her happy, we obliged her. We planned our hikes to include as many waterfalls as possible. As anyone who's ever lived with a Border Collie knows, they mold you with their iron Border Collie will; they order the household as they wish. Even when it seems they are at rest, Border Collies are always herding.

I know New Hampshire's White Mountains from hikes with my wife and Sally, and the rescued dog that preceded her, Tess, another beautiful Border Collie with a rough past. Tess was the perfect trail dog, racing back and forth all day, keeping track of our group. She worked hard; if asked, she could have accounted for everyone. For me the White Mountains are also clear blue days above the tree line walking the stony ridge trails when it feels as if you are flying, aloft in a big country, not at all walking. And it's days when the winds almost blow you off the summit and days when the clouds come down suddenly, until you can only see a few paces ahead. And it's Mount Washington, the range's highest peak, glowing in a winter sunset after a snowy day, and the starry dark skies at 1:00 a.m. It's big country. Not too long ago it was a ruin.

This is why I love the Weeks Act. I love the story of how Congress finally passed the Act in 1911, establishing national forests east of the Mississippi, including the one in the White Mountains. I regard the

Act as my lucky inheritance, a gift I will never ever be able to repay. I enjoyed the act's centenary celebration with its photos of green mountains that were once scarred by clear-cut logging and the scorecard of land acquired under the act for our eastern national forests, which is about 22 million acres (the size of Maine). But the more I read, the more I saw that this celebration was a precarious victory lap. Our national forests are a patchwork of private and public lands, cut up with roads and private mineral rights. They represent, all too well, our faith in private property.

Even with the celebration, the Weeks Act is overlooked. "Although the initiative is one of the most significant pieces of environmental legislation in modern American history, few Americans have ever heard of it," says historian Char Miller. The Weeks Act is worthy of its own holiday.

The millions of acres that we all own in common were saved to protect private property. The forests had been destroyed for profit; runoff from the bare mountains threatened the mills downstream. By appealing to the gospel of private wealth, the reformers of that era were able to add to the commonwealth, or, rather, to return the land to the public. New Hampshire had owned vast tracts of the White Mountains but sold off 172,000 acres in 1867 for the meager sum of about $25,000 ($6.88 an acre or about $115 in today's money for virgin forest). The "lumber kings" who came to rule were frank in their avarice. "I never see the tree yit that doesn't mean a damned sight more to me goin' under the saw than it did standin' on a mountain," said J. E. Henry, whose aggressive logging would start a fire that consumed a valley seven miles long. (The muckraking press called Henry the "Mutilator of Nature.") More than 600 timber companies cut the mountains hard. One, the New Hampshire Land Company, became a leading villain, "a corporation chartered to depopulate and deforest," Rev. John E. Johnson charged in his jeremiad "The Boa Constrictor of the White Mountains, or the Worst 'Trust' in the World." Johnson, an Episcopal missionary, wrote late into the night "for the chance of hitting the devil with an ink bottle."

The company bought public land "for a song," and took over other land for back taxes, until it had almost total control. They refused to sell timber to local loggers, forcing them out. "Robbed of their winter employment," Johnson wrote, they "took no longer to the woods, but to the cities, leaving the old folks to fall slowly but surely into the clutches of the company which took their farms from them or their heirs, in most cases for a dollar or two an acre." The company refused to sell land to innkeepers, to doctors trying to start a sanatorium, and to a town for public waterworks. In the once-prosperous village of Thornton Gore, the company was driving out the last thirty poor families.

The lumber companies clear-cut the mountains, in places felling and leaving behind 200-year-old spruces just to make it easier to roll the largest trees to the logging roads. The barren mountainsides slid into the rivers; the slash caught fire. Before the logging era, fires were rare. In the drought year of 1903, 78,000 acres burned. Laundry hanging out to dry more than one hundred miles away in Manchester would turn gray with ash.

The White Mountains had been spoken of in holy terms by the first white visitors. The artists followed and created paintings that won a national audience and sent a generation to the mountains to stand where they had stood and see what they had seen. The great nineteenth-century landscape paintings were encounters with a little holy terror. They implied that God was near. They were about a vast land, about wonder that bled off the canvas. The huge mountains suggested mountains without end.

The paintings were exhibited in Boston and New York. They were news; they were an advertisement of a great find, a scenic Gold Rush. The tourists followed by the trainload, filling the big wooden arks of the hotels. They stayed for weeks. They came back to the same hotel year after year. The hotels expanded, burned, and were rebuilt ever larger.

But by the end of the nineteenth century many mountains had been reduced to a "dreary waste," said the New Hampshire Forestry Commission. The black-and-white photos of the cut-over and burned mountains remind me of the battlefield photos of Antietam and Gettysburg. The commission was powerless to stop the destruction: "All the mountain forests in New Hampshire are private property . . . and we have no more control over their owners' treatment of them than we have over the condition of life on the moons of Mars."

American law favors "property in motion"—commerce—not property that is locked up, it is said. But how much motion can we live with before we bring the whole house down?

A small band of reformers took on the seemingly impossible task of bringing our connection to the mountain forests closer than the "moons of Mars." After the turn of the century eight men and one woman formed one of the nation's first conservation organizations, the Society for the Protection of New Hampshire Forests, to restore the mountains, and campaigned for ten years before they prevailed. They tried to create national forests in the East by appealing to national pride, forest management, aesthetics, recreation, and tourism. That didn't carry the day. That wasn't the federal government's business; the government had no right buying private property from willing sellers.

The first bill, introduced in Congress after serious drought fires, was rebuked by Speaker of the House Joe Cannon, who said adamantly, "Not one cent for scenery." It was also opposed by senators from the Rocky Mountain states who loathed the federal lands out west. In ten years, forty bills seeking federal money to buy eastern forests would fail.

Cannon was the most powerful Speaker of the House in history. No legislation was discussed or voted on without his approval; he chose the issues and who got to address them. (When a representative was asked by one of his constituents for a copy of the House rules, he sent only a photo of Joe Cannon.) State legislatures and trade groups like the National Association of Manufacturers were petitioning Congress to create national forest reserves in the East, but Cannon wouldn't listen. In an era of reformers, he was a "standpatter." He thought America worked well. "I am god-damned tired of listening to all this babble for reform. America is a hell of a success," he said. It had worked for him, a poor boy from a frontier log cabin along the Wabash who, after his father's death, was left in charge of his family at age fourteen. "We had no eight hour law, no child labor law, no maternity law, no compulsory school law in that settlement. We all worked from morning till night in the woods and fields and then did the chores afterwards." He had been working since he could "hold a hoe or swing an axe or grasp the handles of a plow." He had grown up in a community desperate to clear the forest to grow food. He thought that a man who had saved the last of the walnut forests was a fool who had "buried his talent." Walnut was the main hardwood of the Wabash Valley. He had seen "thousands and thousands of walnut logs split into fence rails" and "millions of feet of walnut timber burned to get rid of," he said. "It's all very well to bewail this sacrifice now when walnut is rare and valuable, but those people 70 years ago were making the country fit for civilization and the walnut and butternut trees were in the way of civilization."

Support grew for forest reserves in the East, adding the president, the Senate, and two House committees. Cannon was opposed. He did not "believe that the social millennium is just one lap ahead and we can catch up with it if only Congress will enact some new fangled law." Cannon believed that 50 percent of all proposals for change were "harmful, the rest useless," said historian Mark Sullivan. He buried the bill. His opposition made the cause famous.

The bill was stalled for years. He only relented at the end of his reign as he was losing control. Cannon appointed Representative John Wingate Weeks, a fellow "standpatter" conservative, to floor-manage the bill. Weeks had grown up in the White Mountains; his ancestors were early settlers, arriving in 1787, and he summered there. Still, he was an unlikely champion. Weeks had made his fortune as a stockbroker. As an alderman and later mayor of the small Massachusetts city of Newton, he was a practical visionary. He welcomed novel solutions, married them to his steady business habits, and adeptly compromised to win support. Cannon told Weeks that if he could craft a forest bill that business would accept, he would support it. Weeks made it an economic bill, adopting the argument that saving forests was about protecting watersheds. There was not one cent for scenery.

Since the White Mountains had been clear-cut, terrible floods had torn through New Hampshire's main cities, Concord and Manchester, shutting down the textile mills. Commerce had destroyed the forests and commerce would save them. In a society with property as its religion, the appeal only succeeded when it was cast as a property issue. And, constitutionally, interstate commerce was the way the federal government could act without impinging upon the states' powers. The bill was amended on the floor to read that forests would be bought "for the purpose of preserving the navigability of navigable streams." The bill was now on its way, but slowly—it would be delayed another two years by filibusters and amendments.

"*The navigability of navigable streams*"—in this lumpy phrase can be found the Act's success. It didn't call for saving and restoring the White Mountains for conservation or scenery. It was what we'd call today a "jobs bill." Runoff from the bare mountains was flooding the Merrimack River, closing the mighty Amoskeag Mills in Manchester, one of the largest cotton mills in the world. These "navigable streams" affected interstate commerce and thus were a federal concern.

Navigability, utility, that's our property creed. Everything is owned; everything has its use. "All land should have an owner," said an 1890 Supreme Court ruling. With the Weeks Act the lawmakers, Weeks and company, got right in step with our property religion to create a national commons, our eastern national forests. They dressed the Act in property concerns to get Congress to surrender property. Millions of acres would no longer be private. The first national forest established under the Act was North Carolina's Pisgah in 1916. The White Mountain National Forest followed two years later. In all, the Weeks Act created fifty-two national forests in the East.

Forests were useful, as it now says on the signs to every national forest welcoming visitors to the "Land of Many Uses."

Our public lands come to us from a tangle of laws, from moments of political will, pushback, and bureaucratic fights. Any land that's been set aside has been marked by contention. How set aside is it? Will it still be open to grazing, hunting, logging, and mining? And who gets to decide?

The new Forest Service was just two years old when its chief, Gifford Pinchot, was heckled at a Colorado convention in 1907 during the first Sagebrush Rebellion. Other such rebellions followed in the 1920s, 1950s, the Reagan 1980s, and today. The Utah state legislature has called for all federal lands to be "returned." (And they don't mean returned to the Indians.) Arizona voters rejected a similar ballot initiative by two to one, but five other western states have passed or introduced legislation to take over federal lands. In 2006, the Bush administration proposed selling 300,000 acres of national forests and grasslands to pay for the reauthorization of the Secure Rural Schools and Community Self-Determination Act of 2000. This proposal was a nonstarter. Meanwhile, California Representative Richard Pombo, the anti-environmental chair of the House Natural Resources Committee, pushed legislation to sell some national parks in lieu of not drilling in the Arctic, and Colorado Representative Tom Tancredo wanted to sell some Department of Interior lands to pay for the cleanup after Hurricane Katrina. Those notions failed. In 2016 a group of heavily armed men took over a National Wildlife Refuge in Oregon and demanded that all federal lands be "returned." (Again, they were not calling for the land to be returned to the Northern Paiute, whose native land it is.)

It's very difficult in a nation in which "property in motion" is our creed to let anything rest. Land of Many Uses is a gentle way of stating our real attitude: use it, use it hard, and let someone else clean up the mess. Land can be hallowed with signs and laws, but it doesn't mean that it's protected. It's still in motion, subject to trespass, and to the deal made with the seller. Fifty percent of our eastern national forests are privately owned. Out west the checkerboard of private and public lands is the result of congressional bills that awarded alternating square miles to a railroad on either side of a proposed rail line. (Between 1850 and 1870 the railroads were gifted with 223 million acres, about 7 percent of the nation.) In the last few years land trusts have put together complicated purchases and swaps of hundreds of thousands of acres to begin to erase the checkerboard.

Even publicly held land is fragmented by roads and power lines. On federal land that is currently logged there are about two miles of road for every square mile. Biologists say that fish and larger species like bears are imperiled by a much lower ratio than that: three-quarters of a mile of road per square mile. And studies have shown that the dramatic drop in the migratory songbird population is due to the loss of unbroken forests.

This is the story hiding behind the celebration of the Weeks Act—and hiding underneath. The Act bought the "surface estate," which left intact the private mineral rights in many places. This is known as a "split estate" and it's a sizeable estate: There are six million acres of privately held mineral rights in the national forests. The owners don't need a permit from the Forest Service to use their mineral rights. (The Forest Service also can't prohibit underground, hard-rock mines, or even strip-mining sometimes, but it can control access.) The same rules apply to national wildlife refuges. Private oil and gas rights have been executed in 155 of 575 national wildlife refuges. The effects on wildlife are not known; they're not studied.

When the Forest Service became the landlord of the Allegheny

National Forest in 1923, it took charge of 800 square miles of barren hillsides, the "Allegheny Brush patch" as some locals called it. The forest was gone, many oil wells were dry, and the deer had been hunted until few remained. The Forest Service bought the land, but it could not afford the mineral rights. Today the Allegheny National Forest contains two wilderness areas, two Wild and Scenic rivers, and, incongruously, hundreds of miles of roads in the forest and 9,000 working oil and gas wells (more than the 154 other national forests combined). At the start of the twenty-first century, Pennsylvania's boom in drilling and fracking for natural gas in the Marcellus Shale increased the value of the "reserved mineral rights" in the forest. Thousands of new wells were proposed, so the Forest Service instituted a temporary moratorium to review the drilling as required by the National Environmental Policy Act. The owners of the mining rights, led by the Minard Run Oil Company, took the Forest Service to federal court.

Under Pennsylvania law, the owners of the "subsurface estate" must show "due regard" for the surface owner, but they don't need the surface owner's permission to enter the land to mine. The mineral estate is "dominant." The Third Circuit Court of Appeals followed this precedent, ruling against the Forest Service. The proposed wells were not subject to environmental regulations because the Weeks Act does not apply to mineral rights, the court ruled. Forest Service approval was not needed to access drilling sites. Requiring mineral rights holders to obtain permission would wipe the national forests "clean of any and all easements," said the court. The moratorium could cause the driller to shut down. In the Act's centennial year, during the victory lap celebration, the court had downsized the Weeks Act. It's as if the ghosts of Joe Cannon's frontier neighbors burning down virgin walnut groves to grow turnips had risen to have the last word.

Our undiminished land hunger makes the Weeks Act all the more miraculous. Reading about the court rulings and the pressures from logging, roads, visitors, and climate change, it gets so that you can't see the forest *or* the trees. That's when it's a good idea to take a hike. We'll celebrate the Weeks Act with our own holiday. We'll head to the White Mountains. The White Mountain National Forest is not a fragmented checkerboard of public and private lands. For more than a century, the Society for the Protection of New Hampshire Forests has been a spirited steward, buying up parcels of land, and even in the 1970s doing something that had never been done: stopping an interstate highway from paving Franconia Notch into a disposable landscape. After twenty-one intense years, they got the interstate squeezed down to a two-lane parkway. Any legislation, no matter how wise, needs constant defense.

"A century ago, it is doubtful John Weeks or anyone else fully anticipated the profound impact of the law they passed authorizing the purchase of land within the watersheds of navigable rivers," writes James B. Snow, a senior fellow with the Pinchot Institute for Conservation. "Today, most people take for granted the millions of acres of national forest in the eastern United States, probably assuming these forest lands always existed as they appear now. There are many lessons to be learned from the legacy of the Weeks Act, particularly at a time when we cope with the challenges of climate change and environmental pollution," writes Snow. And chief among those lessons is that "enlightened public policy" is worth the fight. We can change the future for the better. "Thank you, John Weeks."

For our Weeks Act holiday, we'll pick a favorite mountain and hike with our new Border Collie, Thurber, and somewhere on the trail a few lines from Henry David Thoreau may come to me, as it has on other trips. On June 22, 1853, Thoreau wrote in his journal: "I long for wildness, a nature which I cannot put my foot through, woods where the wood thrush forever sings, where the hours are early morning ones, and there is dew on the grass, and the day is forever unproved, where I might have a fertile unknown for a soil about me. . . . A New Hampshire everlasting and unfallen." More than one hundred years later, here's to the visionaries who, when looking upon charred and slashed mountains, never lost sight of the everlasting and unfallen. May we one day catch up with them.

THE LAST MEDIEVAL CLAIM

The inquiry is sometimes made, what is the value "of erecting these
old stones in out of the way places?"

—*Thirty-Fourth Annual Report of the City of Keene . . . For 1907*

IN A SMALL TOWN NEARBY THERE'S A CURIOUS monument that I
visit every so often. It's not much, a granite obelisk, all of five feet tall
and, the last time I visited, heaved to one side by the frost. To see it you
have to look carefully. It's off in the woods, 140 feet from the road, or
about nine rods as they would have said then.

It was dedicated on August 27, 1907, with solemn ceremony during the annual town picnic in Sullivan, New Hampshire. The monument celebrates a vast property claim, one that shaped New England.
It stands in the woods because it is set on one of the oddest lines ever
surveyed in North America. The day's speakers tied this stubby monument "back to the discovery of the New World." I think of it as marking the end of the Middle Ages.

We have to begin with some big claims. In 1629 Captain John
Mason claimed a sweep of land that he named after his English
home county, Hampshire. Mason's land was granted by a council
chartered by King James I. Captain Mason wanted to establish a
lordship for himself and collect rents on this land. New Hampshire
would be like feudal Hampshire. His grant, like many others, was
vague, extending sixty miles inland from the mouths of two rivers
on the seacoast. In time his inland reach came to be defined by a
curving boundary, and this curve and this claim would afflict the
province for 150 years. This lone curve can still be seen on maps

today looking like a ripple on a pond, the last wash of a feudal sea.

The king established the Council of Plymouth in 1620 to "stretch out the bounds of our Dominions." The council was charged with the "planting, ruling, ordering and governing of New England in America." This was a super-sized New England. The council's forty men were granted a big swath of North America. They claimed possession of all the land lying between the 40th and 48th degrees of northern latitude, from the St. Lawrence River south to present-day Philadelphia, from "sea to sea," from the Atlantic to the Pacific—the "South Sea" as they then said. Captain John Smith had named this place New England in 1614, but the English lacked even one settlement north of Plymouth. When Captain Smith showed his map of New England to Prince Charles, the fifteen-year-old prince wrote in the names of the English places he fancied (hence the Charles River in Boston). Smith kept a separate list of the Indian names he had learned, but most of these were already off the English map.

The English declared that North America was deserted. "Within these late years there hath by Gods visitation reign'd a wonderfull plague together with Many Terrible slaughters & Murders committed amongst the savage & brutish people there heretofore Inhabiting in a manner to the utter devastation destruction & depopulation of that whole Territory," said the council's charter. The Indians' first contact and trade with the fishing fleets off the coast had brought mysterious diseases, possibly smallpox. "The Great Dying," as the Wampanoag called it, had reduced some tribes by as much as 90 percent, but they had not vanished as the charter claimed: "Almighty God in Great goodness & bounty towards us & our people hath Thought fit & determined that those Large & goodly Territories deserted as it were by their natural Inhabitants should be possessed & Enjoyed by such of our subjects & people."

The council set about granting land. They granted Mason and Sir Fernando Gorges land the two men called Maine. They granted Mason land he called Mariana, which bordered on the "Great River Naumkeek," a mythical river recorded on only one English map. They leased land for 3,000 years, and reserved all gold and silver that might be found there for themselves and their king. The deeds were fanciful and confused; they deeded the same land to different people. The council's forty men did not know this land and had never been there. They were Englishmen. There were no maps as we know them. Before they knew where the rivers ran, and if the mountains were more than rumors, they claimed it all.

Here are the mechanics of dominion. This is ours. This can be bought and sold. This land will be called by the name of an English county. We few men—"persons of honor and gentlemen of blood"—sitting in council in Plymouth or London can trade and sell this land we have not seen, in any quantity we choose.

Mason's curve is one of our founding stories. Other cultures begin with stories of creation, the shaping of the earth, the creatures, the first people. America begins with claims, with land deals, litigation, and speculation. In some creation stories, God, or the creator, forms people out of dust, out of stone, in his own image. In this creation story, America is formed out of boundaries and legal deeds, rents and mortgages. The council and its heirs and successors claimed "forever any Lands, Mannors, Tenements, Rents, royalties, privileges, Immunities, Reversions, annuities, hereditaments, Goods & Chattles whatsoever" in this land stretching from the Atlantic to the South Sea.

Property, we learn early, is "nine-tenths of the law." What if it is also nine-tenths of our history?

∾

The strangest thing about these kinds of proclamations claiming lands almost without end, with mythical rivers, misplaced mountains, and their poetic reach to the South Sea, is what is left unsaid: Why do they get to own this land? *Sez who?*

You'd think that this question—How do things come to be owned?—would be easily answered. After all, everything around us, except the birds and wild animals, is owned. But an answer is elusive. Philosophers have debated the origins of property for centuries and have essentially decided on one of two stories: private property is God's plan, or private property is created by the state. The first story says that owning land is a "natural right" predating the state, existing among people in a primitive condition. The other story says that private property is a "conventional right" existing *because* of the state. This is a schematic history, of course. There's every possible mix of these two stories. There was a lot of hand-wringing over just how "natural" property was until John Locke wrote his *Second Treatise* in 1689, which reasons that private property is our natural condition, that God gave us the world to use, and that government exists to protect property. The state exists to keep mine and thine sorted out. After Locke, life and liberty are bound up with property.

But if owning things is God's natural plan, why does questioning this seem, well, a little rude, like something we are taught not to say when company is visiting? Consider a cornerstone of property law, William Blackstone's *Commentaries on the Laws of England*.

"There is nothing which so generally strikes the imagination, and engages the affections of mankind, as the right of property; or that sole and despotic dominion which one man claims and exercises over the external things of the world, in total exclusion of the right of any other individual in the universe," writes Blackstone. A great deal of legal writing about property proceeds from Blackstone's book, published in the 1760s. Blackstone's statement about "sole and despotic dominion" is like a big old bulky piece of furniture that's handed down through the family. The house could be stripped to a beige and concrete minimalism, but there sits the great-great's walnut rolltop desk, like the last of its kind in some modernist zoo.

But as Yale law professor Carol Rose points out, few readers ever follow Blackstone's quote to its conclusion, which has the directness of a child asking, But *why* can we own things? "There are very few, that will give themselves the trouble to consider the original and foundation of this right," says Blackstone. "Pleased as we are with the possession, we seem afraid to look back to the means by which it was acquired, as if fearful of some defect in our title. We do not really want to learn too much about such matters." We don't want to face that "there is no foundation in nature or in natural law, why a set of words upon parchment should convey the dominion of land . . . or why the occupier of a particular field or of a jewel, when lying on his deathbed and no longer able to maintain possession, should be entitled to tell the rest of the world which of them should enjoy it after him."

Blackstone has brought up "ownership anxiety," says Rose. Legal scholars shoo this anxiety out of the room with "just-so stories" about the origins of property, she says. In short: It's always been this way, it's meant to be this way, and what's more, there's centuries of laws built on ownership. Blackstone, she notes, ducks his own question by taking a deep dive into English common law, discussing the feudal rights of "advowsons, tithes, commons, ways, offices, dignities, franchises, corodies or pensions, annuities, and rents." He unloads an entire moving van of old furniture. And still, says Rose, "the institution of property is as mysterious as it was at the beginning."

The best way to understand these continental land grabs is to think of them as fiction, that with brute force becomes fact.

Captain John Mason had served his king well, and for his service he was rewarded with a large debt. The king had dispatched Mason to put down a rebellion in the Hebrides Islands. He gave Mason two warships and two smaller vessels to provision at his own expense with the understanding that the crown would repay him. It did not. Mason owed 2,238 pounds in 1612, which at a 10 percent interest rate had grown to a staggering 12,489 pounds by 1629. That was the year he sought to establish his lordship in the land he named New Hampshire.

Mason founded two plantations with the support of other investors. They sent men, food, clothing, tools, arms, and, it is said, many head of Danish cattle (whose lineage was still visible in New England livestock nearly 300 years later). His workers built one of the first sawmills in New England. As he was planning to make his first visit to New Hampshire, Mason died at age forty-nine, in 1635.

He had wanted to found an empire. He had named two of his other holdings Masonia and the Isle Mason. He had plans to rule all of New England. His ambitious friend, Fernando Gorges, had convinced King Charles I to put New England under one government. Gorges would serve his king as the Lord Governor, and Mason would serve as Vice-Admiral. With civil war in Scotland and England, that plan was abandoned. Mason did not leave a kingdom to his heirs, but a legacy of court cases in England and America.

Mason left the vast majority of his real estate to his eldest grandson. His widow received the estate's rents and profits until the grandson came of age. If that grandson died without children, the estate would pass to the boy's brother. Either brother could inherit the realm only if he took Mason as his surname. The eldest grandson died, and his brother, Robert Tufton, adopted the Mason name. In the many years before Robert Tufton Mason came of age, Captain Mason's plantations went to ruin. His widow did not look after them and provided no financial support. The servants declared themselves free and seized what they could. One man drove the oxen to market in Boston, receiving twenty-five pounds a head, the best rate then paid. Others took over the mills and Mason's house. A few wealthy families owed their start to their theft of Mason's property.

There were new claims for this land in the years while the Mason family was absent. New towns were established, and Massachusetts claimed New Hampshire. Settlers were clearing the land, oblivious of their feudal obligations.

Having come of age, Robert Tufton Mason worked tirelessly for nearly thirty years to recover his ancestral estate. He petitioned the royal courts. The courts upheld his title and awarded him the overdue land rents. Across the ocean, the colonists, still Englishmen at this point, refused to pay. Some of them had been living there fifty years without challenge, and had cleared the land, planted crops, and built houses and barns. They, too, had legal deeds and royal patents.

When Mason petitioned the crown for relief in the 1670s, the king evicted Massachusetts from New Hampshire, creating a separate provincial government. Mason's dogged pursuit of rent set New Hampshire on a course to stand by itself. He had the king appoint him to the council that governed New Hampshire. The "Lord Proprietor," as he called himself, sailed to New Hampshire and summoned his fellow councilors to appear before the king. They responded by issuing a warrant for Mason's arrest. He sailed home and once more petitioned the king for help. (He had petitioned the crown sixteen times in the 1670s alone.)

Back in England, Mason was busy. He had his own man, Edward Cranfield, appointed "lieutenant-governor and commander-in-chief" of New Hampshire. To insure the lieutenant governor's devotion, he promised him a share of the future land rents and his daughter in marriage. Mason also mortgaged New Hampshire to the lieutenant governor for twenty-one years. Cranfield's interests would be Mason's.

This was only a small part of Mason's campaign. He had several royal commissions appointed (one of which was headed by his cousin and a hand-picked staff). He had commissions sent to New Hampshire to hear his claims. He offered the king a profit-sharing plan: one-fifth of "all rents, revenues and profits" from the province. He tried to evict the man he claimed was living on his grandfather's plantation. He won many judgments and collected nothing. Robert Mason was "a Pragmaticall, and Indefatigable person . . . very Conversant at Court," said his legal opponents in Puritan Massachusetts. He was also, says one historian, a tireless liar who knew "that his claims were highly misleading if not completely invalid."

Mason's agents "rendered themselves obnoxious by demanding rents of several persons and threatening to sell their houses for payment," wrote Jeremy Belknap, one of America's first historians. They tried to seize estates, forbade the cutting of timber and firewood, and advertised that all landholders must take out a lease. Lieutenant Governor Cranfield picked judges, juries, and sheriffs to award Mason back rent and much of the settled land in the province. He dismissed the assembly when it refused to rig the judicial system for Mason. He jailed a Puritan minister for not following Anglican liturgy; harassed other ministers; jailed, fined, and sued other royal subjects; instituted heavy taxes; revalued the silver currency, causing hardship; and barred ships from Massachusetts to enter the port. One historian judges Cranfield "the most reckless and tyrannical" of any appointee of the crown in the American colonies. Charges were brought against him, and he left the province for a profitable posting in the Caribbean.

The colonists would not be moved. If Mason had papers to prove his ownership, they had papers, too. If Mason had the king's blessings, so did they. Mason could declare dominion, but they were living on the land. "Keep to your patent. Your patent was a royal grant indeed; and it is instrumentally your defense and security," Pastor Urian Oakes lectured the new session of the Massachusetts colonial legislature in his Election Day sermon for 1673. "Recede from that, one way or the other, and you will expose yourself to the wrath of God and the rage of man."

This was their land. They were not tenants of this "pretended proprietor." They denounced Mason in town meetings, and petitioned the king to protect their liberties. "Now, the plaintiff, nor ancestors, never planted this province, nor expended any thing upon it, to the upholding of it, in peace nor war," they said. But the present inhabitants "in the late Indian war, did defend it against the enemy, to the loss of many of their lives, and considerable part of their estates, without any assistance from Mr. Mason, who now claims not only what poor people have purchased and labored hard upon, but also conquered or relieved from cruel attempts of the barbarous heathen, and we conceive we were under no obligations to run such adventures to make ourselves slaves to Mr. Mason."

~

Mason and his heirs had missed an essential point: They couldn't export feudalism. Today we'd say that they had a bad business plan. They were peddling feudalism in a democratic age.

In their break with England, Americans broke with the old notions of ownership. "Indeed, the entire Revolution could be summed up by the radical transformation Americans made in their understanding of property," writes historian Gordon S. Wood. Landed property had been "a part of a person's identity and the source of his authority." It was the "guarantee" of "one's gentility and independence from the caprices of the market." But after the Revolution, land became another "commercial, dynamic and unpredictable" commodity. Anyone could acquire land.

The young republic shed all the finely sliced feudal privileges and rights that defined property. No more knight service, escuage, burgage, frankalmoign, advowson, tithes to the clergy, fee tail, corody, primogeniture, dignities. . . . All gone. These "incorporeal hereditaments" vanished. Land became a product—buy it, use it, trade it, sell it. It's not something that gives you a say in who the next minister is; it's not required to stay in your family as long as there are heirs; it's not required to be passed to the eldest son; and most importantly, it doesn't belong to the king. Americans wanted a market. They would own their land "free, clear and absolute"— once the Indians were dispossessed. Flux will be the common fate. Americans were leaving behind the stability of the feudal land economy for what we know today, writes legal scholar Morris R.

Cohen, a "money economy" that "seems a state of perpetual wars."

In the Middle Ages, a family and its land were one. Under the entail laws, land could be inherited, but not sold. It was a static world. The many generations of the family were, in a sense, contemporaries, says Alexis de Tocqueville. Everyone knew their role, and the obligations of each class held everything in its orbit. But in a democratic era, Tocqueville writes, "the fabric of time is forever ripped and vestiges of the generations disappear." New families rise and fade away. Land is sold and divided. Ohio had been a state only thirty years when Tocqueville visited, and already people were dashing for the West. He returns many times in his book, *Democracy in America*, to the upheaval and restlessness he saw on his 1831 tour. It was as if Tocqueville had crossed the centuries to arrive in America. The French aristocrat was surprised and troubled to find a society without an aristocracy. It was as "though his kind had never been," says the political theorist Sheldon S. Wolin. "Aristocracy had disappeared with as little trace as the once busy settlements that the relentless wilderness had reclaimed."

"People in those days believed in the immortality of families," Tocqueville writes of the Middle Ages. But "in centuries of equality the human mind takes on a different cast. It is easy to imagine that nothing stays put. The mind is possessed by the idea of instability."

If property is liberty, why enslave yourself to a feudal overlord? Mason's claim was large, but the country was larger, and the appetite for land as large as the continent. "In democratic centuries," says Tocqueville, "the most mobile thing of all is the human heart."

Robert Tufton Mason died in 1688, at age fifty-three, leaving two sons, but no empire. His "unwearied renewal" of his claims had succeeded only in ensuring that New Hampshire was independent from the power-grabbing Puritans of Massachusetts. In 1691 his sons sold New Hampshire, and Mason's other royal land grants, to a London merchant, Samuel Allen, for 2,750 pounds. Allen, short of funds, ended up paying less than half of the agreed-upon price. Allen had purchased generations of legal grief for his family. There were twenty-two towns squatting on Allen's purchase—twenty-two towns whose entire populations were, by his deed, trespassers. His realm, in all legal papers, was still bordered by the mythical river Naumkeek. (And as a further fiction, for the purposes of the sale, all this land was said to be in Greenwich, England.)

Allen continued Robert Mason's legal pursuit as if he were his son. He even sued the son of a man that Mason had sued. After Allen died, his own son continued the suit. By the time that case was decided against the Allens, it had been in and out of the courts, brought by Masons and Allens, back and forth across the ocean, in England and the colonies, for twenty-four years.

The colonists resisted a revival of Mason's claims, and the return of a proprietor who would harass them for rent. Four New Hampshire towns petitioned the king asking to rejoin Massachusetts, but they were denied. Allen had himself commissioned governor of New Hampshire, and his son-in-law, John Usher, lieutenant governor. While Allen was in London, Usher ruled, or tried to—he was a loud, rude, "very disagreeable" man whose "public speeches were always incorrect," says historian Belknap.

Allen pressed his claim, demanding rent. Like Mason he won in the courts but collected little. He won a big case when the Crown's attorney general ruled that he was entitled to all "waste lands," including the land that the towns held in common for grazing cattle, farming, and firewood. Now the colonists owed him rent for land they depended upon for their livelihood. Allen published an order that no one was to use any "waste land" without first leasing it from him. When the colonists refused, he prepared to go to court and sought the superior court record of Mason's cases. He had the legal precedent of Mason's many legal victories. But those cases, twenty-four pages of judgments for Mason, had been cut out of the book. (The jurors had been bribed to rule for Mason in about 160 land cases.)

A special convention of representatives from each New Hampshire town proposed a compromise. If Allen surrendered his claim to land in the older towns, he would receive several thousand acres in these towns, all the unsettled land in the province, and a payment of 2,000 pounds. All legal suits would be withdrawn and the province would be free of his claim. But Allen died the next day. His only son resumed the fight and sold half of the province to another man. The son died and the Allens' claim lapsed in court.

Mason's grant appeared to be dead. Two generations of the Mason family lived without ever entering court or pleading with the king for the rent on this land. But then John Tufton Mason, the captain's great-great-great-grandson, recovered possession of the grant 103 years after the captain's death. He was back in the family business.

The Massachusetts Bay Colony took a keen interest in John Tufton Mason's renewed defense of his ancestor's land. In the late seventeenth century, the Bay Colony was a thriving empire with Maine as its northernmost possession. Pushing its border north, the Bay Colony claimed a good portion of New Hampshire, a scruffy place with a few striving towns. If the Puritans could not rule New Hampshire, they would make sure it was a small province. And they would use every precedent they could find, including Mason's grant.

The two provinces fought for dominion town by town. In the disputed territory, Massachusetts granted twenty-one new towns and New Hampshire established thirteen towns. To this day, the main street of New Hampshire's state capital is the one laid out by Massachusetts when it claimed the town—now thirty-five miles north of the state line. In some places the two governments granted overlapping towns, and residents were pursued by tax collectors from both provinces. "The borderers on the lines live like toads under a barrow, being run into gaols on the one side and the other as often as they please to quarrel," said Governor Jonathan Belcher, who was in charge of both provinces. "They pull down one another's houses, often wound each other, and I fear it will end in bloodshed."

To end this strife, the king appointed a commission to establish the boundary. The Bay Colony countered by purchasing part of Mason's claim. With the grant in hand, they were prepared to argue that New Hampshire could extend no farther than that curved line sixty miles from the sea. They dispatched John Tufton Mason to London, at their expense, to press their case. But once he was there, Massachusetts changed course, and summarily dismissed him. He was angry and embarrassed. New Hampshire's representative quickly signed him up and brokered a deal for John Tufton Mason to sell his grant for 1,000 pounds within twelve months after New Hampshire had its own provincial governor.

In London, the two provinces competed for advantage "in the same spirit of intrigue which has frequently influenced the conduct of princes, and determined the fate of nations," writes Belknap. Massachusetts may have had a good case, but they were outplayed by New Hampshire's clever representative. He portrayed "the poor, little, loyal, distressed province of New Hampshire" as defending the king's interests against "the vast, opulent, overgrown province of Massachusetts," writes Belknap.

The boundary ruling in 1740 surprised both sides. New Hampshire won far more land than it had ever claimed. Massachusetts lost twenty-eight new towns, parts of six older towns, and much land. (When the boundary was surveyed six years later, it was done incorrectly due to a mistake in calculating true north. New Hampshire and Vermont lost almost 195,000 acres, and despite another 150 years of squabbling, there it stands today. New Hampshire did not accept the line as surveyed until 1901.)

John Tufton Mason waited a long time to be paid. After six years he told the legislature that he could wait no longer, hinting that other buyers were interested. The legislature finally honored the sales agreement, appointing a committee to negotiate some restrictions. But on the same day Mason, a mariner who was under order to sail within twenty-four hours, sold his grant, his ancestors' design for empire, to a group of twelve men. For 1,500 pounds they had bought one-sixth of the province. The legislature angrily denounced the sale. This group of men, the most influential, wealthy men in the province, had stolen the deal, they said.

The purchasers called themselves the Masonian Proprietors and they proceeded to do what generations of Masons and Allens had failed to do: survey the land for towns, and grant land for settlers to own. They immediately renounced any claim to land in the established towns, quieting fears there. They were willing to compromise with settlers who had land granted from Massachusetts and now found themselves in a new town. In short, they modernized Mason's 117-year-old grant, recognizing that these Englishmen owed no rent to any "lord proprietor." They would create an orderly real estate market.

But the land was still unknown to them. They thought they had purchased about 200,000 acres, but it was closer to two million acres. They hired a surveyor to discover the westernmost reach of their purchase, the curving boundary believed to mark Captain Mason's 1629 grant. How would the surveyors mark a curve in the wilderness?

Gunter's chain was the primary tool of eighteenth century surveyors for measuring the land. One chain was made of one hundred links which equaled sixty-six feet, or four rods. The rod, also called a pole or perch since the Middle Ages, was the unit of measure that had defined the land. A small New England road was one rod wide, or sixteen-and-a-half feet. Roads between towns might be three rods wide, and turnpikes four rods wide. The measure of one acre was 160 square rods or ten square chains. An Anglo-Saxon acre—as much land as a yoke of oxen could plow in a day—was always 160 square rods.

In 1751, Joseph Blanchard Jr., twenty-one years old, and his crew of nine men set out to measure a curve with a compass and a chain. Blanchard was the Surveyor General of the Kings Lands in New Hampshire. His crew carried the chain over mountains, through swamps, and across nearly half the eight-mile length of Lake Sunapee by raft. At each mile Blanchard recorded their progress on a tree with a marking iron with the letters J. B. and the date. They surveyed the curve by running straight lines for five miles (a chord of the curve) and measuring the angle of each chord before proceeding. "All the hands Labourd Very hard & Were Diligent & Exact," two of the crew reported. Blanchard and his men made it sixty-seven miles, about four-fifths of the way, before "provisions failed" and they were "obliged to return." The Masonian Proprietors sent out more surveyors to complete and re-measure the curve. Fifty years later, Blanchard testified that he could still find his marked trees and follow the curve. At the twentieth mile he found his mark on a fallen beech tree, and farther north he had to chop into several large trees to find his mark now covered by fifty years of growth.

Nowhere in Captain Mason's original 1629 grant is the western boundary described as a curve. It is a line that crosses over the land sixty miles from the sea. To maintain that sixty-mile distance from the curving seacoast it was long thought that the line had to be a curve to be measured from the mouth of the Merrimack, or the Piscataqua, or both, depending on how you read the grant. No one knows who decided it was a curve. Blanchard surveyed his curve with the Merrimack as the focus. To extend the territory, Blanchard began his survey sixty *nautical* miles from the sea.

A new surveyor, Robert Fletcher, was dispatched eighteen years later with new instructions. Fletcher surveyed an ellipse with the mouths of both rivers as foci. An ellipse would cover many more square miles than the already surveyed circle. The Proprietors grabbed more land with each interpretation—that the boundary was a curve, that it be measured in nautical miles, and finally that their territory was an ellipse. It was a good old American land grab.

∾

Captain John Mason's paper kingdom outlived eight monarchs, civil war in England, and revolution in America. Property that King James I had sanctioned and that the royal courts had upheld, was honored in a new country that had declared itself free of kings. John Mason and his heirs saw little gain from this property—it was never to be feudal holding—but in defending their dominion they had defined the provincial boundaries that would become state lines. The Masonian Proprietors, a dozen well-organized Portsmouth elites, made the claim pay. They established towns and sold plots of land, each as uniform as possible, but some part of the boundaries of dozens of towns and three counties curve. Many other town lines were drawn in reference to the curve, so that on a map they seem to be tilting. On a modern map, the curve is like one of those illusions—which do you see: a wine goblet or two faces in profile? Squint and the curve jumps out from all those town lines. Or as one surveyor told me, it stands out like a sore thumb. Many surveyors in its vicinity come across the curve, and it's sometimes the cause of odd boundaries they must untangle.

The Masonian Curve is one of the strangest lines drawn on the continent. There is only one other curving boundary like this in the United States. Delaware's northern boundary is a curve twelve miles long, centered on a courthouse cupola. West of the Ohio River, it is the great rectangular survey of 1785, the six-mile-square grid, that is our distinctive mark on the land.

The Masonian Proprietors met regularly to administer their real estate venture until 1807, and met for the last time in 1846—more than 200 years after Mason made his claim. With Mason's grant settled, New Hampshire contested other boundaries. New Hampshire claimed all of what would become Vermont and Governor Benning Wentworth would enrich himself laying out 129 towns, such as Bennington, Vermont. New Hampshire's boundary disputes continued into the twenty-first century. In 2001, New Hampshire and Maine met in the US Supreme Court contesting ownership of the Portsmouth Naval Shipyard. (Maine won.) The shipyard sits on an island in the Piscataqua River, right on the border of Mason's 1629 grant.

∽

Each colonial power had its own rituals for claiming territory, says historian Patricia Seed. The French would stage a grand procession to plant a cross with the natives taking part to show that they had accepted Christ and the French king. The Spanish would read The Requiruimiento, a formal speech commanding the natives to accept Christ and the Spanish king—if they refused, they would be attacked and enslaved. The Portuguese justified their discovery by recording how they had determined latitude from the sun's position. The Dutch relied on maps to make their claim. The English believed in "improving" land by enclosing it. Preoccupied with boundaries, they planted hedges, built fences, cleared land, and gardened. They were following their interpretation of Genesis 1:28 to "replenish the earth and subdue it and have dominion . . . over every living thing." This was "the grand charter given to Adam and his posterity," said Puritan leader Robert Cushman in his justification for the voyage of the *Mayflower*.

On August 27, 1907, a small caravan of automobiles made its way over the dirt roads to the little town of Sullivan. A half-dozen cars on the road at one time was notable, especially in a town of just over 250 people. The men—the mayor and city clerk of nearby Keene, the clergy, and other invited guests—had come to honor Captain John Mason. With an extensive ceremony, echoing older rituals of conquest, they dedicated their small monument to Mason.

After patriotic music by the East Sullivan town band and a prayer, there were six speeches recounting the history of Mason's curve. The men spoke of the debt "we owe to the old pioneers" and of all the claims that were joined, link by link, since "Columbus took possession of all the lands, islands, and seas which he had discovered." All ownership descended from Columbus, and later John Cabot off Newfoundland in 1497, and then King James I, Plymouth Colony, and the council that granted New Hampshire to Mason, and so on to the most recent deeds that had been filed at the courthouse that week. This memorial was like Gunter's chain measuring the distance from Columbus to the family farm. To these men, Columbus and King James I were, in a sense, their contemporaries.

But why memorialize this? Mason's fiefdom could have been onerous, with rents and a long list of feudal obligations. This is a marker of the future that we didn't have. Mason's ambition was to leave an empire to forever carry his name, and here's his memorial: a granite stump. This would be like memorializing the job you didn't get, the school that rejected you, the spurned marriage proposal. The monument would say: On this spot (*your name here*) failed. He didn't get the great job; he didn't get the part in the play. He got back on the bus and went home to Nebraska. This ceremony seems to claim possession long after such an act is possible. It would be like sailing into New York Harbor today and planting your flag in Battery Park to declare yourself the Landlord-in-Chief.

The hill farms of Sullivan are a long way from the South Sea, the far shore of King James I's dream of a new England in America. The short granite post is inscribed with a few words noting the three town lines that once met at this spot, and the curve surveyed by Blanchard. In the true Yankee manner it is reticent even in celebration. What the monument could say is: Here the Middle Ages end, and here starts the America of westward migration, displaced Indians, suburban backyards, malls and highways and dams. All of it. No wonder the small marker is leaning east, falling back toward England and the feudal past.

2

What Am I Going to Lose?

The quiet citizen must keep out of the way of
the exuberantly active one.

—Thomas Beven, *Principles of the Law of Negligence*, 1895

Whenever a plan for a new project is brought
before a neighborhood meeting, people first think:
"What am I going to lose?"

—Cristina Prochilo, Field Officer, National Trust for Historic Preservation

THE BALLAD OF ROMAINE TENNEY

IN THE SUMMER OF 1964 ROMAINE TENNEY was a bachelor farmer. He milked twenty-five cows by hand on his farm in Ascutney, Vermont. He had no electricity in his house, used no gas-powered machinery. He cut his firewood with an axe and a saw; cut his hay with workhorses. He didn't own a tractor or drive a car. When he went to the nearby big town of Claremont, across the river in New Hampshire, he'd walk the six miles—except that he probably never walked all the way. People always picked him up. Everyone knew Romaine. With his long beard, felt hat, and overalls, he was a familiar sight. Romaine enjoyed visiting on these rides and all his neighbors liked him. His farm was right on the major road between Ascutney and Claremont—the road hugged his cow barn and neighbors would often stop to chat. He rose late and worked late into the night. "You could drive by at midnight and there he would be in his barn fixing some harnesses or just puttering about," said Deputy Sheriff Robert Gale. It was as if Romaine held the office of Bachelor Farmer in town.

His house, trimmed under the eaves with Gothic gingerbread, stood behind a row of majestic maples. Tourists loved to take pictures of the house, and he'd sometimes pose for them. If they wanted a true, old-time Yankee, he'd oblige them. He was the real thing, happy to play the part for a moment, sending a tourist on his way with his prize catch: *Look at this old farmer I found in Vermont. Milks his cows by hand. No electricity, no car, no tractor.* Romaine Tenney was the Vermont they wanted to find.

Romaine looked good in every picture. "What I remember is his beautiful blue eyes and his eternal smile," said his niece, Rosemary Safford. "He was always smiling." And that's what everyone said. "He had a wonderful twinkle in his eye," said his neighbor, Rolly Cann.

Romaine was born on the farm and spent his life there. He loved his family—his many brothers and sisters, nephews and nieces. He loved his animals. He was a happy man—until his farm was destroyed to build Interstate 91.

~

The first six miles of interstate highway in Vermont opened in 1958. It ran from the Massachusetts border to just south of Brattleboro and drivers marveled at what we now take for granted: It was straight and smooth. It was the shape of things to come and they couldn't wait. When a new section opened up near Montpelier in 1960, 300 cars lined up to drive the six miles to Middlesex. The interstate was more than just another road; it was a belief in progress. The highway would rescue Vermont, take the state "out of the sticks" and put it "right in the economic mainstream of the country," said Elbert Moulton, the state's economic development chief under four governors. "The interstate was seen [as] the answer to many, if not most, of Vermont's problems," said Paul Guare, the executive secretary of the state Transportation Board at the time. "It was universally applauded." When the state condemned the houses and farms in the way, filled wetlands and leveled hills, "people were mostly happy to settle with the state." Progress as a religion permitted everything. It was the gravity of America; it was the force that held everything in its course.

Dedicating a new section of the interstate in 1961, Senator George Aiken said, "We're on the verge of the greatest development Vermont has ever seen." That section of highway had buried the senator's boyhood home.

~

Romaine's farm was ninety acres of good pasture and woods with southern exposure and plenty of water. There was a spring up the hillside that almost never went dry, a brook, and a hand pump in the kitchen sink. The fields were good for three hay crops a year. There was an orchard and a ten-acre woodlot. In the farm's prime, in the 1950s, Romaine milked fifty or sixty cows and had about one hundred head of livestock total. He kept two teams of large workhorses and a couple of dogs to bring the cows home. Prince and Spot are the dogs everyone remembers as always being at his side.

Romaine's parents bought the farm in 1892. Myron and Rosa Tenney came over Mendon Mountain in a wagon with all they owned in a trunk or two, the family story goes. He was forty-five and she was twenty-five. The house had been built around 1843. The family that sold it to the Tenneys had dressed up the house and barns in the latest fashion, the Gothic Revival, giving the house leaded windows in a diamond-paned pattern, gingerbread trim, two false dormers, and a big porch. The Tenneys liked to sit out on the flat porch roof to enjoy the long view down the Connecticut River Valley. It "was a real show place when my people came there," recalled Ruth Tuttle, the oldest child. "My father was very proud of the big meadow beside the house and he used to sing and whistle as he worked there."

Myron and Rosa had nine children. Romaine was their fourth, born in 1900. His father died when he was fourteen, leaving his mother alone to raise the large family and run the farm. At times all they ate was oatmeal. All the children left the farm, except for Romaine. He lived there with his mother until her last years when she moved in with a daughter nearby.

Romaine was the closest to his brothers Myron and particularly Emerson, who was the youngest and lived in Claremont. Emerson saw Romaine all the time, took him off to Thanksgiving dinner, and brought his children by to play and work on the farm.

Emerson's children loved the farm. It was a thrill. "I remember all of us piling into the car and just so excited," said Rosemary. "And I remember stretching my neck and I could see the train trestle and I knew we were almost there. And then the big metal bridge that crossed the river. You'd look down—the floor of the bridge had holes in it so you could see the water—and then you knew the farm was right up there. And then the beautiful trees with the house just nestled in those trees. And we were there. And I think perhaps before Daddy even got the car in park, those back doors were open, and we were gone.

"When we got there Daddy would say, 'Don't go near the horses.' We always rode the horses." They rode them through the fields for hours, ran barefoot all over, played in the barn, drank from the cow-watering trough, rolled down the hills. "Can you imagine four or five little kids running in and out of the barn, in the house and up the fields?" she asked. "And he's just smiling. We'd run by and he'd squirt the milk toward us.

"I can remember running behind the hay wagon, and thinking we were really helping," she said. "It took two of us to lift a bale of hay. We'd pick it up and drop it, and pick it up and drop it, but they were so patient. They were all very kind, gentle people—all the brothers and sisters."

Her sister, Gerri Dickerson, remembers her uncle cutting the tall grass by the house with his carefully honed scythe. "His broad shoulders swung in harmony with the scythe," she wrote. "The muscles in his forearms moved like liquid in a repetitious pattern. It was difficult to tell whether it was the man or the tool that led and controlled the motion. This exercise would have me spellbound. Watching and listening would soothe me as I sat in the soft swath of grass left by Uncle Romaine's rhythm."

Thinking of those days, Gerri said, she can smell the freshly cut grass and hear the "swishing cadence of the scythe." "I am brought back to a safe and happy childhood. It is not just an image of Uncle Romaine; it is as well an image of Vermont."

The new highway was inserting itself into that image of Vermont. Rod Tenney, the oldest of Emerson's children, worked on the farm for several summers. He had a friend who was on the crew surveying the route for the highway down the valley. "They were coming from the north. And it was towards the end of the day and people were ready to call it a day and go home. The guy in charge said, 'We'll just get one more site. We'll just shoot it on that barn down there,'" picking out Romaine's cow barn. "And that's how the interstate highway ended up going right through the middle of the property. If it had gone five degrees one way or the other, you know, things would have been very different. But highways have to go somewhere."

By 1964, the Tenney farm had seen better days. "The barn and parts of the house was just coming down around him. He wasn't a carpenter at all," said another nephew, Ron Tenney. The woodshed toward the end of the ell was falling in and the porch was gone. Tourists who liked their Vermont picture-perfect would ask his neighbors why someone didn't buy that house and fix it up.

Romaine lived in the ell, which had a large kitchen with a wood cookstove, a soapstone sink ,waist-high piles of newspapers and magazines—agricultural journals, *National Geographic*—and a transistor radio by any chair he might sit in. He cooked simple meals on the cookstove, things like oatmeal, biscuits, or beans. He didn't have a garden, except for a large rhubarb patch. On hot days when he was haying he drank switchel, a homemade mix of ginger, vinegar, water, and molasses or maple syrup. He slept in a room upstairs over the kitchen. He followed his own routine; he didn't change his clocks for Daylight Savings Time. Why bother? The cows didn't know the difference.

He never went in the main part of the house, which sat just as it had been forty years earlier, a dusty museum with lace tablecloths, curtains, kerosene lamps, portraits, sepia family photos, mirrors, and an old organ, all sitting in the dark because there was no electric light and the big maples shaded out the sun. No one ever entered the house, except when his nieces and nephews would sneak in, and just once when he invited his neighbors Rolly and Lois. "He showed us his mother's dresses still hanging up in the closet. He never wanted to give them away or anything," said Rolly. "He hung on to his earlier life."

Romaine's dairy operation was in decline. Small dairy farms like his were closing in record numbers. The creameries were no longer picking up milk cans. They required bulk tanks, a setup too expensive to be supported by the average herd, which at the time was just seventeen cows. All across Vermont, families had to face the end of their dairy farms. In the ten years after the first bulk tanks in 1953, one-third of the state's dairy farms closed. Romaine's town, Weathersfield—which included the village of Ascutney—went from having more than fifty farms that shipped milk to thirteen in 1970. It has just one dairy farm today.

He did attempt to modernize. He had some arthritis in his hands, so a neighbor helped him install electricity in the barn for milking machines. But Romaine didn't like the machines and soon went back to milking by hand. He also hired a hay baler; before that he had stacked the hay loose in the barn.

The highway was pressing in. He was supposed to be out by April 1, 1964, but he didn't move. Romaine's old house and barns were an island in the midst of piles of dirt and boulders. By June they were dynamiting within one hundred yards of the house. Rocks from one blast had gone through a wall. They leveled the rolling meadows, removing 100,000 cubic yards of earth on both sides of the house. Bulldozers crossed his front yard; the diamond-paned windows were covered in dirt.

Neighbors pleaded with him to auction his antiques and buy a trailer to live in. They'd look after him. They offered him a room and a barn for his animals. They offered to raise money around town for a new house. Romaine refused. He'd take care of himself.

The Tenney family looked into moving the old brick house, but a construction company told them it couldn't be done. And anyway, it wasn't the house alone that Romaine cared about, it was the farm. The state offered $10,600 for the land and buildings, and then a jury increased the offer to $13,600. Romaine owned the farm with his eight siblings and his mother's estate; that amount would be split ten ways. But it didn't matter. He wasn't leaving his land, he said repeatedly. He'd been away only once, for military service.

Under eminent domain, the government has the power to take private property for public use without the owner's consent. The Constitution says the owner must receive "just compensation" and this has usually been defined as market value. But you couldn't put a market value on Romaine's love for his land.

On the afternoon of Friday, September 11, one week after Romaine Tenney's sixty-fourth birthday, Sheriff Melvin Moore and his deputies arrived with a court order. They emptied the horse barn and the sheds of tools, plows, harnesses and bridles, wagons, and a sleigh. The men "moved gingerly," said one report, looking at Romaine who watched from a side porch for a few minutes before going inside. They stacked it all in two piles under a big elm on the hillside—beautiful old bridles with a gold T on the blinders that he took pride in, and the old sleigh. "He loved that sleigh—he used to spend hours polishing it, painting it," said Emerson's wife, Peggy.

The first call came in at 2:50 a.m. on Saturday, September 12. The alarm could be heard for two miles; the night sky glowed orange. Romaine's house, sheds, and barns were blazing. The cow barn across the road was on fire, as were the two piles of harnesses and tools.

Rod Spaulding was one of the first of thirty volunteer firemen on the scene. The whole ell, where Romaine lived, was burning. Spaulding and the fire chief tried to enter the house. The front door was spiked shut. They knocked the door down. Romaine's dog, Spot, charged in after them. But a few feet inside there was another door—the one that led to where Romaine slept—and it was either nailed shut or blocked. "We just couldn't go any further. At the time we didn't have any breathing apparatus. That entryway, the hall, was all filling up with smoke. There's nothing more we could have done. The fire was crackling and setting to burn overhead. Just too far gone," said Spaulding. The fire chief ordered everyone out, knowing the worst—Romaine was in there. They had to hold Spot back.

Sometime after midnight, Romaine had let his horses and cows free, and set the barns on fire. Neighbors believed that he had timed the fire so that no one would see it. The local machine tool companies were between shifts; no one was on the road. He put his beloved dogs outside, barricaded himself in the house, and set it on fire. Then, probably, he shot himself.

It was an extremely hot fire, so hot that it melted the plastic light on top of the fire chief's car, which was parked about eighty feet away. "With our little engine with only 100 gallons of water on it, we couldn't do a thing," said Spaulding. The nearby town of Windsor had also sent ten men and an engine, but it was no use. The firemen rounded up the cows and spent a half-hour reviving one calf. Neighbors would take the cows and Spot; his niece, Rosemary, would take his other dog, Prince.

More than fifty years later, that night still upsets Spaulding. "It's a terrible, helpless feeling. Been through it a couple of times. It's just terribly hopeless. What do I do now? You go because that's what we do and you get there and there's nothing you can do."

On the solemn morning after, Romaine's family stood by the smoldering ruins with hundreds of others. They had been there all night. "This can't be true. This can't be true," Rosemary thought. "I didn't know where the animals were. I didn't know where he was. It was just a total, total loss. And this beautiful farm and this beautiful man and those animals loved him as much as he loved them. And it was gone. All over a ridiculous highway—which could have, and should have, taken a big old turn."

She had just seen Romaine hours before, around midnight. She and her sister, Joan Newcity, and her brother Ron had gone over to the farm with her father to move some of his things back to their house. They thought he was going to move in with them. Romaine and her father always took a long time to part, chatting by the car. "And that night Romaine cried," Rosemary recalled. "I remember him saying, 'I didn't even milk the cows today.' He was very emotional."

"Don't worry," Emerson had told him. "We'll fix you up. There's got to be some way out of this."

The firemen searched the woods for Romaine all night and the next day hoping that Romaine was "sitting up in the back somewhere" having a good laugh. Rolly and Lois also thought he might be hiding. They put out food in the woods, like other cultures do for the souls of the departed, and they called out his name. "We hiked all over the area, calling his name and telling him that there was food and where it was."

Joan kept seeing Romaine in her dreams. She would dream that he was hiding in a small cave-like space among the ledges up behind the house where they used to play. "We used to crawl in. In my dreams he was in there, hiding. I always used to think that."

"For the longest time we all just expected him to come out of the woods," said Rosemary.

~

He had said that he was going to burn down his farm. "I was born here and I will die here," Romaine had said. When his closest neighbor heard the fire siren, she knew; "he has done what he said he was going to do," she said. Rolly had offered him big boxes to pack his things to move. Rolly was on the school board and they had just got a shipment of new desks that had come in big cardboard boxes. "'Well, yeah,' Romaine said slowly after a minute, 'and if I don't use them for that, they burn well.'"

He had said goodbye many times, but no one really believed him. "Toward the end you'd ask him how he was when you saw him and he'd say: 'Living now, but I won't be living long.' But it was hard to know if he meant it," said Deputy Sheriff Gale, who knew him well. One minute he spoke like that and the next he'd be joking and laughing

On Friday night, in his last hours, Romaine had visited one of his sisters in Claremont, Lena Simpson. "You won't be seeing me anymore," he told her. He had said the same thing a month earlier to Emerson's oldest son. Rod had gone off to college and the Army. He was home on leave in August and he went to see Romaine with his father. "His closing comment was, 'This will be the last that I see you.' At that time I said, 'Oh no, I'll be back, in six months or so, and I'll see you then.' But it was the last time I did see him."

"Maybe none of us knew him as much as we thought we did," said Spaulding. We didn't know "just how he felt about progress." It was like watching someone drown close to shore.

All weekend thousands of people drove out to the farm. They stood silently, staring at the smoking ruins, and then they drifted away. The state fire marshal arrived on Sunday. Spaulding and a few others worked in the cellar hole all day carefully moving bricks. Emerson sat nearby on a brick wall, chain-smoking. The fire marshal would pull some bit out of the rubble, and hold it up to Emerson, asking, "Know it?" and then return to the grim task. "Along about two, three o'clock in the afternoon we came across an iron bed with a rifle that had been fired and underneath the iron bed we found some bones," said Spaulding. They wrapped the blackened bones in brown paper and put them in a metal box to be sent to the state pathologist.

"How do you know why he would do such a thing?" Emerson said in answer to a reporter's question. "Pride," he said. "Progress." That was the collision.

Emerson "just aged terribly after that," said Joan. "You could almost see his hair turn white." Rosemary agreed, "You did see it in his eyes."

No one in town had ever seen anything like the huge earth-moving machines that were building the highway. They could be heard wherever you went. In the summer heat, in the grip of a long drought, windows were shut against the dirt drifting everywhere.

The town had just lived through another huge project. A few years before the interstate arrived, the Army Corps of Engineers had built a large flood-control dam and reservoir, submerging for all time Lower Perkinsville. When they put the reservoir in, they took six farms, four covered bridges, and forty houses. They moved two cemeteries.

Spaulding didn't understand it when he fought the fire—he was twenty-four years old—but as he got older, he came to understand that this had been a "traumatic time for Weathersfield." . . . "It seemed to be upheaval there for four or five years with jobs, the interstate, the building of the dam. What in the heck is going on here?"

When Romaine killed himself, some felt guilty that he had been left alone to fight for his land; others were angry with the state for not finding a way to accommodate him. "You can't treat all old men the same," said Deputy Sheriff Gale. "They don't make old men like machines. Each old man is different. You can't just move them out like you would a younger man. They didn't have to do it. Think of it: Here's a highway that's costing a million dollars a mile and they can't find the money to take care of an old man."

The construction crew and state highway officials were shaken up. They respected and admired Romaine's stubborn stand, even as he was in their way. "He was a damned good old man," said the construction superintendent. The workers "go about it as if they didn't really want to anymore," said one news report. "Do you think we wanted this?" said a state highway official. "I'm sick about it." And yet no state official ever called on the Tenneys to offer their condolences.

Tim Murphy, a deputy sheriff in Ascutney, was seventy-four that summer. He had known Romaine since Romaine was a boy and he had seen his town changing. "It's like chess, this progress thing," Murphy said. "It gets so you don't run the game anymore. It runs you—or over you."

~

The suicide of the old Yankee farmer was national news, picked up by newspapers from New York to Los Angeles. Romaine rapidly became a symbol: the stubborn Yankee, the farmer who wouldn't be bought, the man who chose his own death. This independent farmer person-ified Vermont. He became the man who stood up against Progress, a lone dissenter against the rush to build 41,000 miles of interstate highway. The Tenneys got letters from all over the country, from every state. "Lengthy letters, short letters, condolences. Oh, my goodness. Yes, hundreds of letters," said Rosemary.

Shortly after the fire, the big machinery covered over the cellar hole, dumping thousands of yards of fill to raise the road level, and finished laying out Exit 8 through the blasted rock to meet a rerouted state road. It cost about $5 million to build this 6.8-mile stretch of the interstate. Three hundred and fifty dollars had been budgeted to tear down the house and barns, but that was no longer needed. Af-ter the fire, bulldozing the ruins showed up as line item 586 in the contract: $4. Romaine's ashes were removed and "placed outside the limits of R.O.W. [the right-of-way]." If the investigation of Romaine's death were reopened, they wouldn't have to tear up the interstate to sift through his remains. His suicide had delayed the road for seven days.

The old farm is cut up and buried. To figure out where it once was you need to use the construction plans like a treasure map. On the plans, Romaine's house, sheds, and horse barn trail out like geese trying to cross the road, heading from the southbound lanes and the grass median toward an on-ramp. His cow barn sits a couple hundred feet on, square on across the southbound lanes. A couple of parcels of the farm remain. The Tenney family owns about twenty-five acres that the highway missed, and there are two maples. One of them is marked on another plan for a recent improvement of the Park & Ride lot by Exit 8: "Existing 36-inch Tenney farm maple."

In the years since, Romaine's death has become an iconic tale—

there's a song (*The Ballad of Romaine Tenney*); his story is told in the *Vermont Book of Days* and is covered in the *Rutland Herald* and the *Times Argus* roundup of the "Top Twenty Stories of the 20th Century" in Vermont. In a way Joan and Rolly were right: Romaine is still there. His farm is gone, but his spirit won't leave people alone.

Romaine's story stands as a regret. He stands as the lost and the last. He's the lost authentic life, the unrecoverable past. He's as vanished as the road under our wheels at 65 mph. We know that "all is change"; but we don't know that. It's the truth we don't want to acknowledge. We want Romaine to be there on his farm forever. He is the Vermont we want to believe in. As his niece Gerri wrote, "He not only . . . represented what Vermont stood for, but also unwittingly took so many of us to task to do the same." We wish he were still there. We want the old life, accessible to be visited, *and* we want the new things. Why do we have to give up one for the other? Regret is the literature of progress.

Romaine's nephews and nieces—the living generation of witnesses to his life—will tell you that Romaine was not a protestor. He never waved his pitchfork, or shook his fists in anger. That just wasn't Uncle Romaine. He wasn't an agitator, or even someone who wrote letters to the editor. He didn't go to Town Meeting and speak out. Romaine was a farmer minding his own business when the world came to his door. In that way, he's like most of us. History comes and finds us. A terrorist's bomb announces his grievances and kills people we love. Our own government says we have to fight a war in a distant country and our son or daughter dies there, thousands of miles from home. Or the factory closes and we lose our job and our house. History comes for us. As Americans, we think that we are masters of our fate, until the truth humbles us. The great interstate highways were built to speed us along, built to leave things behind, but you can't outrun history.

Romaine Tenney had the misfortune of living right in the path of the largest peacetime construction project in history. In fact, the surveyors laying out the highway sighted the peak of his barn and aimed Interstate 91 right at it. All of us can end up in the crosshairs of some surveyor, some big project in the public interest, our house sliced in two by the dotted line on someone's plan. Romaine may belong more to our future than our past. There are more of us and we're in the way of ever-bigger projects.

Look at the scale of the projects now facing down small towns and rural places: 200-foot-tall cellphone towers and 400-foot-tall windmills, power lines, gas lines, road widenings, oil depots, airport runway extensions. The scale of these developments, unimagined just a decade ago, overpowers the old houses and towns we love. All the easy places have already been built in. We're all in the way of something, and we're all as married to our routine as a bachelor farmer. As *The New York Journal-American* wrote just one week after the fire, "One has the fleeting suspicion that . . . there may just be a little of Romaine Tenney in each of us, too."

The loss of this one farm more than fifty years ago should be a distant event, but the story seems to be coming at us head on. An online story about Romaine catches its immediacy. The writer mistakenly called eminent domain, *imminent* domain. But that's exactly right. Imminent—impending, threatening, "in imminent danger of being run over." It captures the looming threat, the way the state suddenly appears on the horizon and bears down on you—and only then do you discover that your little ship—*The Pursuit of Happiness*—is just a rowboat. By calling it Imminent Domain, the writer caught the mismatch and how the act of taking is completed at the moment of its announcement. It's like being told you have cancer—you are now imprisoned in a story you did not choose.

For us today, Romaine Tenney's death is the story of *Imminent* Domain. It's a story that's being told hundreds of times a year as towns and neighbors face some new road, behemoth big-box store, cell tower. Somewhere this week or next, someone will approach the microphone in a tense public meeting and say, "I was born here and I'm going to die here. I won't sell out. All I ask is to live in peace on this small plot of land. Why can't you let me live in peace?"

At those times, Romaine Tenney is in the room.

But that's not the whole story, either. This talk of Romaine's death overlooks his life, obscuring his gentleness. What people don't see about this bachelor farmer is that this is a love story. "It's all centered around love and dedication," Rosemary said. "He was born there. He loved the land. He loved farming. He *loved*—I'm talking *truly loved*. He worked hard every day, seven days a week, and it was all out of love. Not out of duty or commitment or need."

"We put that on his gravestone and it's true," said Joan: "'Guardian of his land & friend to all.' And that's what he was."

"It was the love of the land. That's all it was," said Rosemary.

And that makes Exit 8 on Interstate 91 in Vermont just about the saddest spot along thousands of miles of highway.

MY ROOTS ARE DEEPER THAN YOUR POCKETS

LYNNE PLACEY IS A SIXTY-FIVE-YEAR-OLD WIDOW who loves teaching piano in her small house in Stewartstown, New Hampshire. She gives lessons on her mother's piano, which was made the year her mother was born in 1920. "I have enjoyed every minute of what I have done for thirty years," Lynne says. "People say: 'How do you listen to all those sour notes?'" She replies, "Because I know what's coming. I can see down the road." She feels "very blessed" to teach piano.

Lynne used to have about forty-five students a week. She was able to just get by on that, but lately her students have dwindled to about eighteen a week. They're too busy playing soccer, and these days not every home has a piano. She doesn't have any other income, except for Social Security. Her late husband, Donald, was ill for more than ten years, confined to a hospital bed in their living room. Lynne would look after him between giving lessons. Then she broke her back and two months later, on October 8, 2009, her husband died. Like many people, she didn't have health insurance, and the long illness had wiped out their small savings.

A year later, her nephew Landon Placey came by to tell her how she could make a half-million dollars, just as he had done. Her money worries would be over. He asked her not to tell anyone else about his visit; this was just between them. Landon had sold his land, 114 acres, to a group of utilities, Northern Pass Transmission, that wanted to build big, high-voltage transmission towers across his land. He was one of the first to sell, and the contract he'd signed required him to secretly recruit others, he said. Lynne's husband had left her seventy-eight acres on Holden Hill, about nine miles from her home. Her land was right next to her nephew's and it was in the path of the proposed power lines. Landon stayed quite a while trying to convince her to sell. Lynne told him what she had told a real estate agent who had called a month earlier: "I'll listen to what you have to say, but I'm not selling." Northern Pass is a $1.1 billion joint venture of Hydro-Quebec and Eversource Energy to build a 180-mile transmission line through New Hampshire. (Eversource is a large utility serving New Hampshire, Massachusetts, and Connecticut.) To do this they wanted to cut forty new miles of right-of-way across the North Country for transmission towers as tall as 140 feet. Since it was announced in October 2010, the project had angered and divided the North Country.

On one of their first dates, Donald took Lynne to see his land on Holden Hill. "I think he was trying to impress me," she says. Donald was one of eight children who had inherited land from his father, Guy Placey, Sr., once the largest landowner in Stewartstown. Donald was proud of that land; he hunted and fished there. Later, they took their three small girls there for cookouts and he'd take each daughter off by herself and teach her to fish.

"It was a beautiful piece of land," Lynne says. "I grew to have a love for it like my husband did." There are good views in three directions: east to the Balsams ski area, west to Vermont's Monadnock Mountain, and north to Pittsburg. And there is a thirty-five-acre field that was hayed up until twenty-five years ago. Lynne would like to bring that field back.

Shortly after her nephew's visit, Lynne wrote a letter to the local newspapers telling about her "secret" conversation: "Can you imagine what half a million dollars would do for me? I won't tell you I didn't give some thought to all that money. The gold-plated carrot was dangled in my face. Would I bite?" she wrote.

No. "On principle, the idea of a foreign corporation coming in to our pristine North Country to ruin it for their personal gain went against everything I believe in." She was not for sale. Against all that money, she put up "my conscience, my ethics, my devotion to New Hampshire's beauty, the memory of my husband, the love for my children and grandchildren, my concern for the health of those living near the towers, and more. . . ." She asked that everyone stand together: Don't believe them when they tell you Northern Pass is a done deal, that your land will be worthless if you don't sell. Don't let them isolate you; don't let them scare you. Don't sell out your neighbors.

She concluded, "I know in my heart I am doing what is best for my beloved North Country." And she signed it, "Yours truly, a devoted native."

The response was overwhelming. The letter made Lynne Placey a North Country hero.

"I got cards from all over the state thanking me for taking my stand. And I got so many phone calls I recorded the phone calls on to a cassette player because I wanted to be able to listen to them again."

One of those calls was from Atta Girl Records in Thornton, New Hampshire. They invited her to Plymouth, where outside on the sidewalk, posing for a photo, they presented her with an oversized check for $2,000 that they had raised and another $650 from the Alliance Against Northern Pass and others. One man stopped to ask what was going on and wrote her a check on the spot.

It was a timely gift. Because of her nephew's sale, Lynne had to have her land surveyed. "I really didn't know how I was going to make ends meet because I used any spare cash that I had to pay this. Still I knew I had to do it to protect myself on Holden Hill." The money covered the survey and some of her lawyer's fees.

"The thing is: you know material things are going to eventually rust out, break. They're going to end up in the garbage or in the dump, whatever. Or the recycling plant," she says. "I think it's more important to leave my children and my grandchildren the inheritance of land. Because land is something that you can pass from one generation to the other. And they can enjoy working on the land just the same as we have and Donald's family before him." Her grandchildren are talking about farming again, using that land, and she's agreed to sell a conservation easement to the Society for the Protection of New Hampshire Forests.

~

A stranger has come to town. He has suitcases full of money. He wants your land. Will you sell? Does everyone have a price? Is everything for sale—every last piece of land, every rock, mineral, pond, and mountain? Around kitchen tables, families divide: sell or don't sell? And then your neighbor sells—he may be your nephew or the cousin you grew up with—and now he's estranged. There's a new stranger in town.

Strangers' money has drawn a line across the land, sowing discord. It has divided the Placey family. They no longer talk to those who sold out. "We pleaded with them; we asked them not to sell," says Lynne. But they wouldn't listen; they wanted the money. Her sister-in-law is distraught. Lynne was close to her nieces; she can't believe that they would do this. "It just breaks my heart to see these kids toss away the Placey land," she says.

This division is repeated all over town, straining the North Country ethic of looking out for your neighbor. I talked with people who were painfully sorting it out: He's my neighbor—I've known him my whole life—but selling to Northern Pass is a grievous wound. I won't shun him on the street, but I'll avoid his business if I can. They mention David Hicks, who owns Hicks Hardware on Main Street in Colebrook. He had a sign against Northern Pass in the window, but then he sold his land. Talk of boycotting his business was quickly put down—that wasn't the North Country way. One of his friends went to him, in private, and said: *How could you do that?* It was despair beyond anger, the bitter taste of disappointment that parents sometimes feel. *You know better than that, don't you?*

There's nothing left to say. It's a difference of belief. Land is an asset; land is allegiance. That's two different units of measure: money and devotion. What is land worth? What is love worth? Families debated, talking past each other as if they were speaking different languages. Love and money, and love of money. Try having a calm family discussion about that. How do you answer when your brother says, "This place is who I am—how can we sell it?"

~

Rod McAllaster could have sold his dairy farm for four million dollars. But where would he be? He would have sold himself off the earth. This is his place; he was born here. At age sixty, he's a man who knows what he's about. He loves this land. When a real estate agent showed up unannounced at his farm, Rod said, "I'm not interested at all. I don't even have to think about it." There's no amount of money they could offer. "My roots are deeper than your pockets," Rod told him.

His Stewartstown farm is right in the middle of where Northern Pass wants to run its power lines. He describes each route that they have tried to take near his farm, or through his farm. It's like a Gettysburg of real estate maneuvers. They move here and are countered and move again. He lays out the routes: this mountain, that cousin, this stream, that other cousin. Power lines and bloodlines, land and family and money all mixed together. Northern Pass has bought land or rights-of-way on land adjoining his on several sides.

Rod has seen his cousins around him sell. He's grown up with them. Has that led to arguments? I ask. "Well, it don't really help anything," he says diplomatically. "Even though I don't agree with them selling, they've got the right to sell. You know it's America. But at the same time I've got a right *not* to sell." Like his cousin Lynne, he has agreed to sell a conservation easement to the forest society.

I ask whether there is any place he would show visitors to try to get them to see the land as he does. He smiles and asks if I have time to see the farm. Because the small dirt road uphill is blocked by some drainage work, we get on a four-wheeler to ride through the fields. Even though he takes care as we cross the gullies and muddy cow paths, the four-wheeler pitches and yaws like a small boat at sea. We pass cows, some of his 150-head dairy herd.

As we bounce along, he tells me the history. His 967 acres were once five smaller farms. He shows me the house where his grandfather was raised, the site of the house where his father and uncle were born, and where many old barns once stood. He shows me fields that have grown up, and woods they've cut that have come back. Each story is a map with a history. He's in no hurry. "This land here—trapping porcupines when I was a kid, hunting whatever—I know every inch of it. It never gets old. Not for me. I don't care if I'm up here in a blizzard—what I'm up here in—I like being here."

We stop in a field by the house where his grandfather was raised. There are magnificent views to Bunnell Mountain, Mount Washington, and Vermont's Monadnock Mountain. It's a big landscape—lots of sky and green hills. It's panoramic. Farther up the hill you can see Canada.

How would he describe this view? "Spectacular. Breathtaking. I mean, I've seen it every day for sixty years and I'm not sick of it. That's the way I feel about it. I'd rather be here looking at this view than I would somewhere else doing something that actually made money. I don't make any money here, but we've been able to stay here. That's all I ask for. Just to get by and hold on to this property that's been in the family. It's important and there's a lot of history here. And you start wrecking it and the history goes with the wreckage."

He tells me about his older brother, who had no interest in farming and left to become a career Army man. Rod looks around, taking in the view as if it were all new to him. "He's seen the world," he says. "I've seen this."

❧

"There are very few places left like this," John Harrigan says. It's "wild country that you could travel as far as the eye could see and maybe not see anybody. A landscape that's largely untouched by any great scars. . . . You can plant a foot, pivot like a hoop-star, and gaze at a landscape uncluttered by anything but the Milky Way." He's been called "the voice of the North Country." He's owned three newspapers in northern New Hampshire, and today writes a twice-weekly column that runs in thirteen papers. For thirty-eight years he has written an outdoors column for the *New Hampshire Sunday News*. He knows what some from "down below" say: "We've got big power lines all around the place down here. We don't mind them. What's your problem?"

"Is there any place you could show that person to change his mind?" I ask.

"That guy represents to me the far edge of any hope, of making anybody understand the deep-rooted attachment to the land that's being so evidenced here," John replies. "It's almost like a religion. Really you have a hard time beginning to describe to somebody why you have a religion. It's just there."

"So you're saying it's like faith?" I ask. "Yes," he answers.

Faith has been described as "the evidence of things not seen." But for people with strong ties to the land, their faith is the evidence of the things they can see—and that they wish big companies could see as well. John explains it this way: "I've got some meat in my refrigerator that came from a deer that a guy shot up on my first meadow, my first hayfield. I've watched that deer grow up. My mother's and father's ashes are in that hayfield." He chokes up. "As are my younger brother's. I'm eating ashes and microbes that grew into grass that the deer ate. It's just the way it is. I'm from the land, I'm on the land, I love the land, and eventually I'll go back to the land."

In one of his first columns about Northern Pass, in December 2010, he wrote: "We here in the North Country are at rope's end. Having lost about all our industry and not having [help from the state] we have only the landscape left, which is our definition, our heritage, our livelihood, and our meager future."

～

Land is all we have. I was haunted by the echo I heard in what John Harrigan said. It's been said by others who have found themselves in the way of Hydro-Quebec. About six hundred miles north of Rod McAllaster and Lynne Placey, the Innu—the native people of northeast Quebec and parts of Labrador—have been protesting Hydro-Quebec's installation of power lines through their ancestral lands without their permission. In 2012 they blocked the road to a hydroelectric complex for five days. "Our land is the last thing we have left," said one. "It's our identity."

In the early 1970s few in the United States had heard of Hydro-Quebec—now North America's largest power producer—until other native people, the Cree, began to protest that they were about to be flooded out of their homeland near Hudson Bay. And since Hydro-Quebec wanted to sell some of that power to Americans, they were about to lose their ancient way of life so we could plug in our televisions.

The Cree have seen thousands of square miles of old hunting grounds, sacred burial grounds, and villages drowned by Hydro-Quebec. Wild rivers have been dammed; forests clear-cut and sliced up with roads. They have negotiated long, complex agreements with Quebec. It has won them some compensation: jobs, investment, autonomy in local governments and schools. And it has lost them much of their old way of life and brought high mercury levels in fish and people from the newly flooded land, alcoholism, drugs, and suicide. It's a complicated ledger sheet of loss and gain. They have bravely attempted to meet modern times on their own terms.

"It's very hard to explain to white people what we mean by "Land is part of our life," Chief Robbie Dick of the Great Whale community said in 1990 when Hydro-Quebec was looking to dam more wild rivers and flood another several hundred square miles ("actions conservationists say would cut out the ecological heart of a rocky region the size of Massachusetts, Vermont, and New Hampshire," reported *The*

New York Times). A few years later, Grand Chief Matthew Coon Come said, "It's always the case that we are asked to give up a way of life. We are asked to compromise."

You could substitute what the native peoples have said over the last forty years for what the opponents of Northern Pass say today. The Cree, the Innu, and now a North Country piano teacher and a farmer are united by the losses they face.

∽

The face-off in the North Country brings us full circle. Our history on this continent began with taking land. We're reenacting the first encounters in North America between the Europeans and the natives: the Europeans arrive and take the Indians' land. The first things the Pilgrims did when they landed on Cape Cod were to shoot at the Indians, steal a store of corn, dig up graves to take the beads, and steal the most beautiful decorations from empty wigwams. The brilliant Abenaki scholar and storyteller Marge Bruchac asks, "What kind of people would do that?" Us. We would. In fact, we're still at it. Only now we play both parts in this drama. We are the raiders and the raided. We are a people who began with a huge land hunger, and there was a lot to grab. But the era of wide-open spaces is gone and now, says one conservative scholar, we are "cannibalizing each other's land."

We are witnessing a new land rush. In the nineteenth century the settlers would rush to the frontier, occupying land in advance of completed surveys or laws, squatting, as it was known. They would claim more land than they could farm, fell forests, deplete the soil, and move on. In today's land rush, citizens and corporations battle over big utility projects. These confrontations, played out through lawyers and legislation, often have the push and shove of a street fight.

What was once remote—the Cree's homeland, the North Country—may now be taken by someone else. It's going on all over the country. The small Hawaiian island of Lanai'i—population 3,135 on

141 square miles—faced a huge wind power project. A billionaire, Larry Ellison, founder of the computer company Oracle, bought 98 percent of the island. Ellison wanted to cover a quarter of the island with 45-story-tall windmills. "It's awful, just awful," said Robin Kaye, a leading opponent. "There are families who won't talk to each other anymore. It has really ripped us up."

Against this new land rush we have only a piano teacher and a dairy farmer—and fortunately thousands of others who can see across boundaries and generations. When I visit they walk me around their place naming the distant mountains, or host me on their porch, hoping that I can see what they see before them, near and far. Their house is a ship in the land moving across the seasons. This is what they wish the power companies and regulators could see. They know the practical case can be made to protect their land, but that's not why they act. They tell me, "There's a feeling in your soul about this place that I can't really express." They say what John Harrigan says: *You can't explain it. It's just there. You feel it in your soul.* Forget for a moment about looking out for miles at mountains; forget all the practical talk of forest management. The most important view is hidden at first. It's how the land lives inside that person.

Back in the 1920s the writer Sherwood Anderson saw this in the old men around him:

"Is it not likely that when the country was new and men were often alone in the fields and forest they got a sense of bigness outside themselves that has now in some way been lost?" Anderson wrote in a letter. "I can remember old fellows in my home town speaking feelingly of an evening spent on the big empty plains. It had taken the shrillness out of them. They had learned the trick of quiet. It affected their whole lives. It made them significant."

In the North Country and around New Hampshire, I meet many people like this. They know what they are about and where they live is a big part of this. That's why Lynne Placey can say no to $500,000 and why Rod McAllaster will not sell his farm at any price. They have something of the reach of the land within themselves. It has made them significant.

THE PIPELINE IN THE NEIGHBORHOOD

ONE MORNING IN JANUARY 2014 VINCE PREMUS was leaving for work when he met a man in the driveway of his home in Pepperell, Massachusetts. Did Vince have a minute? There was something he wanted to discuss. He laid out a map on the hood of his car and made it seem like a routine matter: a thirty-six-inch-diameter, high pressure, natural gas pipeline would be going right *here*: crossing the Premuses' small field—right where Vince, his wife, Denene, and their two children play soccer, lacrosse, and fly kites. It's where they've had Halloween party sack races for fifty children. "It's just a wonderful place to be outside," Vince told me. "In the fall it's just incredible how beautiful it is." The pipeline would flow 200 feet from his house.

The company representative wanted Vince to sign permission to survey his land, and failing that, to sign a form saying he didn't give permission. Vince refused. He was late for work (he's an engineer studying ocean acoustics), but he went straight to Pepperell's town hall to find out what was going on.

They didn't know much more than he did: Kinder Morgan, North America's largest energy infrastructure company, based in Houston, Texas, wanted to build a natural gas pipeline called Northeast Energy Direct across New York State and then 180 miles across northern Massachusetts. This scene was playing out in town after town: a stranger knocking, saying something about a pipeline, rumors and questions all over town. Scant information.

It seemed unreal to Vince and Denene. How could some Texas company lay claim to their land, and as they saw it, threaten their safety? There had been more than twenty serious accidents, including fires and explosions, on Kinder Morgan's pipelines in the eleven years before they showed up at his house. But Kinder Morgan claims it "consistently outperforms the industry averages in virtually all environmental, health and safety measures." One look shows that "industry averages" can be discomforting. On other companies' pipelines, three major accidents in five years had killed eight, injured more than fifty, destroyed forty-one houses, and damaged many others.

In April the Premuses received a letter with a brace of legal citations saying they couldn't refuse: Kinder Morgan could survey their land anyway. "It was pretty aggressive," said Vince. "It's when we knew we were really into something." And that was for openers: if the pipeline won federal approval, Kinder Morgan could take the Premuses' land by eminent domain. Almost all proposed pipelines are approved by the Federal Energy Regulatory Commission (FERC), say industry observers. And further, the Premuses and everyone else plugged into a utility in New England would pay for the pipeline's construction with a "tariff" added to the utility bills of everyone in New England. The Premuses, by law, had been reduced to serfs on their own land.

After that January morning, the pipeline wedged itself into their life. They studied energy policy, organized their neighbors, picketed in front of the statehouse. The pipeline became their unpaid job. They were out late at countless meetings. Denene addressed so many letters to legislators that her son, Duncan, age ten, memorized their names—he helped her address the envelopes. She wanted her children to see that "you have a voice as well as a vote."

"Can they really do this?" Duncan asked; can Kinder Morgan take our land? "I thought the government was supposed to make sure stuff like this didn't happen to anyone. Because we're kind of small and we can't really fight this on our own."

When I visited, Denene brought out a box of papers and Vince did the same, pulling out a thick file that he had stacked as neatly as a deck of cards. "We have boxes of stuff; we have piles of stuff," said

Denene. "We put it away because we wanted to live normally."

The pipeline had put part of their life on hold. They were going to redo their kitchen, fix a few other things. Their home is a gracious, two-story Georgian built in 1730, a big welcoming farmhouse, decorated with Denene's quilts, appliqué, stenciling, and a few Shaker reproductions. It's an airy, cheery household. On a small chalkboard in the kitchen is a brochure: "Getting Good Grades in Twelve Steps." Theirs is a serious homework house.

They had looked at a lot of other houses when they were house hunting, but this one felt like home from the moment they walked in the door. They could see themselves raising a family here; they could see living here forever. Now, fifteen years later, Denene was cleaning out closets. They may have to move. "We don't want to move. We shouldn't have to move," said Vince. "If they build this thing, we probably will. But we shouldn't be forced to have to face that decision. It's just not right. We love this place. We love this town. Everything about it is just perfect." Pepperell reminds them of small-town New Hampshire where they grew up. They didn't just move into an old house; they adopted the town. Each fall they help collect about 150 old bikes in their barn to repair to raise money for town recreation projects. The Premuses are the kind of people you'd recruit for your town: energetic parents, volunteers, good neighbors.

It's like grieving, Denene said of the way she now feels about their home. "It's amazing how life was going along until we got this knock at the door. There's this world out there and decisions are being made. I find it frightening. My life has been totally turned upside down. I haven't slept well."

I met many people at the pipeline meetings who are united by the same story. The pipeline had become the immovable object against which they throw letter after letter, emails, phone calls, meetings, rallies. It upends normal life. They stay up late at their computers in a crash course on energy policy. The rest of us, obliviously turning on lights, knowing little of how electricity gets to us, are in the dark.

One day you're taking your son to Cub Scouts, commuting to work, thinking of redoing your kitchen, and the next day your house is full of boxes of papers and your dinner talk is all about gas pipelines, the electric grid, and the dizzying array of governing agencies and regulations: FERC, ISO-NE, NESCOE, NEPOOL. From the moment someone draws a line across your land, you're already living with the pipeline. Once it's proposed, your life has changed—maybe forever.

~

If Kinder Morgan's pipeline had been sketched across the map of your town, sooner or later you would meet Allen Fore, the company's vice president for public affairs, and its public face for the pipeline in New England.

Fore is a skilled spokesman, delivering a soundbite for the TV news, or giving a lengthy answer at a public meeting that tends to hold off follow-up questions. One-on-one he seems to enjoy challenging questions as others enjoy returning a serve in tennis. He has a smile that says, go ahead, ask me another question.

The pipeline "is transportation, servicing customers" he has said at forty town meetings in Massachusetts. Kinder Morgan will only build it if they have "firm contracts" with customers—in fact, this is required by federal regulators to approve the project. "We don't pursue projects that we aren't confident in securing both" customers and approval, Fore said.

Fore wants the public to understand why the Northeast Energy Direct pipeline was proposed in the first place: In New England, the amount of electricity generated by natural gas has increased from 15 percent in 2000 to 44 percent in 2014. In the winters of 2011 to 2013 the cost of energy spiked during the morning and evening hours on anywhere from ten to twenty-seven days. There's a "pipeline bottleneck," resulting in the nation's highest natural gas prices, said the nonprofit agency responsible for operating the grid, ISO-NE (Independent System Operator for New England). "The region is challenged by a lack of natural gas pipelines," said ISO-NE. The pipeline's critics countered that this pipeline is not the solution.

The pipeline company has consultants, lawyers, money, and a regulatory review process primed to approve projects. And the opposition only has a ridge-backed determination, an ability to find new recruits, and the burden of being painted as selfish NIMBY naysayers. Immediately they are cast as obstructionists just for living where they do. They're in the way and they're called upon to provide alternatives. Asking someone who is facing a pipeline cutting across her land to solve our energy future is like asking someone with cancer to find a cure for her own disease. But the pipeline opponents are serious students of energy policy.

To answer those who ask: *Well, what would you do?* pipeline opponents said: repair and expand existing pipelines, change the way utilities buy gas so it's cheaper, use liquefied natural gas which can be stored and delivered for far less than what a pipeline costs, reduce demand during peak hours, and increase solar power and energy efficiency programs. The winter shortfall was caused by poor management, they said—the pipelines were only 75 percent full—and was largely due to how the region contracts to buy gas. "The incremental expansions of existing natural gas pipelines, as well as the current supplies of liquefied natural gas that we have on the system, would be plenty to make up the shortfalls that we've been seeing in the winter months and reduce those price spikes," said Shanna Cleveland, a senior attorney then with the Conservation Law Foundation.

Across the nation, pipelines run at only slightly more than half their capacity. The interstate pipeline system used 54 percent of its capacity between 1998 and 2013, according to the US Department of Energy. "Given the cost of building new pipelines, finding alternative routes utilizing available capacity on existing pipelines is often less costly than expanding pipeline capacity," the department concluded in a 2015 report.

But still the industry pushes for new pipelines. It's "definitely an overbuild," says Het Shah, who is in charge of natural gas market research at Bloomberg L.P., a consulting firm. In the race to move gas out of the shale fields, the flow of older pipelines is being reversed all toward one goal—exports, says Shah.

"The build-out of pipelines is a true climate disaster," says Robert Howarth, a professor of ecology and environmental biology at Cornell University and one of *Time Magazine's* fifty "People Who Matter" in 2011. Howarth estimates that methane emissions produced by shale gas from wellhead to delivery could add up to a 12 percent leak rate, causing substantially more warming in the short term than coal. "Methane is an incredibly powerful greenhouse gas that is one-hundred-fold greater in absorbing heat than carbon dioxide, while both gases are in the atmosphere, and eighty-six-fold greater when averaged over a twenty-year period following emission. When methane emissions are included, the greenhouse gas footprint of shale gas is significantly larger than that of conventional natural gas, coal, and oil," writes Howarth. His assessment is vigorously contested by other scientists and the gas industry which says gas is a green alternative to coal. If Howarth is right, the United States will not meet the goals it set to limit global warming. He says that the rapid pace of gas development contributed to the recent spike in global temperatures, including record-breaking heat waves in 2015 and 2016.

Toward the end of my visit with the Premuses I looked at a five-page letter Vince has written to the assistant attorney general of Massachusetts. "You've written a letter with thirteen footnotes," I said. The letter takes issues point by point. It's like something a consultant would have prepared. In fact, Vince is a consultant; he has a PhD in electrical engineering. He can spot an inconsistency in numbers or practices a mile away. And while he's as rational as you'd expect an engineer to be, he's also impassioned. At one point while we were talking, he choked up telling me about the feeling of solidarity that comes over him at the homeowner meetings as he looks at his neighbors. "We're all in this together. Your home is just as important as my home." He stopped, eyes red. "If they build this thing and I didn't fight it, I won't be able to sleep. I won't be able to live with myself. You have to fight this fight. You have to be true to yourself. You have to do what makes sense in your own heart. And if you don't do that, you'll regret it for the rest of your life."

He sat with his head in his palm for a moment and after finishing his thought, he left the table.

This is just one household among thousands of others sitting in the way of a 180-mile-long section of one natural gas pipeline. The map is crisscrossed with plans for new pipelines, power lines, and windmills. By the estimate of one physicist who has studied wind power, to make a significant contribution, windmills will have to occupy three thousand miles of ridgelines in the Northeast, not counting the access roads, transmission towers, or the acres cleared and covered with gravel for the temporary storage of equipment. The way electricity has been generated for a century is changing rapidly. Anyone may be in the way of the next big project slicing across the countryside.

Our power grid is like the "tangled maze of poorly maintained roads" that we had before the interstates were built, President Obama said in 2009, calling for a "clean energy superhighway." The National Renewable Energy Laboratory is advocating ten major power line corridors to deliver electricity from big solar and wind projects. The windiest areas are where population is sparse and the only room for big solar projects is out in the last wide-open spaces. (Renewable energy is not a Get Out of Jail card. It's not that simple.)

In New England, Kinder Morgan's Northeast Energy Direct pipeline was competing with at least four other planned natural gas pipelines, some of them in existing right-of-ways. Nationally, 100,000 miles of natural gas pipeline—main lines, laterals, and gathering lines—have been built in the new century and more than 300,000 miles may be built in the next twenty years. New pipelines leading to planned gas liquefaction plants are radiating out of Pennsylvania's shale gas fields. While the 1,179-mile Keystone XL pipeline to bring Canadian tar sands oil to the Gulf Coast starred in the national news, all these other projects make waves locally. But Americans have yet to realize the large scale of these industrial projects, often inserted into the last rural landscapes. There is no place left that is beyond the reach of industrial development. With each new pipeline and transmission tower, we're choosing winners and losers. Corporations are battling citizens, and citizens are in a dog fight with their fellow citizens. It's a turf war.

As many letters as the opponents like the Premuses wrote, as many marches as they organized, they knew that the power to decide if the pipeline would get built was concentrated in one federal agency: the Federal Energy Regulatory Commission—FERC. It is the acronym that commands center stage. The pipeline's opponents pronounce it sharply; it's the F word.

FERC issues a Certificate of Public Convenience and Necessity to a project it approves. The commission's "statement of policy" says it must consider "the enhancement of competitive transportation alternatives, the possibility of overbuilding, the avoidance of unnecessary disruption of the environment, and the unneeded exercise of eminent domain." FERC requires separate examination of the impacts on landowners and surrounding communities, and the environmental impact.

In a speech before the National Press Club, Cheryl LaFleur, then the FERC chairman, offered some insight into how this works in practice. "Market demand and . . . contractual commitments for pipeline capacity," she said, "determine what pipeline infrastructure is needed." Only after judging market demand do they "look at the environmental and safety aspects."

How much do public comments matter? Noting that "pipelines

are facing unprecedented opposition," LaFleur said, "We have a situation here. We take the views of all stakeholders seriously and try as hard we can to thoroughly consider issues that are relevant to the decisions we're required to make. But FERC's responsibility under the National Gas Act, because we're a creature of Congress, is to consider and act on pipeline applications after insuring that they can be built safely and with limited environmental impact."

Critics assail FERC for being too ready to approve projects. US Senator Elizabeth Warren has said, "I am very concerned about a regulatory agency that is only able to say 'yes, yes, yes.' That's not the job of a regulatory agency."

Chase any one of these fights about a planned pipeline or electric transmission line and you eventually wind up in a thicket of technical details, up late at your computer like the pipeline opponents. Battles are waged over the "Low Demand Scenario" and "forward capacity auctions," but the anguish of these confrontations is lost in the technical talk. What's at stake can be seen by looking at another power line war.

In the 1970s in Minnesota, some farmers were told that a high-voltage power line was going right across their farms. If they refused to cooperate, the power company would take their land by eminent domain. They asked their county planning commission to stop the power line. Just as the commissioners were going to agree, the power company changed the rules for reviewing the project and it went to the state. Hearings were held and the power companies won. After their defeat some farmers turned to sabotage. Men and women who set great store by their neighbors' opinions and living properly, toppled at least fourteen high-voltage transmission power line towers. Some farmers backed a reform candidate for governor, Alice Tripp, a farm wife who had never run for any office. As Tripp campaigned, she summarized the great challenge of these large projects, asking, "Who sacrifices, who benefits, and who decides in America today?"

"What happened to the farmers of western Minnesota is happening to all of us," Tripp said in 1978. "Partnerships of large corporations and entrenched bureaucracies dominate the vital decisions that affect our lives and shape our future. . . . This is not democratic." (Tripp made a good showing but lost, as did the farmers.)

Who sacrifices, who benefits, who decides? That's being asked all over the country. These three questions are a clarifying lens. Homeowners along the ever-shifting proposed pipeline route were being asked to sacrifice, but for whose benefit? And what say do they have?

None of the pipeline's planners would call families like the Premuses "stakeholders." That term is reserved for the pipeline company and the agencies that oversee them. "The ironic thing to me," said Vince, "is that the real stakeholder—the one who is investing, against their will by the way—but the one who's really investing, who has the skin in this game: it's us. They're going to take our property. If you measure the size of our stake as a percentage of our net worth, we have far more invested in this process than even Richard Kinder himself," he said, naming Kinder Morgan's CEO. "And yet we have no say—no real say. We're not invited to the table as a stakeholder. The time has come to actually update the definition of stakeholder and give people like us a seat at that table. There should be representation.

"You should never ever come to some homeowner and say we're going to take your land whether you like it or not because we think that this is needed, unless it's an absolute last case. When you say you have to put a pipeline 200 feet from my children's bedrooms, you'd better have tried every other option first," he said. "There is nothing that they can do to me that's more serious and more threatening than what they're proposing to build outside of our doorstep."

∽

Kinder Morgan, of course, does not think that it is threatening Vince's family. Pipelines, Fore will tell you, are the safest way to move gas, but that's misleading. He is referring to a US Department of Transportation (DOT) study which says that pipelines are safer for moving oil and gas than trucks and trains. That's not a reassuring comparison. Railroads spilled oil more often in 2013 than in the preceding forty years, due in part to a 4,000 percent increase in the number of railcars carrying oil in the previous five years, according to DOT.

There's another way to look at this comparison. There were twice as many railroad spills as pipeline spills in the eight years from 2004 to 2012. But pipelines spilled three times as much crude oil in that same period, according to an International Energy Agency study of DOT's own statistics. Pipeline accidents are less frequent, but each spill is much larger. During this period 800,000 gallons of Canadian tar sands crude spilled in and around the Kalamazoo River, and another 63,000 gallons of crude poured into the Yellowstone River.

Though pipelines have fewer accidents, "when they blow, they really blow," said Carl Weimer, executive director of the independent, nonprofit Pipeline Safety Trust. "They have huge potential for wide-ranging explosions." In ten years there have been an average of seventy-seven "significant" accidents each year on natural gas pipelines, according to DOT, resulting in two deaths, ten injuries, and $141 million in property damage.

"Kinder Morgan is proud of its safety record," the company said. "We have consistently outperformed industry averages." They publish their safety statistics on their website. No other company does that, said Fore. But there have been more than twenty serious accidents on Kinder Morgan's pipelines from 2003 to 2014. One Wall Street analyst reviewing Kinder Morgan's capital expenditures believes they spend too little to maintain their 80,000 miles of pipelines. "We struggle to understand how [Kinder Morgan] can safely operate the largest portfolio of transmission and storage assets in the industry for just a fraction of its peers' expenditures," the analyst wrote. Compared with

Spectra Energy Partners, another big pipeline operator, Kinder Morgan spends about half the maintenance capital of Spectra per mile of pipeline, he said.

Kinder Morgan dismisses this analysis as "uninformed and irresponsible." Its capital expenditures represent only a portion of what it spends on maintenance, they said. Many of those costs are in its operating expenses. "We are safe and efficient operators," they said.

Maintaining pipelines—gas and oil—is crucial since most accidents occur on older pipelines. "There's more than one leak, failure, or rupture involving an oil or gas pipeline every day in the United States," said an investigative report by *Politico*.

These older pipelines—gas and oil—are not properly inspected. There's minimal regulation, and many of the regulations are written by the industry. "Tens of thousands of miles of pipeline go completely unregulated by federal officials, who have abandoned the increasingly high-pressure lines to the states," reported *Politico*. The agency in charge, the Pipeline and Hazardous Materials Safety Administration, "lacks the manpower to inspect the nation's 2.6 million miles of oil and gas lines" and it "has stubbornly failed to take a more aggressive regulatory role, even when ordered by Congress to do so."

Two months after I visited the Premuses, Vince and four other landowners did get a seat at the table as stakeholders—for one meeting. For three hours they met with Gordon Van Welie, the CEO of ISO-NE, and four of his staff, in the conference room overlooking the operations floor where ISO-NE manages New England's electric grid. (It looks like Mission Control.)

Vince's group pressed Van Welie to look harder at using renewable energy and efficiency programs to reduce peak demand. The public is receiving a message that gas pipelines are the only option, they said. Van Welie conceded that they could present a more balanced range of options, but he emphasized that as older power plants retire, the grid managers are "painted into a corner." They are "right at the edge" during the peak winter demand hours. Only an "incremental" increase in the supply of natural gas could meet the shortfall in the shortest time, Van Welie said. He doesn't think renewable energy or promoting efficiency would be enough in the short term. Van Welie sees a future in which New England will rely on natural gas to generate as much as 88 percent of its electricity, and he is not concerned about this increasing dependence on gas.

Inserting a pipeline into the landscape of rural towns rewrites thousands of stories. It changes everything it touches. Townsend, Massachusetts, lies six miles west of the Premuses' home. I toured the town with Carolyn Sellars, a busy volunteer, and Leslie Gabrilska, the town's conservation agent. Leslie is the paid staff for Townsend's conservation commission. Carolyn is jovial. She talks fast—she kind of percolates—and she seems relentlessly happy. Leslie is serious and quieter, with the demeanor of a deliberating official.

They showed me the planned pipeline route. I immediately saw why Leslie was one of my guides. A tour of the proposed pipeline route is a tour of conservation land. It looks like Kinder Morgan had stitched together the green spaces on the map. Each time the car paused, they pointed to a state forest, a brook, or land conserved in a town trust, each with its own Eagle Scout project to mark the walking trails. I may have seen a decade's worth of Eagle Scout projects. Kinder Morgan picks bigger open spaces because it's easier than dealing with many small landholders, said Carolyn. Fore, the company spokesman, denied this, explaining that they are trying to balance environmental concerns and the interests of private landowners.

This is hard-won land, assembled with land deals to build clustered and affordable houses, put together with state money and private donations. Townsend was one of the first towns to allow denser housing developments in exchange for setting aside open space. Massachusetts has designated two-thirds of Townsend as an Area of Critical Environmental Concern. There's a state park and forest, two state wildlife areas, and the closest cold-water fishery to Boston. The state has heavily invested to protect Townsend.

At one of our stops, they showed me the possible location of a huge compressor station which is required at intervals along the pipeline to keep the gas moving. Rated at 120,000 horsepower of continuously running gas engines, the station would be one of the largest in Massachusetts. Because of its noise, the station is required to be set off on fifty to seventy-five acres. It's hard to imagine the intrusion of a big industrial installation, lit twenty-four hours, on this wooded, rural backroad, but the neighbors of compressor stations in other towns say that they've had their lives ruined, the country quiet they love and the dark night skies vanquished. The noise can be overwhelming. It "sounds like a truck in my driveway, twenty-four seven," said the neighbor of one Pennsylvania compressor station. He couldn't sleep at night. "I don't even look forward to going home anymore," he said. "Inside my house, I can't get away from the noise." That compressor station was just one-fourth the size of the one planned for Townsend.

Thousands of hours of volunteer work, committee meetings, time drafting state laws, and public investment lie behind the landscape that Townsend has curated. The pipeline and the compressor station—if it sticks to this route—would wipe out all this work. It would nullify the community's and the state's will. The pipeline would be a veto delivered out of nowhere. No to your ideas of what land should be conserved. Overruled.

Carolyn Sellars has been here before; she's upset, but she doesn't despair. Carolyn has defended her family's farm in Winchendon, Massachusetts. Her family has owned the 125-acre farm since her great-grandparents, millworkers, bought it in 1901. She showed me photos of the farm first thing when we met, the way others might show off photos of grandchildren, starting with the house, ca. 1815, with the ancient maple in front. The arborists said "'you really should take that tree down,' but we said, 'We can't: This tree is like a member of this family.'" The lawn by that maple is where her mother, Carolyn, and her own two daughters celebrated their weddings; she spoke of the walks on trails and forgotten dirt roads to a small beaver dam and an early mill dam and of "The Spot"—a place where Bailey Brook meanders through a hemlock grove. Everyone in the family calls it "The Spot" as in, "Let's go for a walk to 'The Spot.'" The pipeline's one-hundred-foot-wide clear cut would cross right near "The Spot." The pipeline would bisect her family's farm.

Twenty-five years ago Carolyn helped fight off a regional garbage dump (1,200 truckloads a day) which would have been built next to her land. That was followed by a proposal to build a major international airport, and then a dragstrip with seating for 30,000 spectators. She's been a part of the efforts that beat them all back.

"This is a marathon," she said. "You do what you can each week, but you have to keep your life going or you'll burn out." During the earlier fights, she was raising her children, working part-time, and helping her mother face cancer. She learned to take a Saturday off to play with her kids, to pace herself, take a breath. Carolyn Sellars is a jolly warrior.

Almost a year after Vince Premus found that company man in his driveway, all those months that they had lived on Pipeline Time, they heard rumors that a big announcement was coming. Then on November 7, 2014, Kinder Morgan announced it was "seriously considering" a new "preferred" route for part of the pipeline. The pipeline had jumped the border into New Hampshire. It would still enter Massachusetts from New York and cross the Connecticut River in the western part of the state, but then turn north and cross seventeen New Hampshire border towns—and 155 wetlands and 116 bodies of water—before turning south to meet the pipeline terminal in Dracut, Massachusetts. But this would not be its final route. Kinder Morgan was keeping its options open. It hadn't said that it had given up on the old route through Carolyn Sellarses' farm, through Townsend, or through the Premuses' field. They had been spared—or had they?

When Denene heard that the route was being moved, she started to cry. "There was going to be a whole new set of people who were going to have to go through everything we had gone through and it was going to start all over for them. There was no relief. I just felt terrible."

No official reason was given for the move. But the project's opponents cited the 50 percent of landowners who refused to give permission to survey, the opposition of both US senators from Massachusetts, and Article 97 of the state constitution, which requires a two-thirds legislative vote to change the use of conservation land.

In New Hampshire the pipeline became *the* news. Few people had ever heard about it, even though its previous route was just a dozen miles south—across the state line. Borders can camouflage things, hide all sorts of connections.

On a Saturday in mid-December, a meeting at an elementary school in Mason, New Hampshire, was packed, maybe 175 in the room. No one had put "learn about gas pipeline" on their Christmas to-do list, but now here they were.

The crowd was patient, almost holding its breath. It was like the first day of school—Energy School, Power Grid 101. The pipeline opponents from Massachusetts helped organize this meeting; no one from Kinder Morgan was here. They would be soon enough. The opponents were all over the room in their yellow T-shirts, this army's uniform. They had sold out of lawn signs ($6); they had petitions, signup lists, and cards on which to mail comments to FERC. There were maps of possible routes across Mason on the walls. "The map has been changing as fast as you can print them out," said the first speaker.

The next speaker dove into the economics and regulation of gas pipelines, and the power grid. When he said, "It's complicated," some laughter rippled in the room. He asked, "Aren't you overwhelmed? Yes. And guess what? They're not. This is their day job; it's our night job."

A man followed who showed a short film he's making about compressor stations—they release methane and other toxins into the air to relieve pipeline pressure. Kinder Morgan disagrees, saying that "under normal operations" methane is not released. "When gas is vented, it is done under controlled conditions," they say. (But just what are "controlled conditions"?) On other pipelines, Kinder Morgan has returned to add more stations so there's one every fifteen to twenty-three miles or so. The proposed pipeline plans now showed compressor stations every thirty-nine to fifty miles. That can change, the man said. It can all change.

Here it was once again, the "quiet citizen" being told to yield to the "exuberantly active one." Real estate was becoming more unreal as they sat in the meeting. The pipeline was an earthquake that set their property in motion, and left their property rights unsure.

At the Mason meeting I spotted Vince and Denene Premus standing on one side of the room. This is like the first meetings in Townsend eight or nine months ago, Vince told me. And then I saw Carolyn Sellars. She said that the pipeline was off her farm, but a smaller lateral pipeline was now across the street from her Townsend home. The pipeline had left Vince and Denene and Carolyn, but they were still fighting this project. They were marathoners. They were here to help. They know what it's like to show up in the crosshairs of somebody else's plans.

I would run into them again at other meetings that year and I would see the same questions being asked as the pipeline, like a wave, hit each community. The proposed pipeline's arrival in each town was met with a civic version of the stages of grieving: disbelief, anger, organizing. I sat late into the night at the formal, federal meetings, as speaker after speaker tried everything to drive off the monster at the city gates: reasoned, factual testimony; fiery Patrick Henry oration;

and the sending forth of eloquent children to speak for the earth.

At one meeting on a hot summer night in a stuffy auditorium, Holly Lovelace, her voice quavering, got down on her knees and begged for her home, her deck where they loved to play cribbage, and the twenty-two acres with its woodlands, fish pond, stream, and the night sky "so dark you can see the Milky Way with all its brilliant clarity." "Our home life is simple but it is awesome and it is our American dream come true," she told the state agency. But a letter with a map had arrived in the mail; their home would be just 1,500 feet from a compressor station. "Our home has been effectively condemned," she said as she struggled not to cry. "We cannot refinance. We cannot sell because no bank will approve a mortgage in an impact zone [from a possible pipeline explosion]. Our homeowner's insurance provider will keep us on until the compressor station is built, but after that we can't get homeowner's insurance anymore. When we asked our attorney if [Kinder Morgan] would buy our now unsellable and uninsurable home, he said there's a long, expensive process for this. He has given us the name of an attorney in Washington, DC, who is a FERC law specialist. . . . We don't have the money to retain a lawyer. So next week I will cash out my retirement account and I will pay her to try to get back what was mine only a week ago." She started to cry. "I am begging you," she said, falling to her knees. "I am begging you."

And then, just as suddenly as the pipeline had arrived, it was over. Almost two and a half years after that man had appeared in the Premuses' driveway, and ten months after Holly Lovelace had begged for her home, Kinder Morgan announced that they were suspending the project. On April 20, 2016, they said they had failed to find enough customers. The competing pipelines were further along and there was stiff opposition. The Premuses, Carolyn Sellars, and the rest of the opposition took in the news cautiously with a kind of exhausted disbelief, wary that the pipeline could be revived. Holly Lovelace hugged her friends and celebrated.

The death sentence on her American dream had been revoked. A month after that announcement, Kinder Morgan withdrew their application. The Northeast Energy Direct pipeline was dead.

Gordon Van Welie, who runs ISO-NE, the agency in charge of the grid, also changed his mind about the need for new pipelines. "We've got as much pipe as we're ever going to see in New England," Van Welie said at a power industry seminar in October 2017. "There's hardly any congestion left in New England." The "pipeline bottleneck," the chief justification for building the Northeast Energy Direct pipeline, seemed to have vanished.

The pipeline battle had left scars. On some day during all this when it was snowing—was it late fall or early spring?—I was once again visiting with the Premus family, when Vince asked his daughter McKenna to bring her notebook to the table. She was in the ninth grade at Bishop Guertin High School. This spiral notebook was from her civics class. From time to time, Vince and Denene tutor her in algebra and go over her notes from her other classes.

Vince read aloud from his daughter's neat handwriting, looking at that moment oversized for his own dining room table, as if he had wedged himself into his daughter's desk for parents' night at school. He summarized the first classes which covered the different forms of government: "There's some mutual benefit by coalescing. You might forgo certain individual rights for the benefit, but the definition of democracy is: the people hold sovereign power, and the government is conducted by people.

"There were distinctions drawn here for direct democracy, pure democracy: the will of the people translates directly into public policy. And then there's indirect democracy, representative democracy, where a group of persons is chosen by people to express the will of the people. But ultimately the thing that differentiates or distinguishes democracy is that people are sovereign.

"And I remember at the time, sitting with her, and just thinking, like, Oh, my gosh, what has happened here?"

"It's gotten muddled. It's gotten muddied," said Denene.

"But worse yet," said Vince. "If it were muddled or muddied that would be fine. Because it's like, alright, well, we're starting to lose our way. But this is like the impact is real and it's on such an enormous, incomprehensible scale. And many people aren't even aware that they've lost their sovereignty and people in other parts of the country are taking advantage of that. These are Americans doing this to Americans. And I feel like I'm being goofy talking at a den meeting trying to teach the kids about what's special about being an American." Vince is a den leader for his son Duncan's troop.

"But at some point you're teaching Duncan, making sure Duncan has his pledge of allegiance down when he goes to the next Boy Scout meeting because it's one of the things he's going to have to demonstrate that he can do. And of course he knows it by heart. He's known it for years. But then you start to really think about the meaning of those words that you recite by rote since you were six years old. And it's like: Do those words really mean anything? What the hell happened here?"

He sat quietly. On my tape of that interview all I hear is my pen scratching the paper.

THREE BEAUTIFUL DAYS
ON A WARMING PLANET

Great Duck

IN THE FOG, CAPTAIN TOBY STEPHENSON MOORS his boat, the *Osprey*. We can't see Great Duck Island, which is only 250 feet away. The forty-six-foot-long boat belongs to the College of the Atlantic. Stephenson is bringing Professor John Anderson and four students back to the island this morning, sixteen miles from Bar Harbor, Maine, to continue their bird research.

Two students get into an orange inflatable landing boat that's tied to the *Osprey*'s stern. It's packed with gear—watertight boxes and plastic jugs and backpacks and two long oars in case the outboard motor quits. Everything has to come to the island on this rubber craft: peanut butter, potable water, gasoline, lumber—three planks on this trip—and propane tanks. We'll take away two empties. As they push off, Stephenson jokes, "You know which direction to go?"

The boat's silhouette disappears in the fog and moments later the fog thins so that the day is both occluded and bright at the same moment. I can see the little white boathouse, the only building on this shore, and the long ramp down to the water. The Coast Guard built the boathouse back in 1890 and the house that the college uses. "This is a good day," says Stephenson. The sea is calm. He doesn't care about fog or rain as long as it's calm. He's been in seas that have slapped the rubber craft into the *Osprey*, and knocked the craft among the rocks as they tried to land.

I board for the next landing. The orange craft runs right up the slippery ramp, which looks like a wet rollercoaster track. A student jumps out to help the others ashore pull the boat forward. We get out and carry the supplies up the steep ramp.

Professor Anderson immediately walks me out about a quarter of a mile, no more, from the boathouse. This is no country ramble; he's determined. Great Duck is a mile and a half long and all of 220 acres. The island has a fascinating history that is representative of the settlement of coastal New England—pirates who left a treasure that has been found, the first lone farmer, and an era when the population peaked at forty with the lighthouse keepers and their families (and one schoolhouse in a barn). But most importantly, at a time when gulls were hunted to near extinction, Great Duck was home to the last big gull colony on the East Coast. The Nature Conservancy acquired the island in 1998. They share it with the college, the State of Maine, and a private summer resident. The college owns about twelve acres, which includes the lighthouse, the Head Keeper's House, and two boathouses.

We walk through a wind-thrown spruce forest on an old track until we come to a clearing, a low meadow sitting down before the next rise. Like a good teacher Anderson lets me look things over, saying nothing, until I ask him for an explanation.

"You're basically at sea level where you're standing right now," he says. "This is very typical of a lot of the islands up and down the coast." We are standing in a "surge channel" between two small granite domes. A natural seawall of cobblestones keeps the sea out—for now. The sea has rushed in twice in the recent past, during a hurricane in the 1930s and when a big winter storm in the 1960s filled the channel for weeks.

"With a six-foot sea-level rise"—one current forecast for the end of the century—"we're not going to have Great Duck anymore. We're going to have little dinky duck and medium-sized duck," he says. "And this is going to be a sea channel." This will likely be underwater in his lifetime, says Anderson, who is fifty-nine. Born in New Zealand and raised in Britain, he has taught ecology and natural history at the college for twenty-nine years. He'll tell you, "I find much of the post-1914 world in extremely bad taste, and deeply resent having missed Charles Darwin by less than a century." There will be other changes on the island. "All of the present guillemot habitat is going to get flooded. In some areas they'll be able to move to higher ground," he says. "They're in better shape here than on some other islands."

He reviews some of the islands he has studied for Acadia National Park: Shabby Island is a marsh surrounded by a rocky berm. "End of island." Great Spoon Island has some higher places for the gulls to nest, but all the prime eider duck habitat will flood out. Schoodic Island, Acadia's prime seabird island, "will be reduced to three chunks of rock. No eiders, no guillemots. Little bit left for cormorants and gulls, but all pressed together in a very small space. I don't think they'll stay. The low islands are in trouble."

Walk a few hundred yards on this one small island and you can see that the rising sea is about more than seawater and shorelines. Maine's islands are home to about 50 percent of all nesting seabirds in the eastern United States. And that population is in turmoil.

On our way out to Great Duck we had passed close to Thrumcap, a vertical-sided stone ship of an island. It's part of Acadia. All granite; no flourish of greenery. Twenty years ago Thrumcap was busy with cormorants and gulls. "There were over 800 pairs of cormorants nesting on it. The last time we counted there were only forty left," he says. On Great Duck's neighboring island, Little Duck, there was a big gull colony ten years ago. "The last time we landed there, two summers ago, we found twenty-four nests. There should have been 800." (But on Great Duck itself, the population is stable.)

Seabirds are "declining very, very rapidly. Depending on who's doing the numbers we've had between a 40- and 70-percent decline in gulls here in the Northeast in the last ten years. If this were any other group of birds, we'd have headlines and national boards of inquiry, but because it's just gulls, nobody cares. But we care. And part of what makes us very nervous is that everybody is seeing a big decline in gulls all around the northern North Atlantic." Bird by bird, we are losing the world.

Anderson is an impressive teacher. In our short walk he has set the Gulf of Maine spinning. It's a bit much to take in. This is only a glancing tour of the natural and unnatural history of the islands, but the story is all about change. The natural changes and the ones we've caused are as knotted up as the storm-wracked fishing lines that wash up alongshore. For a moment I wish I were a tourist at a lobster pound sitting there soaking in "unchanging Maine," maybe telling children and grandchildren that it's just as it ever was and so it will be for them. But that's a fairy tale Maine, a fairy tale earth. The Maine coast that we see today, the stunning place of deep pine green rimmed with granite, is only 5,000 years old. When you begin to see that 5,000 years is, in the earth's time, really but a weekend getaway, you feel the earth spinning, you feel everything moving—including yourself—which is closer to the true condition.

Forget for a moment the distant forecasts, and all the arguments about climate change. We first have to accept that the earth is changing and it has always changed. Now it is changing very fast due to our carbon-producing ways. If we can learn to see that marshes move, that barrier beaches belong more to the sea than the land, then we can begin to listen to what the scientists have been trying to tell us.

We walk to the old Head Keeper's House, the college's headquarters for its island research. The students are busy unpacking the groceries, getting ready to go to the lighthouse where they'll observe the herring gull colony. Later they'll suit up in black trash bags (for obvious reasons) and walk among the gulls to do a "chick check," weighing the young in the nests. One student is reviewing the images from game cameras she has set out to watch the burrows of Leach's Storm Petrels. They are the most numerous birds on the island—about 5,000 —making this the largest studied petrel colony in the eastern United States, a point of pride for Anderson and his students.

Captain Stephenson is everywhere; he's a busy man on land. I can't count the number of things that he fixes. Each time I see him he's off on another task, carrying a long ladder with a student, repairing the boat ramp with another student, fixing the hot water heater, then the internet connection. He's calm about it all. He pretty much can't walk forward without stopping to fix a waiting project. By day's end his light tan shorts are dirty and he has a bloody cut below one knee that he doesn't bother to look at.

Anderson is sitting at a computer doing what scientists do; he's checking the data his students have gathered on the birds. When he did his three-year-long study for Acadia of the future of the major seabird islands, he had to keep revising his forecast. The climate scientists kept increasing their estimate of how much the sea will rise—it's the fastest rise on record for the last twenty-eight centuries. By mid-century Boston could see a "one-hundred-year flood" every two to four years, and every year by the century's end. Scientists are shocked at how fast the earth is warming with each year's temperatures topping the previous years. The ocean science is complicated. The ice melts, the sea rises; the sea warms, water expands, the sea rises more; the water becomes 30 percent more acidic; the Gulf Stream may slow with dire consequences. We're conducting an irreversible, earth-sized experiment.

Acadia is on the advance edge, seeing greater changes than nearly every other national park. Native trees and plants are dying; Lyme disease is surging. Acadia is spending more than $200,000 yearly to suppress invasive plants like Morrow's honeysuckle, purple loosestrife, barberry, and glossy buckthorn. They are crowding out native orchids, lilies, and asters. Fir, spruce, aspen, and paper birch are declining. "Our forests, coastlines, wildlife, and iconic views are already very different than they were when the park was created one hundred years ago, and will change even more in coming years," says Abe Miller-Rushing, the park's science coordinator.

"The Gulf of Maine is warming faster than any other body of its size in the world. When you have that sort of shift in temperature, all kinds of other things happen," says Anderson. The temperature had been rising 1 degree Celsius every forty years; it's now warming up one degree every *four* years. In water just a couple of degrees warmer, the oxygen content is reduced and some fish, like tuna, can't survive. Algae blooms and pathogens increase. "One of the immediate problems is softshell, which is a fungal infection which basically makes lobsters look like Frankenstein's monster," Anderson says. One third of lobsters in southern New England have softshell disease and it's creeping north. Lobsters make up three-quarters of the value of Maine's fishery, and while there have been record catches recently, there are signs of trouble with recent declines in the number of the youngest lobsters. Scientists know that it's difficult to predict the lobster population. And it's getting more difficult as climate change is creating what they are calling a "brave new ocean," an ocean they don't understand.

At the same time, the fishery is threatened by a new development, rockweed harvesting, which is sold for fertilizer or to eat as nori. "As people lose their fisheries, they're looking for other things to live off, and there's rockweed," he says. But "rockweed is a very important nursery for an awful lot of species of marine organisms. So you're matching nineteenth-century greed with twenty-first-century technology. You've got a problem. And I don't know how to solve that. Because people are used to the idea of 'let's go get it.' And, yeah, I was brought up with the idea the seas would feed the world. And we finished that off in twenty-five years." Anderson is sitting on an old couch under some maps of the island. And while his crisp British accent never loses its bite, he is weighed down by what he's saying. He slumps deeper into the couch. It's one of those moments when people speak directly and say here, this is the burden I carry.

What's happening in the Gulf of Maine is a part of a great dying. In the history of life on earth there have been five mass extinctions when thousands of species were extinguished. We're living in the midst of the sixth great extinction, say scientists, the greatest since the dinosaurs vanished, and we're causing it. "No other creature has ever managed this, and it will, unfortunately, be our most enduring legacy," writes Elizabeth Kolbert in *The Sixth Extinction*.

The earth is on course to exceed the limits agreed upon in the international climate agreements: holding warming to a 2 degree Celsius (3.6 F) increase. Life on an earth hotter than that offers a grim future that could bring down civilization, say the climate scientists. It's widely accepted that there's going to be a significant loss of ice at the poles. "That's not actually the worst-case situation. Worst-case situation is the Antarctic ice shelf goes," Anderson says. "If the Antarctic ice shelf goes, Bar Harbor goes with it. It's not just going to be seabirds. It's going to be New York. And honestly I don't worry about wildlife and global warming. All of the species you're looking at have survived massive periods of global warming, global cooling. The thing that I'm really terrified of is how these guys are going to deal with global warming and seven-and-a-half billion humans not dealing with it very well. So we're going to lose New York," he says with a frankness that is jarring. He has pointed out the elephant in the room.

Superstorm Sandy, which caused $65 billion in damage in 2012, making it second only to Hurricane Katrina, was just a modest storm. Sandy, many say, is the preview. "Imagine when we get a real superstorm when we've had six feet of sea level rise. The whole of lower Manhattan is going to go under. What are we going to do with six million New Yorkers? That's the question. We're going to lose New Orleans. And we've got two choices at this point: We can evacuate New Orleans carefully and slowly and deliberately, finding new places for people to live, finding new jobs for them, settling them in better places. Or we can evacuate them in the rain on busses with what they can carry. The one thing we don't get, is we don't get New Orleans. But we don't want to think about it. That's the next one hundred years. My students are going to live to see this."

And yet here we are. It's a glorious day. The fog gathers, thins, and gathers. The students are busy counting chicks, studying the gulls. We go to see the petrel burrows a student has flagged. "Think of how lucky we are," says Anderson. "Isn't this a wonderful place? Isn't it insane that we're getting paid to go drive around the coast and look at beautiful birds and beautiful islands?"

～

For months I've been doing this, showing up along the coast and saying: show me what we're going to lose, show me what will be underwater, show me the shape of things to come. In each place I am standing at the corner of Loss and Change. And in each place I have the same conversation: This is changing—*fast*. This conversation is going on all along the East Coast and, in fact, the world over.

I've been rowing against my own belief in rock-steady traditions, in wanting to believe that the iconic places that we all love—the islands of Maine, the stony ridges of the Presidentials—you name it—that they will abide. That it will all be the same as it ever was, the same when the next generations discover this place. We want to be the benefactors of all that is graceful; we want heirs. But this is no longer our future.

Think about the coast as a long necklace of beloved places, coves and marshes, peninsulas, docks and bays, clam shacks and lobster pounds and ice cream stands, old cabins and places people return to year after year. Summer lives here. Summer stories are about the joy of returning to the same treasured places every year. They're about the joy of constancy, the comfort of routine. We get older, children leave home, parents face illness and die, careers flourish or stall, but come summer people can head to their favorite beach, or B & B, or family summer place. They can exhale. They can do as they have done as far back as they can remember.

When just one of those places closes, it's like a thread snaps. It's like a rude shove in the back, a push to keep on moving along your life's small timeline. The yearly pilgrimage, the eternal return, is cut off and people miss it deeply. Now magnify that a thousand-fold up and down the coast from Maine to Connecticut. Goodbye to beaches, marinas, beach houses, marshes, lighthouses. . . . Who can face that? I know that I can't.

Doggerland

In the nineteenth century, fishermen in the North Sea off England would sometimes pull up odd things in their nets. Clumps of peat. Deer, bison, bear, and horse bones. How strange, they thought. How did these animals get sixty miles from shore? As trawling increased in the twentieth century, they found more things. In 1931 a trawler brought up a barbed harpoon that was carved from an antler. A beautiful object, like something the Inuit would make.

With the boom in North Sea oil, the sea floor was extensively mapped, revealing the valleys, rivers, lakes, and hills of a lost land. The archaeologist Bryony Coles named this place "Doggerland" in a 1998 paper. Part of it was located on what today is the Dogger Bank. Doggerland connected England to Norway, Denmark, the Netherlands, and northern Germany. There was no North Sea. The English Channel was a tidal estuary, a river snaking through a marsh to meet the sea. Then as the great ice sheets up north began to melt, the sea rose, the English Channel formed, and Doggerland became, after many years, an island. It was a chilly, rainy place of marshes, inlets, and fens. Small bands of hunter-gatherers lived off the salmon run for months. Survival required precise local knowledge of the fish, birds, deer, and plants. This knowledge was your inheritance. From one generation to the next, Doggerland changed as old fishing grounds were lost. Some of that inherited local wisdom became obsolete. Each generation had to relearn their native home. As the sea rose, the marshes flooded, the land flooded, and at last on a day lost in prehistory, about 7,600 years ago, Doggerland's people left for England. Today there are people in Britain and Australia who can trace their lineage to Doggerland. DNA tests link them all the way back to that North Sea Atlantis.

Doggerland is not something most people know about, but it is now squarely in our future.

Doggerland is our story. Under the current forecast from the scientific research organization Climate Central, about 760 million people—more than 10 percent of the world's population—will be left homeless in the next century by the rising sea. About 40 percent of the US population lives along the coasts. Many of our revered landmarks are close to sea level, like the Statute of Liberty, Faneuil Hall, the Kennedy Space Center, and the location of the country's first permanent English settlement. Almost all the archaeological sites on the historic Virginia island of Jamestown will be lost in fifty years. The "birthplace of America" will be underwater. "If you're within two to seven feet of sea level today, then saltwater *is* in your future this century," says one museum consultant. Many coastal cities are "close to a tipping point," says the National Oceanic and Atmospheric Administration. At high tide flooding will be routine. The rising seas will upend our notions of property. Each bit of warming today locks in the changes for centuries to come.

Old Lyme: Another Beautiful Day

When Evan Griswold was seven years old in 1954, his mother led him out into the eye of Hurricane Carol. The sky was bright blue, the surf raged. They ran all the way down to the beach. He saw the front of a house disappear into the waves. He's never forgotten it, a lesson that's with him more than sixty years later when I visit.

On a sunny day I walk out to Griswold Point with him and Adam Whelchel, the director of science for the Nature Conservancy in Connecticut. Griswold Point, at the mouth of the Connecticut River, is one of those places where you feel yourself pinned to the map, where you can imagine yourself as seen from far above. We look south to a marsh, a barrier beach, and Long Island Sound, and west across the Black Hall River to more marsh and the white stone lighthouse commissioned by George Washington that marks the entrance of the broad Connecticut River. Piping plover nest on the barrier beach. The *yip yip yip* of osprey fills the air; a light breeze keeps the mosquitoes moving. Osprey have a made a spectacular recovery since DDT was banned in 1972; pesticides had wiped out 90 percent of the birds. This is some of the most productive habitat for osprey in the country. Old Lyme even has an Osprey Festival. Griswold's old Lab, Maya, follows us, looking like a small black bear in the tall grass, until she's distracted by the lone elm in the hayfield. Osprey land there and drop fish parts, an irresistible treat for her.

Evan Griswold's family has lived here since 1645, back to the day Matthew Griswold was granted 10,000 acres, land that stretched from Old Lyme to New London. Growing up, Evan spent his summers here, and in 1983 he and his wife, Emily, had a post-and-beam house built under the trees. He sited it back from the marsh, quite far back, he thought. When Superstorm Sandy hit, the tide kept coming in for three days. Long Island Sound rose until it was about fifty feet from his house, just down a small slope. Griswold is a proper host as he shows us around, but also a bit of the real estate agent that he is, efficiently walking a prospective buyer through a house.

The changes he has seen in the marsh during his twice-daily walks over the last thirty-three years are quietly persistent. He shows us where the stone wall his ancestors built falls right into the low-tide mudflats of the cove. This was once pasture. He shows us where the bright green of salt meadow hay (*Spartina patens*) has retreated and almost disappeared as the rougher, duskier, salt marsh cordgrass (*Spartina alterniflora*) advances. The salt marsh cordgrass marks the advance of the tides. It can live where it is submerged daily.

And the changes are also large. We tour a neighborhood that was flooded in Sandy and was without municipal water for six months. A rebuilt house stands awkwardly on stilts thirteen feet high. There are signs up around Old Lyme announcing meetings to discuss "resiliency"—about preparing for the new post-Sandy world. The three lectures were poorly attended by the beach communities. The lectures were lost in the festivity of early summer—tennis, tag sales, and all the rest. But the future is hurrying in anyway. Old Lyme's Main Street is just twenty feet above sea level. Most of the town's residents depend on wells and septic systems. The rising sea is starting to salt the wells and carry nitrogen from the septic systems into the Sound. This will be very expensive to fix.

The post-Sandy reality was vividly presented to Old Lyme in the new flood maps drawn by FEMA. People are angry at FEMA: *"How dare you tell me I'm in a flood zone? Do you know what this will do to my property value?"* Griswold is uniquely positioned to see both sides. Before he began selling real estate, he trained as a forester, and he has served as the executive director of the Nature Conservancy's Connecticut Chapter. When the sea rises to take land, it ignores property lines.

"It's a creeping crisis," he says. But it's as if climate change is happening in an invisible spectrum, a bandwidth of light that we won't see until it's too late. The water rises a few inches, one kind of grass dies off over decades, the days heat up. We're dying by inches. The world changes inch by inch, and when a big storm blows the sea inland we are finally forced to see that we are living in a new world.

We talk about Hurricane Carol, Sandy, and Hurricane Irene in 2011, which also flooded the marsh and brought water close to Griswold's house. It's hard to imagine, says Whelchel. Climate change can seem to be a gentle thing on such a fine day. The osprey are thriving, Long Island Sound sparkles, the long tenure of the Griswold family here suggests a comforting continuity. "So how do you convince people on a beautiful day like today that there was ever anything like a nor'easter or hurricane?" Whelchel asks. And as for "sea level rise—what's that?"

"It may seem subtle—an inch to five inches over a ten-year period or more," he says, but it raises the stage; the water from the next storm is going to push farther inland. And while the flooding from a hurricane is temporary, "sea level rise is permanent. It's permanent inundation. There is no going back from a sea level rise increase."

Adam Whelchel is cofounder of a pioneering program to get people to see the coming changes. He works with all twenty-four of Connecticut's coastal communities, home to about a third of the state's population, running workshops to help towns plan for the rising sea. Since starting the Coastal Resilience Program in 2007—the first in the nation—he has brought it to fourteen states and seven countries. Whelchel is boyish, with a sunny disposition. He enjoys our short walk, and he seems eager to throw on a backpack and disappear into the back-country. He has trekked through the Himalayas, climbed ice crevasses in Alaska, and explored river caves in Belize and Thailand.

One of his tools is a bit of shock therapy: aerial photo maps showing in red the houses, schools (often used as emergency shelters), and roads that will be "in daily conflict with the tides." "Where's the road going to be? Where's the neighborhood going to be? Where's the boat dock going to be? Where's the beach going to be?" Whelchel makes it local, even personal. That's your living room going out to sea.

The photos are marked to show where the marshes are going to move. Saving marshes is essential to protect the coast. Healthy marshes absorb storm surges, purify the water, and serve as a nursery for most of the fish we eat. Marshes naturally respond to the changes in the ocean. As it gets too salty, they move back. Essentially, over time, they walk away. The Nature Conservancy's study has determined how much room Connecticut's marshes need to move (using a conservative forecast of 4.3 feet of sea level rise). The Old Lyme Marsh, near Griswold Point, is likely to move 936 acres by the 2080s. There's room for 86.7 percent of that move on protected and unprotected land, but for the rest, 13.3 percent (124 acres), roads and buildings are in the way.

"Marsh Advancement"—the marsh walking—is the central metaphor we lack: we have to move. The sea is going to move; animals and plants are going to move. We have to move our thinking and our planning and our politics. The usual story is about anguished efforts to hold seaside cottages back from the sea with seawalls, sandbags, dune restoration, and millions of dollars of sand. But we need a new approach: We have to flex, to bend. To call it retreat disguises the new thinking required. We have to imagine a changing world. We have to imagine the world without us. We have to admit that every action we take has consequences, even if they cannot be neatly tallied. We have to live with uncertainty.

We need to "be prudent," advises Whelchel. "That is such a hall-mark of being a Yankee, right? Thinking forward and being prudent. That's what it's all about. And by not considering sea level rise, it's just not prudent. Why wouldn't you, if it's going to affect you? Why wouldn't you turn that to your advantage somehow?"

He's guardedly optimistic. Before Sandy hit, attendance at his workshops was sparse, and usually included a few climate change de-niers looking for an argument. But these days, he says, "I think we have hit an inflection point where people are asking the right ques-tions now, and maybe even making some smarter decisions because of that, but the transformation to a sea-level-rise-ready community or a resilient community, whatever you want to call it, is a little beyond us at this point." He notes that Old Lyme was one of the first places to be settled in New England. The land is marked by nearly 400 years of decisions "about what goes where and why. People are not going to change immediately."

We haven't really begun to reckon with this, Whelchel says. "We spent close to $14 billion post-Katrina on New Orleans. There are roughly 350,000 people who live there. So you start figuring out the per capita investment to maintain that city where it is and the key question for me would be: Can we sustain that level of investment when we add on a Houston, a Miami, a Los Angeles, a San Francisco, a Boston, a New York, a Seattle? And that's not even the Mississippi river communities." How are we going to afford this? How are we going to choose which cities and towns to protect? "What would it cost to secure a network of cities—name them: Boston, Portsmouth, Hull, Chatham, Falmouth, Providence, Gloucester, Portland?" His roll call hangs in the air, each city a possible disaster. We've faced disasters one at a time, and we have not done well. It's a prescient roll call: Since he and I spoke, hurricanes have flooded Houston and paralyzed Puerto Rico.

As we stand there talking about great changes, Griswold points to an island in the middle of the Connecticut River. "In the middle of Great Island—this great mass of marsh—there are hilltops—old hilltops which 10,000 years ago were hills. So the sea level rose. Long Island Sound was a freshwater lake. Then the oceans rose enough to break through the glacial moraines that were left so the seawater came in and made Long Island Sound a brackish water estuary. And then over the centuries these marshes have migrated inland and risen. So now you have these little forested hummocks in the middle of the marsh that were hilltops 10,000 years ago." I look at the sparkling water trying to see thousands of years of changes as if they were happening in a day. Great changes had created this spot where we stood, and great changes were coming in a hurry.

Whelchel continues the story beyond the next decades. "At some point in the future we're going to stop thinking about Connecticut's coast," he says, "and we'll start thinking about the Archipelago of Connecticut," a chain of islands where today there are beaches, roads and towns. Call them after the places we know: Fairfield Island, Westport Island, Branford Island. And somewhere under there will be Griswold Point, a New England Doggerland.

I think about that day 7,600 years ago when the inhabitants of the sinking Doggerland Island finally left. Were there those who refused to go, who tried to turn back the rising waters with their disbelief? Did they stay behind blinded by their own denial?

The Earth as We Knew It

A nother beautiful day. I've been out for an afternoon paddle in my kayak at an Audubon sanctuary watching two loons feed their chick. As I pull my boat out at the landing, a father is trying to coax his daughter, maybe two years old, to go home. She's wearing inflatable water wings and a beseeching look that says, *Why leave? This is the best.*

"We'll come back tomorrow," he says. "It will still be here."

That's the promise that could be posted at every nature sanctuary, land easement, state and national park, and that's the promise that we have broken.

Epilogue

Have all ages been like ours and have men always dwelt, as in our day, in a world where nothing is connected?

—Alexis de Tocqueville, *Democracy in America*, 1835

FINDING THE GOOD LIFE IN BROKEN TOWN

ONCE LONG AGO, THE ENGLISH AND THE ABENAKI sat down to make a treaty of peace. The Abenaki had allied with the French. Many had removed to Canada. There the Abenaki of St. Francis met with Captain Phineas Stevens who was representing the Governor of Boston. He was a trusted friend who spoke their language. "Were it not for the French it would be easy to live at peace with the Indians," Stevens said. Also at that meeting on July 5, 1752, were the Governor of Montreal, and Iroquois from Sault-Saint-Louis and Lake of the Two Mountains. The English and the French had warred sporadically for North America, but this was two years before they were fighting what we call the French and Indian War. In that war, the Abenaki's village at St. Francis would be set afire in an attack. There is a record in English from that day's meeting of what was said in several languages and it is stilted and formal, as if Alfred Lord Tennyson were writing a cowboy and Indian movie. But at the center of their talk was a single question: Why do you want our land? asked the Abenaki.

Property may be the air we breathe, as I said earlier. It may be everywhere and invisible, but this meeting of cultures makes property visible.

Atiwaneto, a sachem, speaking for the Abenaki, said:

"We hear on all sides that this Governor and the Bostonians say that the Abenakis are bad people. 'Tis in vain that we are taxed with having a bad heart; it is you, brother, that always attack us; your mouth is of sugar but your heart of gall. . . .

"Brothers, we tell you that we seek not war, we ask nothing better than to be quiet, and it depends, Brothers, only on you English, to have peace with us.

"We have not yet sold the lands we inhabit, we wish to keep the possession of them. Our elders have been willing to tolerate you, brothers Englishmen, on the seaboard as far as Sawakwato; as that has been so decided, we wish it to be so.

"But we will not cede one single inch of the lands we inhabit beyond what has been decided formerly by our fathers. . . .

"We acknowledge no other boundaries of yours than your settlements whereon you have built, and we will not, under any pretext whatsoever, that you pass beyond them. The lands we possess have been given us by the Master of Life. We acknowledge to hold only from him."

The English reply was brief. Why do you attack us? they asked. "Are you satisfied with the death of your people on account of your attacks on the English?"

The Abenaki answered by explaining their attacks. One was to avenge the killing of a man and a woman from their village by the English. Two other Englishmen were killed for hunting beaver on the Abenaki's lands. "On this point we repeat to you, with all the firmness we are capable of, that we will kill all the Englishmen we shall find on the lands in our possession." But now they had settled the score. "Our heart is good," Atiwaneto said, "and since we struck the blow our thirst for vengeance is extinguished."

Peace, of a sort. But we know how the rest of the story goes: Many wars, many treaties, more fighting. In the years after this council, the Indians will lose almost all of their land in the East. They will remain, many hiding in plain sight, keeping their heritage quiet.

This is the part of America's founding that we know and forget at the same time: America begins by taking the land of the natives. This taking sets up two characteristics of American land law that will become indelible in the nineteenth century—One: Americans believe in "property in motion," and not at rest. Property is an investment, to be risked and sold. And, Two: The "quiet citizen" must step aside for the "exuberantly active one." The great churning of land deals and development starts in treaty councils like this. It moves forward through a raft of case law giving big water mills the legal right to stop rivers and destroy smaller mills along with fishing, and with laws giving railroads the power of eminent domain and the impunity to offer little or no settlement to landowners whose rights have been extinguished. In our time, the battles over power lines and pipelines forcing their way through a settled landscape moves to the same script, one that would be recognized by the St. Francis Abenaki. Why do you want to take our land? ask the homeowners. And the big corporations answer, Do you realize the damage you are causing by interfering with our property and our notions of progress?

If you seek a manual for turning the world upside down, you can find it in the preceding pages: draw an arbitrary boundary through a desert, cutting an ancient world in two; enslave Africans to drain an immense swamp; clear-cut mountains until the slash catches fire and the runoff from the rains floods rivers; march through the wilderness attempting to claim a feudal realm; flatten hills and farms to make a highway that runs straight and fast; push gas pipelines and transmission towers across the land in a blind march toward profit; load the atmosphere with carbon dioxide until winter is summer and the seas rise to take away your great cities.

It's our habit to turn the world upside down, to make night into day, to unbend rivers and blast through mountains. Our nation is founded upon this habit we have called by many names, including Manifest Destiny and The Pursuit of Happiness. A habit this ingrained will not be remedied by a reform movement or by a new set of laws. To break this habit we need to change how we view ourselves and our property. Ownership defines us. We are owned by what we own. As Lewis Hyde said, "How we imagine property is how we imagine ourselves." Can we re-imagine ourselves?

We'd go a long way to freeing ourselves if we can imagine living the good life in Broken Town.

~

Researching a seaside town I was going to visit for a magazine assignment, I came across some videos made by a land trust showing a local salt marsh and beach. The video was all piping plovers and waves. Peace. Then I chanced upon a rap video by some of the town's teens who hated that one of the town's other beaches was overrun by outsiders in the summer. This video was ugly and racist, without using a single racist word. It was a straight-up shot of us vs. them. And then a third video: *Our Town*. It's a perfect place, just perfect, perfect. This video was made by one of the real estate brokerages in town. It's all culture and history and green trees and well-dressed women in exquisite kitchens saying: "It's perfect." And I thought: *This glass is ready to break*. No place is perfect. The contrast between the teens and the piping plovers and the piping real estate ladies was strange until I realized it all represents the mentality of a gated community. But most of us, I'd say, live in some sort of emotional or mental gated community, walling ourselves off from different ideas, facts, and people.

Instead of trying to live in Perfectville, wouldn't it be better to live in Broken Town? In Broken Town we admit that the community is *im*perfect, is home to suffering as well as joy, failure as well as success, unhappiness as well as happiness. If we admit this maybe we could come together to do what we can to alleviate suffering, to be a compassionate community.

In Broken Town we would understand the lesson of the broken glass, a lesson that Mark Epstein, a psychiatrist and Buddhist, brought home with him from his travels. Epstein was on the road in search of answers, one of many pilgrims in the 1970s "pillaging the wisdom of the East, trolling for some seed or shoot that might take root in American soil," he says. He and his three friends were deep in the rainforest of Thailand, visiting a Buddhist monk, Ajahn Chah. The monastery looked fragile. It "wasn't just in the tropical forest; it was constructed out of it. Ruddy wooden buildings rose on stilts over neatly swept dirt strewn paths. Bird calls mixed with the ringing of chimes, the murmurs of the monks and the faint residue of incense from the temple."

They were shown in to meet the monk. "A severe-looking man with a kindly twinkle in his eyes, he sat patiently waiting for us to articulate the question that had brought us to him from such a distance. Finally, we made an attempt: 'What are you really talking about? What do you mean by eradicating craving?' Ajahn Chah looked down and smiled faintly. He picked up the glass of drinking water to his left. Holding it up to us, he spoke in the chirpy Lao dialect that was his native tongue: 'Do you see this glass? I love this glass. It holds the water admirably. When the sun shines on it, it reflects the light beautifully. When I tap it, it has a lovely ring. Yet for me, this glass is already broken. When the wind knocks it over or my elbow knocks it off the shelf and it falls to the ground and shatters, I say, 'Of course.' But when I understand that this glass is already broken, every minute with it is precious.'"

What did he mean? *This glass is already broken.* To Americans that sounds wrong. Take the glass back! Let's get it fixed; let's perfect it. Let's fund research to make better, unbreakable glasses. Let's install alarms to warn of falling glasses. Let's engineer, redesign, plan, legislate, until we make the last, best glass of mankind.

In this story of the broken glass, notes Epstein, can be found the central teachings of Buddhism: Life is fleeting and we suffer when we try to hold on to some idea of unchanging reality, some perfect, static world. We can find a measure of peace in accepting that we can't control things. "Whoever sees the uncertainty in things, sees the unchanging reality of them," says Ajahn Chah.

Property is our attempt to make a steady place for ourselves. But we've drawn our lines in a fluid world. We can see this where the land meets the sea. At the shore, barrier islands are not fixed. They move with with the winds and the tides. Over thousands of years they slowly move landward. That is until we have come along, interfering with this dance by building and dredging. Similar changes are taking place on the land you may own, but we are often not willing to see the changes, just as we are blind to our land's true history.

If you are fortunate enough to own a beach house, enjoy it, but know that it is a broken glass. Like Ajahn Chah, you could say, "See this beach house? I love this beach house. I love having my children and grandchildren here. I love sitting here in the morning with my wife reading the morning paper as the sun has just risen over the beach. But this house is already gone, already smashed by a hurricane yet to be named, already enveloped by the rising seas. This beach house is a broken glass."

And: "See this house in the woods? This house on its own suburban quarter-acre, this row house on the quiet side-street in the good school district? I love this house, but everything will change—the neighborhood, my sense of myself, my spouse, our marriage itself, our nation, the climate." All our houses are founded on barrier islands. "Like the glass, this world is already broken," says Epstein. "And yet when you drop your fear and open your heart, its preciousness is there too."

The world is broken, it is fragile. But this doesn't mean we shouldn't fight to save a beloved farm from being lost to a highway or cut up with transmission towers. When we protect our property, if we see the broken glass in what we love, we understand that we are *not* protecting a static thing. We are caretakers. What we hold is valuable because it is transitory. It's no easy thing to hold on and to let go.

See your home here and gone. See other families living in it; see them foolishly changing the qualities you love—cutting down beloved trees, paving lawns, knocking down walls. See your house broken by neglect or storm, or the ravages of fashionable renovations. See this and know, still, that you will hand off this baton, and that you will trust the next runner to take the baton. We hold property for the time being. Our grasp should be more like a runner's on a relay team, ready to hand off. If you've ever tried this—quickly handing off the slick aluminum cylinder without dropping it—you understand how tricky it can be to let go at the right moment.

This is how to hand off a baton in a relay race: hold the baton at the bottom, stay to the outside of the lane, don't slow down. When you are about two meters away, call out, "*Stick!*" Place it in the palm of the waiting runner. *Let go.* You need to be in sync with your teammate; you need to let go. That's how we should handle the land and houses we own, as part of a relay race, our part done with a quick letting go.

It's right at that moment when you have to let go, but hold firm, hand off, but not too soon—it's at that moment we are failing. We are in such a moment now as the planet heats up. We have to learn to hand off the baton, to open our grip, to accept change. What's keeping us from dealing with climate change is that we don't understand the word "change." We need to accept a contradiction: We own a changing thing; we've bound a boundless world. We need to see past the boundary lines. We need to see the world in our property. We really have our hands on a paradox—the attempt to permanently hold on to the impermanent. We are trying to hold on to running water.

"We must have beginner's mind, free from possessing anything, a mind that knows everything is in flowing change. Nothing exists but momentarily in its present form and color. One thing flows into another and cannot be grasped," says Zen teacher Shunryu Suzuki. Once we draw boundaries, "flowing change" is a problem. Fearing the glass breaking, we'll do anything to keep it. We want certainty in the world. We exaggerate our own reality.

Our conception of property is like our sense of self. We conceive of ourselves as separate from the world. Our property is an extension of our self. We see our property in opposition to the world rather than as part of it. Property is about separation. It's the rock-solid part of our creed of individualism. Property is the self spread over the land, marked and measured, paid for and defended. It's the flag we wave, the one saying, Here I am.

We are reaching for permanence—for iron borders, for an idea of possession that doesn't allow for shadings, overlapping claims, changes in the land itself. We have a fortress mentality. Legal deeds are our castle walls. "Keep out" and "No trespassing" is our creed.

Under attack we tighten our hold on our property. We have to. It's as if the property tightens its hold on us. Faced with a big corporation's lawyers pulling at our land, we can only dig in and pull back. It's a war; all the delicacy of peacetime ambiguity is lost. At the moment we feel the glass slipping, we grasp it tighter. There is no way to linger on feelings that the world is abiding and passing as the day passes, that our homes are "both broken and whole, intensely alive" and also "burning with the trauma of impermanence," as Epstein says.

In a property fight, eminent domain not only aims to separate us from our home, it also breaks apart the paradox of a place being both "broken and whole"; it snatches us from the poignancy of impermanence. We have to wall off any thought that this property is a passing thing, and fight for it with absoluteness. The fear of losing our property, our house, is a match set to dry tinder. It ignites our fears of losing family, love, life itself. This is what makes eminent domain a blowtorch. Eminent domain is a fire set by the state.

Anyone who has ever been swept along by a serious illness may understand Ajahn Chah. Illness slows us down, revealing that our life rests upon what we take for granted—a foundation of continued health and modest prosperity. We assume that the most intangible things will be tangible. But, as my friends the disability rights activists remind us, we're all temporarily able-bodied.

"Temporary" is a word we struggle to understand. In *Blue Nights*, Joan Didion's memoir about her daughter's early death, she's forced to face the evanescence of our lives. She recalls a toast at the wedding day lunch for her daughter and son-in-law: Happiness, health, love, luck, and beautiful children. We counted these as "ordinary blessings," Didion says.

Just as empires are built on hubris, our ordinary days are built on hubris. Health and happiness are due to us, and if not due, then certainly accepted without question, "ordinary blessings." We have to be jostled by tragedy and loss to see how extraordinary the ordinary is. Loss is a great sorting out. *Pathei mathos*, Aeschylus wrote in *Agamemnon*, notes a classical scholar: we "suffer into knowledge."

We all may face a long illness at some point. If we'd accept that likelihood, it would transform the healthcare debates. Of course we should all pay for health care; we're all broken glasses. We all suffer.

〜

I wonder if we could learn from the broken glass, accept it in our lives and our republic, and resume the work of making sure the experiment of democracy continues another generation. I don't know what a politics of the broken glass would look like, but I'd accept my political opposite and they would accept me. We'd disagree, at times harshly, but we'd still be neighbors. We'd never forget that we're all in this together.

The politics of the broken glass knows that we are the refugees; we are the battered wives, the drug addicts. Boats adrift, boats under sail, boats taking on water, boats to be envied for their grace and seaworthiness. The politics of the broken glass accepts this all. It would accept imperfect outcomes and imperfect people. It would steadily aim to mend the world, avoiding the great hurrahs and disappointments of pushes for reform; it would accept that we won't have a swift, tidy outcome.

It would require patience. We offer help, assistance, charity, aid, education, love, knowing that most efforts fail or cannot even be measured. It would require commitment to the long haul.

And it would require an admission that we're all mortal, and broken. So how can our institutions and ideals be anything else but that?

All reforms should aim to build Broken Town, not Perfectville.

〜

The United States is the broken glass. It is not the "citty upon a hill," in the oft-quoted words that John Winthrop wrote aboard ship in 1630, leading his Puritan believers to establish a New Jerusalem. Our nation's founding is flawed—it proclaims liberty as it enslaves, it sings of the pursuit of happiness as it dispossesses native peoples. The good and bad in the American idea are tightly folded together. This is our inheritance as Americans, and it is the work of each generation to prise that apart.

The founders are great men, and Madison is brilliant, no doubt. They managed to establish a stable government and a workable civil society. They took us from the Enlightenment to today. But even as they believed that all men are created equal, they owned people and were willing for the sake of compromise to count slaves as three-fifths of a person. So which is true? That "conscience is the most sacred of all property," as Madison said, or that they were slaveholders who were committed to a limited, male-only suffrage? Both are true. If we judge them by our standards today, it is because they made that possible. They set our course.

One truth doesn't eclipse the other. This is our dual inheritance. The founders laid the tracks for great things, high ideals, and terrible things. The "citty upon a hill," and the slaves to build it, the civic forum and the urban riot, the act of charity and the dead weight of winner-take-all greed. All of this is America.

We have to acknowledge one more part of our inheritance: we can't ignore the founders' rapacious hunger for land. In each confrontation about the big projects that I witnessed, I saw this land hunger on both sides. Grasping for land, we are still pioneers, strangers come ashore on this continent. We reenact the frontier, leveling and reshaping places that we don't pause to understand. This land hunger is who we are. It's how we began in this land, taking and taking. This is the root of our ecological blindness and our destruction of native people and places. Alexis de Tocqueville's observation still applies: "There is no country in the world where the feeling for property is keener or more anxious."

Tocqueville was a tourist in search of the exotic when he visited a young America in 1831. With his friend, Gustave Beaumont, he set out for the wild frontier: Saginaw, Michigan, a rude gathering of five or six cabins in a deep forest that ran, or so he imagined, to the Pole and to the Pacific. "You want to go to Saginaw!" said one pioneer they met on their way. "Do you know that Saginaw is the last inhabited point until you come to the Pacific Ocean?"

Tocqueville found what he was looking for in Saginaw. It was a place in which "the buzzing of the mosquitoes was the only breathing of this sleeping world," he wrote. In the wild forest he found an "immensity" like the ocean with a "sense of isolation and abandonment" that "astonishes and overwhelms the imagination."

He was moved by a kind of anticipatory nostalgia for its coming loss. "In but few years these impenetrable forests will have fallen. The noise of civilization and of industry will break the silence of the Saginaw. Its echoes will be silent. Embankments will imprison its sides, and its waters which today flow unknown and quiet through nameless wilds, will be thrown back in their flow by the prows of ships. Fifty leagues [more than 150 miles] still separate this solitude from the great European settlements and we are perhaps the last travelers who will have been allowed to see its primitive splendor, so great is the force that drives the white race to complete conquest of the New World.

"It is this consciousness of destruction, this *arriere-pensee* of quick and inevitable change that gives, we feel, so peculiar a character and such a touching beauty to the solitudes of America. One sees them with a melancholy pleasure; one is in some sort of hurry to admire them."

Our New World is the world turned upside down.

BIBLIOGRAPHY

Desert Teachings

Fontana, Bernard L. *Of Earth & Little Rain: The Papago Indians*. Tucson: University of Arizona Press, 1989.

Johnson, Terrol. Interview with the author, March 18, 2000.

Johnson, Tony. Interview with the author, March 19, 2000.

Lopez, Tony. Interview with the author, March 18, 2000.

Mansfield, Howard. "Desert Walk." *DoubleTake*, Summer 2001.

Montgomery, Sy. "A Return to Native Foods." *The Boston Globe*, March 7, 2000.

Nabhan, Gary. *Coming Home to Eat*. New York: W. W. Norton, 2001.

———. *Cultures of Habitat*. Washington, DC: Counterpoint, 1997.

———. *The Desert Smells Like Rain: A Naturalist in Papago Indian Country*. San Francisco: North Point Press, 1982.

———. Interview with the author, March 18, 2000.

Raver, Anne. "Finding Desert Blooms That Heal." *The New York Times*, March 30, 2000.

Reader, Tristan. Interview with the author, March 18, 2000.

Tolan, Sandy. "The Walk" (from the National Public Radio program *Living on Earth*). Broadcast the week of April 21, 2000.

Land Rush: A Brief History of Turning the World Upside Down

Alexander, Gregory S. *Commodity and Propriety: Competing Visions of Property in American Legal Thought, 1776–1970*. Chicago: University of Chicago Press, 1997.

Banner, Stuart. *American Property*. Cambridge, MA: Harvard University Press, 2011.

———. *How the Indians Lost Their Land*. Cambridge, MA: Harvard University Press, 2005.

Baude, William. "Rethinking the Federal Eminent Domain Power." *The Yale Law Journal* 122, no. 7 (2013).

Behn, Richard J. "The Founders and the Pursuit of Land." From the website of the Lehrman Institute. http://lehrmaninstitute.org/history/founders-land.html.

Benedict, Jeff. *Little Pink House*. New York: Grand Central Publishing, 2009.

Bodenhamer, David J. *The Revolutionary Constitution*. Oxford: Oxford University Press, 2012.

Boorstin, Daniel. *The Americans: The Colonial Experience*. New York: Vintage Books, 1958.

———. *The Americans: The National Experience*. New York: Vintage Books, 1965.

———. *The Lost World of Thomas Jefferson*. New York: Henry Holt, 1948.

Brown, Frederick, ed. and trans. *Letters from America: Alexis de Tocqueville*. New Haven, CT: Yale University Press, 2010.

Byrd, William, II. *Description of the Dismal Swamp and a Proposal to Drain the Swamp*. Charles F. Heartman Manuscripts of Slavery Collection. Xavier University of Louisiana, New Orleans, 1922.

———. *William Byrd's Histories of the Dividing Line Betwixt Virginia and North Carolina*. The North Carolina Historical Commission, Raleigh, NC, 1929.

Carpenter, David. *Magna Carta*. London: Penguin Books, 2015.

Chernow, Ron. *Washington*. New York: Penguin Press, 2010.

Clay, Grady. *Real Places: An Unconventional Guide to America's Generic Landscape*. Chicago: University of Chicago Press, 1994.

Cohen, Morris R. *Law and the Social Order*. New York: Harcourt, Brace, 1933.

Commager, Henry Steele. *The American Mind*. New Haven, CT: Yale University Press, 1950.

Cronon, William. *Changes in the Land: Indians, Colonists and the Ecology of New England*. New York: Hill and Wang, 1983.

Damrosch, Leo. *Tocqueville's Discovery of America*. New York: Farrar, Straus and Giroux, 2010.

Degen, Julie E. "The Legislative Aftershocks of *Kelo*: State Legislative Response to the New Use of Eminent Domain." *Drake Journal of Agricultural Law* 12, no. 1 (2007).

Demsetz, Harold. "Toward a Theory of Property Rights." *The American Economic Review* 57, no. 2. (May 1967).

Diouf, Sylviane. *Slavery's Exiles: The Story of American Maroons*. New York: New York University Press, 2014.

Dodd, Edwin Merrick. *American Business Corporations until 1860*. Cambridge, MA: Harvard University Press, 1954.

Ely, James W., Jr. *The Guardian of Every Other Right: A Constitutional History of Property Rights*. 3rd ed. Oxford: Oxford University Press, 2008.

Foner, Eric. *The Story of American Freedom*. New York: W. W. Norton, 1998.

Freyfogle, Eric T. *Bounded People, Boundless Lands: Envisioning a New Land Ethic*. Washington, DC: Island Press, 1998.

———. "Property and Liberty." *Harvard Environmental Law Review*, 34 (2010).

———. *The Land We Share: Private Property and the Common Good*. Washington, DC: Island Press, 2003.

Friedman, Lawrence M. *A History of American Law*. New York: Simon & Schuster, 1973.

———, and Harry N. Scheiber, eds. *American Law and the Constitutional Order*. Cambridge, MA: Harvard University Press, 1978.

Gardner, Andrew J. "How Did Washington Make His Millions?" *Colonial Williamsburg Journal* (Winter 2013).

Grandy, Moses. *Narrative of the Life of Moses Grandy; Late a Slave in the United States of America*. London: C. Gilpin, 1843.

Grant, J. A. C. "The 'Higher Law' Background of the Law of Eminent Domain." *Wisconsin Law Review* 6 (1930–1931).

Grant, Richard. "Deep in the Swamps, Archaeologists Are Finding How Fugitive Slaves Kept Their Freedom." *Smithsonian*, September 2016.

Grobbel, Christopher. *Summary of Property Takings Case Law*. East Lansing, MI: Michigan State University, December 16, 2002.

Hall, Kermit L. *The Magic Mirror: Law in American History*. Oxford: Oxford University Press, 2009.

Hart, John F. "Colonial Land Use Law and its Significance for Modern Takings Doctrine." *Harvard Law Review* 109, no. 6 (1995–1996).

———. "Land Use Law in the Early Republic and the Original Meaning of the Takings Clause." *Northwest University Law Review*, 94 (1999–2000).

Henretta, James A., and Gregory H. Nobles. *Evolution and Revolution: American Society, 1600–1820*. Lexington, MA: D. C. Heath, 1987.

Hofstadter, Richard. *The Age of Reform*. New York: Vintage Books, 1955.

Holt, J. C. *Magna Carta*. Cambridge: Cambridge University Press, 1965.

Horwitz, Morton J. *The Transformation of American Law, 1780–1860*. Cambridge, MA: Harvard University Press, 1977.

Hurst, J. Willard. *Law and the Conditions of Freedom in the 19ᵗʰ Century United States*. Madison, WI: University of Wisconsin, 1967.

Hyde, Lewis. *Common as Air: Revolution, Art and Ownership*. New York: Farrar, Straus and Giroux, 2010.

Jackson, J. B. *American Space: The Centennial Years, 1865–1876*. New York: W. W. Norton, 1972.

Ketcham, Ralph. *Framed for Posterity: The Enduring Philosophy of the Constitution*. Lawrence, KS: University Press of Kansas, 1993.

Kratovil, Robert, and Frank J. Harrison, Jr. "Eminent Domain—Policy and Concept." *California Law Review* 42, no. 4 (1954).

Kutler, Stanley I. *Privilege & Creative Destruction: The Charles River Bridge Case*. Philadelphia: J. B. Lippincott, 1971.

Levy, Leonard. *The Law of the Commonwealth and Chief Justice Shaw*. Cambridge, MA: Harvard University Press, 1957.

Linklater, Andro. *Measuring America*. New York: Walker & Company, 2002.

———. *Owning the Earth: The Transforming History of Land Ownership*. New York: Basic Books, 2013.

Livermore, Shaw. *Early American Land Companies*. New York: Octagon Books, 1968.

Lobingier, Charles Sumner. "Rise and Fall of Feudal Law." *Cornell Law Review* 18, no. 2 (1933).

Massey, Calvin R. "An Assault upon 'Takings' Doctrine: Finding New Answers in Old Theory." *Indiana Law Journal* 63, no. 1 (Winter 1987).

Matthews, Richard K. *If Men Were Angels: James Madison and the Heartless Empire of Reason*. Lawrence, KS: University Press of Kansas, 1995.

Maxwell, William B. *West Virginia Historical Society* 24, no. 1 (Spring 2010).

McKean, Dayton D. "The Constitutional Limits Upon the Power of Eminent Domain." *Rocky Mountain Law Review* 6 (1933–1934).

Miceli, Thomas J., and Kathleen Segerson. "Takings." *Encyclopedia of Law and Economics, Volume IV. The Economics of Public and Tax Law*. Boudewijn Bouckaert et al., eds. Cheltenham, UK: Edward Elgar, 2000.

Mitchell, John Hanson. *Trespassing: An Inquiry into the Private Ownership of Land*. New York: Perseus Books, 1998.

Nedelsky, Jennifer. *Private Property and the Limits of American Constitutionalism*. Chicago: University of Chicago Press, 1990.

Paul, Ellen Frankel. *Property Rights and Eminent Domain*. Piscataway, NJ: Transaction Books, 1987.

Perkel, Jeffrey M. "Gene Patents Decision: Everybody Wins." *The Scientist*, June 18, 2013.

Plater, Zygmunt J. B. "The Three Economies: An Essay in Honor of Joseph Sax." *Ecology Law Quarterly* 25 (1998).

Plotkin, Sidney. *Keep Out: The Struggle for Land Use Control*. Berkeley, CA: University of California Press, 1987.

Potter, David M. *People of Plenty*. Chicago: University of Chicago Press, 1954.

Price, Edward T. *Dividing the Land: Early American Beginnings of Our Private*

Property Mosaic. Chicago: University of Chicago Press, 1995.

Rakove, Jack N., ed. *James Madison Writings*. Library of America, 1999.

Reps, John. *The Making of Urban America*. Princeton, NJ: Princeton University Press, 1965.

Reynolds, Susan. *Before Eminent Domain: Toward a History of Expropriation of Land for the Common Good*. Chapel Hill, NC: University of North Carolina Press, 2010.

Richter. Daniel K. *Before the Revolution: America's Ancient Pasts*. Cambridge, MA: Belknap Press, 2011.

Rose, Carol M. "A Dozen Propositions on Private Property, Public Rights, and the New Takings Legislation." *Washington & Lee Law Review* 53 (1996).

———. "Joseph Sax and the Idea of the Public Trust." *Faculty Scholarship Series*. Paper 1805, January 1, 1998.

———. "Possession as the Origin of Property." *Faculty Scholarship Series*. Paper 1830, January 1, 1985.

———. "Property and Expropriation: Themes and Variations in American Law." *Faculty Scholarship Series*. Paper 1800, January 1, 2000.

———. *Property and Persuasion*. Boulder, CO: Westview Press, Inc., 1994.

———. "Public Property, Old and New." *Faculty Scholarship Series*. Paper 1832, January 1, 1984.

Royster, Charles. *The Fabulous History of the Dismal Swamp Company*. New York: Alfred A. Knopf, 1999.

Ryan, Alan. *On Politics—Book One & Book Two*. W. W. Norton, 2012.

Sax, Joseph L. "The Constitutional Dimensions of Property: A Debate." *Loyola of Los Angeles Law Review* 26 (1992).

———. "Some Thoughts on the Decline of Private Property." *Washington Law Review* 58 (1982).

———. "Why America Has a Property Rights Movement." *University of Illinois Law Review* (January 2005).

Sayers, Daniel O. *A Desolate Place for a Defiant People: The Archaeology of Maroons, Indigenous Americans, and Enslaved Laborers in the Great Dismal Swamp*. Gainesville, FL: University Press of Florida, 2013.

Scheiber, Harry N. "The Road to Munn: Eminent Domain and the Concepts of Public Purpose in the State Courts." *Perspectives in American History, Vol. 5*. Donald Fleming and Bernard Bailyn, eds. Cambridge, MA: Harvard University Press, 1971.

Schlatter, Robert. *Private Property: The History of an Idea*. London: George Allen and Unwin, 1951.

Schwartz, Marie Jenkins. *Ties That Bound: Founding First Ladies and Slaves*. Chicago: University of Chicago Press, 2017.

Scott, William B. *In Pursuit of Happiness: American Conceptions of Property from the 17th to the 20th Century*. Bloomington, IN: Indiana University Press, 1978.

Simpson, Bland. *The Great Dismal: A Carolinian's Swamp Memoir*. Chapel Hill, NC: University of North Carolina Press, 2000.

Steinberg, Theodore. *Slide Mountain, or the Folly of Owning Nature*. Oakland, CA: University of California Press, 1995.

Stoebuck, William B. "A General Theory of Eminent Domain." *Washington Law Review* 47, no. 4 (1972).

Taylor, Alan. *American Colonies: The Settling of North America, Vol. 1*. New York: Viking, 2001.

———. *The Divided Ground: Indians, Settlers and the Northern Borderland of the American Revolution.* New York: Alfred A. Knopf, 2006.

———. *Liberty Men and Great Proprietors.* Chapel Hill, NC: University of North Carolina Press, 1990.

Tocqueville, Alexis de. *Democracy in America.* 1835 & 1840: Library of America, 2004.

Treanor, William Michael. "The Original Understanding of the Takings Clause." *Georgetown Environmental Law & Policy Institute Papers & Reports,* June 2010.

———. "The Origins and Original Significance of the Just Compensation Clause of the Fifth Amendment." *The Yale Law Journal* 95 (1985).

Underkuffler, Laura S. "On Property: An Essay." *The Yale Law Journal* 100, no. 1 (1990).

Vandevelde, Kenneth J. "The New Property of the Nineteenth Century: The Development of the Modern Concept of Property." *Buffalo Law Review* 29 (1980).

Washington, George. Letter to Jacob Read, November 3, 1784. http://founders.archives.gov/documents/Washington/04-02-02-0105.

Wiecek, William M. *The Lost World of Classical Legal Thought: Law & Ideology in America, 1886–1937.* Oxford: Oxford University Press, 1998.

Wilentz, Sean. *The Rise of American Democracy.* New York: W. W. Norton, 2005.

Wills, Gary. *Inventing America: Jefferson's Declaration of Independence.* New York: Vintage Books, 1978.

Winthrop, John. "Reasons for the Plantation in New England." ca. 1628. http://www.winthropsociety.com/doc_reasons.php.

Wolin, Sheldon S. *Tocqueville Between Two Worlds.* Princeton, NJ: Princeton University Press, 2001.

Wood, Gordon S. *The Creation of the American Republic 1776–1778.* Chapel Hill, NC: University of North Carolina Press, 1969.

———. *Empire of Liberty: A History of the Early Republic, 1789–1815.* Oxford: Oxford University Press, 2009.

———. *The Idea of America.* New York: Penguin Press, 2011.

Woodard, Calvin. "Law and the Condition of Freedom in the Nineteenth-Century United States, by James Willard Hurst. Madison: The University of Wisconsin Press, 1956." *Loyola of Los Angeles Law Review* 19, no. 2 (1956).

Wythe, Holt. "Morton Horwitz and the Transformation of American Legal History." *William & Mary Law Review* 23 (1982).

Youssef, Sharif, producer. "The Great Dismal Swamp," Episode 271, *99% Invisible,* August 15, 2017.

Land of Many Uses

American Forestry. "The Passage of the Appalachian Bill." March 1911.

———. "The Weeks Bill in Congress." August 1910.

Bennett, Randall H. *The White Mountains: Alps of New England.* Charleston, SC: Arcadia Publishing, 2003.

Bolles, Blair. *Tyrant from Illinois: Uncle Joe's Experiment with Personal Power.* New York: W. W. Norton, 1951.

Busbey, L. White. *Uncle Joe Cannon.* New York: Henry Holt, 1927.

Conrad, David E. *The Land We Cared For: A History of the Forest Service's Eastern Region.* Jay H. Cravens, George Banzhaf & Company for the US Forest Service, 1997.

Conroy, Rosemary G., and Richard Ober. *People and Place: Society for the Protection of New Hampshire Forests, The First 100 Years.* Society for the Protection of New Hampshire Forests, 2001.

Diaz, Jessica. "A Forest Divided: *Minard Run Oil Co. v. U.S. Forest Service* and the Battle over Private Oil and Gas Rights On Public Lands." *Ecology Law Quarterly* 40, no. 2 (2013).

Dickerman, Mike. *The White Mountain Reader.* Littleton, NH: Bondcliff Books, 2000.

Dumanoski, Dianne. "Northern Habitat Loss Blamed for Songbird Decline." *The Boston Globe,* October 21, 1991.

Fair, Sally K. *Buying Nature: The Limits of Land Acquisition as a Conservation Strategy, 1780–2004.* Cambridge, MA: MIT Press, 2005.

Forest Notes, "Saving the White Mountains: The Weeks Act Then and Now." Weeks Act Centennial Issue. Society for the Protection of New Hampshire Forests, Summer 2011.

Goodale, Christine. "Fire in the White Mountains: A Historical Perspective." *Appalachia,* December 2003.

Gwinn, William Rea. *Uncle Joe Cannon: Archfoe of Insurgency.* New York: Bookman Associates, 1957.

Hall, William L. "The White Mountain Forest and How it is to Be Made Useful." *American Forestry,* September 1913.

Heffernan, Nancy Coffey, and Ann Page Stecker. *New Hampshire: Crosscurrents in Its Development.* Grantham, NH: Tompson & Rutter, 1986.

Historical New Hampshire. Beauty Caught and Kept: Benjamin Champney in the White Mountains. New Hampshire Historical Society, Fall/Winter 1996.

———. *Consuming Views: Art & Tourism in the White Mountains 1850–1900.* New Hampshire Historical Society, 2006.

———. *The Grand Resort Hotels and Tourism in the White Mountains.* New Hampshire Historical Society, Spring/Summer 1995.

———. *Nature's Nobleman: Edward Hill and His Art.* New Hampshire Historical Society, Spring/Summer 1989.

———. *A Suburb of Paradise: The White Mountains and the Visual Arts.* New Hampshire Historical Society, Fall/Winter 1999.

Johnson, Christopher. *This Grand & Magnificent Place: The Wilderness Heritage of the White Mountains.* Durham, NH: University of New Hampshire Press, 2006.

Johnson, Christopher, and David Govatski. *Forests for the People: The Story of America's Eastern National Forests.* Washington, DC: Island Press, 2013.

Johnson, John E. "The Boa Constrictor of the White Mountains, or the Worst 'Trust' in the World." *The New England Homestead,* December 8, 1900.

Jones, Bradley R. "*Minard Run Oil Company v. United States Forest Service.*" *Public Land and Resources Law Review* (Fall 2011).

Mansfield, Howard. "Chasing Eden." *Beyond the Notches.* John R. Harris et al., eds. Littleton, NH: Bondcliff Books, 2010.

Miller, Char. *Public Lands, Public Debates: A Century of Controversy.* Corvallis, OR: Oregon State University Press, 2012.

Novak, Barbara. *Nature and Culture: American Landscape and Painting 1825–1875.* Rev. ed. Oxford: Oxford University Press, 1995.

Sears, John F. *Sacred Places: American Tourist Attractions in the Nineteenth Century.* Oxford: Oxford University Press, 1989.

Shands, William E. "The Lands Nobody Wanted: The Legacy of the Eastern National Forests." *Origins of the National Forests: A Centennial Symposium.* Harold K. Steen, ed. Forest History Society, 1992.

Soule, Michael. "Forget About Building the Road to Nowhere." *Christian Science Monitor,* October 16, 2000.

Strand, Ginger. "A Win for the West." *Nature Conservancy,* April/May 2015.

Sullivan, Mark. *Our Times: The United States 1900–1925. Vol. IV. The War Begins 1909–1914.* New York: Charles Scribner's Sons, 1932.

Tolles, Byrant F., Jr. *The Grand Resort Hotels of the White Mountains: A Vanishing Architectural Legacy.* Jaffrey, NH: David R. Godine, 1998.

Walker, Joseph P. "Our New Hampshire Forests. An Address. . . ." 1891.

Washburn, Charles G. *The Life of John W. Weeks.* Boston: Houghton Mifflin, 1928.

Weeks, John W. Letter to Gifford Pinchot, June 18, 1912. http://whitemountainhistory. org.

Weeks Act Centennial Issue, *Forest History Today Magazine,* Spring/Fall 2011.

Wolraich, Michael. *Unreasonable Men: Theodore Roosevelt and the Republican Rebels Who Created Progressive Politics.* Basingstoke, UK: Palgrave Macmillan, 2014.

The Last Medieval Claim

Akagi, Roy Hidemichi. *The Town Proprietors of the New England Colonies: A Study of Their Development, Organization, Activities, and Controversies, 1620–1770.* Philadelphia: Press of the University of Pennsylvania, 1924.

Batchellor, Albert Stillman, ed. *Documents Relating to the Masonian Patent, 1630–1846.* Provincial and State Papers, Vol. 29. Edward N. Pearson, Public Printer, 1896.

Belknap, Jeremy. *The History of New-Hampshire. Vol. 1.* 1831: Johnson Reprint Corp., 1970.

———. *The History of New-Hampshire. Vol. III.* 1812: Johnson Reprint Corp., 1970.

Blanchard, Joseph. "Affidavit of Joseph Blanchard: Running the Masonian Patent Line." *The Benchmark,* Summer/Fall 1985.

Breckenridge, John A., ed. "The First Work on Surveying Mason's Curve." *The Benchmark,* April 1976.

———. "The First Work on Surveying Mason's Curve, Part II." *The Benchmark,* July 1976.

City of Keene. *Thirty-Fourth Annual Report of the City of Keene . . . For 1907.* Keene, NH: Sentinel Printing, 1908.

Clark, Charles E. *The Eastern Frontier: The Settlement of Northern New England, 1610–1763.* New York: Alfred A. Knopf, 1970.

Daniell, Jere R. *Colonial New Hampshire.* Millwood, NY: KTO Press, 1981.

Dean, John Ward. *Capt. John Mason, The Founder of New Hampshire.* 1887: New York: Burt Franklin, 1967.

Fry, William H. *New Hampshire as a Royal Province.* 1908: New York: AMS Press, 1970.

Garvin, James L. "The Range Township in Eighteenth-Century New Hampshire." *New England prospect: maps, place names, and the historical landscape.* The Dublin Seminar for New England Folklife Annual Proceedings, Vol. 5, 1980.

Hammond, Otis Grant. "The Masonian Title and Its Relations to New Hampshire and Massachusetts." American Antiquarian Society, *Proceedings* (New Series) 27, 1916.

Johnson, Richard R. "Robert Mason and the Coming of Royal Government to New England." *Historical New Hampshire,* Winter 1980.

Keene Evening Sentinel. "Masonian Line Monument." August 27, 1907.

Laskey, Mark. "The Great Dying: New England's Coastal Plague, 1616–1619." http://www.cvltnation.com.

Lord, G. T., ed. *Belknap's New Hampshire: An Account of the State in 1792.* 1812: Portsmouth, NH: Peter E. Randall, Publisher, 1973.

Marr, J. S., and J. T. Cathey. "New Hypothesis for Cause of Epidemic among Native Americans, New England, 1616–1619." *Emerging Infectious Diseases,* February 2010. (Internet serial) http://www.cdc.gov/EID/content/16/2/281.htm.

New Hampshire Land Surveyors Association. "Proceedings of the Seminar on the History of New Hampshire Relating to Land Surveying," February 24, 1979. Jointly Sponsored by The Lifelong Learning Center/Merrimack Valley College and New Hampshire Land Surveyors Association.

———. "Proceedings of History Seminar II: New Prospective on History of New Hampshire Pertaining to Land Surveying." January 31, 1981.

Page, Elwin L. "The Case of Samuel Allen of London, Esqr., Governor of New Hampshire." *Historical New Hampshire,* Winter 1970.

———. *Judicial Beginnings in New Hampshire, 1640–1700.* Concord, NH: New Hampshire Historical Society, 1959.

———. "The Validity of John Mason's Title to New Hampshire." *Historical New Hampshire,* August 1953.

Richeson, A. W. *English Land Measuring to 1800: Instruments and Practices.* Cambridge, MA: MIT Press, 1966.

Roberts, E. N. *Land Surveying in New Hampshire.* State of New Hampshire, State Planning Project, 1965.

Rose, Carol M., "Canons of Property Talk, or, Blackstone's Anxiety." *Faculty Scholarship Series.* Paper 1802, 1999. http://digitalcommons.law.yale.edu/fss_papers/1802.

Seed, Patricia. *Ceremonies of Possession in Europe's Conquest of the New World, 1492–1640.* Cambridge, UK: Cambridge University Press, 1995.

Upham, George B. "The 'Great River Naumkeek' Once the Southern Boundary of New Hampshire." *The Granite Monthly,* May 1920.

———. "New Hampshire Town Boundaries Determined by Mason's Curve." *The Granite Monthly,* January 1920.

Van Deventer, David E. *The Emergence of Provincial New Hampshire, 1623–1741.* Baltimore, MD: Johns Hopkins University Press, 1976.

Wadsworth, Samuel. "The Masonian Patent Line." *The Benchmark,* Winter/Spring 1985.

Wood, Gordon S. *The Radicalism of the American Revolution.* New York: Alfred A. Knopf, 1992.

The Ballad of Romaine Tenney

Adams, Nathan M. "He Burned Himself to Death to Protest Superhighway." *New York Journal-American,* September 19, 1964.

Albers, Jan. *Hands on the Land: A History of the Vermont Landscape.* Cambridge, MA: MIT Press, 2000.

Allsbee, Roger. Interview with the author, September 21, 2011.

Cann, Roald. "The Humane Thing." *Daily Eagle,* Claremont, NH, October 1, 1964.

Dickerson, Gerri. Letter to Weathersfield Historical Society, April 30, 2008.

Graff, Christopher. "New Interstates Pushed Vt. to Modern Age." http://www.ver-monttoday.com/century/topstories/newinterstate.html.

Gutfreund, Owen D. *Twentieth-Century Sprawl: Highways and the Reshaping of the American Landscape.* Oxford: Oxford University Press, 2004.

Harrison, Blake. *The View from Vermont: Tourism and the Marketing of an American Rural Landscape.* Burlington, VT: University of Vermont Press, 2006.

Haupt, Kathryn. Interview with the author, May 21, 2012.

Highway Board Meetings and Compensation Hearings, Docket #4543: "Rose Tenney Estate vs. State Highway Board," "Informal Summary of Hearing . . . December 19, 1963." "Informal Summary of Continued Hearing . . . January 20, 1964." Vermont State Archives.

Huffman, Benjamin L. *Getting Around Vermont: A Study of 20 Years of Highway Building in Vermont with Respect to Economics, Automotive Travel, Community Patterns, and the Future.* The Environmental Program, University of Vermont, 1974.

Hunter, Edith. "Man Who Would Not Be Moved—Romaine Tenney." Unpublished paper, Weathersfield Historical Society.

Jarvis, Joe. Interview with the author, September 13, 2011.

Jarvis, Paul. Interview with the author, September 13, 2011.

Jarvis, Victor. Interview with the author, September 19, 2011.

Laas, William, ed. *Freedom of the American Road.* Dearborn, MI: Ford Motor Co., 1956.

Lewis, Tom. *Divided Highways.* London: Penguin Books, 1999.

Mansfield, Howard. "I Will Not Leave." *Yankee,* March/April 2013.

Meeks, Harold A. *Time & Change in Vermont: A Human Geography.* Guilford, CT: Globe Pequot Press, 1986.

Mountain, Archie. "Vt. Farmer Who Defied Highway May Have Set His Own Funeral Pyre." *Daily Eagle,* Claremont, NH, September 14, 1964.

Newcity, Joan. Interview with the author, September 12, 2011.

O'Donnell, Maureen. Interview with the author, September 20, 2011.

Patton, Phil. *Open Road.* New York: Simon & Schuster, 1987.

Richard Stanton Haupt Papers, Norwich University Archives & Special Collections, Northfield, VT.

Rose, Mark H. *Interstate: Express Highway Politics, 1941–1956.* Lawrence, KS: Regents Press of Kansas, 1979.

Safford, Rosemary. Interview with the author, September 12, 2011.

SAPA TV (Springfield, VT). "The Farmer Wouldn't Be Bought: Romaine Tenney. A program from the Weathersfield Historical Society." Broadcast February 10, 2009.

Sherman, Joe. *Fast Lane on a Dirt Road: Vermont Transformed 1945–1990.* Woodstock, VT: Countryman Press, 1991.

Shulins, Nancy. "He Lies Buried Beneath Road He Hated." Associated Press, July 29, 1979.

Singh, Dev. Interview with the author, September 23, 2011.

Spaulding, Rod. Interview with the author, September 20, 2011.

Springfield Reporter. "Ascutney Man Who Didn't Want to Leave Farm to Make Way For Interstate Missing After Fire." September 16, 1964.

Stoughton, Roger. Interview with the author, September 18, 2011

Tenney, Rod. Interview with the author, August 18, 2011.

Tenney, Ron. Interview with the author, September 12, 2011.

Textual Records from the Dept. of Transportation, Federal Highway Administration, Bureau of Public Roads, VT District Office (04/01/1967–08/10/1970), Record Group 30, F. A. Project Files 1966–68. National Archives, Northeast Region, Waltham, MA.

Weathersfield Annual Report, Year Ending 1963.

Weathersfield, VT, Historical Society. Romaine Tenney scrapbook.

Whitney, Don. Interview with the author, September 23, 2011

Wood, Willis. Interview with the author, September 18, 2011.

My Roots Are Deeper than Your Pockets

Alexander, Laura. "Coos at the Crossroads." *Beyond the Notches.* John R. Harris et al., eds. Littleton, NH: Bondcliff Books, 2010.

Baker, Bob. Interview with the author, August 23, 2012.

Beauregard, Denis. "Hydro-Quebec." *Economic Development Journal,* Winter 2003.

Came, B. "Colliding Cultures." *Maclean's,* August 12, 1991.

The Colebrook Chronicle. "Lynne Placey Honored as Opposition Hero." January 20, 2012.

Cowan, Edward. "Quebec Said to Plan Huge Hydroelectric Project." *The New York Times,* April 29, 1971.

Crossette, Barbara. "Indian Chief Miffs Canada, Not Sparing Indians." *The New York Times,* October 31, 2001.

Dillon, John. "Big Hydro: Changes to Cree Culture." Vermont Public Radio, broadcast August 19, 2010.

Farnsworth, Clyde M. "Toughest Fight Yet for Hydro-Quebec." *The New York Times,* October 6, 1991.

Garner, Fradley. "View from Abroad: The Northern Frontier." *Environment,* March 1975.

Giniger, Henry. "Cree Indians on Canadian Island are Facing Disruption." *The New York Times,* June 7, 1977.

Gold, Allan R. "Quebec Hydropower Battle Entangles Utilities in U.S." *The New York Times,* October 17, 1990.

———. "Quebec Indians Ponder True Cost of Electricity." *The New York Times,* October 12, 1990.

Harrigan, John. Interview with the author, August 23, 2012.

———. Letter to the N. H. Site Evaluation Committee, March 17, 2009.

———. "Wrecking What Little We Have Left." https://nonorthernpassnh.blogspot.com. *New Hampshire Union Leader,* December 18, 2010.

Heindel, Naomi. "The Cree and the Crown." *Northern Woodlands,* Winter 2012.

Indian Country Today Media Network. "James Bay Cree: Hunting as Way of Life and Death." http://indiancountrytodaymedianetwork.com, January 20, 2011.

The James Bay and Northern Quebec Agreement. http://www.gcc.ca/, 1975.

Jensen, Chris. "Saying No to Northern Pass and Hundreds of Thousands of Dollars." New Hampshire Public Radio, broadcast October 27, 2011.

Johnsen, Rick. Interview with the author, August 23, 2012.

Jones, Howard Mumford, and Walter B. Rideout, eds. *Letters of Sherwood Anderson.* Boston: Little, Brown & Co., 1953.

Jordan, Donna. "Forest Society Easements Hope to Block Northern Pass." *The Colebrook Chronicle,* August 24, 2012.

Journeyman Pictures. *Cree Freedom—Canada*. http://www.youtube.com/watch?v=7GdegbTQGAo, July 1996.

Kaufman, Marty. Interview with the author, August 24, 2012.

Kaye, Robin. "The Island of Lana'i Has Been Remade into Ellisonia." *The Huffington Post*, September 10, 2013.

Krauss, Clifford. "Mistissini Journal: Will the Flood Wash Away the Crees' Birthright?" *The New York Times*, February 27, 2002.

Lathem, Alexis. "Hydro Quebec is Back." *Earth Island Journal*, Summer 1995.

Linden, Eugene. "Bury My Heart at James Bay." *Time*, July 15, 1991.

Lewandowski, René. "What 8,000 Studies Reveal." *Canadian Geographic*, November/December 2005.

Main, Carla T. "How Eminent Domain Ran Amok." *Policy Review*, October 1, 2005.

Mann, Brian. "Thirst for Energy Fuels Controversial Power Project." *All Things Considered* (National Public Radio), November 9, 2007.

Mansfield, Howard. "My Roots Are Deeper than Your Pockets." *Yankee*, March/April 2013.

McAllaster, Rod. Interview with the author, August 23, 2012.

Nagourney, Adam. "Tiny Hawaiian Island Will See if New Owner Tilts at Windmills." *The New York Times*, August 22, 2012.

Niezen, Ronald. *Defending the Land: Sovereignty and Forest Life in James Bay Cree Society*. Boston: Allyn and Bacon, 1998.

Placey, Lynne. Interview with the author, August 31, 2012.

———. Letters to the Editor. *The Colebrook Chronicle*, October 14, 2011, June 15, 2012.

Powers, John M. "Hydro-Quebec CEO has Good Reason to Smile." *Electric Light & Power*, July/August 2007.

Rappaport, Larry. Interview with the author, August 24, 2012.

Salisbury, Richard F. *A Homeland for the Cree*. Kingston, ON: McGill-Queen's Press, 1986.

Shulgan, Christopher. "The Price of Peace." *Canadian Geographic*, November/December 2005.

Society for the Protection of New Hampshire Forests. *Power of the People: A Guide to Public Decisions Needed by Northern Pass and How to Make Your Voice Heard*. 2011.

Stanton, John. "In Canada, an Environmental Disaster Looms." Letter to the Editor, *The New York Times*, July 14, 1990.

Timmons, Sara. Interview with the author, September 17, 2012.

The Toronto Star. "Marching to save identity, way of life; 'Our land is the last thing we have left,' say Innu protesting hydro project." April 14, 2012.

Tripp, Nathaniel. "Drowning Quebec for Hydropower." *Country Journal* 18, no. 6 (November/December 1991).

US Department of Energy. *Northern Pass Transmission Line Draft EIS. Public Scoping Hearing*. Colebrook Elementary School, March 19, 2011.

Williams, Ted. "Hydro-Quebec Hits Granite." *Audubon*, May/June 2014.

Winslow, Edward, et al. *Mourt's Relation: A Journal of the Pilgrims in Plymouth*. 1622: Carlisle, MA: Applewood Books, 1986.

Wray, Elizabeth. "Travelers between Two Worlds." *Sierra*, May/June 1992.

Wright, Kristen. "Electric Light & Power Utility 2010 Utility of the Year: Hydro-Quebec." *Electric Light & Power*, November/December 2010.

Yakabuski, Konrad. "New England Lusts for Quebec's Power." *The Toronto Globe & Mail*, December 30, 2008.

The Pipeline in the Neighborhood

Atkin, Emily. "Data: Oil Trains Spill More Often, But Pipelines Spill Bigger." https:// thinkprogress.org, February 18, 2015.

Bade, Gavin. "Beyond the Substation: How Five Proactive States are Transforming the Grid Edge." UtilityDive.com, March 2, 2015.

Bary, Andrew. "Kinder Morgan: Trouble in the Pipelines?" *Barron's*, February 22, 2014.

Bernstein, Hattie. "Protests Grow Over Proposed Pipeline Through Mass." *The Boston Globe*, July 17, 2014.

Black & Veatch. *New England Natural Gas Infrastructure and Electric Generation: Constraints & Solutions, Phase II*, April 16, 2013.

Casper, Barry M., and Paul David Wellstone. *Powerline: The First Battle of America's Energy War*. Amherst, MA: University of Massachusetts Press, 1981.

Cathles, Lawrence M., et al. "A Commentary on 'The Greenhouse-Gas Footprint of Natural Gas in Shale Formations' by R. W. Howarth, R. Santoro, and Anthony Ingraffea." *Climatic Change*, October 21, 2011.

Cleveland, Shanna, and N. Jonathan Peress. Conservation Law Foundation letter to Heather Hunt, Executive Director, New England States Committee on Electricity (NESCOE). May 30, 2014.

Conservation Law Foundation. "Preliminary Documents: New England Governors' Pipeline and Power Line Plan Shaped by Utility and Gas Pipeline Companies, Incomplete Analysis." www.clf.org, June 2014.

Douglas, Mary. "Anatomy of a Pipeline Decision: A Scheme of 'Dubious' Legality." theberkshireedge.com, October 20, 2014.

Driscoll, Andy. *Power Play*. Twin Cities Public Television, 1979.

Feely, Paul. "Pipeline Basics: The Players, The Project." *New Hampshire Union Leader*, January 15, 2015.

Feldstein, Michael, and Kathy Kessler. "Burden of Proof: The Case Against the Proposed Northeast Energy Direct (NED) Fracked Gas Pipeline." http://www. nofrackedgasinmass.org, August 27, 2014.

Fisher, Joe. "Northeast Natural Gas Pipe Projects Under Fire, But It's Never Been Easy." NGI: Natural Gas Intelligence. www.naturalgasintel.com, January 23, 2015.

Fore, Allen. Interview with the author, February 23, 2015.

Gabrilska, Leslie. Interview with the author, October 24, 2014.

Howarth, Robert W. "Methane Emissions and Climatic Warming Risk from Hydraulic Fracturing and Shale Gas Development: Implications for Policy." *Energy and Emission Control Technologies*, October 8, 2015.

INGAA. "North American Midstream Infrastructure Through 2035: Capitalizing on Our Energy Abundance. An INGAA Foundation Report, Prepared by ICF International." www.ingaaa.org, March 18, 2014.

ISO New England. "New England 2030 Power System Study." www.iso-ne.com, February 2010.

———. "State of the Grid: Managing a System in Transition." www.iso-ne.com, January 21, 2015.

Karanian, Edna M. "Regional Market Trends: The Gas Utility Perspective, May 1, 2014." www.northeastgas.org.

Kinder Morgan. "Additional Kinder Morgan Partners Responses to Barron's." http:// ir.kindermorgan.com/sites/kindermorgan.investorhq.businesswire.com.

———. "Fixed Income Investor Presentation." Kindermorgan.com, March 5, 2015.

Lafleur, Cheryl. National Press Club Luncheon. January 27, 2015.

Living on Earth. "Unprecedented New England Pipeline Proposal" (National Public Radio broadcast). Broadcast the week of July 25, 2014.

Lombardi, Kristen, and Jamie Smith Hopkins. "Natural Gas Building Boom Fuels Climate Worries, Enrages Landowners." Center for Public Integrity, www.publicintegrity.org, July 17, 2017.

Lovelace, Holly. Interview with the author, December 2, 2015.

———. Public hearing testimony, Massachusetts Department of Public Utilities, Greenfield, MA, June 11, 2015.

Luce, Benjamin. "Issues with Windpower in New England." Lecture for New Hampshire Windwatch. http://www.nhwindwatch.org, January 18, 2015.

Mansfield, Howard. "Power Struggles." *Yankee,* January/February 2016.

Massachusetts Pipeline Awareness Network. "Kinder Morgan Natural Gas Pipeline: Wrong Choice for Massachusetts." www.massplan.org, September 2014.

McKenna, Phil. "Pipeline Payday: How Builders Win Big, Whether More Gas Is Needed or Not." https://insideclimatenews.org, February 8, 2017.

Moskowitz, Peter. "With the Boom in Oil and Gas, Pipelines Proliferate in the U.S." Environment360. e360.yale.edu, October 6, 2014.

Northey, Hannah. "Energy Policy: FERC Faces Heightened Scrutiny as Gas Projects Proliferate." http://www.eenews.net, November 3, 2014.

Obama, Barack. "Remarks by the President on Recovery Act Funding for Smart Grid Technology." http://www.whitehouse.gov/the-press-office, October 27, 2009.

Pipeline Safety Trust. "Briefing Paper #6: Considering Risk." pstrust.org.

———. "Briefing Paper #9: Pipeline Routing and Siting Issues." pstrust.org.

Premus, Vince. "Gordon Van Welie, Do Your Job." http://commonwealthmagazine.org, April 3, 2015.

———. Letter to Massachusetts Assistant Attorney General Jesse Reyes, June 16, 2014.

Premus, Vince and Denene. Interviews with the author, October 24, 2014, March 21, 2015.

Restuccia, Andrew, and Elana Schor. "'Pipelines Blowup and People Die.'" Politico.com, April 21, 2015.

Sellars, Carolyn. Interview with the author, October 24, 2014.

Shattuck, Peter, et al. "The Missing Energy Crisis & What it Tells Us About the Energy System of the Future." Acadiacenter.org, May 26, 2015.

Solomon, Dave. "Kinder Morgan Pipeline Safety Record Fuels N.H. Opposition." *New Hampshire Union Leader,* February 7, 2015.

Southwest Pennsylvania Environmental Health Project. "Summary on Compressor Stations and Health Impacts." www.environmentalhealthproject.org, February 24, 2015.

Stockman, Lorne, et al. *A Bridge Too Far: How Appalachian Basin Gas Pipeline Expansion Will Undermine U.S. Climate Goals.* Oil Change International, July 2016.

Subra, Wilma. "Human Health Impacts Associated with Chemicals and Pathways of Exposure from the Development of Shale Gas Plays." Earthworks Oil and Gas Accountability Project. n.d.

Sweeney, Rory D. "Shale Gas Impacts Cut a Wide Swath, Panelists Say." www.rtoinsider.com, October 30, 2017.

Tennessee Gas Pipeline Co. "Franklin Regional Council of Governments: Northeast Energy Direct Project Questions and Answers." Kinder Morgan.

———. "Northeast Energy Direct Project, Docket No. PF14-22-000. Draft Environmental Report. Resource Report 1." March 2015.

Thimble Creek Research. "Madison County, New York Department of Health Comments to the Federal Energy Regulatory Committee Concerning Docket No. CP14 -497-000, Dominion Transmission, Inc." http://www.newipswichpipeline. info, September 30, 2014.

UtilityDive.com. "2015 State of the Electric Utility."

Van Welie, Gordon. Press conference, January 21, 2015.

Wald, Matthew L. "How to Build the Supergrid." *Scientific American*, November 2010.

Three Beautiful Days on a Warming Planet

Abel, David. "In Maine, Scientists See Signs of Climate Change." *The Boston Globe*, September 21, 2014.

Anderson, John. Interview with the author, June 22, 2016.

———. *The Potential Impact of Sea Level Rise on Seabird Nesting Islands in Acadia National Park.* Natural Resource Report NPS/ACAD/NRR—2015/1055.

Anderson, L., P. Glick, S. Heyck-Williams, and J. Murphy. *Changing Tides: How Sea-Level Rise Harms Wildlife and Recreation Economies along the U.S. Eastern Seaboard.* National Wildlife Federation, 2016.

Coles, B. J. "Doggerland: The Cultural Dynamics of a Shifting Coastline." *Geological Society, London, Special Publications*, vol.175, 2000. http://sp.lyellcollection.org/content/175/1/393.abstract.

———. "Doggerland: A Speculative Survey." *Proceedings of the Prehistoric Society*, vol. 64. Prehistoric Society, London, 1998.

College of the Atlantic. "Great Duck Island." https://coa.edu/islands/great-duck-island/.

Connecticut Audubon Society, "Osprey Nation 2016." http://www.ctaudubon.org

DeConto, Robert M., and David Pollard. "Contribution of Antarctica to Past and Future Sea-Level Rise." *Nature,* March 31, 2016.

Dennis, Brady, and Chris Mooney. "Scientists Nearly Double Sea Level Rise Projections for 2100, because of Antarctica." *The Washington Post*, March 30, 2016.

Fagan, Brian. *The Attacking Ocean.* New York: Bloomsbury Press, 2013.

Flannery, Tim. *The Weather Makers.* New York: Atlantic Monthly Press, 2005.

Gaffney, Vincent, et al., eds. *Mapping Doggerland: The Mesolithic Landscapes of the Southern Northern Sea.* Oxford: Archaeopress, 2007.

Gedan, Keryn B., et al. "Uncertain Future of New England Salt Marshes." *Marine Ecology Progress Series*, vol. 434, 2011.

Gessner, David. *Return of the Osprey.* Chapel Hill, NC: Algonquin Books of Chapel Hill, 2001.

Gillis, Justin. "Climate Model Predicts West Antarctica Ice Sheet Could Melt Rapidly." *The New York Times,* March 30, 2016.

Griswold, Evan. Interview with the author, June 1, 2016.

Griswold, Wick. *Griswold Point.* Charleston, SC: History Press, 2014.

Gulf of Maine Council on the Marine Environment. *Salt Marshes in the Gulf of Maine: Human Impacts, Habitat Restoration, and Long-Term Change Analysis,* 2008.

Henson, Robert. *The Thinking Person's Guide to Climate Change.* Boston: American Meteorological Society, 2014.

Hirji, Zahra. "Flood Damage Costs Will Rise Faster Than Sea Levels, Study Says." Insideclimatenews.org, March 1, 2016.

Holtz, Debra, et al. *National Landmarks at Risk: How Rising Seas, Floods, and Wild-fires Are Threatening the United States' Most Cherished Historic Sites.* Union of Concerned Scientists, May 2014.

Jacobson, G. L., et al., eds. *Maine's Climate Future: An Initial Assessment.* http://www.climatechange.umaine.edu/mainesclimatefuture/. University of Maine, 2009.

Kahn, Brian. "Climate Change is Decimating Cod in the Gulf of Maine." Climate-Central.org, October 29, 2015.

Kelley, J. "Sea Level Rise along the Coast of Maine." Climate Change Institute. http://climatechange.umaine.edu/Research.

Kolbert, Elizabeth. *The Sixth Extinction.* New York: Henry Holt, 2016.

Laffoley, D., and J. M. Baxter, eds. *Explaining Ocean Warming: Causes, Scale, Effects and Consequences.* International Union for Conservation of Nature and Natural Resources, 2016.

Marshall, George. *Don't Even Think About It: Why Our Brains Are Wired to Ignore Climate Change.* New York: Bloomsbury, 2014.

Massachusetts Office of Coastal Zone Management. *Sea Level Rise: Understanding and Applying Future Trends for Analysis and Planning.* December 2013.

McKibben, Bill. "Global Warming's Terrifying New Math." *Rolling Stone,* July 19, 2012.

Merritt, Elizabeth. "Unwanted Water." Center for the Future of Museums Blog, May 31, 2016. http://futureofmuseums.aam-us.org/2016/05/unwanted-water.html.

The Nature Conservancy. *Adapting to the Rise: A Guide for Connecticut's Coastal Communities.* 2015.

———. *Developing a Regional Framework for Assessing Coastal Vulnerability to Sea Level Rise in Southern New England.* September 2010.

New Hampshire Coastal Risk and Hazards Commission. *Preparing New Hampshire for Projected Storm Surge, Sea-Level Rise and Extreme Precipitation.* 2016.

Rinaldi, Jessica. "The Seas are Rising Fast—And Even Faster in Mass." *The Boston Globe,* February 25, 2016.

Ryan, A., and A. W. Whelchel. *The Salt Marsh Advancement Zone Assessment of Connecticut.* The Nature Conservancy, Coastal Resilience Program, 2015.

———. *A Salt Marsh Advancement Zone Assessment of Old Lyme, Connecticut.* The Nature Conservancy, Coastal Resilience Program. Publication Series #1-U, New Haven, CT. 2015.

Salisbury, Edward E. *The Griswold Family of Connecticut.* New Haven, CT: Press of Morehouse, Tuttle & Taylor, 1884.

Schwartz, Michael. "The Age of Extremes." *Ag Professional.* http://www.commodities-now.com/reports/, October 29, 2015.

Steneck, Robert S., and Richard A. Wahle. "American Lobster Dynamics in a Brave New Ocean." *Canadian Journal of Fisheries and Aquatic Sciences* 70, no. 11 (2013).

Stephenson, Toby. Interview with the author, June 22, 2016.

Strauss, Benjamin H., et al. *Mapping Choices: Carbon, Climate, and Rising Seas—Our Global Legacy.* ClimateCentral.org, November 2015.

Sweet, William V. *Patterns and Projections of High Tide Flooding Along the U.S. Coastline Using a Common Impact Threshold.* NOAA Technical Report NOS CO-OPS 086. National Oceanic and Atmospheric Administration, February 2018.

USGCRP, 2017: *Climate Science Special Report: A Sustained Assessment Activity of the U.S. Global Change Research Program.* U.S. Global Change Research Program, 2017.

Weiss, Judith S., and Carol A. Butler. *Salt Marshes: A Natural & Unnatural History.* New Brunswick, NJ: Rutgers University Press, 2009.

Whelchel, Adam. Interview with the author, June 1, 2016.

Whittle, Patrick. "With Young Lobsters in Decline, Concern for Maine Fishery Rises." *Portland Press Herald*, April 22, 2014.

Woodard, Colin. "Mayday: Gulf of Maine in Distress." A six-part series. *Portland Press Herald*, October 25–30, 2015.

Epilogue: Finding the Good Life in Broken Town

Bruchac, Marge. "Sokoki Homeland from Monadnock (K'namitobena Sokwaki)." *Where the Mountain Stands Alone: Stories of Place in the Monadnock Region.* Howard Mansfield, ed. Rindge, NH: Monadnock Institute of Nature, Place and Culture/University Press of New England, 2006.

Calloway, Colin G. *New Worlds for All: Indians, Europeans, and the Remaking of Early America.* Baltimore, MD: Johns Hopkins University Press, 1997.

———, ed. *After King Philip's War: Presence and Persistence in Indian New England.* Hanover, NH: Dartmouth College/University Press of New England, 1997.

Chah, Ajahn. *The Teachings of Ajahn Chah.* The Sangha, Wat Nong Pah Pong, 2007. http://www.accesstoinsight.org/lib/.

Didion, Joan. *Blue Nights.* New York: Alfred A. Knopf, 2011.

Epstein, Mark. "Freud and Buddha." The Network of Spiritual Progressives. Spiritual-progressives.org

———. *Going to Pieces without Falling Apart.* New York: Broadway Books, 1998.

———. *Psychotherapy without the Self.* New Haven, CT: Yale University Press, 2007.

———. *Thoughts without a Thinker.* New York: Basic Books, 1995.

———. *The Trauma of Everyday Life.* New York: Penguin Press, 2013.

Mayer, J. P., ed. *Journey to America: Alexis de Tocqueville.* New Haven, CT: Yale University Press, 1959.

Mendelsohn, Daniel. "Battle Lines: A Slimmer, Faster Iliad." *The New Yorker*, November 7, 2011.

Morgan, Edmund S., ed. *Puritan Political Ideas.* Indianapolis: Bobbs-Merrill Educational Publishing, 1965.

Morrison, Kenneth M. *The Embattled Northeast: The Elusive Ideal of Alliance in Abenaki-Euramerican Relations.* Berkeley, CA: University of California Press, 1984.

Nhat Hanh, Thich. *Answers from the Heart.* Berkeley, CA: Parallax Press, 2009.

———. *No Death, No Fear.* New York: Riverhead Books, 2002.

O'Callaghan, E. B., ed. *Documents Relative to the Colonial History of the State of New York*, Vol. X. Albany, NY: Weed, Parsons, and Co., 1853–1887.

Tocqueville, Alexis de. "A Fortnight in the Wilderness." http://oll.libertyfund.org/titles/2435, August 1831.

ACKNOWLEDGMENTS

My thanks to Jane Eklund and Joni Praded for reading and commenting on the manuscript. "The Land of Many Uses" chapter benefited from Jane Difley's review. Donna-Belle and James Garvin subjected "The Last Medieval Claim" chapter to a thoughtful reading. Thanks to Sandy Tolan for his help on the desert walk, Rick Johnsen for acting as my North Country guide, and to Cristina Prochilo Miedema for her insights into the public reaction to proposed neighborhood projects. Thanks once again to Hancock Town Library Director Amy Markus for hunting down an eclectic zoo of books through Interlibrary Loan.

Quite a few people took the time to answer my questions, sometimes in many hours of interviews or during days of hanging around. I am indebted to: Roger Allbee; John Anderson; Bob Baker; John D. Bridges; Brian Burford; Meade Cadot; Gerri Dickerson; Sally Fairfax; Leslie Gabrilska; John Harrigan; Kathryn Haupt; Joe, Paul, and Victor Jarvis; Terrol Johnson; Tony Johnson; Marty Kaufman; Bob King; Holly Lovelace; Danny Lopez; Dick Martin; Rod McAllaster; Joan Newcity; Lynne Placey; Denene and Vince Premus; Larry Rappaport; Rodney Rowland; Rosemary Safford; Carolyn Sellars; Cookie Shand; Dev Singh; Rod Spaulding; Toby Stephenson; Roger Stoughton; Rod and Ron Tenney; Sara Timmons; Adam Whelchel; Don Whitney; and Willis Wood.

Several of these chapters appeared in different versions in *Yankee* magazine. Writing for *Yankee* has given me the chance to test out my ideas by reporting on some complex issues. My thanks to *Yankee*'s editor Mel Allen.

It's a pleasure to work with the crew at Bauhan Publishing: Sarah Bauhan, Henry James, Mary Ann Faughnan, and Jocelyn Lovering.

As always, my deepest thanks to my wife and editor, the octopus whisperer, Sy Montgomery.

PRAISE FOR HOWARD MANSFIELD'S BOOKS:

Summer Over Autumn

At a time of change, when common ground is becoming less common, it helps to have a tour guide who can navigate the American landscape. That's a job for Howard Mansfield, who offers a unique lens on our history, customs and habits. . . .

—Joan Silverman, *Portland Press Herald*

A wonderful book . . . goes straight to the heart of the inscrutable nature of small-town life, of New England life. We are left with no choice but to love this book.

—Edie Clark, author of *The Place He Made*

Dwelling in Possibility

I was only halfway through this book when I began to quote from it. It is strong stuff and goes deep. It should be on every thoughtful citizen's "must read" list.

—Karen Dahood, *BookPleasures.com*

Mansfield is a perceptive writer about an American sense of place, and how people pass through and how they settle down. . . . [He] is an eloquent master of the discursive essay.

—*Valley News,* West Lebanon, New Hampshire

Turn and Jump

Mansfield takes the reader deeper into worlds familiar and strange, illuminating both this historical process and its human cost. . . .

—Eric Miller, associate professor of history, Geneva College,
Books & Culture: A Christian Review

As an excavator and guardian of our living past, Howard Mansfield is unmatched.

—John Heilpern, Contributing Editor, *Vanity Fair.*

The Bones of the Earth

[T]here's nobody else in this generation who writes so meaningfully about what really matters in the American landscape.

—Beth Kephart, author of *A Slant of Sun*

The Same Ax, Twice

The Same Ax, Twice is filled with insight and eloquence . . . a memorable, readable, brilliant book on an important subject. It is a book filled with quotable wisdom.

—*The New York Times Book Review*

The Same Ax, Twice is one of those quiet books that foments revolution.

—*ArchitectureBoston*

In the Memory House

. . . the best of [books], the deepest, the widest-ranging, the most provocative is Howard Mansfield's *In the Memory House*."

—*Hungry Mind Review.*

Mr. Mansfield gets beneath the patina of the tangible and intangible relics of our history to locate the emotional core of our past. . . . A wise and beautiful book.

—*The New York Times Book Review.*

ELEMENTS
OF SIMULATION

ELEMENTS
OF SIMULATION

Byron J. T. Morgan

MATHEMATICAL INSTITUTE
UNIVERSITY OF KENT, CANTERBURY, UK

LONDON   NEW YORK
CHAPMAN AND HALL

First published 1984 by
Chapman and Hall Ltd
11 New Fetter Lane, London EC4P 4EE
Published in the USA by
Chapman and Hall
29 West 35th Street, New York NY 10001
Reprinted 1986

Printed in Great Britain at
The University Press, Cambridge

ISBN 0 412 24580 9 (cased)
ISBN 0 412 24590 6 (Science Paperback)

British Library Cataloguing in Publication Data

Morgan, Byron J. T.
  Elements of simulation
  1. Simulation methods
  I. Title
  001.4'24   T57.62

  ISBN 0-412-24580-9
  ISBN 0-412-24590-6 Pbk

Library of Congress Cataloging in Publication Data

Morgan, Byron J. T., 1946-
  Elements of simulation.

  Bibliography: p.
  Includes indexes.
  1. Mathematical statistics—Data processing.
2. Digital computer simulation.   I. Title.
QA276.4.M666   1984   519.5'0724   83-25238
ISBN 0-412-24580-9
ISBN 0-412-24590-6 (pbk.)

CONTENTS

PREFACE

The use of simulation in statistics dates from the start of the 20th century, coinciding with the beginnings of radio broadcasting and the invention of television. Just as radio and television are now commonplace in our everyday lives, simulation methods are now widely used throughout the many branches of statistics, as can be readily appreciated from reading Chapters 1 and 9. The rapid development of simulation during this century has come about because of the computer revolution, so that now, at one extreme, simulation provides powerful operational research tools for industry, while at the other extreme, it is also taught in schools. It is employed by a wide variety of scientists and research workers, in industry and education, all of whom it is hoped will find this book of use. The aim is to provide a guide to the subject, which will include recent developments and which may be used either as a teaching text or as a source of reference.

The book has grown out of a fifteen-hour lecture course given to third-year mathematics undergraduates at the University of Kent, and it could be used either as an undergraduate or a postgraduate text. For elementary teaching the starred sections and Exercises may be omitted, but most of the material should be accessible to those possessing a knowledge of statistics at the level of *A Basic Course in Statistics* by Clarke and Cooke (1983). This book, henceforth referred to as '*ABC*', is widely used for teaching statistics in schools. Most of the standard statistical theory encountered that is not in the revision Chapter 2 is given a page reference in *ABC*.

Simulation may either be taught as an operational research tool in its own right, or as a mathematical method which cements together different parts of statistics and which may be used in a variety of lecture courses. In the last three chapters indications are made of the varied uses of simulation throughout statistics. Alternatively, simulation may be used to motivate subjects such as the teaching of distribution theory and the manipulation of random variables, and Chapters 4 and 5 especially will hopefully be useful in this respect. To take a very simple illustration, a student who finds subjects such as random

ix

variables and probability density functions difficult is greatly helped if the random variables can be simulated and histograms drawn that may be compared with density functions, as is done in Chapter 2.

Inevitably simulation requires algorithms, and computer programs for the operation of those algorithms. A seemingly complicated algorithm can often be programmed in a simple way, and so the approach has been adopted here of occasionally presenting illustrative computer programs in the BASIC computer language. BASIC is chosen because of its current wide use, from home microcomputers upwards, and the programs are written in accordance with the American Standard for what is called 'Minimal BASIC', so that they need no modification for running on a variety of different implementations of BASIC. Graduate students and research workers in industry and elsewhere will frequently use pre-programmed computerized algorithms in their simulation work. Ready supplies of these algorithms are provided by the NAG and IMSL libraries, and Appendix 1 describes the relevant routines from these libraries, indicating also which methods are being used. Computer simulation languages are discussed in Chapter 8.

There are 232 Exercises and complements, some of which are designed to be of interest to advanced readers and contain more recent and more difficult material. These, and more advanced sections are marked with a *. In each chapter a small number of straightforward exercises are marked with a †symbol, and it is recommended that all readers should attempt these exercises.

ACKNOWLEDGEMENTS

This book was originally planned as a joint venture with Barrie Wetherill, but because of his involvement in several other books, it became mine alone. I am therefore particularly grateful to Professor Wetherill for his early motivation, advice and enthusiasm. Many others have been extremely helpful at various stages, and I am especially indebted to Mike Bremner, Tim Hopkins, Keith Lenden–Hitchcock, Sophia Liow and my wife, Janet Morgan, for discussions and advice. Tim Hopkins's computing expertise proved invaluable. Christopher Chatfield, Ian Jolliffe and Brian Ripley were kind enough to comment in detail on an earlier draft, and I have benefited greatly from their perceptive comments. Finally, it is a pleasure to record my appreciation for all the hard work of Mrs Mavis Swain who typed the manuscript.

Canterbury
May, 1983.

LIST OF NOTATION

$(a, b; m)$	a congruential, pseudo-random number generator, with multiplier a, constant b, and modulus m (p. 144)
$B_e(a, b)$	beta distribution with parameters a and b (p. 29)
$B(n, p)$	binomial distribution with parameters n and p (p. 18)
c.d.f.	cumulative distribution function (p. 13)
χ_v^2	chi-square distribution with parameter v
Corr (X, Y)	product-moment correlation between random variables X and Y (p. 16)
Cov (X, Y)	covariance between random variables X and Y (p. 16)
$\mathscr{E}[X]$	expectation of the random variable X (p. 15)
$F(x), F_X(x)$	cumulative distribution functions (p. 13)
$F(x, y), F_{X,Y}(x, y)$	joint cumulative distribution functions (p. 14)
$f(x), f_X(x)$	probability density functions (p. 13)
$f(x, y), f_{X,Y}(x, y)$	joint probability density functions (p. 14)
$f_{Y\vert X}(y\vert x)$	conditional probability density function of the random variable Y, given the random variable X (p. 15)
FIFO	queue discipline of 'first-in-first-out' (p. 190)
$\Gamma(n)$	gamma function (p. 26)
$\Gamma(n, \lambda)$	gamma distribution with parameters n and λ
$G(z), G_X(z)$	probability generating functions (p. 42)
$G(u, v), G_{X,Y}(u, v)$	bivariate probability generating functions (p. 87)
$M(\theta), M_X(\theta)$	moment generating functions (p. 41)
m.g.f.	moment generating function (p. 41)
M/D/1	single-server queue with exponentially distributed inter-arrival times and constant service time (p. 206)
M/E$_2$/1	single-server queue with exponentially distributed inter-arrival times and service times with a $\Gamma(2, \mu)$ distribution (p. 206)
M/M/1	single-server queue with exponentially distributed inter-

arrival times and exponentially distributed service times (p. 49)

$M_\mathbf{x}(\theta)$	multivariate moment generating function (p. 44)
$N(\mu, \sigma^2)$	normal distribution with mean μ and variance σ^2 (p. 23)
$N(0, 1)$	standard normal distribution
$N(\mu, \Sigma)$	multivariate normal distribution, with mean vector μ and variance/covariance matrix Σ (p. 40)
p.d.f.	probability density function (p. 13)
p.g.f.	probability generating function (p. 42)
$\phi(x)$	probability density function of a standard normal random variable (p. 23)
$\Phi(x)$	cumulative distribution function of a standard normal random variable (p. 23)
$\Pr(A)$	probability of the event A (p. 12)
r.v.	random variable
$U(a, b)$	uniform distribution over the range $[a, b]$ (p. 20)
U, U_i	$U(0, 1)$ random variables
$\mathrm{Var}(X)$	variance of a random variable X (p. 15)

1

INTRODUCTION

1.1 Simulation

'Simulation' is a word which is in common use today. If we seek its definition in the *Concise Oxford Dictionary*, we find:

> '**simulate**, *verb transitive*. Feign, pretend to have or feel, put on; pretend to be, act like, resemble, wear the guise of, mimic, ... So **simulation**, *noun*.'

Three examples are as follows:

(a) Following the crash of a DC-10 jet after it had lost an engine, the *Observer* newspaper, published in Britain on 10 June 1979, reported that, 'A DC-10 flight training simulator is being programmed to investigate whether the aircraft would be controllable with one engine detached.'
(b) Model cars or aeroplanes in wind-tunnels can be said to simulate the behaviour of a full-scale car, or aeroplane, respectively.
(c) A British television 'Horizon' programme, presented in 1977, discussed, with the aid of a simulated *Stegosaurus*, whether or not certain dinosaurs were hot-blooded. In this case the simulating model was very simply a tube conducting hot water; the tube was also equipped with cooling vanes, the shape of which could be changed. Different shapes then gave rise to differing amounts of cooling, and the vane shapes of the *Stegosaurus* itself were shown to provide efficient cooling. Thus these shapes *could* have developed, by natural selection, to cool a hot-blooded creature.

For statisticians and operational-research workers, the term 'simulation' describes a wealth of varied and useful techniques, all connected with the mimicking of the rules of a model of some kind.

1.2   What do we mean by a model?

It frequently occurs that we find processes in the real world far too complicated to understand. In such cases it is a good idea to strip the processes of some of their features, to leave us with *models* of the original processes. If we can understand the model, then that may provide us with some insight into the process itself. Thus in examples (a) and (b) above, it proves much cheaper and easier to investigate real systems through simulated models. In the case of (b) a physical scale model is used, while, in the case of (a), the model would most likely have been a computer simulation. In example (c) one can only employ simulated models because dinosaurs are extinct!

Subjects such as physics, biology, chemistry and economics use models to greater and lesser extents, and the same is true of mathematics, statistics and probability theory. Differential equations and laws of mechanics, for instance, can be viewed as resulting from models, and whenever, in probability theory, we set up a sample-space (see for example $ABC^\dagger$, p. 62, who use the term 'possibility space') and assign probabilities to its elements, we are building a model of reality. Some particular models are described later. When the models are given a mathematical formulation, but analytic predictions are not possible, then quite often simulation can prove to be a useful tool, not only for describing the model itself, but also for investigating how the behaviour of the model may change following a change in the model; compare the situation leading to example (a) above. Abstract discussion of models is best accompanied by examples, and several now follow.

1.3   Examples of models which may be investigated by simulation

(a) Forest management

A problem facing foresters is how to manage the felling of trees. One possible approach is to 'clear fell' the forest in sections, which involves choosing a section of forest, and then felling all of the trees in it before moving on to a new section. An alternative approach is to select only mature, healthy trees for felling before moving on to the next section. A disadvantage of the former approach is that sometimes the trees felled will not all be of the same age and size, so that some will only be useful for turning into pulp, for the manufacture of paper, say, while others will be of much better quality, and may be used for construction purposes. A disadvantage of the latter approach that has been encountered in the Eucalypt forests of Victoria, in Australia, is that the resulting tree stumps can act as a food supply for a fungal disease called *Armilleria* root-rot. Spores of this fungus can be transmitted by the wind,

---

† *ABC: A Basic Course in Statistics*, by Clarke and Cooke (1983).

alight on stumps and then develop in the stumps, finally even proceeding into the root-system of the stump, which may result in the transmission of the infection to healthy neighbouring trees from root-to-root contact.

While one can experiment with different management procedures, trees grow slowly, and it could take a lifetime before proper comparisons could be made. One can, however, build a model of the forest, which would include a description of the possible transmission of fungal disease by air and root contact, and then simulate the model under different management policies. One would hope that simulation would proceed very much faster than tree growth. For a related example, see Mitchell (1975).

(b) Epidemics

Diseases can spread through animal, as well as plant populations, and in recent years there has been much interest in mathematical models of the spread of infection (see Bailey, 1975). Although these models are often extremely simple, their mathematical solution is not so simple, and simulation of the models has frequently been employed (see Bailey, 1967).

(c) Congestion

Queues are a common feature of everyday life. They may be readily apparent, as when we wait to pay for goods in a shop, cash a cheque at a bank, or wait for a bus, or less immediately obvious, as, for example, when planes circle airports in holding stacks, waiting for runway clearance; time-sharing computers process computer jobs; or, in industry, when manufacturers of composite items (such as cars) await the supply of component parts (such as batteries and lights) from other manufacturers. The behaviour of individuals in the more apparent type of queue can vary from country to country: Mikes (1946) wrote that the British, for instance, are capable of forming orderly queues consisting of just single individuals!

Congestion in queues can result in personal aggravation for individuals, and costly delays, as when planes or industries are involved. Modifying systems with the aim of reducing congestion can result in unforeseen secondary effects, and be difficult and costly. Here again, models of the systems may be built, and the effect of modifications can then be readily appreciated by modifying the model, rather than the system itself. An example we shall encounter later models the arrival of cars at a toll; the question to be answered here was how to decide on the ratio of manned toll-booths to automatic toll-booths.

Another example to be considered later deals with the queues that form in a doctor's waiting-room. Some doctors use appointment systems, while others do not, and a simulation model may be used to investigate which system may be better, in terms of reducing average waiting times, for example. (See also Exercise 1.6.)

(*d*) *Animal populations*

Simulation models have been used to mimic the behaviour of animal populations. Saunders and Tweedie (1976) used a simulation model to mimic the development of groups of individuals assumed to be colonizing Polynesia from small canoe-loads of migrants, while Pennycuick (1969) used a computer model to investigate the future development of a population of Great Tits. Gibbs (1980) used simulation to compare two different strategies for young male gorillas. The background to this last example is the interesting sociology of the African mountain gorilla, which lives in groups headed by a single male, who is the only male allowed to mate with the group females. In such a society young males who do not head groups have to choose between either biding their time in a hierarchy until the head male and other more dominant males all die, or attempting to set-up a group themselves, by breaking away and then trying to attract females from established groups. Both of these strategies involve risks, and the question to be answered was 'which strategy is more likely to succeed?'

(*e*) *Ageing*

The age-structure of the human populations of many societies is changing, currently with the fraction of individuals that are elderly increasing. The prospect for the future is thus one of progressively more elderly individuals, many of whom will be in need of social services of some kind. Efficient distribution of such services depends upon an understanding of the way in which the population of elderly is composed, and how this structure might change with time.

Harris *et al.* (1974) developed categories which they called 'social independence states' for the elderly, the aim being to simulate the changes of state of elderly under various assumptions, and thereby to evaluate the effect of those assumptions. See also Wright (1978), and Jolliffe *et al.* (1982), who describe a study resulting in a simple classification of elderly into a small number of distinct groups.

1.4 The tools required

Models contain parameters, such as the birth and death rates of gorillas, the rate of arrival of cars at a toll-booth, the likelihood of a car-driver to possess the correct change for an automatic toll-booth, and so forth. In order for the models to be realistic, these parameters must be chosen to correspond as closely as possible to reality, and typically this means that before a model can be simulated, the real-life process must be observed, and so sampled that parameter estimates can be obtained; see, for example, *ABC*, chapter 10. In the case of modelling forests, the forests must first of all be surveyed to assess the extent of fungal infection. In the case of a doctor's waiting room, variables such as consultation times and inter-arrival times must be observed; see Exercises 1.6 and 1.7. With epidemics, one needs reports of the spread of disease, possibly within households on the one hand, or on a larger scale from field observations, as occurred with the recent spread of rabies in Europe, or of myxomatosis in Australia in the early 1950s. For useful comments on data collection see *ABC*, chapter 11, and Barnett (1974).

The models of Section 1.3 all involve random elements, so that the predictions of the models are not known or calculable with certainty. In some cases one encounters models which do not involve random elements, and then simulation is simply a case of enumerating the certain predictions of the model. Such models can occur in the area of manpower planning, for example; see Bartholomew and Forbes (1979). Models with random elements are the concern of this book, and the simulation of such models then necessitates the generation of such random variates as waiting times, service times and life times. In some cases longer times will be more likely than shorter times, while in others the converse will be true, and in general statistical terms what we need to be able to do is simulate random variables from a whole range of different statistical distributions; see for example *ABC*, chapters 7, 9, 13, 14 and 19. This, then, is the motivation for the material of Chapters 3, 4 and 5 of this book, while Chapter 2 provides a résumé of some of the results regarding random variables and distributions that we shall need in the remainder of this book.

An additional feature of the simulation of models is how to assess the performance of the model. Measures such as expected waiting times have already been mentioned, and this is a topic to which we shall return later. (See also Exercise 1.9).

1.5   Two further uses of simulation

As we shall see later, not only may we use simulation to mimic explicitly the behaviour of models, but also we can use simulation to evaluate the behaviour of complicated random variables whose precise distribution we are unable to evaluate mathematically. An early example of this also provides us with an interesting footnote to the history of statistics. Let us suppose that $(x_1, x_2, \ldots, x_n)$ form a random sample from a normal distribution of mean μ and variance σ^2 (details are given in Section 2.9). We can estimate the parameter σ^2 by

$$s^2 = \frac{1}{(n-1)} \sum_{i=1}^{n} (x_i - \bar{x})^2$$

$$\tag{1.1}$$

where

$$\bar{x} = \sum_{i=1}^{n} x_i/n.$$

The quantity $z = \sqrt{n}(\bar{x} - \mu)/s$ is itself a realization of a random variable with a t-distribution on $(n - 1)$ degrees of freedom.

These t-distributions were first studied by W. S. Gossett, who published his work under the pen-name of 'Student'. In 1908, he wrote:

'Before I had succeeded in solving my problem analytically, I had endeavoured to do so empirically [i.e. by simulation]. The material used was a . . . table containing the height and left middle finger measurements of 3000 criminals . . . . The measurements were written out on 3000 pieces of cardboard, which were then very thoroughly shuffled and drawn at random . . . each consecutive set of 4 was taken as a sample . . . [i.e., $n = 4$ above] . . . and the mean [and] standard deviation of each sample determined. . . . This provides us with two sets of . . . 750 z's on which to test the theoretical results arrived at. The height and left middle finger . . . table was chosen because the distribution of both was approximately normal . . .'

This use of simulation has increased over the years, and is remarkably common today. We have seen that simulation can be used to assess the behaviour of models, and also of certain random variables. Similarly, it can be used to gauge the performance of various techniques.

Examples can be provided from the area of multivariate analysis (see Everitt, 1980; Jolliffe, 1972), time-series analysis (see Anderson, 1976), and also ecology (see Jeffers, 1972). In the area of cluster analysis, many different techniques have been devised for detecting groups in populations of individuals. One way of assessing the performance of these techniques, and of comparing different techniques, is to simulate populations of known structure and then to see to what extent the techniques identify that structure. This is done, for example, by Everitt (1980, p. 77) and Gardner (1978). In the ecological case, sampling techniques exist for assessing the size of mobile animal populations, such as sperm whales, for example. Frequently these techniques involve marking captured animals, releasing them, and then considering the proportions of marked individuals in further samples. One can examine the performance of such techniques by simulating the behaviour of animal populations of known size, and then considering how well the techniques recover the known population size. This is another application of simulation in common use today, and, like the investigation of sampling distributions, it also requires the material of Chapters 3, 4 and 5.

Both of the uses considered here are still, fundamentally, simulating models, though perhaps not quite so obviously as before, and in the second use above the model is only of secondary interest. We shall return to the examples of this section in Chapter 9.

1.6   Computers, mathematics and efficiency

Most simulation in statistics is carried out today with the aid of a computer. Interesting exceptions do exist, however. One example is provided by Fine (1977), in which we find a description of a mechanical model which mimics the rules of an epidemic model. Another is provided by Moran and Fazekas de St Groth (1962), who sprayed table-tennis balls with paint in a particular way, to mimic the shielding of virus surfaces by antibodies.

As computers become more powerful and more widely available, one might even ask whether simulation might render mathematical analysis of models redundant. The fact is that, useful though simulation is, an analytical solution is always preferable, as we can appreciate from a simple example. Exercise 1.4 presents a simple model for a queue. If $p < \frac{1}{2}$ in this model then, after a long period of time since the start of the queueing system, we can write the probability that the queue contains $k \geq 0$ customers as

$$p_k = \left(\frac{q-p}{2pq}\right)\left(\frac{p}{q}\right)^k \qquad \text{for } k \geq 1$$

with
$$p_0 = \left(\frac{q-p}{2q}\right), \qquad \text{where } q = 1 - p.$$

Mathematical analysis has, therefore, provided us with a detailed prediction for the distribution of queue size, and while we might, for any value of $p < \frac{1}{2}$, use simulations to estimate the $\{p_k, k \geq 0\}$ distribution, we should have to start again from scratch for every different value of p.

Analytical solutions could be mathematically complex, in which case simulation may provide a reassuring check, as was true of Gossett's simulations mentioned earlier. In some cases, approximate analytical solutions are sought for models, and then simulations of the original model can provide a check of the accuracy of the approximations made; examples of this are provided by Lewis (1975) and Morgan and Leventhal (1977).

The relative power of an analytical solution, as compared with a simulation approach to a model, is such that even if a full analytical solution is impossible, such a solution to part of the model, with the remainder investigated by simulation, is preferable by far to a simulation solution to the whole model. We find comments to this effect in, for example, Estes (1975), Brown and Holgate (1974), Mitchell (1969) and Morgan and Robertson (1980). Mitchell (1969) puts the point as follows:

'It is well to be clear from the start that simulation is something of a sledgehammer technique. It has neither elegance nor power as these terms would be understood in a mathematical context. In a practical context it can, however, be an extremely powerful way of understanding and modelling a system. More importantly from the practical point of view, it can be the only way of deriving an adequate model of a system.'

Mitchell goes on to emphasize a major disadvantage of simulation, which is that it could well be time-consuming and expensive. Even with powerful computers, poor attention to efficiency could make a simulation impractical. Efficiency in simulating is therefore very important, and is a common theme underlying a number of the following chapters, especially Chapter 7.

1.7   Discussion and further reading

We have seen from this chapter that simulation can be a powerful technique of wide-ranging applicability. Of course, the examples given here comprise only a very small part of a much larger and more varied picture. The current interest in simulation is reflected by the papers in journals such as *Mathematics and Computers in Simulation, Communications in Statistics*, Part B: *Simulation and Computation*, and the *Journal of Statistical Computation and Simulation*.

Further examples of models and their simulation can be found in Barrodale, Roberts and Ehle (1971), Gross and Harris (1974), Hollingdale (1967), Kendall (1974), Smith (1968) and Tocher (1975). In addition, many books on operational research contain useful chapters dealing with simulation, as do also Bailey (1964), Cox and Miller (1972) and Cox and Smith (1967).

1.8   Exercises and complements

†**1.1**   Toss a fair coin 500 times and draw a graph of
  (i) r/n vs. n, for $n = 1, 2, \ldots, 500$, where n is the number of tosses and r is the number of heads in those n tosses;
  and
  (ii) $(2r - n)$ vs. n, i.e. the difference between the number of heads and the number of tails.
  Here you may either use a coin (supposed fair), the table of random digits from Appendix 2 (how?) or a suitable hand-calculator or computer if one is available. A possible alternative to the table of random digits is provided by the digits in the decimal expansion of π, provided in Exercise 3.10.

  Comment   (a) on the behaviour of r/n
  and   (b) on the behaviour of $(2r - n)$.

1.2   In 1733, George Louis Leclerc, later Comte de Buffon, considered the following problem:

  'If a thin, straight needle of length l is thrown at random onto the middle of a horizontal table ruled with parallel lines a distance $d \geq l$ apart, so that the needle lies entirely on the table, what is the probability that no line will be crossed by the needle?'

You may like to try to prove that the solution is $1 - 2l/\pi d$, but this is not entirely straightforward.

Dr E. E. Bassett of the University of Kent simulated this process using an ordinary sewing needle, and obtained the following results from two classes of students:

(i) 390 trials, 254 crossings;   (ii) 960 trials, 638 crossings.

Explain how you can use these data to estimate π; in each of these simulations, d was taken equal to l, as this gives the greatest precision for estimating π, a result you may care to investigate. We shall return to this topic in Chapter 7.

†**1.3**   Give five examples of processes which may be investigated using simulation.

***1.4**   The following is a *flow diagram* for a very simple model of a queue with a single server; time is measured in integer units and is denoted by $T \geq 0$. T changes only when the queue size changes, by either a departure or a new arrival. $Q \geq 0$ denotes queue size. Explain how you would use the table of random digits in Appendix 2 to simulate this model

(i) when $p = 0.25$ and   (ii) when $p = 0.75$,

in each case drawing a graph of Q vs. T for $0 \leq T \leq 50$. For the case (i) estimate p_0, the proportion of time the queue is empty over a long period, and compare your estimate with the known theoretical value for p_0 of $p_0 = (q - p)/(2q)$.

Modify the flow diagram so that (a) there is limited waiting room of r individuals; (b) there are two servers and customers join the smaller queue when they arrive.

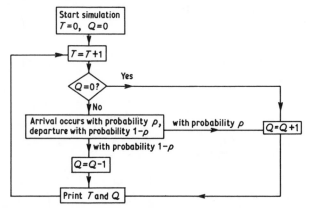

1.5   A canoe containing three men and three women lands on a previously uninhabited island. Discuss the information you require in order to model the society of these individuals, and how their population size changes with time.

†**1.6** Consider how you would simulate the following model of a doctor's surgery operating an appointment system:
   (i) Patients are scheduled to arrive every 5 units of time.
   (ii) Independently of all other patients, each patient fails to turn up with probability 0.1.
   (iii) Independently of all other patients, each patient has arrival times with the following distribution:

	Early		On time	Late	
	2 units	1 unit		1 unit	2 units
probability:	$\frac{1}{10}$	$\frac{1}{5}$	$\frac{2}{5}$	$\frac{1}{5}$	$\frac{1}{10}$

   (iv) Consultation times have the following distribution:

Time units	2	3	4	5	6	7	8	9
Probability	$\frac{1}{10}$	$\frac{1}{10}$	$\frac{1}{10}$	$\frac{1}{5}$	$\frac{1}{5}$	$\frac{1}{10}$	$\frac{1}{10}$	$\frac{1}{10}$

   (v) Patients are seen in the order in which they arrive.

Discuss how you might use this model to investigate the efficiency of the appointment system.

1.7 Criticize the assumptions made in the last exercise. It may help to consider the following data collected from a British medical practice by Keith Gibbs as part of his mathematics undergraduate dissertation at the University of Kent:

		Patient			
		Male		Female	
	Consulting time	Adult	Child	Adult	Child
Morning	≤ 5 minutes	8	2	24	11
	> 5 minutes	8	2	12	8
Afternoon	≤ 5 minutes	21	9	11	6
	> 5 minutes	6	4	6	2

1.8 Construct suitable histograms (*ABC*, p. 8) of $z = (\bar{x}\sqrt{n}/s)$, using samples of size $n = 2$ from the table of standard normal variables given in Appendix 2, and, separately, using samples of size $n = 4$ from the same table. Repeat this, but now for $z = (\bar{x} - 1)\sqrt{n}/s$, using the table of exponential random variables, and comment on the results.

1.9   One way of comparing different queueing systems is to compare mean waiting times. Can you envisage a situation in which other features of the entire distributions of waiting times might also be important?

1.10   Simulate English text, using the word frequencies of Dewey (1923). Examples can be found in Shannon and Weaver (1964).

2

SOME PROBABILITY
AND STATISTICS
REVISION

We have seen from Chapter 1 that in many uses of simulation, statisticians need to simulate discrete and continuous random variables of different kinds, and techniques for doing this are provided in Chapters 4 and 5. The aim of this chapter is to specify what we mean by random variables, and generally to provide revision of the material to be used later.

It will be assumed that the reader is familiar with the axioms of probability and the ideas of independence, and conditional probability. The material assumed is covered in, for example, chapters 1–6 of *ABC*. In the following we shall write $\Pr(A)$ for the probability of any event A. We shall begin with general definitions, and then proceed to consider particular important cases.

2.1 Random variables

Underlying all statistical investigations is the concept of a random experiment, such as the tossing of a coin k times. The set of all possible outcomes to such an experiment is called the *sample-space*, introduced by von Mises in 1921 (he called it a *Merkmalraum*), and we can formally define random variables as functions over the sample-space, about which it is possible to make probability statements. In the simple model of a queue given in Exercise 1.4, the change in queue size forms a random experiment, with just two possible outcomes, namely, an arrival or a departure. With this random experiment we can associate a random variable, X, say, such that $X = +1$ if we have an arrival, and $X = -1$ if we have a departure. The model is such that $\Pr(X = +1) = p$, and $\Pr(X = -1) = 1 - p$. In detail here, the sample-space contains just two outcomes, say ω_1 and ω_2, corresponding to arrival and departure, respectively, and we can write the random variable X as $X(\omega)$, so that $X(\omega_1) = +1$ and $X(\omega_2) = -1$. However, we find it more convenient to suppress the argument of $X(\omega)$, and

12

simply write the random variable as X in this example. We shall adopt the now standard practice of using capital letters for random variables, and small letters for values they may take. When a random variable X is simulated n times then we obtain a succession of values: $\{x_1, x_2, \ldots, x_n\}$, each of which provides us with a *realization* of X.

Another random experiment results if we record the queue size in the model of Exercise 1.4. From Section 1.6 we see that if $p < \frac{1}{2}$, then after a long period of time since the start of the queueing system we can denote the queue size by a random variable, Y, say, such that Y may take any non-negative integral value, and $\Pr(Y = k)$ for $k \geq 0$ is as given in Section 1.6.

2.2 The cumulative distribution function (c.d.f.)

For any random variable X, the function F, given by $F(x) = \Pr(X \leq x)$ is called the *cumulative distribution function of* X. We have

$$\lim_{x \to \infty} F(x) = 1; \qquad \lim_{x \to -\infty} F(x) = 0$$

$F(x)$ is a nondecreasing function of x, and $F(x)$ is continuous from the right (i.e. if $x > x_0$, $\lim_{x \to x_0} F(x) = F(x_0)$).

The nature of $F(x)$ determines the type of random variable in question, and we shall normally specify random variables by defining their *distribution*, which in turn provides us with $F(x)$. If $F(x)$ is a step function we say that X is a *discrete* random variable, while if $F(x)$ is a continuous function of x then we say that X is a *continuous* random variable. Certain variables, called *mixed* random variables, may be expressed in terms of both discrete and continuous random variables, as is the case of the waiting-time experienced by cars approaching traffic lights; with a certain probability the lights are green, and the waiting-time may then be zero, but otherwise if the lights are red the waiting-time may be described by a continuous random variable. Mixed random variables are easily dealt with and we shall not consider them further here. Examples of many common c.d.f.'s are given later.

2.3 The probability density function (p.d.f.)

When $F(x)$ is a continuous function of x, with a continuous first derivative, then $f(x) = dF(x)/dx$ is called the *probability density function* of the (continuous) random variable X. If $F(x)$ is continuous but has a first derivative that is not continuous at a finite number of points, then we can still define the probability density function as above, but for uniqueness we set $f(x) = 0$, for instance, when $dF(x)/dx$ does not exist; an example of this is provided by the c.d.f. of the random variable Y of Exercise 2.25.

The p.d.f. has the following properties:

(i) $f(x) \geq 0$

(ii) $\displaystyle\int_{-\infty}^{\infty} f(x)\,dx = 1$

(iii) $\Pr(a < X < b) = \Pr(a \leq X < b) = \Pr(a < X \leq b) = \Pr(a \leq X \leq b)$

$$= \int_{a}^{b} f(t)\,dt$$

EXAMPLE 2.1

Under what conditions on the constants α, β, γ can the following functions be a p.d.f.?

$$g(x) = \begin{cases} e^{-\alpha x}(\beta + \gamma x) & \text{for} \quad x \geq 0 \\ 0 & \text{for} \quad x < 0 \end{cases}$$

We must verify that $g(x)$ is non-negative, and that $\int_{-\infty}^{\infty} g(x)\,dx = 1$. If $\alpha \leq 0$, this integral cannot be finite, and so we must have $\alpha > 0$.

$$\int_{-\infty}^{\infty} g(x)\,dx = \int_{-\infty}^{\infty} e^{-\alpha x}(\beta + \gamma x)\,dx = \frac{\beta}{\alpha} + \frac{\gamma}{\alpha^2}$$

Thus we must have $\gamma = \alpha^2 - \alpha\beta = \alpha(\alpha - \beta)$ resulting in the p.d.f.

$$g(x) = e^{-\alpha x}(\beta + \alpha(\alpha - \beta)x) \qquad \text{for} \quad x \geq 0$$

In order that $g(x) \geq 0$ for $x \geq 0$, we must have $\beta \geq 0$, and $\alpha \geq \beta$. Hence set $\beta = \theta\alpha$ and $\gamma = \alpha^2(1 - \theta)$, for $\alpha > 0$ and $0 \leq \theta \leq 1$.

We sometimes abbreviate 'probability density function' to just 'density'.

2.4   Joint, marginal and conditional distributions

In the case of two random variables X and Y we can define the *joint* c.d.f. by $F(x, y) = \Pr(X \leq x \text{ and } Y \leq y)$, and then the univariate distributions of X and Y are referred to as the *marginal* distributions. If

$$f(x, y) = \frac{\partial^2 F(x, y)}{\partial x\, \partial y}$$

is a continuous function, except possibly at a finite number of points, then $f(x, y)$ is called the *joint* p.d.f. of X and Y, and in this case the marginal p.d.f.'s are given by:

$$f_X(x) = \int_{-\infty}^{\infty} f(x, y)\,dy$$

and

$$f_Y(y) = \int_{-\infty}^{\infty} f(x, y)\,dx$$

Here we have adopted a notation we shall employ regularly, of subscripting the p.d.f. of a random variable with the random variable itself, so that there should be no confusion as to which random variable is being described. The same approach is adopted for c.d.f.'s, and also, sometimes, for joint distributions.

The *conditional* p.d.f. of the random variable Y, given the random variable X, may be written as $f_{Y|X}(y|x)$, and is defined by

$$f_{Y|X}(y|x) = f_{X,Y}(x, y)/f_X(x) \qquad \text{if} \qquad f_X(x) > 0$$

For two *independent* continuous random variables X and Y, with joint p.d.f. $f_{X,Y}(x, y)$, we have

$$f_{X,Y}(x, y) = f_X(x)\,f_Y(y) \qquad \text{for any } x \text{ and } y$$

The above definitions of independence and joint, marginal and conditional p.d.f.'s have straightforward analogues for discrete random variables. For example, the marginal distribution of a discrete random variable X may be given by:

$$\Pr(X = x) = \sum_y \Pr(X = x, Y = y)$$

Furthermore, while we have only discussed the bivariate case, these definitions may be extended in a natural way to the case of more than two random variables.

2.5 Expectation

The *expectation* of a random variable X exists only if the defining sum or integral converges absolutely. If X is a continuous random variable we define the expectation of X as:

$$\mathscr{E}[X] = \int_{-\infty}^{\infty} xf(x)\,dx \qquad \text{if} \qquad \int_{-\infty}^{\infty} |x|\,f(x)\,dx < \infty$$

where $f(x)$ is the p.d.f. of X. Similarly, if X is a discrete random variable which may take the values $\{x_i\}$, then

$$\mathscr{E}[X] = \sum_i x_i \Pr(X = x_i) \qquad \text{if} \qquad \sum_i |x_i| \Pr(X = x_i) < \infty$$

The *variance* of a random variable X is defined as

$$\mathrm{Var}(X) = \mathscr{E}\big[(X - \mathscr{E}[X])^2\big]$$

and the *covariance* between random variables X and Y is defined as

$$\mathrm{Cov}(X,Y) = \mathscr{E}[(X - \mathscr{E}[X])(Y - \mathscr{E}[Y])]$$

The expectation of a random variable X is frequently used as a measure of location of the distribution of X, while the variance provides a measure of spread of the distribution. Independent random variables have zero covariance, but in general the converse is not true.

The *correlation* between random variables X and Y is defined as

$$\mathrm{Corr}(X,Y) = \frac{\mathrm{Cov}(X,Y)}{\sqrt{[\mathrm{Var}(X)\,\mathrm{Var}(Y)]}}$$

2.6 The geometric, binomial and negative-binomial distributions

Consider a succession of independent experiments, such as tosses of a coin, at each of which either 'success' (which we could identify with 'heads' in the case of the coin) or 'failure' ('tails' for the coin) occurs. This rudimentary succession of experiments, or trials, provides the framework for three important discrete distributions.

$$\text{Let } p = \mathrm{Pr}(\text{success}) \quad \text{and} \quad q = 1 - p = \mathrm{Pr}(\text{failure})$$

The simplest is the *geometric* distribution, which results if we let X be the discrete random variable measuring the number of trials until the first success. We have

> Geometric distribution:
> $$\mathrm{Pr}(X = i) = q^{i-1}p \quad \text{for} \quad 1 \le i \le \infty$$

$$\mathscr{E}[X] = 1/p \quad \text{and} \quad \mathrm{Var}(X) = q/p^2$$

Figure 2.1(a) gives a bar-chart illustrating the geometric distribution for the case $p = 0.5$. Figure 2.1(b) demonstrates the result of simulating such a geometric random variable 100 times.

The *binomial* distribution results if we fix a number of trials at $n \ge 1$, say, and

---

Figure 2.1 (a) Bar-chart illustrating the geometric distribution with $p = 0.5$. (b) Bar-chart illustrating the results from simulating a random variable with the distribution of (a) 100 times. Here i is observed n_i times, $i \ge 1$.

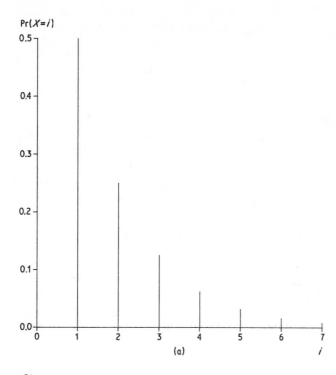

(a)

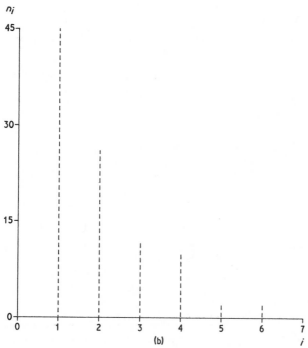

(b)

count the number, X, of successes. This gives

> Binomial distribution:
>
> $$\Pr(X = i) = \binom{n}{i} p^i q^{n-i} \quad \text{for} \quad 0 \leq i \leq n$$

$$\mathscr{E}[X] = np \quad \text{and} \quad \text{Var}(X) = npq$$

Figure 2.2(a) gives a bar-chart illustrating the binomial distribution for the case $p = 0.5$ and $n = 5$. Figure 2.2(b) demonstrates the result of simulating such a binomial random variable 100 times. We shall refer to such a random variable as possessing a $B(n, p)$ distribution, thus specifying the two parameters, n and p.

A geometric random variable provides the *waiting-time* measured by the number of trials until the first success. The random variable X which measures the waiting-time until the nth success has a *negative-binomial* distribution. When $X = n + i$, for $i \geq 0$, then the $(n + i)$th trial results in success, and the remaining $(n - 1)$ successes occur during the first $(n + i - 1)$ trials, and we can write

> Negative-binomial distribution:
>
> $$\Pr(X = n + i) = \binom{n + i - 1}{i} p^i q^n \quad \text{for} \quad 0 \leq i \leq \infty$$

$$\mathscr{E}[X] = n/p \quad \text{and} \quad \text{Var}(X) = nq/p^2.$$

As is shown in Exercise 2.18, there is a simple relationship between the binomial and the negative-binomial distributions.

2.7   The Poisson distribution

A random variable X with a *Poisson* distribution of parameter λ is described as follows:

> Poisson distribution:
>
> $$\Pr(X = i) = \frac{e^{-\lambda} \lambda^i}{i!} \quad \text{for} \quad 0 \leq i \leq \infty$$

$$\mathscr{E}[X] = \lambda \quad \text{and} \quad \text{Var}(X) = \lambda$$

Figure 2.2 (a) Bar-chart illustrating the binomial distribution for the case $n = 5$, $p = 0.5$. (b) Bar-chart illustrating the results from simulating a random variable with the distribution of (a) 100 times. Here i is observed n_i times, $0 \leq i \leq 5$.

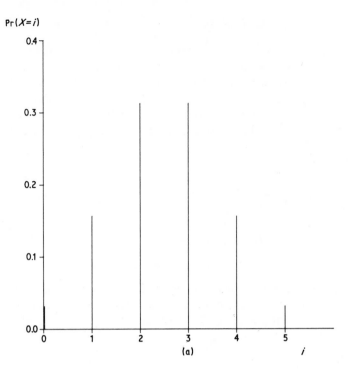

(a)

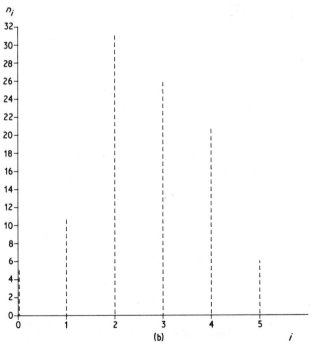

(b)

Named after the French mathematician, S. D. Poisson, who derived the distribution in 1837, the distribution had been obtained earlier by De Moivre. The Poisson distribution is often useful as a description of data that result when counts are made of the occurrence of events, such as the occurrence of telephone calls in fixed intervals of time, or the numbers of plants within areas of a fixed size. This is because the real-life processes giving rise to the data approximate to a model called a *Poisson process*, which predicts a Poisson distribution for the data. We shall discuss the Poisson process in detail in Section 4.4.2.

Figure 2.3(a) gives a bar-chart illustrating the Poisson distribution for $\lambda = 5$, and Fig. 2.3(b) describes the results of simulating such a Poisson random variable 100 times.

2.8   The uniform distribution

The simplest continuous random variables have *uniform* (sometimes called rectangular) distributions. As we shall see later, uniform random variables form the basis of most simulation investigations. A uniform random variable over the range $[a, b]$ has the p.d.f.

Uniform p.d.f. over $[a, b]$:

$$f(x) = \frac{1}{(b-a)} \qquad \text{for } a < x < b$$

$$f(x) = 0 \qquad \text{for } x < a \text{ and } x > b$$

We shall frequently refer to this as the $U(a, b)$ p.d.f., the most important case being when $a = 0$ and $b = 1$.

The c.d.f. of a $U(0, 1)$ random variable X is given by

$$F(u) = \int_0^u 1 \, dx = u \qquad \text{for } 0 \le u \le 1$$

and so for any $0 \le \alpha \le \beta \le 1$,

$$\Pr(\alpha \le X \le \beta) = \Pr(0 \le X \le \beta) - \Pr(0 \le X \le \alpha)$$

$$= F(\beta) - F(\alpha) = (\beta - \alpha)$$

Figure 2.3   (a) Bar-chart illustrating the Poisson distribution for $\lambda = 5$. (b) Bar-chart illustrating the results from simulating a random variable with the distribution of (a) 100 times. Here i is observed n_i times, for $i \ge 0$.

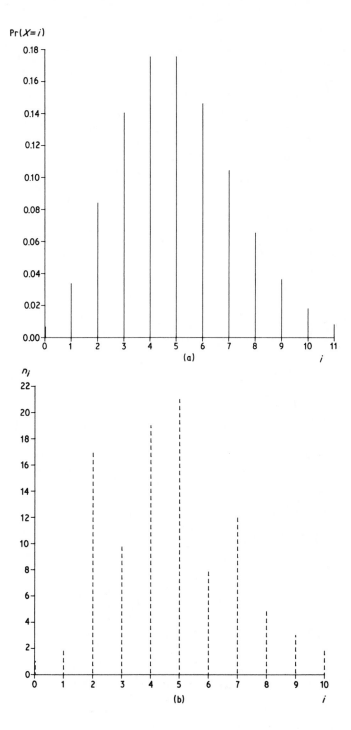

(a)

(b)

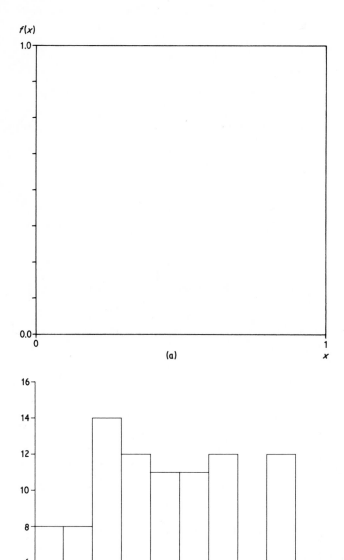

(a)

(b)

a result which is much used in later chapters. For a $U(0, 1)$ random variable X, $\mathscr{E}[X] = \frac{1}{2}$; $\mathrm{Var}(X) = 1/12$. Figure 2.4 illustrates the $U(0, 1)$ probability density function, and also a histogram resulting from a random sample of size 100 from this density.

2.9   The normal distribution and a central limit theorem

A continuous random variable with a *normal* distribution, and mean μ and variance σ^2 has the p.d.f.

Normal probability density function:

$$f(x) = \frac{1}{\sigma\sqrt{(2\pi)}} \exp\left[-\frac{1}{2}\left(\frac{x-\mu}{\sigma}\right)^2\right] \quad \text{for } -\infty \leq x \leq \infty$$

Early work on this distribution was by such pioneers as De Moivre, Laplace and Gauss, towards the end of the 18th century and at the start of the 19th century. The normal distribution is so called because of its common occurrence in nature, which is due to 'central limit theorems', which state that, under appropriate conditions, when one adds a large number of random variables, which may well not be normal, the resulting sum has an approximately normal distribution. A formal statement of the commonest central limit theorem is that:

if $X_1, X_2, \ldots, X_n$ are independent, identically distributed random variables, with $\mathscr{E}[X_i] = \mu$ and $\mathrm{Var}(X_i) = \sigma^2$, then for any real x,

$$\lim_{n \to \infty} \mathrm{Pr}\left\{\frac{1}{\sqrt{n}} \sum_{i=1}^{n} \left(\frac{X_i - \mu}{\sigma}\right) \leq x\right\} = \Phi(x)$$

where $\Phi(x)$ is the c.d.f. of a normal random variable with zero mean and unit variance.

For a more general central limit theorem, and historical background, see Grimmett and Stirzaker (1982, p. 110).

We shall use the notation $N(\mu, \sigma^2)$, to denote the distribution of a normal random variable with mean μ and variance σ^2. The $N(0, 1)$ case is frequently called the *standard* normal, when the p.d.f. is denoted by $\phi(x)$. Figure 2.5 illustrates $\phi(x)$ and also presents a histogram resulting from a random sample of size 100 from this density.

Figure 2.4   (a) The $U(0, 1)$ probability density function. (b) Histogram summarizing a random sample of size 100 from the density function of (a).

2.10 Exponential, gamma, chi-square and Laplace distributions

We say that a continuous random variable X has an *exponential* distribution with parameter λ when we can write the p.d.f. as

> Exponential probability density function:
>
> $$f(x) = \lambda e^{-\lambda x} \qquad \text{for } 0 \le x \le \infty$$

$$\mathscr{E}[X] = 1/\lambda \qquad \text{and} \qquad \text{Var}(X) = 1/\lambda^2$$

Some authors (see for example Barnett, 1965) call this the 'negative exponential' p.d.f.

The Poisson process, mentioned in Section 2.7, is often used to model the occurrence of events in time. It predicts that

$$\text{Pr}(k \text{ events in a time interval of length } t) = \frac{e^{-\lambda t}(\lambda t)^k}{k!} \quad \text{for} \quad 0 \le k \le \infty$$

where $\lambda > 0$ is the *rate* parameter for the model, and is equal to the average number of events per unit time.

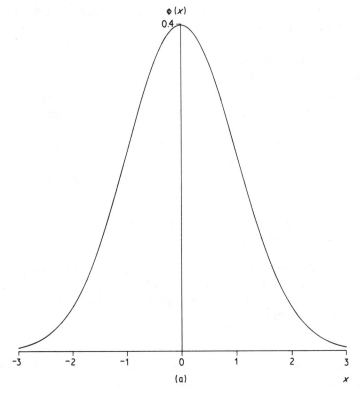

(a)

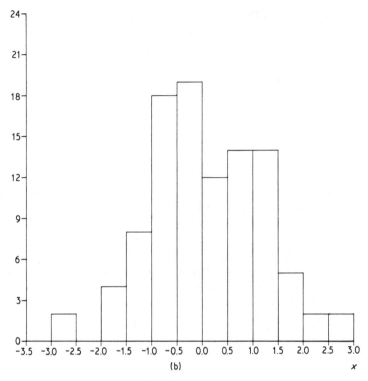

Figure 2.5 (a) The standard normal probability density function,

$$\phi(x) = \frac{1}{\sqrt{(2\pi)}} \exp\left(\frac{-x^2}{2}\right)$$ over the range $-3 \le x \le 3$. (b) Histogram summarizing a random sample of size 100 from the density function of (a).

If T is a random variable denoting the time to the next event in the Poisson process, measuring time from some arbitrary time origin, then

$$\Pr(T \ge t) = \Pr(\text{no events in the time interval } (0, t))$$

$$= e^{-\lambda t}$$

i.e. $$f(t) = \lambda e^{-\lambda t}$$

and so times between events in a Poisson process have an exponential distribution.

If we form the sum

$$S = \sum_{i=1}^{n} X_i$$

in which the X_i are independent random variables, each with the above exponential distribution, then (see Exercise 2.6 and Example 2.6) S has a *gamma* distribution with the p.d.f.

Gamma probability density function:

$$f(x) = \frac{e^{-\lambda x}\lambda^n x^{n-1}}{\Gamma(n)} \qquad \text{for} \qquad 0 \leq x \leq \infty$$

$$\mathscr{E}[X] = n/\lambda \quad \text{and} \quad \text{Var}(X) = n/\lambda^2$$

We shall refer to such a gamma distribution by means of the notation $\Gamma(n, \lambda)$. In this derivation, n is a positive integer, but in general gamma random variables have the above p.d.f. in which the only restriction on n is $n > 0$.

Figure 2.6 presents an exponential p.d.f., and two gamma p.d.f.'s, and a histogram summarizing a random sample of size 100 from the exponential p.d.f.

A random variable with a $\Gamma(v/2, \tfrac{1}{2})$ distribution is said to have a *chi-square* distribution with parameter v. For reasons which we shall not discuss here, the parameter v is usually referred to as the 'degrees-of-freedom' of the distribution. A random variable X with a $\Gamma(v/2, \tfrac{1}{2})$ distribution is also said to have a χ_v^2 distribution, with the p.d.f.

Chi-square probability density function with v degrees of freedom:

$$f(x) = \frac{e^{-x/2}x^{v/2-1}}{\Gamma(v/2)2^{v/2}} \quad \text{for } x \geq 0$$

$$\mathscr{E}[X] = v \qquad \text{and} \qquad \text{Var}(X) = 2v$$

The exponential and gamma distributions describe only non-negative random variables, but the exponential distribution forms the basis of the *Laplace* distribution, discovered by Laplace in 1774 and given below:

Laplace probability density function:

$$f(x) = \frac{\lambda}{2}e^{-\lambda|x|} \qquad \text{for } -\infty \leq x \leq \infty$$

$$\mathscr{E}[X] = 0 \qquad \text{and} \qquad \text{Var}(X) = 2/\lambda^2$$

Figure 2.6 (a) The $\Gamma(1, 1)$ (i.e., exponential), $\Gamma(2, 1)$ and $\Gamma(5, 1)$ probability density functions. (b) Histogram summarizing a random sample of size 100 from the exponential density function of (a).

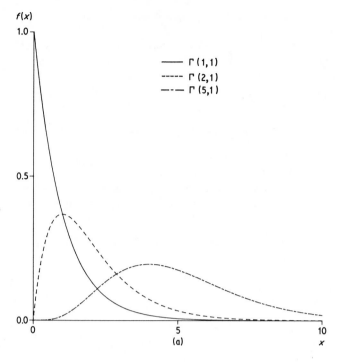

(a)

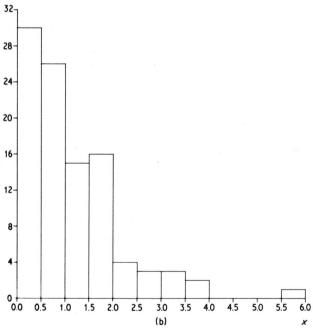

(b)

Just as the geometric and negative-binomial distributions describe waiting times when time is measured in integer units, the exponential and gamma distributions describe waiting times (in a Poisson process) when time is a continuous quantity, and we shall return to this point again later.

2.11   Distributions of other continuous random variables

Here we shall simply list the standard forms of the p.d.f. of a number of other common continuous random variables to which we shall refer later.

A probability density function that has the same qualitative shape as the normal p.d.f. is the *logistic* p.d.f., given in standard form below:

The standard logistic probability density function:

$$f(x) = \frac{e^{-x}}{(1+e^{-x})^2} \qquad \text{for } -\infty \leq x \leq \infty$$

$$\mathscr{E}[X] = 0 \qquad \text{and} \qquad \text{Var}(X) = \pi^2/3$$

We shall later make use of the logistic c.d.f., which in standard form is

$$F(x) = (1+e^{-x})^{-1} \qquad \text{for} \qquad -\infty \leq x \leq \infty$$

A unimodal symmetric p.d.f. with more weight in the tails than either the normal or logistic is the *Cauchy* p.d.f., so called because of its appearance in a paper by Cauchy in 1853. In its standard form the Cauchy p.d.f. is as follows:

The standard Cauchy probability density function:

$$f(x) = \frac{1}{\pi(1+x^2)} \qquad \text{for } -\infty \leq x \leq \infty$$

Because of the large weight in the tails of this p.d.f., a random variable with this distribution does not possess a finite mean or variance.

Finally, we give below the p.d.f. of a random variable with a *beta* distribution over $[0, 1]$.

The beta probability density function over $[0, 1]$:

$$f(x) = \begin{cases} \dfrac{x^{\alpha-1}(1-x)^{\beta-1}\,\Gamma(\alpha+\beta)}{\Gamma(\alpha)\,\Gamma(\beta)} & \text{for } 0 < x < 1 \\[2mm] 0 & \text{for } x < 0 \text{ and for } x > 1 \end{cases}$$

$$\mathscr{E}[X] = \alpha/(\alpha + \beta) \quad \text{and} \quad \text{Var}(X) = \frac{\alpha\beta}{(\alpha + \beta)^2 (\alpha + \beta + 1)}$$

The beta distribution contains the uniform distribution as a special case: $\alpha = \beta = 1$. If the random variable X has this distribution, then such a beta random variable will be said to have a $B_e(\alpha, \beta)$ distribution.

Figures 2.7–2.9 provide examples of these p.d.f.'s together with histograms summarizing random samples of size 100 from the respective p.d.f.'s

The logistic and Cauchy p.d.f.'s are given in standard, parameter-free form, but we can simply introduce location and scale parameters β and α respectively by means of the transformation $Y = \alpha X + \beta$. This is an example of transforming one random variable to give a new random variable, and we shall now consider such transformations in a general setting.

2.12   New random variables for old

Transforming random variables is a common statistical practice, and one which is often utilized in simulation. The simplest transformation is the linear transformation, $Y = \alpha X + \beta$. In the case of certain random variables, such as uniform, logistic, normal and Cauchy, this transformation does not change the distributional form, and merely changes the distribution parameters, while in other cases the effect of this transformation is a little more complicated.

In the case of single random variables, a general transformation is $Y = g(X)$, for some function g. In such a case, if X is a discrete random variable then the distribution of Y may be obtained by simple enumeration, using the distribution of X and the form of g. Thus, for example, if $Y = X^2$, $\Pr(Y = i) = \Pr(X = -\sqrt{i}) + \Pr(X = \sqrt{i})$. Such enumeration is greatly simplified if g is a strictly monotonic function, so that in the last example, if X were a non-negative random variable then we simply have

$$\Pr(Y = i) = \Pr(X = \sqrt{i})$$

The simplification of the case when g is a strictly monotonic function applies also to the case of the continuous random variables X. Two possible examples are shown in Fig. 2.10.

For the case (a) illustrated in Fig. 2.10, the events $\{Y \leq y\}$ and $\{X \leq x\}$ are clearly equivalent, while for case (b) it is the events $\{Y \geq y\}$ and $\{X \leq x\}$ that are equivalent, so that

$$\left. \begin{array}{l} \text{for case (a),} \quad F(y) = \Pr(Y \leq y) = \Pr(X \leq x) = F(x) \\ \text{and for case (b),} \ 1 - F(y) = \Pr(Y \geq y) = \Pr(X \leq x) = F(x) \end{array} \right\} \quad (2.1)$$

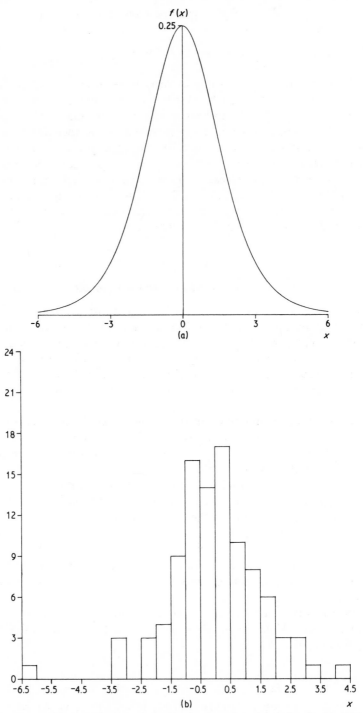

Figure 2.7 (a) The standard logistic density function for $|x| \leq 6$. (b) Histogram summarizing a random sample of size 100 from the density function of (a).

30

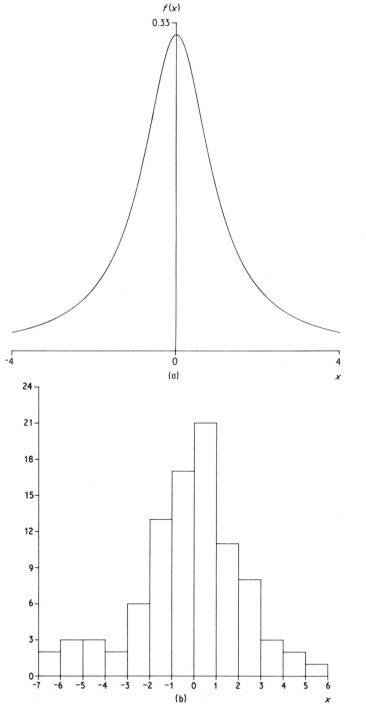

Figure 2.8   (a) The standard Cauchy density function for $|x| \leq 4$. (b) Histogram summarizing a random sample of size 100 from the density function of (a).

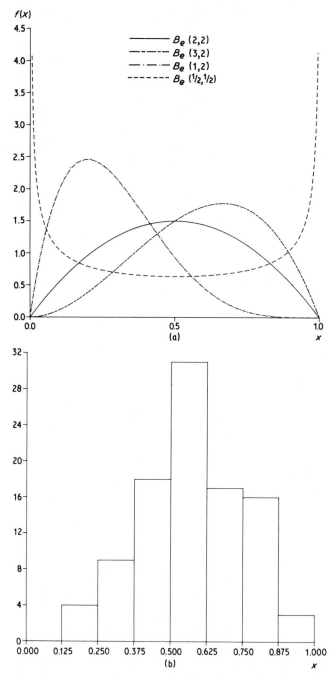

Figure 2.9 (a) The $B_e(2, 2)$, $B_e(3, 2)$, $B_e(1, 2)$ and $B_e(\frac{1}{2}, \frac{1}{2})$ density functions. (b) Histogram summarizing a random sample of size 100 from the $B_e(3, 2)$ density function.

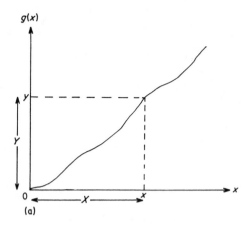

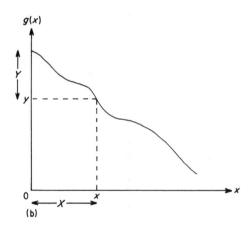

Figure 2.10   Illustrations of $y = g(x)$ where g is strictly monotonic, (a) increasing and (b) decreasing.

leading naturally to:

$$f(y) = f(x)\frac{dx}{dy} \qquad \text{for case (a)}$$

and

$$f(y) = -f(x)\frac{dx}{dy} \qquad \text{for case (b)}$$

(2.2)

Two examples of case (a) now follow.

EXAMPLE 2.2

$$y = x^2$$

$$f_X(x) = \lambda e^{-\lambda x} \quad \text{for } x \geq 0$$

$$f_Y(y) = \frac{\lambda e^{-\lambda x}}{2x} = \frac{\lambda e^{-\lambda \sqrt{y}}}{2\sqrt{y}}$$

See Fig. 2.11 for the case $\lambda = 1$.

EXAMPLE 2.3

$$y = \sqrt{x}$$

$$f_X(x) = \lambda e^{-\lambda x} \quad \text{for } x \geq 0$$

$$f_Y(y) = 2\lambda y e^{-\lambda y^2}.$$

See Fig. 2.12 for the case $\lambda = 1$.

We can see from Figs 2.11 and 2.12 how the two different transformations have put different emphases over the range of x, resulting in the two different forms for $f_Y(y)$ shown. Thus in Fig. 2.12, 'small' values of x are transformed into larger values of y (for $0 < x < 1$, $\sqrt{x} > x$), with the result that the mode of $f_Y(y)$ is to be found at $y = \sqrt{(1/2\lambda)} > 0$. However, in Fig. 2.11, for $0 < x < 1$, $x^2 < x$, and the mode of $f_Y(y)$ remains at 0.

The aim in the above has been to obtain $f(y)$ as a function of y alone, and to do this we have substituted $x = g^{-1}(y)$. Cases (a) and (b) in Equation (2.2) are both described by:

$$f_Y(y) = f_X(g^{-1}(y)) \left| \frac{dx}{dy} \right| \tag{2.3}$$

If g does not have a continuous derivative, then strictly (2.3) does not hold without a clear specification of what is meant by dx/dy. In practice, however, such cases are easily dealt with when they arise (see Exercise 2.25), since the appropriate result of Equation (2.1) always holds, giving $F(y)$.

The result of (2.3) is very useful in the simulation of random variables, as we shall see later. It may be generalized to the case of more than one random variable, when the derivative of (2.3) becomes a *Jacobian*. Thus, for example, if

$$w = g(x, y)$$

and

$$z = h(x, y)$$

provide us with a one-to-one transformation from (x, y) to (w, z), then the Jacobian of the transformation is given by the determinant

$$J = \begin{vmatrix} \dfrac{\partial w}{\partial x} & \dfrac{\partial w}{\partial y} \\[2ex] \dfrac{\partial z}{\partial x} & \dfrac{\partial z}{\partial y} \end{vmatrix}$$

and if $J \neq 0$ and all the partial derivatives involved are continuous, we can write the joint density of W and Z as:

$$f_{W,Z}(w, z) = f_{X,Y}(x, y)|J^{-1}| \qquad (2.4)$$

As with the case of a single random variable, we express the right-hand side of (2.4) as a function of w and z only. It is sometimes useful to note that

$$J^{-1} = \begin{vmatrix} \dfrac{\partial x}{\partial w} & \dfrac{\partial x}{\partial z} \\[2ex] \dfrac{\partial y}{\partial w} & \dfrac{\partial y}{\partial z} \end{vmatrix}$$

It often occurs that we require the distribution of the random variable, $W = g(X, Y)$. Introduction of some suitable function, $Z = h(X, Y)$, may result in a one-to-one transformation, so that (2.4) will give the joint density function of W and Z, from which we may then derive the required density of W as the marginal density:

$$f_W(w) = \int f_{W,Z}(w, z)\, dz$$

(See Exercises 2.14 and 2.15 for examples.) We shall now consider an example of the use of (2.4).

EXAMPLE 2.4
Let N_1 and N_2 be independent $N(0, 1)$ normal random variables. The pair (N_1, N_2) defines a point in two dimensions, by Cartesian co-ordinates. The transformation, from Cartesian to polar co-ordinates given by

$$N_1 = R \cos \Theta$$

$$N_2 = R \sin \Theta$$

is one-to-one, and all the partial derivatives involved are continuous, so that we

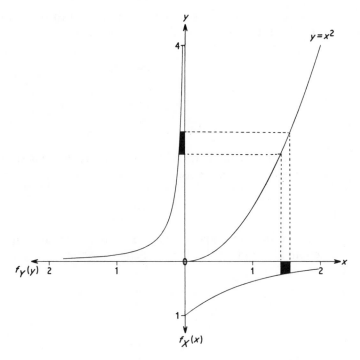

Figure 2.11 An illustration of the transformation $y = x^2$ and the densities $f_X(x) = e^{-x}, f_Y(y) = \dfrac{e^{-\sqrt{y}}}{2\sqrt{y}}$. The shaded regions have the same area.

may use (2.4) to derive the joint density function R and Θ as follows:

$$J^{-1} = \begin{vmatrix} \dfrac{\partial n_1}{\partial r} & \dfrac{\partial n_1}{\partial \theta} \\[2mm] \dfrac{\partial n_2}{\partial r} & \dfrac{\partial n_2}{\partial \theta} \end{vmatrix} = \begin{vmatrix} \cos\theta & -r\sin\theta \\ \sin\theta & r\cos\theta \end{vmatrix} = r$$

Thus $f_{R\Theta}(r, \theta) = \dfrac{r}{2\pi} \exp\left[-\tfrac{1}{2}(n_1^2 + n_2^2)\right]$

$$= \dfrac{r}{2\pi} \exp\left[-r^2/2\right] \qquad \text{for } 0 \le \theta \le 2\pi, \, 0 \le r \le \infty.$$

We thus see that R and Θ are independent random variables, with $f_\Theta(\theta) = 1/2\pi$, i.e. Θ is uniform over $[0, 2\pi]$, and $f_R(r) = r\exp\left[-r^2/2\right]$, i.e. (see Example 2.3), R^2 has an exponential distribution of parameter $\tfrac{1}{2}$.

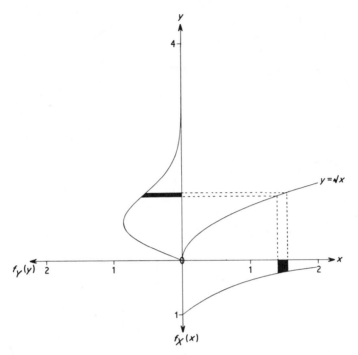

Figure 2.12 An illustration of the transformation $y = \sqrt{x}$ and the densities $f_X(x) = e^{-x}, f_Y(y) = 2ye^{-y^2}$. The shaded regions have the same area.

2.13 Convolutions

We have seen earlier that a further common transformation is a linear combination of a number of independent random variables. Again, in some cases the distributional form of the components of the sum is preserved, as occurs with Poisson, normal and Cauchy random variables, for example, while in other cases the distributional form changes, as when a sum of independent exponential random variables has a gamma distribution, as we have seen above.

The sum of mutually independent random variables is called a *convolution*. Its distribution may be evaluated by a convolution sum, or integral, as appropriate, as can be seen from the two examples that now follow.

EXAMPLE 2.5
Suppose X_1 has a $B(n_1, p)$ distribution, X_2 has a $B(n_2, p)$ distribution, and that X_1 and X_2 are independent.

Let $S = X_1 + X_2$

$$\Pr(S = k) = \sum_{i=0}^{\min(k, n_1)} \Pr(X_1 = i)\Pr(X_2 = k - i)$$

$$= \sum_{i=0}^{\min(k, n_1)} \binom{n_1}{i} p^i (1-p)^{n_1 - i} \binom{n_2}{k-i} p^{k-i}(1-p)^{n_2 - k + i}$$

$$= p^k (1-p)^{n_1 + n_2 - k} \sum_{i=0}^{\min(k, n_1)} \binom{n_1}{i}\binom{n_2}{k-i}$$

which can be shown to equal:

$$\binom{n_1 + n_2}{k} p^k (1-p)^{n_1 + n_2 - k} \qquad \text{for } 0 \le k \le n_1 + n_2$$

Thus S has a $B(n_1 + n_2, p)$ distribution.

EXAMPLE 2.6
Suppose X_1 and X_2 are independent exponential random variables, each with the p.d.f. $\lambda e^{-\lambda x}$ for $x \ge 0$.
Let $S = X_1 + X_2$

$$f_S(s) = \int f_{X_1}(x) f_{X_2}(s - x)\,dx$$

$$= \int_0^s \lambda^2 e^{-\lambda x} e^{-\lambda(s-x)}\,dx$$

$$= \lambda^2 e^{-\lambda s} s \qquad \text{for } s \ge 0$$

i.e. S has a $\Gamma(2, \lambda)$ distribution.

The result of this last example was anticipated in Section 2.10, and further examples of convolutions are given in Exercises 2.5–2.8. An important and often difficult feature in the evaluation of convolution sums and integrals is the correct determination of the admissible range for the convolution sum or integral.

2.14   The chi-square goodness-of-fit test

In the above figures illustrating distributions we can see the good qualitative match between the shapes of distributions and the corresponding shapes of histograms or bar-charts. For larger samples we would expect this match to improve. Whatever the sample size, however, we can ask whether the match between, say, probability density function and histogram is good enough. This is an important question when it comes to testing a procedure for simulating random variables of a specific type.

Special tests exist for special distributions, and we shall encounter some of these in Chapter 6; however, a test, due to K. Pearson, exists which may be applied in any situation. When this test was established by Pearson in 1900 it formed one of the cornerstones of modern statistics. The test refers to a situation in which, effectively, balls are being placed independently in one of m boxes. For any distribution we can divide up the range of the random variable into m disjoint intervals, observe how many of the simulated values (which now correspond to the balls) fall into each of the intervals (the boxes), and compare the observed numbers of values in each interval with the numbers we would expect. We then compute the statistic,

$$X^2 = \sum_{i=1}^{m} \frac{(O_i - E_i)^2}{E_i}$$

where we have used O_i and E_i to denote respectively the observed and expected numbers of values in the ith interval. If the random variables are indeed from the desired distribution then the X^2 statistic has, asymptotically, a chi-square distribution on an appropriate number of degrees of freedom. The rule for computing the degrees of freedom is

degrees of freedom = number of intervals $-1-$ number of parameters,
suitably estimated, if any

This test is useful because of its universal applicability, but simply because it may be applied in general it tends not to be very powerful at detecting departures from what one expects. A further problem with this test is that the chi-square result only holds for 'large' expected values. Although in many cases this may simply mean that we should ensure $E_i > 5$, for all i, we may well have to make a judicious choice of intervals for this to be the case. For further discussion, see Everitt (1977, p. 40), and Fienberg (1980, p. 172). The distribution of X^2 when cell values are small is discussed by Fienberg; this case may be investigated by simulation, and an illustration is given in Section 9.4.1. We shall use this test in Chapter 6 (see also Exercise 2.24).

*2.15   Multivariate distributions

In the last section we encountered the simplest of all multivariate distributions, the multinomial distribution, which results when we throw n balls independently into m boxes, with $p_i = \text{Pr (ball lands in the } i\text{th box)}$, for $1 \leq i \leq m$ and $\Sigma_{i=1}^{m} p_i = 1$.

Here we have a *family* of random variables, $\{X_i, 1 \le i \le m\}$, where X_i denotes the number of balls falling into the ith box, and so $\Sigma_{i=1}^{m} X_i = n$. The joint distribution of these random variables is given below.

Multinomial distribution:

$$\Pr(X_i = x_i, 1 \le i \le m) = \begin{pmatrix} n \\ x_1, x_2, \ldots, x_m \end{pmatrix} \prod_{i=1}^{m} p_i^{x_i},$$

$$\text{where } \sum_{i=1}^{m} x_i = n \text{ and } \sum_{i=1}^{m} p_i = 1$$

Here

$$\begin{pmatrix} n \\ x_1, x_2, \ldots, x_m \end{pmatrix} = \frac{n!}{x_1! x_2! \ldots x_m!},$$

the multinomial coefficient.

An important continuous multivariate distribution is the multivariate normal distribution, also called the multi-normal distribution. In its bivariate form the multivariate normal density function is

---

Bivariate normal probability density function:

$$\phi(x_1, x_2) = \frac{1}{2\pi\sigma_1\sigma_2(1-\rho^2)^{1/2}} \exp\left\{ -\frac{1}{2(1-\rho^2)} \left[\left(\frac{x_1-\mu_1}{\sigma_1}\right)^2 \right.\right.$$

$$\left.\left. -2\rho\left(\frac{x_1-\mu_1}{\sigma_1}\right)\left(\frac{x_2-\mu_2}{\sigma_2}\right) + \left(\frac{x_2-\mu_2}{\sigma_2}\right)^2 \right] \right\}$$

$$\text{for } -\infty < x_1, x_2 < \infty$$

Here ρ is the correlation between the two random variables; Fig. 2.13 illustrates two possible forms for $\phi(x_1, x_2)$. The p-variate density function has the following form:

---

p-variate multivariate normal probability density function:

$$\phi(\mathbf{x}) = (2\pi)^{-p/2} |\mathbf{\Sigma}|^{-1/2} \exp\left(-\tfrac{1}{2}(\mathbf{x}-\mathbf{\mu})' \mathbf{\Sigma}^{-1}(\mathbf{x}-\mathbf{\mu})\right)$$

$$\text{for } -\infty < x_i < \infty, 1 \le i \le p$$

notation used: $N(\mathbf{\mu}, \mathbf{\Sigma})$

Here $\mathbf{\mu}$ is the mean vector, $(\mathbf{x} - \mathbf{\mu})'$ is the transpose (row vector) of the column vector $(\mathbf{x} - \boldsymbol{\mu})$, and $\boldsymbol{\Sigma}$ is the variance/covariance matrix, i.e. $\boldsymbol{\Sigma} = \{\sigma_{ij}, 1 \le i, j \le p\}$, in which σ_{ij} is the covariance between the component random variables X_i and X_j. Thus σ_{ii} is the variance of X_i, for $1 \le i \le p$.

It can readily be shown (see, e.g., Morrison, 1976, p. 90, and cf. Exercise 2.16) that if $\mathbf{Y} = \mathbf{AX}$ and $\mathbf{X}$ has the $N(\boldsymbol{\mu}, \boldsymbol{\Sigma})$ distribution, where $\mathbf{A}$ is a nonsingular $p \times p$ matrix, then $\mathbf{Y}$ has the $N(\mathbf{A}\boldsymbol{\mu}, \mathbf{A}\boldsymbol{\Sigma}\mathbf{A}')$ distribution.

*2.16   Generating functions

The material of this section is not used extensively in the remainder of the book, and many readers may prefer to move on to Section 2.17.

It is often convenient to know the forms of *generating functions* of random variables. For any random variable X, we define the moment generating function (m.g.f.) as

$$M_X(\theta) = \mathscr{E}\left[e^{\theta X}\right]$$

for an appropriate range of the dummy variable, θ. Not all random variables have m.g.f.'s: the Cauchy distribution provides a well-known example. However, if $M_X(\theta)$ exists for a nontrivial interval for θ, then the m.g.f.

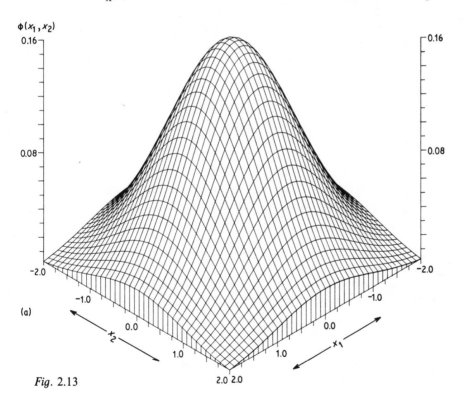

Fig. 2.13

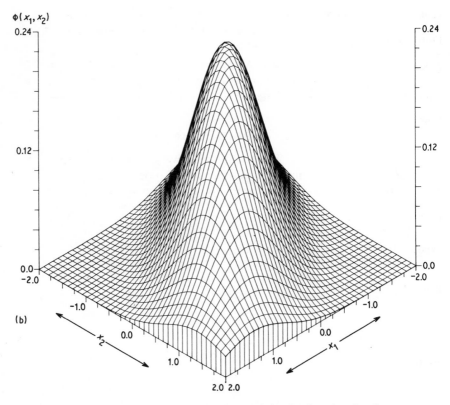

Figure 2.13   Illustration of the bivariate normal density function for the cases

(a) $\mu_1 = \mu_2 = 0$, $\sigma_1 = \sigma_2 = 1$, $\rho = 0$
(b) $\mu_1 = \mu_2 = 0$, $\sigma_1 = \sigma_2 = 1$, $\rho = 0.5$

characterizes the random variable. An alternative generating function is the probability generating function, defined by

$$G(z) = \mathscr{E}[z^X]$$

M.g.f.'s for some of the distributions considered earlier in this chapter are given in Table 2.1.

For the distributions of Table 2.1, the m.g.f. may be used to check the values of means and variances given earlier, since

$$M'(0) = \mathscr{E}[X], \quad \text{and} \quad M''(0) = \mathscr{E}[X^2],$$

illustrating why the m.g.f. is so named.

A glance at the m.g.f.'s of Table 2.1 shows that binomial, negative-binomial and gamma random variables can be expressed as convolutions of identically

Table 2.1   Common distributions and associated moment generating functions

Distribution	m.g.f.
geometric: $\Pr(X = i) = q^{i-1}p$	$pe^{\theta}(1 - qe^{\theta})^{-1}$, for $qe^{\theta} < 1$
binomial: $B(n, p)$: $\Pr(X = i) = \binom{n}{i} p^{i} q^{n-i}$	$(q + pe^{\theta})^{n}$
negative-binomial: $\Pr(X = n + i)$ $= \binom{n+i-1}{i} p^{i} q^{n}$	$p^{n}e^{n\theta}(1 - qe^{\theta})^{-n}$, for $qe^{\theta} < 1$
Poisson: $\Pr(X = i) = \dfrac{e^{-\lambda}\lambda^{i}}{i!}$	$e^{\lambda(e^{\theta} - 1)}$
normal: $N(0, 1)$: $f(x) = \dfrac{e^{-x^{2}/2}}{\sqrt{(2\pi)}}$	$e^{\theta^{2}/2}$
exponential: $f(x) = \lambda e^{-\lambda x}$	$\dfrac{\lambda}{\lambda - \theta}$ for $\theta < \lambda$
gamma: $\Gamma(n, \lambda)$: $f(x) = \dfrac{e^{-\lambda x}\lambda^{n}x^{n-1}}{\Gamma(n)}$	$\left(\dfrac{\lambda}{\lambda - \theta}\right)^{n}$ for $\theta < \lambda$

distributed random variables. We see why this is so as follows:

Let
$$S = \sum_{i=1}^{n} X_{i}$$

then
$$M_{S}(\theta) = \mathscr{E}\left[\exp\left(\theta \sum_{i=1}^{n} X_{i}\right)\right]$$
$$= \mathscr{E}\left[\prod_{i=1}^{n} \exp(\theta X_{i})\right]$$

and if the $\{X_{i}\}$ are mutually independent, then

$$M_{S}(\theta) = \prod_{i=1}^{n} \mathscr{E}[\exp(\theta X_{i})] = \prod_{i=1}^{n} M_{X_{i}}(\theta).$$

Furthermore, if the $\{X_{i}\}$ have the common m.g.f., $M_{X}(\theta)$, say, then

$$M_{S}(\theta) = (M_{X}(\theta))^{n} \qquad (2.5)$$

Thus, for example, a random variable X with the $\Gamma(n, \lambda)$ distribution can be written as

$$X = \sum_{i=1}^{n} E_{i}$$

where the E_i are independent, identically distributed exponential random variables with parameter λ (cf. Exercise 2.6).

Moment generating functions may also be defined for m jointly distributed random variables $X_1, X_2, \ldots, X_m$, as follows:

$$M_{\mathbf{X}}(\boldsymbol{\theta}) = \mathscr{E}\left[\prod_{i=1}^{m} \exp{(\theta_i X_i)}\right]$$

Thus for the multinomial distribution of Section 2.15, we have the multivariate moment generating function

$$M_{\mathbf{X}}(\boldsymbol{\theta}) = \left(\sum_{i=1}^{m} p_i \exp{\theta_i}\right)^n,$$

while the multivariate normal distribution of Section 2.15 has the multivariate m.g.f.

$$M_{\mathbf{X}}(\boldsymbol{\theta}) = \exp{(\boldsymbol{\theta}'\boldsymbol{\mu} + \tfrac{1}{2}\boldsymbol{\theta}'\boldsymbol{\Sigma}\boldsymbol{\theta})}$$

A bivariate Poisson distribution which we shall encounter later is simply defined by its m.g.f.:

$$M_{\mathbf{X}}(\boldsymbol{\theta}) = \exp{\left[\lambda_1 (e^{\theta_1} - 1) + \lambda_2 (e^{\theta_2} - 1) + \lambda_3 (e^{\theta_1 + \theta_2} - 1)\right]} \qquad (2.6)$$

We shall conclude this section with two examples which complement work earlier in the chapter and illustrate further the utility of generating functions.

EXAMPLE 2.7   *Proof of a central limit theorem*
A $B(n, p)$ random variable W can be written as a convolution:

$$W = \sum_{i=1}^{n} X_i$$

where $\mathscr{E}[X_i] = p$ and $\mathrm{Var}(X_i) = pq$, where $q = 1 - p$.

Let $S_n = \dfrac{(W - np)}{\sqrt{(np)}}$

then

$$M_{S_n}(\theta) = \mathscr{E}\left[\exp{\left(\frac{(W - np)\theta}{\sqrt{(npq)}}\right)}\right]$$

$$= \mathscr{E}\left[\exp{\left(\sum_{i=1}^{n}(X_i - p)\phi\right)}\right]$$

where $\phi = \theta / \sqrt{(npq)}$

and so by (2.5), as the $\{X_i\}$ are independent,

$$M_{S_n}(\theta) = (M_Y(\phi))^n, \quad \text{where} \quad Y = (X_i - p)$$

From the above, $\mathscr{E}[Y] = 0$ and $\mathscr{E}[Y^2] = pq$, and so

$$M_{S_n}(\theta) = \left(1 + \frac{\phi^2 pq}{2} + \text{higher order terms in } \phi\right)^n$$

$$= \left(1 + \frac{\theta^2}{2n} + \text{higher order terms in } \left(\frac{\theta}{\sqrt{n}}\right)\right)^n$$

and by a result similar to that of Exercise 2.22,

$$M_{S_n}(\theta) \to \exp(\theta^2/2) \quad \text{as} \quad n \to \infty.$$

Thus as $n \to \infty$ the m.g.f. of $S_n \to$ the m.g.f. of an $N(0, 1)$ random variable, and so the distribution of $S_n \to N(0, 1)$. A similar limiting operation applied to the multinomial distribution results in the multivariate normal distribution.

EXAMPLE 2.8   *Deriving the Poisson distribution from the binomial distribution*

If W has a $B(n, p)$ distribution, then

$$M_W(\theta) = (1 - p + pe^\theta)^n$$

Now let us keep $np = \lambda$, say, fixed, while we let $n \to \infty$ (and consequently $p \to 0$).

Now $$M_W(\theta) = \left(1 + \frac{\lambda(e^\theta - 1)}{n}\right)^n,$$

and as $n \to \infty$, $\quad M_W(\theta) \to \exp[\lambda(e^\theta - 1)]$, $\quad$ (see Exercise 2.22)

i.e. the m.g.f. of a Poisson random variable with parameter λ. Hence under this limiting operation the distribution of W tends to this Poisson form.

It is possible to derive the exponential and gamma distributions by similar limiting processes applied, respectively, to the geometric and negative-binomial distributions (see Exercise 2.23). This approach may be used to provide an heuristic proof that the rules of the Poisson process result in a predicted Poisson distribution (see Parzen, 1960, p. 253).

2.17   Discussion and further reading

While we have dichotomized random variables as usually discrete or continuous, we have not mentioned, for instance, that most discrete random variables simply take integer values. Furthermore, the continuous random variables we have considered are, formally, absolutely continuous random variables. Such discussion is not necessary for the material to follow, but it may be found in books such as Blake (1979) and Parzen (1960). Additional discrete and continuous distributions will arise throughout the book.

In this chapter we have presented only the tip of a very large iceberg. Much more detail can be found in, for example, Haight (1967), Johnson and Kotz (1969, 1970a, 1970b and 1972), Kendall and Stuart (1961) and Ord (1972). Mardia (1970) considers families of bivariate distributions, while Douglas (1980) describes the interesting distributions which can result from special combinations of distributions such as the binomial and Poisson. Cramér (1954) discusses and proves different forms of central limit theorems, and Bailey (1964), Cox and Miller (1965) and Feller (1957) provide the necessary background to the Poisson process. Apostol (1963) is a good reference for the full transformation-of-variable theory, which is also well described by Blake (1979). Further discussion of the chi-square goodness-of-fit test is provided by Cochran (1952) and Craddock and Flood (1970), whose small-sample study is the subject of Section 9.4.1. A more leisurely introduction to some of the material of this chapter is provided by Folks (1981), and Cox and Smith (1967) provide a good introduction to the mathematical theory of queues, relevant to Exercises 2.26–2.28.

2.18   Exercises and complements

(a)   Transforming random variables

†2.1   Derive the density function of the random variable

$$X = -\log_e U, \text{ where } U \text{ is } U(0, 1).$$

2.2   Consider the effect of the transformation $Y = aX$, where a is a fixed constant, and X is, e.g., an exponential, normal, gamma, or Poisson random variable.

2.3   Show that if X has the distribution of Exercise 2.1, and $W = \gamma X^{1/\beta}$ then W has a Weibull distribution with p.d.f.

$$f_W(\omega) = \frac{\beta}{\gamma^\beta} \, \omega^{\beta-1} \exp\left[-(\omega/\gamma)^\beta\right]$$

for $0 \leq \omega < \infty$, $\beta > 0$, $\gamma > 0$.

2.4   Find the distribution of $Y = N^2$, where N is an $N(0, 1)$ random variable.

2.5   If $Y_1, Y_2, \ldots, Y_n$ are all mutually independent $N(0, 1)$ random variables, show, by induction or otherwise, that $\Sigma_{i=1}^n Y_i^2$ has the χ_n^2 distribution.

2.6   If $Y_1, Y_2, \ldots, Y_n$ are all mutually independent exponential random variables with p.d.f. $\lambda e^{-\lambda y}$ for $\lambda > 0$, $y \geq 0$, show, by induction and using the convolution integral, that $\Sigma_{i=1}^{n} Y_i$ has the $\Gamma(n, \lambda)$ distribution.

*2.7   If $Y_1, Y_2, \ldots, Y_n$ are all mutually independent Cauchy random variables with p.d.f., $(\pi(1 + y^2))^{-1}$, derive the distribution of $1/n \Sigma_{i=1}^{n} Y_i$.

†2.8 X, Y are independent random variables. Find the distribution of $X + Y$ when:

(a) X, Y are $N(\mu_1, \sigma_1^2)$, $N(\mu_2, \sigma_2^2)$ respectively
(b) X, Y are Poisson, with parameters λ, μ, respectively
(c) X, Y are exponential, with parameters λ, μ, respectively.

2.9   If X, Y are as in Exercise 2.8(b), find

$$\Pr(X = r | X + Y = n) \qquad 0 \leq r \leq n.$$

2.10   If X and Y are independent random variables, find the distribution of $Z = \max(X, Y)$ in terms of the c.d.f.'s of X and Y.

*2.11 $X_1, X_2, \ldots, X_n$ are independent random variables with the distribution of Exercise 2.1. Prove that the following random variables have the same distribution:

$$Y = \max(X_1, X_2, \ldots, X_n)$$

$$Z = X_1 + \frac{X_2}{2} + \ldots + \frac{X_n}{n}.$$

*2.12   Random variables Y_1 and Y_2 have the exponential p.d.f., e^{-x} for $x \geq 0$. Let $X_1 = Y_1 - Y_2$ and $X_2 = Y_1 + Y_2$. Find the joint distribution of (X_1, X_2).

*2.13   Let X_1, X_2 be two independent and identically distributed non-negative continuous random variables. Find the joint probability density function of $\min(X_1, X_2)$ and $|X_1 - X_2|$. Deduce that these two new random variables are independent if and only if X_1 and X_2 have an exponential distribution. In such a case, evaluate $\Pr(X_1 + X_2 \leq 3 \min(X_1, X_2) \leq 3b)$, where b is constant.

†2.14   Random variables X, Y are independently distributed as χ_{2a}^2 and χ_{2b}^2 respectively. Show that the new random variables, $S = X + Y$ and $T = X/(X + Y)$ are independent, and T has a beta, $B_e(a, b)$ distribution.

*2.15   If N_1, N_2, N_3, N_4 are independent $N(0, 1)$ random variables, show that:

(a) $X = |N_1 N_2 + N_3 N_4|$ has the exponential p.d.f., e^{-x} for $x \geq 0$.
(b) $C = N_1/N_2$ has the Cauchy distribution of Exercise 2.7.

*2.16 $\mathbf{X}$ is a p-dimensional column vector with the multivariate $N(\mathbf{0}, \mathbf{I})$

distribution, in which $\mathbf{0}$ denotes a p-variate zero vector and $\mathbf{I}$ is the $p \times p$ identity matrix. If $\mathbf{Z} = \mathbf{AX} + \boldsymbol{\mu}$, where $\mathbf{A}$ is an arbitrary $p \times p$ matrix, and $\boldsymbol{\mu}$ is an arbitrary p-dimensional column vector, show that $\mathbf{Z}$ has the $N(\boldsymbol{\mu}, \mathbf{AA}')$ distribution.

(b) Manipulation of random variables, and questions arising from the chapter

*2.17 Two independent Poisson processes have parameters λ_1 and λ_2. Find and identify the distribution of the number of events in the first process which occur before the first event in the second process.

*2.18 Random variables X and Y have the related distributions:

$$\Pr(Y = k) = \binom{n+m}{k} (1-\theta)^k \theta^{n+m-k} \qquad \text{for } 0 \le k \le n+m$$

$$\Pr(X = k) = \binom{n+k-1}{k} \theta^k (1-\theta)^n \qquad \text{for } k \ge 0$$

Here n, m are positive integers, and $0 < \theta < 1$, so that Y is binomial, and X is negative-binomial. By finding the coefficient of z^i in $(1+z)^{n+m}/(1+z)^{m+1-i}$, for $0 \le i \le m$, or otherwise, show that

$$\Pr(X \le m) = \Pr(Y \ge n).$$

*2.19 Use a central limit theorem approach to show that

$$\lim_{n \to \infty} e^{-n} \sum_{r=0}^{n} \frac{n^r}{r!} = \frac{1}{2}.$$

*2.20 Show that a random variable with the negative-binomial distribution has the moment generating function

$$M_X(\theta) = p^n e^{n\theta} (1 - qe^\theta)^{-n}.$$

*2.21 Show that a random variable with the gamma $\Gamma(n, \lambda)$ distribution has the moment generating function

$$\left(\frac{\lambda}{\lambda - \theta}\right)^n \qquad \text{for } \theta < \lambda.$$

*2.22 Show that $\lim\limits_{n \to \infty} \left(1 + \dfrac{x}{n}\right)^n = e^x.$

*2.23 Suppose X is a random variable with a geometric distribution of parameter p. Let $Y = aX$. If $a \to 0$ and $p \to 0$ in such a way that $\lambda = a/p$ is a constant, show that the distribution of Y tends to that of a random variable with an exponential distribution with parameter λ^{-1}.

2.24 Use the chi-square goodness-of-fit test to compare the observed and expected values in the intervals: (0, 0.1), (0.1, 0.2), etc., for the example of Fig. 2.4, arising from the $U(0, 1)$ distribution. The grouped data frequencies are, in increasing order: 8, 8, 14, 12, 11, 11, 12, 6, 12, 6.

2.25 X is a random variable with the exponential p.d.f., e^{-x} for $x \geq 0$. We define Y as follows:

$$\text{for } 0 \leq X \leq 1, \quad Y = X$$

$$\text{for } X \geq 1, \quad Y = 2X - 1.$$

Obtain the distribution of Y.

(c) Questions on modelling, continuing Exercises 1.4, 1.6 and 1.7

†**2.26** The simple queue of Exercise 1.4 measured time in integral units. More realistically, times between arrivals, and service times, would be continuous quantities, sometimes modelled by random variables with exponential distributions. Observe a real-life queue, at a post-office, for example, make a record of inter-arrival and service times and illustrate these by means of histograms. What underlying distributions might seem appropriate?

2.27 (*continuation*)   The BASIC program given below simulates what is called an M/M/1 queue (see e.g., Gross and Harris, 1974, p. 8). In this queue, inter-arrival times are independent random variables with $\lambda e^{-\lambda x}$ exponential density function, and service times are independent random variables with $\mu e^{-\mu x}$ exponential density function. There is just one server and $\lambda/(\lambda + \mu)$ plays the rôle of p in Exercise 1.4. Run this program for cases: $\lambda = \mu, \lambda > \mu$ and $\lambda < \mu$, and comment on the results.

NOTE that the statements 100, 150 and 190 below simulate a $U(0, 1)$ random variable. The method used by the computer is described in the next chapter. The function of statements 110 and 160 should be clear from the solutions to Exercises 2.1 and 2.2. An explanation of why this program does in fact simulate an M/M/1 queue is given in Section 8.3.1.

```
10    REM THIS PROGRAM SIMULATES AN M/M/1/ QUEUE, STARTING EMPTY
20    REM AS INPUT YOU MUST PROVIDE ARRIVAL AND DEPARTURE RATES
30    REM NOTE THAT THERE IS NO TERMINATION RULE IN THIS PROGRAM
40    PRINT "TYPE LAMBDA AND MU, IN THAT ORDER "
50    INPUT L,M
60    LET S = L+M
70    LET I = L/S
80    PRINT "QUEUE SIZE.......AFTER TIME"
90    RANDOMIZE
100   LET U = RND
110   LET E = (-LOG(U))/L
120   REM E IS THE TIME TO FIRST ARRIVAL AT AN EMPTY QUEUE
```

```
130  LET Q = 1
140  PRINT Q,E
150  LET U = RND
160  LET E = (-LOG(U))/S
170  REM E IS TIME TO NEXT EVENT, IE., ARRIVAL OR DEPARTURE
180  REM WE MUST NOW FIND THE TYPE OF THAT EVENT
190  LET U = RND
200  IF U > I THEN 250
210  REM THUS WE HAVE AN ARRIVAL
220  LET Q = Q+1
230  PRINT Q,E
240  GOTO 150
250  REM THUS WE HAVE A DEPARTURE
260  LET Q = Q-1
270  PRINT Q,E
280  IF Q = 0 THEN 100
290  GOTO 150
300  END
```

*2.28 (*continuation*) We have seen that exponential distributions result from Poisson processes, and we can consider the parameters λ and μ of Exercise 2.27 to be rate parameters in Poisson processes for arrivals and departures, respectively. In some cases it may seem realistic for λ and μ each to be functions of the current queue size, n, say. For example, if $\lambda_n = 2/(n+1)$ and $\mu = 1$, we have simple 'discouragement' queue, with an arrival rate which decreases with increasing queue size. Modify the BASIC program of Exercise 2.27 in order to simulate this discouragement queue, and compare the behaviour of this queue with that of the M/M/1 queue with $\lambda = 2$, $\mu = 3$. We shall continue discussion of these queues in Chapters 7 and 8.

3

GENERATING UNIFORM
RANDOM VARIABLES

3.1 Uses of uniform random numbers

Random digits are used widely in statistics, for example, in the generation of random samples (see Barnett, 1974, p. 22), or in the allocation of treatments in statistical experiments (see Cox, 1958, p. 72). More generally still, uniform random numbers and digits are needed for the conduct of lotteries, such as the national premium bond lottery of the United Kingdom (see Thompson, 1959).

A further use for random digits is given in the following example.

EXAMPLE 3.1
The randomized response technique

In conducting surveys of individuals' activities it may be of interest to ask a question which could be embarrassing to the interviewee; possible examples include questions relating to car driving offences, sex, tax-evasion and the use of drugs. Let us denote the embarrassing question by E, and suppose, for the population in question, we know the frequency, p, of positive response to some other, non-embarrassing, question, N, say. We can now proceed by presenting the interviewee with both questions N and E, and a random digit simulator, producing 0 with probability p_0, and producing 1 with probability $1 - p_0$. The interviewee is then instructed to answer N if the random digit is 0, say, and to answer E if the random digit is 1. The interviewer does not see the random digit. From elementary probability theory (see *ABC*, p. 85)

$$\Pr(\text{response} = \text{Yes}) = \Pr(\text{response} = \text{Yes}|\text{question is } N)p_0$$

$$+ \Pr(\text{response} = \text{Yes}|\text{question is } E)(1 - p_0)$$

Knowing p_0 and $\Pr(\text{response} = \text{Yes}|\text{question is } N)$, and estimating $\Pr(\text{response} = \text{Yes})$ from the survey, enables one to estimate $\Pr(\text{response} = \text{Yes}|\text{question is } E)$. This illustration is an example of a randomized-

51

response technique (RRT), and for further examples and discussion, see Campbell and Joiner (1973) and Exercises 3.1–3.4.

Uniform random numbers are clearly generally useful. Furthermore, in Chapters 4 and 5 we shall see that if we have a supply of $U(0, 1)$ random variables, we can simulate any random variable, discrete or continuous, by suitably manipulating these $U(0, 1)$ random variables.

Initially, therefore, we must consider how we can simulate uniform random variables, the *building-blocks* of simulation, and that is the subject of this chapter. We start by indicating the relationships between discrete and continuous uniform random variables.

3.2   Continuous and discrete uniform random variables

If U is a $U(0, 1)$ random variable, and we introduce a discrete random variable D such that

$$D = i \text{ if and only if } i \leq 10\,U < i+1, \quad \text{for } i = 0, 1, 2, \ldots, 9$$

then
$$\Pr(D = i) = \Pr(i \leq 10\,U < i+1)$$

$$= \frac{1}{10} \quad \text{for } i = 0, 1, 2, \ldots, 9$$

The random variable D thus provides equi-probable (uniform) random digits.

Conversely, if we write a $U(0, 1)$ random variable, U, in decimal form,

$$U = \sum_{k \geq 1} D(k)\, 10^{-k}$$

Then intuitively we would expect $D(k)$ to be a uniform random digit, for each $k \geq 1$,

i.e.
$$\Pr(D(k) = i) = \frac{1}{10}, \quad \text{for } 0 \leq i \leq 10,$$

$$\text{and } k \geq 1$$

This and further results are proved by Yakowitz (1977, pp. 29–31). We see, therefore, that $U(0, 1)$ random variables can readily give us uniform random digits, while given a means of simulating random digits we can combine them to give $U(0, 1)$ variables to whatever accuracy is required.

3.3   Dice and machines

The simplest random number generators are coins, dice and bags of coloured balls, the very bread-and-butter of exercises in elementary probability theory.

Thus in the RRT example above, the interviewee could be given a well-shaken bag of balls, a proportion p_0 of which are white, with the remainder being black. Without looking, the interviewee then selects a ball from the bag, and answers question N if the ball chosen is white, and answers question E if the ball chosen is black. Similar physical devices are sometimes used in lotteries, and games of chance such as bingo and roulette. Certain countries such as Australia, Canada, France and West Germany televise, once a week, the operation of a complex physical device for selecting winning lottery numbers. West (1955) provides an analysis of the results of a lottery carried out in Rhodesia.

The random digits we usually need are uniform over the 0–9 range, and such digits can be obtained by suitably manipulating simple devices such as coins, as in the following example:

EXAMPLE 3.2

A fair coin is tossed four times. If we record a head as 0 and a tail as 1, then the result of the experiment is four digits, $abcd$, written in order, e.g., 0110. We can interpret $abcd$ as the number, $(a \times 2^3) + (b \times 2^2) + (c \times 2) + d$, so that 0110 is interpreted as 6. If the resulting number is greater than 9 we reject it and start again. If the resulting number is in the 0–9 range then it is a realization of a uniformly distributed random digit over that range. (Based on part of an A-level question, Oxford, 1978.)

We can see this simply by enumerating the possible outcomes to the experiment:

Outcome	Resulting number
0000	0
1000	8
0100	4
0010	2
0001	1
1100	12
1010	10
1001	9
0110	6
0101	5
0011	3
1110	14
1101	13
1011	11
0111	7
1111	15

We are just using the coin to simulate the binary form of the digits 0–15. This method therefore does give rise to uniform random digits over 0–9, but it is rather wasteful, as resulting numbers are rejected 3/8 of the time.

Manipulations of this kind are avoided by the direct use of simple dice to produce 0–9 uniform random digits. Unfortunately, a regular 10-sided figure does not exist, but one can use icosahedral dice (giving regular 20-sided figures), each digit 0–9 appearing separately on two different faces. Further possibilities include rolling a regular 10-faced cylinder, or throwing a 10-faced di-pyramid, with each face being an isosceles triangle of some fixed size. These simple devices are illustrated in Fig. 3.1.

Figure 3.1   (a) Three icosahedral dice. Note the need to distinguish between 6 and 9 (b) A regular, 10-faced cylinder (c) Three 10-faced di-pyramids, with truncated isosceles triangles of the same size as faces. Note that the two pyramids are so attached that when the body is at rest a face is uppermost.

Because of the general demand for random digits, tables, such as those of Appendix 2, are now widely available. A sequence of random digits can be obtained by reading the table by rows, by columns, or by any other rule. The first table of this kind was produced by Tippett in 1927, and it was regarded as a 'godsend' by the statisticians of the day (Daniels, 1982).

In using physical devices such as dice to simulate random digits one is reversing the customary model/reality relationship. As described in Chapter 1, one usually takes a real-life situation, and builds a model of it. Here we start with a model, such as a uniform random digit, seek a real-life mechanism to correspond to that model, and then take observations from the real-life mechanism. There is always a discrepancy between model and reality – coins may not be fair (see, e.g., Kerrich, 1946), dice may be biased, and so on – therefore the numbers produced by physical devices are tested, to ensure that no drastic non-randomness is present. This is simply a form of quality control of random numbers, and one applies only a finite subset of the infinity of tests that are possible. We shall return to the subject of testing of numbers in Chapter 6.

Any process in nature that is thought to be random may be used to try to simulate uniform random numbers. Kendall and Babbington-Smith (1939a) used a rotating disk with ten uniform segments, which was stopped at random. Tippett (1925) used digits read from tables of logarithms. ERNIE, the computer used for selecting winning premium bonds in the British national lottery, uses the electronic 'noise' of neon tubes. The digits of Table 3.1 were obtained from reading the last three digits of successive numbers from the Canterbury region telephone directory. (Cf. Section 6.7 and Exercise 6.8. The relative frequencies of these digits are considered in Example 6.1.)

Student (1908a) drew samples from a set of physical measurements taken on criminals, as described in Section 1.5. In his case we have an illustration of sampling from a non-uniform population, (approximately normal in this case), and similarly exponential and Poisson random variables may be simulated directly if one can observe a process in nature which provides a good approximation to a Poisson process (see Section 4.4.2).

Dice and machines are impractical for all but the smallest simulations, which are now in any case likely to be conducted with the aid of readily available tables (see for instance, Neave, 1981, and Murdoch and Barnes, 1974). Large-scale simulations are usually conducted using computers, and early computers were equipped with built-in random-number generators of the physical kind, using random electronic features, as in ERNIE. Tocher (1975, chapter 5) provides many examples here, and even circuit diagrams. More recently, Isida (1982) presented a compact physical random-number generator based on the noise of a Zener diode. The modern equivalent of this can be found in certain hand-calculators, which have an RND button for simulating $U(0, 1)$ random variables. A problem with all physical devices is the danger that they may

Table 3.1   Digits from telephone numbers

	874	580	873	824	564	663
	478	658	540	561	360	082
	661	839	996	261	052	938
	334	420	356	571	081	866
	569	166	045	091	961	610
(a)	471	378	936	569	107	022
	916	865	961	838	303	826
	665	014	148	764	276	638
	504	776	237	682	634	207
	659	654	774	217	609	684
	423	213	423	002	960	273
	183	059	563	379	252	955
	202	410	451	887	467	427
(b)	207	483	809	265	117	891
	061	658	145	950	135	495
	716	232	955	771	747	699
	693	757	952	053	659	459
	991	876	091	431	316	283
	499	223	743	037	891	729
	611	998	650	527	073	665

become unreliable, through changes to the device in time; thus dice, for instance, could become unevenly worn, resulting in bias. Frequent checks of the generated numbers should therefore be carried out.

The modern approach to large-scale simulation is quite different from that of this section, and it avoids the need for such frequent checking by producing a sequence of numbers that can be shown mathematically to possess certain desirable features. This approach, which is also not without its drawbacks, is described in the next section.

3.4   Pseudo-random numbers

The digits of Table 3.2 superficially have the *appearance* of the digits of Table 3.1, but they have been generated in a blatantly non-random fashion, from the recursion formula

$$u_{n+1} = \text{fractional part of } (\pi + u_n)^5 \qquad \text{for } n \geq 0 \qquad (3.1)$$

where u_0 is some specified number in the range $0 < u_0 < 1$. u_0 is, rather graphically, termed the 'seed'. Knowledge of the formula of (3.1) provides one with complete knowledge of the sequence of numbers resulting in Table 3.2., but in many applications one may find these digits as suitable as those, say, of Table 3.1, and much more easily generated on a calculator or computer. Formula (3.1) can be likened to a 'black box' which takes the place of a physical black-box such as a die. Recursion formulae are most suitable for use on computers

Table 3.2 Digits from the recursion of Equation (3.1).

254	032	329	233	252	444
794	807	600	974	884	454
797	354	440	855	159	290
162	053	737	489	953	381
051	091	224	843	075	513
703	740	755	750	070	002
301	810	903	392	970	915
690	642	767	038	140	051
962	283	420	435	835	150
574	108	551	564	209	788
810	657	491	939	365	537
612	514	020	950	567	239
119	865	638	032	062	491
966	619	460	553	850	096
255	550	872	019	601	282
474	943	141	486	022	074
013	589	023	454	681	854
489	857	712	412	307	910
826	305	753	610	885	458
346	008	309	763	890	300

Each triple is obtained from the first three decimal places of the u_i, when (3.1) was operated using a 32-bit computer and floating-point arithmetic. Successive numbers were obtained moving from left to right across the rows, and down the table.

and calculators, and furthermore the properties of the numbers they produce can be investigated mathematically. If the resulting numbers satisfy a variety of tests, then because of the deterministic nature of a recursion formula, additional application of these tests at a later stage is not necessary, as there is no danger of bias creeping into the black-box, with the progress of time.

In some applications it may be required to re-run a simulation using the same random numbers as on a previous run. Such a requirement may seem unlikely, but we shall see in Chapter 7 that it can be very useful in certain methods for variance-reduction. Knowledge of u_0 for a formula such as (3.1) enables one to do this quite easily, whereas such a 're-run' facility is not possible with physical generators unless a possibly time-consuming record is made of the numbers used, Inoue *et al.* (1983) describe the generation and testing of random digits which may be supplied on magnetic tapes.

3.5 Congruential pseudo-random number generators

An alternative mathematical representation of formula (3.1) is:

$$u_{n+1} = (\pi + u_n)^5 \pmod 1 \qquad \text{for } n \geq 0$$

Currently the recursion formula that is most frequently adopted is:

$$x_{n+1} = ax_n + b \quad (\text{mod } m) \quad \text{for } n \geq 0 \quad (3.2)$$

in which a, b and m are suitably chosen fixed *integer* constants, and the seed is an integer, x_0. Starting from x_0, the formula (3.2) gives rise to a sequence of integers, each of which lies in the 0 to $(m-1)$ range. Because the resulting numbers can be investigated by the theory of congruences, such generators are termed 'congruential'. Although terminology is not always uniform here, we shall call a generator with $b = 0$, 'multiplicative', and one with $b \neq 0$, 'mixed'. Approximations to $U(0, 1)$ variables can be obtained from setting $u_i = x_i/m$, as discussed in Exercise 3.15. For an example, see the solution to Exercise 3.21.

Formulae such as (3.1) are sometimes used to play games involving random elements on hand calculators. We can examine the numbers produced and we may find that they satisfy many criteria of random numbers. However, there is no guarantee, in general, that at some stage the sequence of numbers produced by such formulae may not seriously violate criteria of random numbers, and thus, in general, such formulae are of little use for scientific work. As we shall see, an advantage of the formula (3.2) is that certain guarantees *are* available for the resulting numbers.

The constants a, b and m are chosen with a number of aims in mind. For a start, one wants the arithmetic to be efficient. Human beings do arithmetic to base 10, and so if the formula (3.2) was being operated by hand, using pencil and paper, it would be sensible for m to be some positive integral power of 10. For example, if we have

$$x_0 = 89, \ a = 1573, \ b = 19, \ m = 10^3$$

then from (3.2),

$$x_1 = 140\,016 \quad (\text{mod } 10^3) = 16$$
$$x_2 = \ \ 25\,187 \quad (\text{mod } 10^3) = 187$$
$$\text{etc.}$$

Clearly, if one naturally does arithmetic to number base r, say, then the operation of division by m is most efficiently done if $m = r^k$ for some positive integer k. For most computers this entails setting $m = 2^k$, where k is selected so that m is 'large' (see below) and the numbers involved are within the accuracy of the machine.

A moment's thought shows that the generator of (3.2) can produce no more than m different numbers before the cycle repeats itself, again and again. Thus a second aim in choosing the constants a, b, m is that the cycle length, which could certainly be less than m, is reasonably large. It has been shown (see Hull and Dobell, 1962, and Knuth, 1981, pp. 16–18) that for the case $b > 0$, the maximum possible cycle length m is obtained if, and only if, the following relations hold:

(i) b and m have no common factors other than 1;

(ii) $(a-1)$ is a multiple of every prime number that divides m;
(iii) $(a-1)$ is a multiple of 4 if m is a multiple of 4.

If $m = 2^k$, relation (iii) will imply that $a = 4c + 1$ for positive integral c. Such an a then also satisfies relation (ii). When $m = 2^k$, relation (i) is easily obtained by setting $b = $ any odd positive constant. Proofs of results such as these are usually given in general number-theoretic terms; however, following Peach (1961), in Section 3.9 we provide a simple proof of the above result for the commonly used case: $m = 2^k$, $a = 4c + 1$ and b odd (c, b, and k positive integers).

Although multiplicative congruential generators involve less arithmetic than mixed congruential generators, it is not possible to obtain the full cycle length in the multiplicative case. Nevertheless, if $m = 2^k$ for a multiplicative generator, then a cycle length of 2^{k-2} may be obtained. This is achieved by setting $a = \pm 3$ (mod 8), and now also imposing a constraint on x_0, namely, choosing x_0 to be odd. A suitable choice for a is an odd power of 5, since, for positive, integral q,

$$5^{2q+1} = (1+4)^{2q+1} = (1 + 4(2q+1)) \quad \text{mod (8)}$$
$$= -3 \quad (\text{mod } 8)$$

Five such generators that have been considered are:

a	k
5^{13}	36, 39
5^{17}	40, 42, 43

For further discussion of multiplicative congruential generators, see Exercise 3.31.

When one first encounters the idea of a sequence of 'random' numbers cycling, this is disturbing. However, it is put in perspective by Wichmann and Hill (1982a), who present a generator, which we shall discuss later, with a cycle length greater than 2.78×10^{13}. As they remark, if one used 1000 of these numbers a second, it would take more than 800 years for the sequence to repeat!

Large cycle lengths do not necessarily result in sequences of 'good' pseudo-random numbers, and a third aim in the choice of a, b, m is to try to produce a small correlation between successive numbers in the series; for truly random numbers, successive numbers are uncorrelated, but we can see that this is not likely to be the case for a generator such as (3.2). Greenberger (1961) has shown that an approximation to the correlation between x_n and x_{n+1} is given by:

$$\rho \approx \frac{1}{a} - \frac{6b}{am}\left(1 - \frac{b}{m}\right) \pm \frac{a}{m} \tag{3.3}$$

Greenberger gives the following two examples of sequences with the same full cycle length:

	a	b	m	ρ
(i)	$2^{34} + 1$	1	2^{35}	0.25
(ii)	$2^{18} + 1$	1	2^{35}	$\ll 2^{-18}$

Expressions such as (3.3) are obtained by averaging over one complete cycle of a full-period mixed generator (cf. Exercise 3.13) and exact formulae for ρ, involving Dedekind sums, are presented by Kennedy and Gentle (1980, p. 140). As is discussed by Kennedy and Gentle, and also by Knuth (1981, p. 84), choosing a, b and m to ensure small ρ can result in a poor generator in other respects. For instance, for sequences that are much shorter than the full cycle, the correlation between x_n and x_{n+1} may be appreciably higher than the value of ρ for the complete cycle. Also, higher-order correlations may be far too high; see Coveyou and MacPherson (1967) and Van Gelder (1967) for further discussion. It is sometimes recommended that one takes $a \approx \sqrt{m}$ (see e.g., Cooke, Craven and Clarke, 1982, p. 69). However, this approximate relationship holds for the RANDU generator, originally used by IBM and, as we can see from Exercise 3.25, this generator possesses a rather glaring Achilles heel. Unfortunately, as we shall see in Chapter 6, this is a defect which can be hard to detect using standard empirical tests. As a further example, $a \approx \sqrt{m}$ for the generator of Exercise 3.31 (ii), which passes the randomness tests of Downham and Roberts (1967) yet has since been shown to have undesirable properties by Atkinson (1980). Similar findings for this generator and that of Exercise 3.31 (i) are given by Grafton (1981).

The choice of the constants a, b and m is clearly a difficult one, but the convenience of pseudo-random number generators has made the search for good generators worth while. Ultimately, the properties of any generator will be judged by the use intended for the numbers to be generated, and by the tests applied. A very important feature of congruential generators, which is perhaps inadequately emphasized, is that the arithmetic involved in operating the formula (3.2) is exact, without any round-off error. Thus naïve programming of the formula (3.2) in, say, BASIC can rapidly result in a sequence of unknown properties, because of the use of floating-point arithmetic; this feature is clearly illustrated in Exercise 3.14. This problem is usually solved in computer implementations by machine-code programs which employ integer arithmetic.

In this case the modulus operation can be performed automatically, without division, if the modulus, $m = 2^r$, and r is the computer word size: after $(ax_{i-1} + b)$ is formed, then only the r lowest-order bits are retained; this is the integer 'overspill or carry-out' feature described by Kennedy and Gentle (1980, p. 19).

3.6 Further properties of congruential generators

One might well expect numbers resulting from the formula (3.2) to have unusual dependencies and that this is so is seen from the following illustration:

Let
$$x_{i+1} = 5x_i \pmod{m}$$

Here
$$x_{i+1} = 5x_i - h_i m \tag{3.4}$$

in which h_i takes one of the values, 0, 1, 2, 3, 4. Thus pairs of successive values, (x_i, x_{i+1}) give the Cartesian co-ordinates of points which lie on just one of the five lines given by (3.4), and the larger m is, the longer the sequence of generated numbers will remain on any one of these lines before moving to another line. For example, if $x_0 = 1$, $m = 11$, then

$$x_1 = 5, x_2 = 3, x_3 = 4, x_4 = 9, x_5 = 1$$

and the line used changes with each iteration.
However, if $x_0 = 1$, $m = 1000$, then

$$x_1 = 5, x_2 = 25, x_3 = 125, x_4 = 625, x_5 = 125$$

and the sequence $x_1 \rightarrow x_4$ is obtained from the line

$$x_{i+1} = 5x_i$$

after which the sequence degenerates into a simple alternation pairs of successive values give points which lie on a limited number of straight lines, triplets of successive values lie on a limited number of planes, and so on (see Exercise 3.25).

The mixed congruential generator

$$x_{n+1} = 781 x_n + 387 \pmod{1000} \tag{3.5}$$

has cycle length 1000. Figure 3.2 illustrates a plot of u_{n+1} vs. u_n for a sequence of length 500, where $u_i = x_i/1000$, for $0 \le i \le 999$.

The striking sparseness of the points is because of the small value of m used here, which also allows us to see very clearly the kind of pattern which can arise. Thus many users prefer to modify the output from congruential generators before use. One way to modify the output is to take numbers in groups of size g, say, and then 'shuffle' them, by means of a permutation, before use. The permutation used may be fixed, or chosen at random when required. Andrews *et al.* (1972) used such an approach with $g = 500$, while Egger (1979)

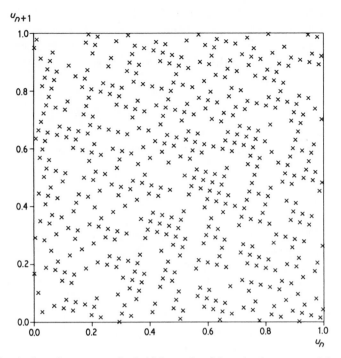

Figure 3.2   A plot of u_{n+1} vs. u_n for half the cycle of the mixed congruential generator of Equation (3.5).

used $g = 100$, a choice also investigated by Atkinson (1980). Page (1967) discusses the construction of random permutations, while tables of these are provided by Moses and Oakford (1963). See also the IMSL routine GGPER described in Appendix 1. An alternative approach, due to MacLaren and Marsaglia (1965) is to have a 'running' store of g numbers from a congruential generator, and to choose which of these numbers to use next by means of a random indicator digit from the range 1 to g, obtained, say, by a separate congruential generator. The gap in the store is then filled by the next number from the original generator, and so on. When this is done for the sequence resulting in Fig. 3.2, we obtain the plot of Fig. 3.3.

For further discussion, see Chambers (1977, p. 173) and Nance and Overstreet (1978). Nance and Overstreet discuss the value of g to be used, and conclude with Knuth (1981, p. 31) that for a good generator, shuffling is often not needed. On the other hand, shuffling can appreciably improve even very poor generators, as demonstrated by Atkinson (1980), a point which is also made in Exercise 3.26. The IMSL routine GGUW employs shuffling with $g = 128$; see Section A1.1 in Appendix 1.

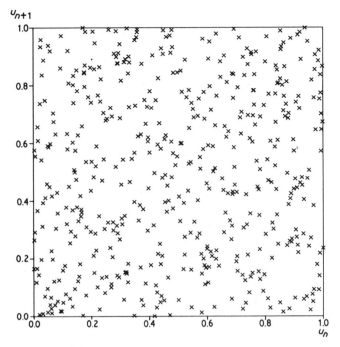

Figure 3.3 The plot resulting from modifying the same sequence that gave rise to Fig. 3.2. The modification entailed choosing the next number 'at random' from a store of length $g = 20$ of numbers from the original sequence, as explained in the text. In this example the random selection was made using Equation (3.1) and a seed of 0.5.

Successive digits in a decimal expansion of a truly random $U(0, 1)$ variable may, as we saw in Section 3.2, be used as uniform random digits. However, this approach is unwise in the case of pseudo-random $U(0, 1)$ variables because of the pattern effects which may arise (see the solution to Exercise 3.21). A disadvantage of congruential generators with $m = 2^k$ is that the low-order bits of generated numbers have short cycles (e.g. Atkinson, 1980). This is not a problem if m is prime (see Exercise 3.31) but then the arithmetic of the method is much more time-consuming on a 'binary' computer than if $m = 2^k$. Ways of reducing computational effort when m is prime are referenced by Law and Kelton (1982, p. 226).

3.7 Alternative methods of pseudo-random number generation

A variety of other methods of pseudo-random number generation exist—see for example, Andrews (1977, p. 170), O'Donovan (1979, p. 33), Law and Kelton (1982, p. 230), Tausworthe (1965) and Craddock and Farmer (1971). Miller and

Prentice (1968), for instance, use the third-order recurrence

$$x_j = x_{j-2} + x_{j-3} \quad (\mathrm{mod}\ p)$$

in which p is a suitable prime.

As with the congruential methods considered above, it is possible here also to examine the theoretical properties of the resulting sequence (cf. Exercise 3.26).

Different computers have different word-lengths (see Kennedy and Gentle 1980, p. 8), which determine the value of the modulus, m, used in congruential generators. This has resulted in machine-dependent generators, which is undesirable, as it makes it difficult to reproduce results, a positive feature of using pseudo-random numbers. *Portable* generators can result from representing 'large' integers by means of a number of 'short' word-length integers; see also Kral (1972) and Roberts (1982). An alternative approach is given by Wichmann and Hill (1982a, b), who combine three simple multiplicative congruential generators in such a way that the overall cycle-length is the product of the individual cycle lengths (see Exercises 3.17 and 3.18). The result is a portable generator with a cycle length greater than 2.78×10^{13}. As well as providing FORTRAN and Ada listings for their algorithm, they also provide an 82-step program for the Hewlett Packard HP-67 hand-calculator.

3.8 Summary and discussion

The building-blocks of simulation are $U(0, 1)$ random variables, and random digits. We have seen that these may be obtained by the use of physical devices, or arithmetic formulae, and that no method is without its drawbacks. Large-scale simulations take place on computers, for which arithmetic formulae provide the most convenient approach. While any formula may seem to be adequate, and produce reasonable-looking numbers, there is always the danger that the formula could break down at some stage. The advantage of congruential generators is that they can be shown to possess certain desirable features, and to give guaranteed cycle lengths. There is always a chance, however, that because the numbers are pseudo-random, and not truly random, unwanted effects could still arise in any particular application. The answer is clearly to proceed with caution, and to make regular checks for oddities. Certain early generators were blatantly unsuitable, and the possibility remains that these generators are still in use. Well-used computer packages, such as MINITAB (see Ryan, Joiner and Ryan, 1976) do not always specify the generator they employ, which is clearly undesirable. (Indeed, different implementations of the same package may use different generators.) The same is true of certain widely used microcomputers. Possible pitfalls, as may occur here, can be avoided by the use of portable generators, which may be used on any machine, even a hand-calculator. Kennedy and Gentle (1980, p. 165) report that as many as about 30% of papers in the *Journal of the American Statistical Association* in 1978 employed simulation. In such a climate it is extremely important for research papers to specify the algorithm used, and the tests for randomness employed in their investigation. At best, simulation results should be verified using a *different* generator.

In minimal BASIC there are two statements which relate to the work of this chapter. These are:

10   RANDOMIZE
20   U = RND

The first statement selects a seed in a random fashion, possibly by reference to the current time. If this statement is omitted, the pseudo-random number sequence that is used will always start from the same seed. In the second statement we obtain a realization of a pseudo-random $U(0, 1)$ variable. Both of these statements will occur in programs in later chapters. While the BASIC instructions are as above, the underlying method used will vary from machine to machine, and on many microcomputers a slightly different form from RND is used.

The bibliographies by Sowey (1972, 1978) reveal that random number generation is a wide field of continuing interest. While new generators of proven improved properties may be developed in the future, congruential generators are likely to continue to prove popular and convenient. The need to test random numbers cannot be stressed too strongly, and this is a subject to which we shall return in Chapter 6. We shall now, in Chapters 4 and 5, proceed to see how uniform random numbers may be changed to give random variables of any kind.

*3.9   Proof of the attainment of a full cycle for a particular mixed congruential pseudo-random number generator

In the following, $a, b, c, k, s, t, \alpha, \gamma, \theta, \phi, h_1, h_2$ and h_3 denote positive integers.

THEOREM 3.1
The mixed congruential generator

$$x_{n+1} = ax_n + b \quad (\text{mod } m)$$

with $a = 4c + 1$, b odd and $m = 2^k$, has cycle length m.

PROOF
The basis of the proof is to show that if $x_i = x_j$, for $i \neq j$, then we cannot have $|i - j| < m$. As the cycle length is $\leq m$, then this will prove that the cycle length is m, and the sequence generated within a single cycle is a permutation of the integers from 0 to $(m - 1)$. Without loss of generality, therefore, we shall take $x_0 = 0$, as this simplifies matters.

First of all, note that $x_n = y_n \pmod{m}$,

where $\qquad\qquad y_{n+1} = ay_n + b, \qquad \text{for } n \geq 0$ $\qquad\qquad$ (3.6)

and, by the above, $y_0 = 0$.

From (3.6) we see that

$$y_n = b(1 + a + a^2 + \ldots + a^{n-1}) \qquad \text{for } n \geq 0$$

Now, $x_i = y_i - h_1 2^k$
and if $x_i = x_j$ for some $i > j$, say, then

$$b(a^j + a^{j+1} + \ldots + a^{i-1}) = h_2 2^k$$

i.e., $$ba^j(1 + a + a^2 + \ldots + a^{i-j-1}) = h_2 2^k \qquad (3.7)$$

Let us write $w_n = 1 + a + \ldots + a^{n-1} \qquad \text{for } n \geq 1.$

In (3.7), by definition, a and b are odd, and so to prove the theorem we must show that:

$$w_{(i-j)} \neq h_3 2^k \qquad \text{for } (i-j) < 2^k \qquad (3.8)$$

and this we shall now do.

THE CASE $(i-j)$ ODD

If $(i-j)$ is odd, we can write $(i-j) = 2t+1$, say, for $t \geq 0$.

$$w_{2t} = \left(\frac{1-a^{2t}}{1-a}\right) = \{(1+4c)^{2t} - 1\}/4c$$

$$= \{(1+4c)^t - 1\}\{(1+4c)^t + 1)/4c\}$$

$$= \{(1+4c)^t + 1\}\left\{\sum_{i=1}^{t} (4c)^{i-1}\binom{t}{i}\right\},$$

which is even, as $$1 + (1+4c)^t = 2 + 4c\sum_{i=1}^{t}\binom{t}{i}(4c)^{i-1} \qquad (3.9)$$

Thus $w_{2t+1} = w_{2t} + a^{2t+1}$ is odd, as a is odd, and so (3.8) is trivially true.

THE CASE $(i-j)$ EVEN

If $(i-j)$ is even, there exists an s such that $(i-j) = \alpha 2^s$, for some odd, positive integral α, and as $(i-j) < 2^k$, then $s < k$.

$$w_{(i-j)} = w_{\alpha 2^s} = 1 + a + \ldots + a^{\alpha 2^{s-1}-1} + a^{\alpha 2^{s-1}} + \ldots + a^{\alpha 2^s - 1}$$

$$= w_{\alpha 2^{s-1}} + a^{\alpha 2^{s-1}} w_{\alpha 2^{s-1}}$$

$$= w_{\alpha 2^{s-1}}(1 + a^{\alpha 2^{s-1}}) \qquad (3.10)$$

$$= w_{\alpha 2^{s-2}}(1 + a^{k_1})(1 + a^{k_2})$$

$$\vdots$$

$$= w_{\alpha}(1 + a^{k_1})(1 + a^{k_2}) \ldots (1 + a^{k_s}),$$

for suitable positive integers, $k_1, k_2, \ldots, k_s$.

Since $a = (1 + 4c)$, we have, from (3.9)

$$w_{\alpha 2^s} = w_\alpha \gamma 2^s,$$

in which w_α and γ are odd positive integers. (We have just proved that w_α is odd if α is odd.) Hence as $s < k$, there does not exist an h_3 such that

$$w_{\alpha 2^s} = h_3 2^k$$

This completes the proof. We note, finally, that it is simple to verify $x_0 = x_m$, since $(x_0 - x_m) = bw_m$, where $m = 2^k$. $w_2 = 1 + a = 2 + 4c$, so for $k = 1$, result (3.7) is true. Let us suppose that

$$w_m = \theta m, \text{ for } m = 2^k \text{ and } k \geq 1 \qquad (3.11)$$

$$w_{2m} = \theta m(1 + a^m), \qquad \text{from (3.10)}$$

and $(1 + a^m)$ is even, from (3.9).

Hence $w_{2m} = \phi(2m)$, and if (3.11) is true for $k \geq 1$, then it is also true for $(k + 1)$. We have seen that it is true for $k = 1$, and so by induction it is true for all $k \geq 1$.

3.10   Exercises and complements

(a) Uses of random numbers

The randomized response technique (RRT) has been much studied and extended. A good introduction is provided by Campbell and Joiner (1973), who motivate the first four questions.

†3.1   Investigate the workings of the RRT when the two alternative questions are:

(i) I belong to group X;
(ii) I do not belong to group X.

3.2   Describe how you would proceed if the proportion of positive responses to the RRT 'innocent' question is unknown. Can you suggest an innocent question for which it should be possible to obtain the proportion of correct responses without difficulty?

3.3   Investigate the RRT when the randomizing device is a bag of balls, each being one of *three* different colours, say red, white and blue, and the instructions to the (female) respondents are:

If the red ball is drawn answer the question: 'have you had an abortion'.
If the white ball is drawn, respond 'Yes'.
If the blue ball is drawn, respond 'No'.

*3.4   In RRT, consider the implications for the respondent of responding 'Yes', even though it is not known which question has been answered. Consider how the technique might be extended to deal with frequency of activity. Consider how to construct a confidence interval (see, e.g., *ABC*, p. 266) for the estimated proportion.

3.5   100 numbered pebbles formed the population in a sampling experiment devised by J. M. Bremer. Students estimate the population mean weight ($\mu = 37.63$ g) by selecting 10 pebbles at random, using tables of random numbers, and additionally by choosing a sample of 10 pebbles, using their judgement only. The results obtained from a class of 32 biology undergraduates are given below:

Judgement sample means	Random sample means
62.63	31.45
35.85	32.12
55.36	51.93
66.43	24.74
34.96	43.32
37.23	29.41
34.45	42.67
60.53	47.94
49.61	28.76
56.07	56.43
59.02	31.21
50.65	32.73
33.34	55.37
58.62	36.65
47.02	22.44
48.34	40.04
28.56	44.65
26.65	41.43
46.34	39.39
27.86	26.39
39.62	23.88
25.45	35.15
48.82	35.88
66.56	28.03
37.25	31.71
45.98	43.98
32.46	61.49
54.03	31.52
51.89	33.99
62.81	33.78
59.74	49.69
14.05	22.97

Discuss, with reference to these data, the importance of taking random samples.

(b) On uniform random digits

3.6   A possible way of using two unbiased dice for simulating uniform random digits from 0 to 9 is as follows: throw the two dice and record the sum. Interpret 10 as 0, 11 as 1, and ignore 12. Discuss this procedure. (Based on part of an A-level examination question: Oxford, 1978.)

3.7   In Example 3.2 we used a fair coin to simulate events with probability different from 0.5. Here we consider the converse problem (the other side of the coin). Suppose you want to simulate an event with probability $\frac{1}{2}$; you have a coin but you suspect it is biased. How should you proceed? One approach is this: toss the coin twice. If the results of the two tosses are the same, repeat the experiment, and carry on like this until you obtain two tosses that are different. Record the outcome of the second toss. Explain why this procedure produces equi-probable outcomes. Discussion and extensions to this simple idea are given in Dwass (1972) and Hoeffding and Simons (1970).

3.8   In a series of 10 tosses of two distinguishable fair dice, A and B, the following faces were uppermost (A is given first in each case): (1, 4), (2, 6), (1, 5), (4, 3), (2, 2), (6, 3), (4, 5), (5, 1), (3, 4), (1, 2).
Explain how you would use the dice to generate uniformly distributed random numbers in the range 0000–9999. (Based on part of an A-level examination question: Oxford, 1980.)

†**3.9**   British car registration numbers are of the form: SHX 792R. Special rôles are played by the letters, but that is not, in general, true of the numbers. Collect 1000 digits from observing car numbers, and examine these digits for randomness (explain how you deal with numbers of the form: HCY 7F).

3.10   Below we give the decimal expansion of π to 2500 places, kindly supplied by T. Hopkins. Draw a bar-chart to represent the relative frequencies of some (if not all!) of these digits, and comment on the use of these digits as uniform random 0–9 digits. Note that Fisher and Yates (1948) adopted a not dissimilar approach, constructing random numbers from tables of logarithms; further discussion of their numbers is given in Exercise 6.8(ii).

3.1415926535 8979323846 2643383279 5028841971 6939937510
5820974944 5923078164 0628620899 8628034825 3421170679
8214808651 3282306647 0938446095 5058223172 5359408128
4811174502 8410270193 8521105559 6446229489 5493038196
4428810975 6659334461 2847564823 3786783165 2712019091
4564856692 3460348610 4543266482 1339360726 0249141273
7245870066 0631558817 4881520920 9628292540 9171536436
7892590360 0113305305 4882046652 1384146951 9415116094
3305727036 5759591953 0921861173 8193261179 3105118548
0744623799 6274956735 1885752724 8912279381 8301194912

9833673362 4406566430 8602139494 6395224737 1907021798
6094370277 0539217176 2931767523 8467481846 7669405132
0005681271 4526356082 7785771342 7577896091 7363717872
1468440901 2249534301 4654958537 1050792279 6892589235
4201995611 2129021960 8640344181 5981362977 4771309960
5187072113 4999999837 2978049951 0597317328 1609631859
5024459455 3469083026 4252230825 3344685035 2619311881
7101000313 7838752886 5875332083 8142061717 7669147303
5982534904 2875546873 1159562863 8823537875 9375195778
1857780532 1712268066 1300192787 6611195909 2164201989

3809525720 1065485863 2788659361 5338182796 8230301952
0353018529 6899577362 2599413891 2497217752 8347913151
5574857242 4541506959 5082953311 6861727855 8890750983
8175463746 4939319255 0604009277 0167113900 9848824012
8583616035 6370766010 4710181942 9555961989 4676783744
9448255379 7747268471 0404753464 6208046684 2590694912
9331367702 8989152104 7521620569 6602405803 8150193511
2533824300 3558764024 7496473263 9141992726 0426992279
6782354781 6360093417 2164121992 4586315030 2861829745
5570674983 8505494588 5869269956 9092721079 7509302955

3211653449 8720275596 0236480665 4991198818 3479775356
6369807426 5425278625 5181841757 4672890977 7727938000
8164706001 6145249192 1732172147 7235014144 1973568548
1613611573 5255213347 5741849468 4385233239 0739414333
4547762416 8625189835 6948556209 9219222184 2725502542
5688767179 0494601653 4668049886 2723279178 6085784383
8279679766 8145410095 3883786360 9506800642 2512520511
7392984896 0841284886 2694560424 1965285022 2106611863
0674427862 2039194945 0471237137 8696095636 4371917287
4677646575 7396241389 0865832645 9958133904 7802759009

9465764078 9512694683 9835259570 9825822620 5224894077
2671947826 8482601476 9909026401 3639443745 5305068203
4962524517 4939965143 1429809190 6592509372 2169646151
5709858387 4105978859 5977297549 8930161753 9284681382
6868386894 2774155991 8559252459 5395943104 9972524680
8459872736 4469584865 3836736222 6260991246 0805124388
4390451244 1365497627 8079771569 1435997700 1296160894
4169486855 5848406353 4220722258 2848864815 8456028506
0168427394 5226746767 8895252138 5225499546 6672782398
6456596116 3548862305 7745649803 5593634568 1743241125

(c) On pseudo-random numbers

3.11   The first pseudo-random number generator was the 'mid-square' proposed by von Neumann (1951). The method is as follows: select a large integer, e.g. 7777. Square it and use the middle four digits as the next integer, square that, and so on. Here we get:

$$7777 \rightarrow 60\,\underline{481}\,729 \rightarrow 4817 \rightarrow 23\,\underline{203}\,489 \rightarrow 2034 \rightarrow 4\,\underline{137}\,156$$
$$\rightarrow 1371 \rightarrow 1\,\underline{879}\,641 \rightarrow \text{etc.}$$

The above sequence illustrates how we proceed when the squared number does not fill the entire possible field-length of 8. Investigate and comment upon this procedure. Further discussion is provided by Tocher (1975, p. 72) and Knuth (1981, p. 3), who explain the problems that can arise with this method. Craddock and Farmer (1971) provide a modification which avoids the obvious degeneration when the process results in zero.

†3.12   Investigate sequences produced by:

$$u_{n+1} = \text{fractional part of } (\pi + u_n)^5.$$

3.13   Show that for a full-period mixed congruential generator the mean and variance of the values produced by dividing each integer element in the full-period sequence by the modulus m are, respectively, $\frac{1}{2}(1 - 1/m)$, and $(1 + 1/m)/12$.

†3.14   The following BASIC program simulates the mixed congruential generator of Equation (3.4), with $a = 781, b = 387, m = 1000$. Run this

```
10   REM MIXED CONGRUENTIAL GENERATOR
20   INPUT U0
30   LET A = 781
40   LET B = 387
50   FOR I = 1 TO 1000
60     LET U1 = (A*U0+B)/1000
70     LET U1 = (U1-INT(U1))*1000
80     LET U0 = U1
90     PRINT U1
100  NEXT I
110  END
```

program with and without the following change (from Cooke, Craven and Clarke, 1982, p. 70):

   75   U1 = INT(U1 + 0.5)

Comment on the results and the reason for using this additional line.

3.15   In pseudo-random number generation using congruential methods we obtain a sequence of integers $\{x_i\}$ over the range $(0, m)$. Approximations to $U(0, 1)$ random variables are then obtained by setting $u_i = x_i/m$. Show that

$$u_{i+1} = (au_i + b/m) \pmod 1$$

and program this in BASIC (note the lesson of Exercise 3.14 with regard to round-off error). Note also the comments of Knuth (1981, p. 525).

3.16 Show that for a mixed congruential generator, with $a > 1$,

$$x_{n+k} = [a^k x_n + (a^k - 1)b/(a-1)] \pmod{m} \qquad \text{for } k \geq 0, n \geq 0$$

This property is useful in distributed array processing (DAP) programming of congruential generators (Sylwestrowicz, 1982).

*3.17 Show that if U_1 and U_2 are independent $U(0, 1)$ random variables, then the fractional part of $(U_1 + U_2)$ is also $U(0, 1)$. Show further that this result still holds if U_1 is $U(0, 1)$, but U_2 has any continuous distribution.

3.18 (*continuation*) Show that if U_1, U_2 and U_3 are formed independently from congruential generators with respective cycle lengths c_1, c_2 and c_3, then we may take the fractional part of $(U_1 + U_2 + U_3)$ as a realization of a pseudo-random $U(0, 1)$ random variable, and the resulting sequence of $(0, 1)$ variables will have cycle length $c_1 c_2 c_3$ if c_1, c_2 and c_3 are relatively prime. For further discussion, see Neave (1972, p. 6) and Wichmann and Hill (1982a).

3.19 In a mixed congruential generator, show that if $m = 10^k$ for some positive integer $k > 1$, then for the cycle length to equal m, we need to set $a = 20d + 1$, where d is a positive integer.

3.20 Show that the sequence $\{x_i\}$ of Section 3.5, for which $x_0 = 89, x_1 = 16$, etc., alternates between even and odd numbers.

†3.21 (a) (Peach, 1961) The mixed congruential generator,

$$x_{n+1} = 9x_n + 13 \pmod{32}$$

has full (32) cycle length. Write down the resulting sequence of numbers and investigate it for patterns. For example, compare the numbers in the first half with those in the second half, write the numbers in binary form, etc.

(b) Experiment with congruential generators of your own.

3.22 (a) Write BASIC programs to perform random shuffling and random replacement of pseudo-random numbers. When might these two procedures be equivalent?

(b) (Bays and Durham, 1976) We may use the next number from a congruential generator to determine the random replacement. Investigate this procedure for the generator

$$x_{n+1} = 5x_n + 3 \pmod{16}; x_0 = 1.$$

3.23   A distinctly non-random feature of congruential pseudo-random numbers is that no number appears twice within a cycle. Suggest a simple procedure for overcoming this defect.

3.24   Construct a pseudo-random number generator of your own, and evaluate its performance.

[†]**3.25**   The much-used IBM generator RANDU is multiplicative congruential, with multiplier 65 539, and modulus 2^{31}, so that the generated sequence is:

$$x_{i+1} = 65\,539 x_i \quad (\text{mod } 2^{31})$$

Use the identity $65\,539 = 2^{16} + 3$ to show that

$$x_{i+1} = (6x_i - 9x_{i-1}) \quad (\text{mod } 2^{31}),$$

and comment on the behaviour of successive triplets (x_{i-1}, x_i, x_{i+1}). See also Chambers (1977, p. 191), Miller (1980a, b) and Kennedy and Gentle (1980, p. 149) for further discussion of this generator. Examples of plots of triplets are to be found in Knuth (1981, p. 90) and Oakenfull (1979).

3.26   (a)   The Fibonacci series may be used for a pseudo-random number generator:

$$x_{n+1} = (x_n + x_{n-1}) \quad (\text{mod } m)$$

Investigate the behaviour of numbers resulting from such a series. See Wall (1960) for an investigation of cycle-length when $m = 2^k$.

(b)   (Knuth, 1981)   In a random sequence of numbers, $0 \le x_i < m$, how often would you expect to obtain $x_{n-1} < x_{n+1} < x_n$? How often does this sequence occur with the generator of (a)? The Fibonacci series above is generally held to be a poor generator of pseudo-random numbers, but its performance can be much improved by shuffling (see Gebhardt, 1967). Oakenfull (1979) has obtained good results from the series

$$x_{n+1} = (x_n + x_{n-97}) \quad (\text{mod } 2^{35})$$

Note that repeated numbers can occur with Fibonacci-type generators (cf. Exercise 3.23).

3.27   (a)   For a multiplicative congruential generator, show that if a is an odd power of $8n \pm 3$, for any suitable integral n, and x_0 is odd, then all subsequent members of the congruential series are odd.

(b)   As we shall see in Chapter 5, it is sometimes necessary to form $\log_e U$, where U is $U(0, 1)$. Use the result of (a) to explain the advantage of such a multiplicative generator over a mixed congruential generator in such a case.

*3.28 Consider how you would write a FORTRAN program for a congruential generator.

*3.29 (Taussky and Todd, 1956)   Consider the recurrence

$$y_{n+1} = y_n + y_{n-1} \qquad \text{for } n \geq 1,$$

with $y_0 = 0$, $y_1 = 1$.
Show that

$$y_n = \left\{ \left(\frac{\sqrt{5}+1}{2} \right)^n - \left(\frac{1-\sqrt{5}}{2} \right)^n \right\} \bigg/ \sqrt{5}$$

and deduce that for large n,

$$y_n \approx \left(\frac{\sqrt{5}+1}{2} \right)^n \bigg/ \sqrt{5}$$

Hence compare the Fibonacci series generator of Exercise 3.26 with a multiplicative congruential generator. Difference equations, such as that above, occur regularly in the theory of random walks (see Cox and Miller, 1965, Section 2.2).

3.30 The literature abounds with congruential generators. Discuss the choice of a, b, m, in the following. For further considerations, see Kennedy and Gentle (1980, p. 141) and Knuth (1981, p. 170).

(i) $\quad a \quad\quad b \quad\quad m$
$\quad\;\; 7^5 \quad\; 0 \quad\; 2^{31} - 1$

Called GGL, this is IBM's replacement for RANDU (see Learmonth and Lewis, 1973). Egger (1979) used this generator in combination with shuffling from a $g = 100$ store, and it is the basis of routines GGUBFS and GGUBS of the IMSL library; see Section A1.1 in Appendix 1.

(ii) $\quad\;\; a \quad\quad\quad b \quad\quad\;\; m$
$\quad\;\;\; 16333 \quad 25887 \quad 2^{15} \quad$ (from Oakenfull, 1979)

(iii) $\quad\; a \quad\quad\;\; b \quad\quad\;\; m$
$\quad\;\;\; 3432 \quad 6789 \quad 9973 \quad$ (see also Oakenfull, 1979)

(iv) $\quad\; a \quad\quad b \quad\quad\;\; m$
$\quad\;\;\; 23 \quad\; 0 \quad\; 10^8 + 1 \quad$ the Lehmer generator

This generator is of interest as it was the first proposed congruential generator, with $x_0 = 47\,594\,118$, by Lehmer (1951).

(v) The NAG generator: GO5CAF:

a b m
13^{13} 0 2^{59}

See Section A1.1

(vi) a b m
171 0 30269

This is one of the three component generators used by Wichmann and Hill (1982a, b).

(vii) a b m
131 0 2^{35} used by Neave (1973).

(viii) a b m
$2^7 + 1$ 1 2^{35}

This generator is of interest as it is one of the original mixed congruential generators, proposed by Rotenberg (1960).

(ix) a b m
397 204 094 0 $2^{31} - 1$

This is the routine GGUBT of the IMSL library—see Section A1.1

3.31 Show that the cycle length in a multiplicative congruential generator is given by the smallest positive integer n satisfying $a^n = 1 \bmod(m)$. (See Exercise 3.16.)

We stated in Section 3.5, that if $m = 2^k$ in a multiplicative congruential generator, only one-quarter of the integers 0–m are obtained in the generator cycle. However, if m is a prime number then a cycle of length $(m - 1)$ can be obtained with multiplicative congruential generators. Let $\phi(m)$ be the number of integers less than and prime to m, and suppose m is a prime number, p. Clearly, $\phi(p) = (p - 1)$.

It has been shown (Tocher, 1975, p. 76) that the n above must divide $\phi(m)$. If $n = \phi(p)$ then a is called a 'primitive root' mod (p), and the cycle length $(p - 1)$ is attained. Ways of identifying primitive roots of prime moduli are given by Downham and Roberts (1967); for example, 2 is a primitive root of p if $(p - 1)/2$ is prime and $p = 3 \bmod (8)$. Given a primitive root r, then further primitive roots r may be generated from: $r = r^k \bmod(p)$, where k and $(p - 1)$ are co-prime.

Use these results to verify that the following 5 prime modulus multiplicative congruential generators, considered by Downham and Roberts (1967), have cycle length $(p - 1)$.

	$m = p$	a
(i)	67 101 323	8 192
(ii)	67 099 547	8 192
(iii)	16 775 723	32 768
(iv)	67 100 963	8
(v)	7 999 787	32

Extensions and further discussion are given by Knuth (1981, pp. 19–22) and Fuller (1976), while relevant tables are provided by Hauptman *et al.* (1970) and Western and Miller (1968).

*3.32 If a fair coin is tossed until there are two consecutive heads, show that the probability that n tosses are required is

$$p_n = y_{n-1}/2^n \qquad \text{for } n \geq 2$$

where the y_n are given by the Fibonacci numbers of Exercise 3.29 (cf. Exercise 3.26). We see from Exercise 3.29 that as $n \to \infty$, the ratio $y_n/y_{n-1} \to$ the golden ratio, $\phi = (1 + \sqrt{5})/2$, so that the tail of the distribution is approximately geometric, with parameter $\phi/2$. Mead and Stern (1973) suggest uses of this problem in the empirical teaching of statistics. Verify that the distribution above has mean 6.

4

PARTICULAR METHODS
FOR NON-UNIFORM
RANDOM VARIABLES

Some of the results of Chapter 2 may be used to convert uniform random variables into variables with other distributions. It is the aim of this chapter to provide some examples of such particular methods for simulating non-uniform random variables. Because of the important rôle played by the normal distribution in statistics, we shall start with normally distributed random variables.

4.1   Using a central limit theorem

It is because of central limit theorems that the normal distribution is encountered so frequently, and forms the basis of much statistical theory. It makes sense, therefore, to use a central limit theorem in order to simulate normal random variables. For instance, we may simulate n independent $U(0, 1)$ random variables, $U_1, U_2, \ldots, U_n$, say, and then set $N = \Sigma_{i=1}^{n} U_i$. As $n \to \infty$ the distribution of N tends to that of a normal variable. But in practice, of course, we settle on some finite value for n, so that the resulting N will only be approximately normal. So how large should we take n? The case $n = 2$ is unsuitable, as N then has a triangular distribution (see Exercise 4.8), but for $n = 3$, the distribution of N is already nicely 'bell-shaped', as will be shown later. The answer to this question really depends on the use to which the resulting numbers are to be put, and how close an approximation is desired.

A convenient number to take is $n = 12$, since, as is easily verified, $\mathscr{E}[U_i] = \frac{1}{2}$ and $\text{Var}[U_i] = 1/12$, so that then

$$N = \sum_{i=1}^{12} U_i - 6$$

is an approximately normal random variable with mean zero and unit variance. Values of $|N| > 6$ do not occur, which could, conceivably, be a problem for large-scale simulations. The obvious advantage of this approach, however, is its simplicity; it is simple to understand, and simple to program, as we can see from Fig. 4.1.

```
10    RANDOMIZE                        10    RANDOMIZE
20    INPUT M                          20    INPUT M
30    REM PROGRAM TO SIMULATE 2*M      30    REM PROGRAM TO SIMULATE 2*M
40    REM APPROXIMATELY STANDARD       40    REM STANDARD NORMAL RANDOM VARIABLES
50    REM NORMAL RANDOM VARIABLES      50    REM USING THE BOX-MULLER METHOD
60    REM USING A CENTRAL LIMIT        60    LET P2 = 2*3.14159265
70    REM THEOREM APPROACH             70    FOR I = 1 TO M
80    FOR I = 1 TO 2*M                 80      LET R = SQR(-2*LOG(RND))
90      LET N = 0                      90      LET U = RND
100     FOR J = 1 TO 12               100      PRINT R*SIN(P2*U),R*COS(P2*U)
110       LET N = N+RND               110    NEXT I
120     NEXT J                        120    END
130     PRINT N-6
140   NEXT I
150   END
```

Figure 4.1   BASIC programs for simulating $2M$ standard normal random variables.

4.2   The Box–Müller and Polar Marsaglia methods

4.2.1   The Box–Müller method

The last method used a convolution to provide approximately normal random variables. The next method we consider obtains exact normal random variables by means of a one-to-one transformation of two $U(0, 1)$ random variables. If U_1 and U_2 are two independent $U(0, 1)$ random variables then Box and Müller (1958) showed that

$$N_1 = (-2 \log_e U_1)^{1/2} \cos(2\pi U_2)$$

and $\qquad\qquad N_2 = (-2 \log_e U_1)^{1/2} \sin(2\pi U_2)$ $\qquad$ (4.1)

are independent $N(0, 1)$ random variables.

At first sight this result seems quite remarkable, as well as most convenient. It is, however, a direct consequence of the result of Example 2.4, as we shall now see.

If we *start* with independent $N(0, 1)$ random variables, N_1 and N_2, defining a point (N_1, N_2) in two dimensions by Cartesian co-ordinates, and we change to polar co-ordinates (R, Θ), then

$$N_1 = R \cos \Theta$$
$$N_2 = R \sin \Theta$$
$\qquad$ (4.2)

and in Example 2.4 we have already proved that R and Θ are then independent random variables, Θ with a $U(0, 2\pi)$ distribution, and $R^2 = N_1^2 + N_2^2$ with a χ_2^2 distribution, i.e. an exponential distribution of mean 2. Furthermore, to simulate Θ we need simply take $2\pi U_2$, where U_2 is $U(0, 1)$, and to simulate R we can take $(-2 \log_e U_1)$, where U_1 is $U(0, 1)$, as explained in Exercises 2.1 and 2.2.

We therefore see that Box and Müller have simply inverted the relationship of (4.2), which goes from (N_1, N_2) of a particular kind to (R, Θ) of a particular kind, and instead move from (R, Θ) to (N_1, N_2), simulating Θ by means of $2\pi U_2$, and an independent R from $(-2 \log_e U_1)^{1/2}$, where U_1 is independent of U_2. You are asked to provide a formal proof of (4.1) in Exercise 4.6. As we can see from Fig. 4.1, this method is also very easy to program, and each method considered so far would be easily operated on a hand-calculator.

If the Box–Müller method were to be used regularly on a computer then it would be worth incorporating the following interesting modification, which avoids the use of time-consuming sine and cosine functions.

*4.2.2   The Polar Marsaglia Method

The way to avoid using trignometric functions is to construct the sines and cosines of uniformly distributed angles directly *without* first of all simulating the angles. This can be done by means of a *rejection* method as follows:

If U is $U(0, 1)$, then $2U$ is $U(0, 2)$, and $V = 2U - 1$ is $U(-1, 1)$.

If we select two independent $U(-1, 1)$ random variables, V_1 and V_2, then these specify a point at random in the square of Fig. 4.2, with polar co-ordinates $(\tilde{R}, \Theta)$ given by:

$$\tilde{R}^2 = V_1^2 + V_2^2$$

and
$$\tan \Theta = V_2/V_1$$

Repeated selection of such points provides a random scatter of points in the square, and rejection of points outside the inscribed circle shown leaves us with a uniform random scatter of points within the circle.

For any one of these points it is intuitively clear (see also Exercise 4.11) that the polar co-ordinates $\tilde{R}$ and Θ are independent random variables, and further that Θ is a $U(0, 2\pi)$ random variable. In addition (see Exercise 4.11) $\tilde{R}^2$ is $U(0, 1)$ and so the pair $(\tilde{R}, \Theta)$ are what are required by the Box–Müller method, and we can here simply write

$$\sin \Theta = \frac{V_2}{\tilde{R}} = V_2(V_1^2 + V_2^2)^{-1/2}$$

$$\cos \Theta = V_1(V_1^2 + V_2^2)^{-1/2}$$

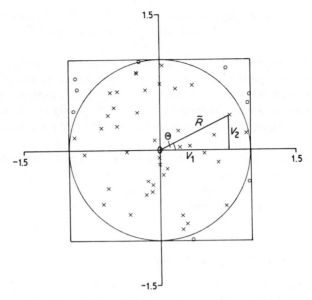

Figure 4.2   An illustration of points (denoted by ∘ and ×) uniformly distributed over the square shown. The points denoted by ∘ are rejected when one simply requires points uniformly distributed over the unit disc, as in the Polar Marsaglia method.

so that a pair of independent $N(0, 1)$ variables, N_1 and N_2, are given by:

$$N_1 = (-2 \log(\tilde{R}^2))^{1/2} V_2 (V_1^2 + V_2^2)^{-1/2}$$

$$N_2 = (-2 \log(\tilde{R}^2))^{1/2} V_1 (V_1^2 + V_2^2)^{-1/2},$$

i.e.

$$N_1 = (-2 \log(V_1^2 + V_2^2))^{1/2} V_2 (V_1^2 + V_2^2)^{-1/2}$$

$$N_2 = (-2 \log(V_1^2 + V_2^2))^{1/2} V_1 (V_1^2 + V_2^2)^{-1/2}$$

resulting in

$$N_1 = V_2 \left(\frac{-2 \log W}{W} \right)^{1/2}$$

$$N_2 = V_1 \left(\frac{-2 \log W}{W} \right)^{1/2}$$

where $W = V_1^2 + V_2^2$.

The philosophy of rejection may seem rather strange at first, as it involves discarding variates obtained at a certain cost and effort, but pairs of variates

(V_1, V_2) are rejected just a proportion $1 - \pi/4$ of the time. The advantage of the rejection method here is that it provides a very simple way of obtaining a uniform scatter of points inside the circle of Fig. 4.2. Another rejection method was described in Example 3.2, and we shall encounter more general rejection methods in the next chapter, which have the same aim and use as here.

A BASIC program for this method is given in Fig. 4.3. Now known as the 'Polar Marsaglia' method, this approach is due originally to Marsaglia and Bray (1964), and is used in the IMSL routine: GGNPM – see Section A1.1.

```
10   RANDOMIZE
20   INPUT M
30   REM PROGRAM TO SIMULATE 2*M STANDARD NORMAL
40   REM RANDOM VARIABLES USING THE POLAR
50   REM MARSAGLIA METHOD
60   FOR I = 1 TO M
70   LET V1 = 2*RND-1
80   LET V2 = 2*RND-1
90   LET R2 = V1*V1+V2*V2
100  IF R2 > 1 THEN 70
110  LET Y = SQR((-2*LOG(R2))/R2)
120  PRINT V1*Y,V2*Y
130  NEXT I
140  END
```

Figure 4.3   BASIC program for simulating $2M$ standard normal random variables, using the Polar Marsaglia method.

4.3   Exponential, gamma and chi-square variates

Random variables with exponential and gamma distributions are frequently used to model waiting times in queues of various kinds, and this is a natural consequence of the predictions of the Poisson process. The simplest way of obtaining random variables with an exponential p.d.f. of e^{-x} for $x \geq 0$ is to set $X = -\log_e U$, where U is $U(0, 1)$, as has already been done in the previous section (see Exercise 2.1).

We have also seen that $Y = X/\lambda$ has the exponential p.d.f. $\lambda e^{-\lambda x}$ for $x \geq 0$ (Exercise 2.2). An alternative approach for simulating exponential random variables will be given later in Exercise 5.35.

We have seen in Section 2.10 that if we have independent random variables, $Y_1, \ldots, Y_n$ with density function $\lambda e^{-\lambda x}$ for $x \geq 0$, then

$$G = \sum_{i=1}^{n} Y_i$$

has a gamma, $\Gamma(n, \lambda)$, distribution. Thus to simulate a $\Gamma(n, \lambda)$ random variable for integral n we can simply set

$$G = -\frac{1}{\lambda} \sum_{i=1}^{n} \log_e U_i$$

where $U_1, \ldots, U_n$ are independent $U(0, 1)$ random variables, i.e.

$$G = -\frac{1}{\lambda} \log_e \left(\prod_{i=1}^{n} U_i \right)$$

Now a random variable with a χ_m^2 distribution is simply a $\Gamma(m/2, \frac{1}{2})$ random variable (see Section 2.10), and so if m is even we can readily obtain a random variable with a χ_m^2 distribution, by the above approach. If m is odd, we can obtain a random variable with a χ_{m-1}^2 distribution by first obtaining a $\Gamma((m-1)/2, \frac{1}{2})$ random variable as above, and then adding to it N^2, where N is an independent $N(0, 1)$ random variable (see Exercise 2.5). Here we are using the defining property of χ^2 random variables on integral degrees of freedom, and use of this property alone provides us with a χ_m^2 random variable from simply setting

$$Z = \sum_{i=1}^{m} N_i^2 \tag{4.3}$$

where the N_i are independent, $N(0, 1)$ random variables. However, because of the time taken to simulate $N(0, 1)$ random variables, this last approach is not likely to be very efficient. Both NAG and IMSL computer packages use convolutions of exponential random variables in their routines for the generation of gamma and chi-square random variables (see Section A1.1). The simulation of gamma random variables with non-integral shape parameter, n, is discussed in Section 4.6.

4.4   Binomial and Poisson variates

4.4.1   Binomial variates

A binomial $B(n, p)$ random variable, X, can be written as $X = \sum_{i=1}^{n} B_i$, where the B_i are independent *Bernoulli* random variables, each taking the values, $B_i = 1$, with probability p, or $B_i = 0$, with probability $(1 - p)$. Thus to simulate such an X, we need just simulate n independent $U(0, 1)$ random variables, $U_1, \ldots, U_n$, and set $B_i = 1$ if $U_i \leq p$, and $B_i = 0$ if $U_i > p$. The same end result can, however, be obtained from judicious re-use of a single $U(0, 1)$ random variable U. Such re-use of uniform variates is employed by the IMSL routine, GGBN when $n < 35$ (see Section A1.1). If $n \geq 35$, a method due to Relles (1972) is employed by this routine: in simulating a $B(n, p)$ variate we simply count how many of the U_i are less than p. If n is large then time can be saved by ordering the $\{U_i\}$ and then observing the location of p within the ordered sample. Thus if we denote the ordered sample by $\{U_{(i)}\}$, for the case $n = 7$, and $p = 0.5$ we might have:

In this example we would obtain $X = 3$ as a realization of a $B(7, \frac{1}{2})$ random variable.

Rather than explicitly order, one can check to see whether the sample median is greater than or less than p, and then concentrate on the number of sample values between p and the median. In the above illustration the sample median is $U_{(4)} > p$, and so we do not need to check whether $U_{(i)} > p$ for $i > 4$. However, we do not know, without checking, whether $U_{(i)} > p$ for any $i < 4$. This approach can clearly now be iterated by seeking the sample median of the sample: $(U_{(1)}, U_{(2)}, U_{(3)}, U_{(4)})$ in the above example, checking whether it is greater or less than p, etc. (see Exercise 4.18). A further short-cut results if one makes use of the fact that sample medians from $U(0, 1)$ samples can be simulated directly using a beta distribution (see Exercise 4.17). Full details are given by Relles (1972), who also provides a FORTRAN algorithm. An alternative approach, for particular values of p only, is as follows.

If we write U in binary form to n places, and if indeed U is $U(0, 1)$, then independently of all other places the ith place is 0 or 1 with probability $\frac{1}{2}$ (cf. Section 3.2). Thus, for example, if $U_1 = 0.10101011100101100$, we obtain 9 as a realization of a $B(17, \frac{1}{2})$ random variable, if we simply sum the number of ones. Here the binary places correspond to the trials of the binomial distribution.

If we want a $B(17, \frac{1}{4})$ random variable, we select further an independent $U(0, 1)$ random variable, U_2, say. If $U_2 = 0.10101101100110101$, then place-by-place multiplication of the digits in U_1 and U_2 gives: 0.10101001100100100, in which 1 occurs with probability $\frac{1}{4}$ at any place after the point. In this illustration we therefore obtain 7 as a realization of a $B(17, \frac{1}{4})$ random variable.

This approach can be used to provide $B(n, p)$ random variables, when we can find m and r so that $p = m2^{-r}$ (see Exercise 4.3). Most people are not very adept at binary arithmetic, but quite efficient algorithms could result from exploiting these ideas if machine-code programming could be used to utilize the binary nature of the arithmetic of most computers. However, as we have seen in Chapter 3, pseudo-random $U(0, 1)$ variables could exhibit undesirable patterns when expressed in binary form.

4.4.2   Poisson variates

Random variables with a Poisson distribution of parameter λ can be generated as a consequence of the following result.

Suppose $\{E_i, i \geq 1\}$ is a sequence of independent random variables, each with an exponential distribution, of density $\lambda e^{-\lambda x}$, for $x \geq 0$. Let $S_0 = 0$ and $S_k = \Sigma_{i=1}^k E_i$ for $k \geq 1$, so that, from Section 4.3, the S_k are $\Gamma(k, \lambda)$ random variables. Then the random variable K, defined implicitly by the inequalities $S_K \leq 1 < S_{K+1}$ has a Poisson distribution with parameter λ. In other words, we set $S_1 = E_1$, and if $1 < S_1$, then we set $K = 0$. If $S_1 \leq 1$, then we set $S_2 = E_1 + E_2$, and then if $S_2 > 1$, we set $K = 1$. If $S_2 \leq 1$, then we continue, setting $S_3 = E_1 + E_2 + E_3$, and so on, so that we set $K = i$ when, and only when, $S_i \leq 1 < S_{i+1}$ for $i \geq 0$.

The BASIC program in Fig. 4.4 shows how easily this algorithm may be programmed, and may also help in demonstrating how it works. Note that we simulate

$$S_k = \sum_{i=1}^{k} E_i \qquad \text{by} \qquad S_k = -\frac{1}{\lambda} \log\left(\prod_{i=1}^{k} U_i\right)$$

```
10   RANDOMIZE
20   INPUT M,L
30   REM PROGRAM TO SIMULATE M RANDOM
40   REM VARIABLES FROM A POISSON
50   REM DISTRIBUTION OF PARAMETER L
60   LET E1 = EXP(-L)
70   FOR I = 1 TO M
80     LET K = 0
90     LET U = RND
100    IF U < E1 THEN 140
110    LET U = U*RND
120    LET K = K+1
130    GOTO 100
140    PRINT K
150  NEXT I
160  END
```

Figure 4.4   BASIC program for simulating M Poisson random variables.

where as usual the U_i are independent $U(0, 1)$ random variables, as explained in Section 4.3. The comparison $S_k > 1$, then becomes

$$-\frac{1}{\lambda} \log\left(\prod_{i=1}^{k} U_i\right) > 1$$

i.e. $\qquad\qquad \log\left(\prod_{i=1}^{k} U_i\right) < -\lambda \qquad$ i.e. $\qquad \prod_{i=1}^{k} U_i < e^{-\lambda}$

and it is this inequality which is being tested in line number 100 of the program.

On first acquaintance, this algorithm has the same 'rabbit-out-of-a-hat' nature as the Box–Müller method. We can certainly show analytically that K thus defined has the required Poisson distribution (see Exercise 4.9), but a consideration of the Poisson process, mentioned in Section 2.7, shows readily the origin of this algorithm, as we shall now see.

In a Poisson process in time (say) of rate λ we have the two important results (see, e.g., *ABC*, chapter 19):

(a) times between events are independent random variables from the exponential p.d.f., $\lambda e^{-\lambda x}$, for $x \geq 0$;
(b) the number of events in any fixed time interval of length t has a Poisson distribution of parameter (λt).

Result (b) tells us that to simulate a random variable with a Poisson distribution of parameter λ, all we have to do is construct a realization of a Poisson process of parameter λ, and then count the number of events occurring in a time interval of unit length. Result (a) tells us how we can simulate the desired Poisson process, by simply placing end-to-end independent realizations of exponential random variables from the $\lambda e^{-\lambda x}$ density. We keep a record of the time taken since the start of the process, and stop the simulation once that time exceeds unity. Figure 4.5 provides an illustration, resulting in $K = 3$, as there have been just three events in the Poisson process in the $(0, 1)$ time interval, occurring at times E_1, $E_1 + E_2$ and $E_1 + E_2 + E_3$ respectively, with the fourth event occurring at time $E_1 + E_2 + E_3 + E_4 > 1$.

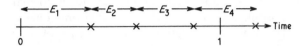

Figure 4.5   Illustration of the simulation of a Poisson process in time, starting at time 0. Four events occur, at times denoted by $\times$. Inter-event times, E_i, are independent random variables from the $\lambda e^{-\lambda x}$ exponential p.d.f., and the value $K = 3$, the number of events in the $(0, 1)$ time interval, is a realization of a random variable with a Poisson distribution of parameter λ.

*4.5   Multivariate random variables

Particular rules may also be exploited to simulate multivariate random variables. Two examples, one discrete and one continuous, will be considered here.

4.5.1   The bivariate Poisson distribution

This distribution was mentioned briefly in Section 2.16. If three independent random variables, X_1, X_2 and X_3, have Poisson distributions with parameters λ_1, λ_2 and λ_3 respectively, then the derived variables $Y_1 = X_1 + X_3$, $Y_2 = X_2 + X_3$ have a bivariate Poisson distribution. This is readily verified by simply writing down the bivariate moment generating function for Y_1 and Y_2, and observing it is of the form given in Equation (2.6). If we simulate X_1, X_2 and X_3 by the method of Section 4.4.2 then this result readily allows us to simulate random variables Y_1 and Y_2 with a bivariate Poisson distribution. An alternative approach is suggested in Exercise 4.7.

4.5.2   The multivariate normal distribution

This distribution was discussed in Section 2.15. We saw there that if the p-variate random variable $\mathbf{X}$ has the multivariate normal, $N(\mathbf{0}, \mathbf{I})$ distribution, then $\mathbf{Z} = \mathbf{AX} + \boldsymbol{\mu}$ has the multivariate normal $N(\boldsymbol{\mu}, \mathbf{AA}')$ distribution. Hence, if we want to simulate random variables from an $N(\boldsymbol{\mu}, \boldsymbol{\Sigma})$ multivariate normal distribution, then we need only find a matrix $\mathbf{A}$ for which $\boldsymbol{\Sigma} = \mathbf{AA}'$. Ways of doing this are discussed in Exercise 4.14 and in the solution to that exercise. $\mathbf{X}$ is readily simulated, as its elements are independent and $N(0, 1)$. We then set $\mathbf{Z} = \mathbf{AX} + \boldsymbol{\mu}$.

4.6   Discussion and further reading

We have seen in this chapter a utilization of formal relationships between random variables, which enables us to simulate a variety of random variables, using only $U(0, 1)$ variates; other illustrations can be found in the exercises. All of these examples are no more than useful tricks for particular cases. Often very simple algorithms result, as we have seen from some of the BASIC programs presented, and these algorithms could be readily implemented for small-scale simulations, using hand-calculators or microcomputers, for example. In a number of cases, however, the algorithms are less efficient than others which may be devised (see Kinderman and Ramage, 1976, for example), some of which will be considered in the next chapter.

The method of Section 4.4.2 for Poisson variates may become very inefficient if λ is large. In this case we would expect large numbers of events in the Poisson process during the $(0, 1)$ interval, resulting in prohibitively many checks. Atkinson (1979a) compares the algorithm of Section 4.4.2 for simulating Poisson random variables with alternative approaches which will be mentioned in the next chapter, while Kemp and Loukas (1978a, b) make similar comparisons for the bivariate Poisson case. More recent work for the univariate Poisson case is to be found in Atkinson (1979c), Kemp (1982), Ahrens and Dieter (1980, 1982) and Devroye (1981).

Atkinson and Pearce (1976), Atkinson (1977) and Cheng (1977) discuss the simulation of gamma $\Gamma(n, \lambda)$ random variables with *non*-integral shape parameter n, and we consider Cheng's method in Exercise 5.22. More recent work is provided by Cheng and Feast (1979) and Kinderman and Monahan (1980).

Neave (1973) showed that when the standard Box–Müller method is operated using pseudo-random numbers from a particular multiplicative congruential generator, the resulting numbers exhibit some strikingly non-normal properties. This finding was taken up by Chay, Fardo and Mazumdar (1975) and Golder and Settle (1976), and we shall return to this point in Section 6.7.

We have in this chapter only scratched the surface of the relations between random variables of different kinds. The books by Johnson and Kotz (1969, 1970a, 1970b, 1972) provide many more such relationships, and the book by Mardia (1970) provides more information on bivariate distributions.

4.7   **Exercises and complements**

4.1   Show that a random variable with a $U(0, 1)$ distribution has mean 1/2 and variance 1/12.

†4.2   Consider how you might simulate a binomial $B(3, p)$ random variable using just one $U(0, 1)$ variate, and write a BASIC program to do this.

4.3   Explain how the approach of using a binary representation of $U(0, 1)$ random variables may be used to simulate random variables with a binomial $B(n, p)$ distribution in which $p = m2^{-r}$, for integral $r > 0$, and integral $0 \leq m \leq 2^r$.

4.4   The following result is similar to one in Exercise 2.14: If the independent random variables X_1 and X_2 are, respectively, $\Gamma(p, 1)$ and $\Gamma(r, 1)$, then $Y = X_1/(X_1 + X_2)$ has the beta density,

$$f(y) = \frac{\Gamma(p+r)}{\Gamma(p)\Gamma(r)} y^{p-1} (1-y)^{r-1} \quad \text{for } 0 \leq y \leq 1$$

Use this result to write a BASIC program to simulate such a Y random variable, for integral $p > 1$ and $r > 1$.

4.5   If X_1, X_2, X_3, X_4 are independent $N(0, 1)$ random variables, we may use them to simulate other variates, using the results of Exercise 2.15 as follows:
(a) $Y = |X_1 X_2 + X_3 X_4|$ has an exponential distribution of parameter 1.
(b) $C = X_1/X_2$ has a Cauchy distribution, with density function $1/(\pi(1+x^2))$.
Use these results to write BASIC programs to simulate such Y and C random variables.

†4.6   Prove that N_1, N_2, given by Equation (4.1) are independent $N(0, 1)$ random variables.

*4.7   When (X, Y) have a bivariate Poisson distribution, the probability generating function has the form

$$G(u, v) = \exp\{\lambda_1(u-1) + \lambda_2(v-1) + \lambda_3(uv-1)\}$$

The marginal distribution of X is Poisson, of parameter $(\lambda_1 + \lambda_3)$, while the conditional distribution of $Y|X = x$ has probability generating function

$$\left(\frac{\lambda_1 + \lambda_3 v}{\lambda_1 + \lambda_3}\right)^x \exp[\lambda_2(v-1)]$$

Use these results to simulate the bivariate Poisson random variable (X, Y).

*4.8 If $X = \Sigma_{i=1}^{n} U_i$, where U_i are independent $U(0, 1)$ random variables, show, by induction or otherwise, that X has the probability density function

$$f(x) = \sum_{j=0}^{[x]} (-)^j \binom{n}{j} (x - j)^{n-1} / (n-1)! \qquad \text{for } 0 \leq x \leq n,$$

$$= 0 \qquad \text{otherwise}$$

where $[x]$ denotes the integral part of x.

*4.9 Prove, without reference to the Poisson process, that the random variable K, defined at the start of Section 4.4.2, has a Poisson distribution of parameter λ.

*4.10 Consider how the Box–Müller method may be extended to more than two dimensions.

*4.11 In the notation of the Polar Marsaglia method, show that Θ and $\tilde{R}$, defined by

$$\tan \Theta = V_1/V_2 \qquad \text{and} \qquad \tilde{R}^2 = V_1^2 + V_2^2$$

both conditional on $V_1^2 + V_2^2 \leq 1$, are independent random variables. Show also that R^2 is a $U(0, 1)$ random variable, and Θ is a $U(0, 2\pi)$ random variable.

*4.12 When a pair of variates (V_1, V_2) is rejected in the Polar Marsaglia method, it is tempting to try to improve on efficiency, and only reject one of the variates, so that the next pair for consideration would then be (V_2, V_3), say. Show why this approach is unacceptable.

*4.13 Provide an example of a continuous distribution, with density function $f(x)$, with zero mean, for which the following result is true: X_1 and X_2 are independent random variables with probability density function $f(x)$. When the point (X_1, X_2), specified in terms of Cartesian co-ordinates, is expressed in polar co-ordinates (R, Θ), then R, Θ are not independent.

*4.14 If S is a square, symmetric matrix, show that it is possible to write $S = VDV'$, where D is a diagonal matrix, the ith diagonal element of which is the ith eigenvalue of S, and V is an orthogonal matrix with ith column an eigenvector corresponding to the ith eigenvalue of S. Hence provide a means of obtaining the factorization, $\Sigma = AA'$ required for the simulation of multivariate normal random variables in Section 4.5.2. More usually, a Choleski factorization is used for Σ, in which A is a lower-triangular matrix. Details are provided in the solution. This is the approach adopted in the IMSL routine GGNSM—see Section A1.1.

*4.15 If X_1, X_2, ..., X_n are independent (column) random variables from a p-variate multivariate normal, $N(0, \Sigma)$ distribution, then $Z = \Sigma_{i=1}^{n} X_i X_i'$ has the *Wishart* distribution, $W(Z; \Sigma, n)$, described, for example, by Press (1972, p. 100). Use this result, which generalizes to p dimensions the result of Equation (4.3), to provide a BASIC program to simulate such a Z. For related discussion, see Newman and Odell (1971, chapter 5).

*4.16 If X_1, X_2 are independent $N(0, 1)$ random variables, show that the random variables X_1 and

$$Y_1 = \rho X_1 + (1 - \rho^2)^{1/2} X_2 \qquad \text{where} \qquad -1 \le \rho \le +1$$

have a bivariate normal distribution, with zero means, unit variances, and correlation coefficient ρ.

†4.17 Let $U_1, U_2, \ldots, U_{2n-1}$ be a random sample from the $U(0, 1)$ density. If M denotes the sample median, show that M has the $B_e(n, n)$ distribution.

*4.18 (*continuation*) Consider how the result of Exercise 4.17 may be used to simulate a $B(n, p)$ random variable (Relles, 1972).

†4.19 We find, from Section A1.1, that the IMSL computer library has routine GGPON for simulating Poisson variables when the Poisson parameter λ may vary from call to call. Otherwise one might use the IMSL routine GGPOS. Kemp (1982) was also concerned with Poisson variable simulation when λ may vary, and one might wonder why one should want to simulate such Poisson variates. An answer is provided by the following exercise.

The random variable X has the conditional Poisson distribution:

$$\Pr(X = k | \lambda) = \frac{e^{-\lambda} \lambda^k}{k!} \qquad \text{for } k \ge 0$$

If λ has a $\Gamma(n, \theta)$ distribution, show that the unconditional distribution of X is

$$\Pr(X = k) = \binom{n+k-1}{k} \left(\frac{1}{\theta+1}\right)^k \left(\frac{\theta}{\theta+1}\right)^n \qquad \text{for } k \ge 0$$

i.e. $X = Y - n$, where Y has a negative-binomial distribution as defined in Section 2.6. As $n \to 0$ the distribution tends to the logarithmic series distribution much used in ecology, and generalized by Kempton (1975). Kemp (1981) considers simulation from this distribution (see Exercise 4.22).

4.20 (*continuation*) Use the result of the last question to provide a BASIC program to simulate negative-binomial random variables, for integral $n > 1$. This result is sometimes used to explain the frequent use of the negative-binomial distribution for describing discrete data when the Poisson distribution is unsatisfactory. The negative-binomial is a 'contagious' distribution, and much more relevant material is provided by Douglas (1980). An algorithm using the waiting-time definition of the distribution is given in the IMSL routine GGBNR (see Appendix 1).

4.21 Use the transformation theory of Section 2.12 to show that the random variable $Y = e^X$, where X has a $N(\mu, \sigma^2)$ distribution, has the density function

$$f_Y(y) = \frac{1}{y\sigma \sqrt{(2\pi)}} \exp\left(-\frac{1}{2}\left(\frac{\log_e(y) - \mu}{\sigma} \right)^2 \right) \qquad \text{for } y \geq 0$$

Y is said to have a *log-normal* distribution. The p.d.f. has the same qualitative shape as that of the $\Gamma(2, 1)$ p.d.f. of Fig. 2.6, and the log-normal distribution is often used to describe incubation periods for diseases (see also Morgan and Watts, 1980) and sojourn times in more general states. We shall, in fact, encounter such a use for this distribution in Example 8.3. Section A1.1 gives IMSL and NAG routines for simulating from this distribution. For full details, see Aitchison and Brown (1966).

4.22 (Kemp, 1981)   The general logarithmic distribution is

$$p_k = -\alpha^k / \{ k \log_e(1 - \alpha) \} \qquad k \geq 1, 0 < \alpha < 1$$

Show that its moment generating function is:

$$\log(1 - \alpha e^\theta)/\log(1 - \alpha)$$

and that successive probabilities can be generated from:

$$p_k = \alpha(1 - 1/k)p_{k-1} \qquad \text{for } k \geq 2$$

Show that if X has the conditional geometric distribution

$$\Pr(X = x | Y = y) = (1 - y)y^{x-1} \qquad \text{for } x \geq 1$$

and if $\qquad \Pr(Y \leq y) = \dfrac{\log(1 - y)}{\log(1 - \alpha)} \qquad \text{for } 0 \leq y \leq \alpha$

then X has the logarithmic distribution. Explain how you can make use of this result to simulate from the logarithmic distribution.

5

GENERAL METHODS
FOR NON-UNIFORM
RANDOM VARIABLES

For many uses, simple algorithms, such as those which may arise from particular methods of the kind described in the last chapter, will suffice. It is of interest, however, to consider also *general* methods, which may be used for any distribution, and that we shall now do. In many cases general methods can result in algorithms which, while they are more complicated than those considered so far, are appreciably more efficient.

5.1   The 'table-look-up' method for discrete random variables

For ease of notation, let us suppose that we have a random variable X that takes the values 0, 1, 2, 3, etc., and with $p_i = \Pr(X = i)$ for $i \geq 0$. Thus X could be binomial, or Poisson, for example.

A general algorithm for simulating X is as follows:
Select a $U(0, 1)$ random variable, U.

Set $X = 0$ if $0 \leq U < p_0$, and

set $X = j$ if $\sum_{i=0}^{j-1} p_i \leq U < \sum_{i=0}^{j} p_i$ for $j \geq 1$.

We can think of the probabilities $\{p_i, i \geq 0\}$ being put end-to-end and, as $\sum_{i=0}^{\infty} p_i = 1$, filling out the interval $[0, 1]$ as illustrated in Fig. 5.1. We can now see that the above algorithm works by selecting a value U and observing in which probability interval U lands. In the illustration of Fig. 5.1 we have

$$\sum_{i=0}^{1} p_i \leq U < \sum_{i=0}^{2} p_i$$

and so we set $X = 2$.

This algorithm is simply a generalization to more than two intervals of the rule used to simulate Bernoulli random variables in Section 4.4.1. The

91

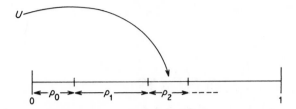

Figure 5.1   An illustration of how the table-look-up method works for simulating a discrete random variable with probability distribution $\{P_i, i \geq 0\}$.

reason the algorithm works is readily explained. We want to simulate the random variable X so that $\Pr(X = i) = p_i$ for $i \geq 0$. As U is $U(0, 1)$, then for any $0 \leq a \leq b$, $\Pr(a \leq U < b) = (b - a)$ (see Section 2.8) and so $\Pr(0 \leq U < p_0) = p_0 = \Pr(X = 0)$, and

$$\Pr\left(\sum_{i=0}^{j-1} p_i \leq U < \sum_{i=0}^{j} p_i\right) = p_j = \Pr(X = j) \qquad \text{for } j \geq 1$$

Thus, for $j \geq 0$, the algorithm returns the value $X = j$ with probability p_j, as required.

The above algorithm is readily modified to cope with discrete random variables with different ranges from that considered above, (i.e. $X \geq 0$) and one such example now follows.

EXAMPLE 5.1
We want to simulate the geometric random variable X, with distribution

$$p_i = \Pr(X = i) = (1 - p)^{i-1} p \qquad \text{for } i \geq 1, 0 < p < 1$$

In order to operate the above algorithm we need successive cumulative sums of the $\{p_i\}$, and in this case, because p_i is of a simple geometric form, then these cumulative sums are also of a simple form. Here,

$$\sum_{i=1}^{j} p_i = \frac{p(1 - (1 - p)^j)}{1 - (1 - p)} = 1 - (1 - p)^j \qquad \text{for } j \geq 1$$

Thus the algorithm becomes:

Set $X = j$ if $1 - (1 - p)^{j-1} \leq U < 1 - (1 - p)^j$ for $j \geq 1$

which is equivalent to: $-(1 - p)^{j-1} \leq U - 1 < -(1 - p)^j$

i.e. $(1 - p)^{j-1} \geq (1 - U) > (1 - p)^j$ (5.1)

Before proceeding, we can here observe that the algorithm entails selecting a $U(0, 1)$ random variable, and then checking the range of $(1 - U)$. Now it is intuitively clear that if U is a $U(0, 1)$ random variable, *then so is* $(1 - U)$, and this result can be readily verified by the change-of-variable theory of Section 2.12 (see Exercise 5.1).

Hence we can reduce the labour of arithmetic slightly if we replace (5.1) by the equivalent test:

$$(1-p)^{j-1} \geq U > (1-p)^j \tag{5.2}$$

Of course, for any particular realization of U, (5.1) and (5.2) will usually give different results; however, the random variable X resulting from using (5.2) will have the same geometric distribution as the random variable resulting from using (5.1). Continuing from (5.2), we set $X = j \geq 1$ if, and only if,

$$(j-1)\log_e(1-p) \geq \log_e U > j\log_e(1-p)$$

so that, recalling that $\log_e(1-p) < 0$, we have $X = j$ if

$$(j-1) \leq \frac{\log_e U}{\log_e(1-p)} < j \qquad \text{for } j \geq 1 \tag{5.3}$$

Finally, we note that we can express (5.3) very simply by setting

$$X = 1 + \left[\frac{\log_e U}{\log_e(1-p)} \right] \tag{5.4}$$

where $[y]$ is used to denote the integral part of y. For further discussion of this result, see Exercise 5.12.

This example therefore uses the general table-look-up algorithm to produce the simple expression of (5.4). This is in contrast to a particular approach which may be used, based upon the definition of the geometric distribution given in Section 2.6. Thus an alternative method would be to test sequentially independent $U(0, 1)$ random variables until one was found to be less than p, and an algorithm using this approach is provided by the IMSL routine GGEOT (see Section A1.1).

This example is unusual in that the cumulative sums of probabilities have a simple form. The next example is far more typical.

EXAMPLE 5.2
If X has a Poisson distribution of parameter 2, its cumulative distribution function is given below to four places of decimals:

i	0	1	2	3	4	5	6	7	8	9
$Pr(X \leq i)$	0.1353	0.4060	0.6767	0.8571	0.9473	0.9834	0.9955	0.9989	0.9998	1.000

Using this table and the table-look-up algorithm, the following eight $U(0, 1)$

random variables can be seen to give rise to the indicated values of X:

U	X
0.0318	0
0.4167	2
0.4908	2
0.2459	1
0.3643	1
0.8124	3
0.9673	5
0.1254	0

This example illustrates why the table-look-up method is so called. Given a table of the cumulative distribution of any discrete random variable, and a supply of $U(0, 1)$ random variables, we can use this method to simulate that random variable. By their very nature, such tables are finite, and if the random variable in question has an infinite range, then the range would have to be truncated for the method to be used. This was done in the above example, where using accuracy of only four decimal places resulted in the range of X being truncated to $[0, 9]$.

Human beings can operate the table-look-up method quite easily, but its implementation on a computer poses some intriguing problems. First of all we can remark that for a computer implementation it is not necessary to store cumulative sums of probabilities—they can be computed each time, as required. Random variables of infinite range need not then have their range truncated, but this approach is usually far too costly in effort because of the repeated duplication of arithmetic each time a new simulation is run. More usually ranges are truncated if necessary, and the resulting finite tables are stored within the computer. The next problem that arises is how to read such stored tables. Computers need specified algorithms which could, for instance, involve reading the table of the cumulative distribution in Example 5.2 from left to right. In such a case, the computer would return $X = 0$ when $U = 0.0318$, with the greatest of ease, but when $U = 0.9673$ it would laboriously check whether $U < 0.1353$, $U < 0.4060$, and so on until it found $0.9473 < U < 0.9834$. Human beings need not be so rigid and have the advantage over computers of being able to change their strategy in the light of superficial evidence on the size of U. By analogy, when looking up a word such as 'wombat' in the dictionary, not many of us would start at the front, with the letter 'A' and then skim through from A to W; rather, we would start from the middle, or somewhere near the end, possibly even working backwards as well as forwards. A more efficient computer algorithm may result if the range of X

were initially subdivided; for example, if $\Pr(X \leq \theta) = p \approx 0.5$, say, for some known θ, then if $U > p$ it would not be necessary to chek U against $\Sigma_{i=0}^{j} p_i$ for $j \leq \theta$. Such an approach is utilized in the IMSL routine GGDT and the NAG routine GO5EYF (see Section A1.1) and was encountered earlier in Section 4.4.1.

5.2   The 'table-look-up', or inversion method for continuous random variables

We shall now consider the analogue of the above method for continuous random variables. Suppose we wish to simulate a continuous random variable X with cumulative distribution function $F(x)$, i.e. $F(x) = \Pr(X \leq x)$, and suppose also that the inverse function, $F^{-1}(u)$ is well-defined for $0 \leq u \leq 1$.

If U is a $U(0, 1)$ random variable, then $X = F^{-1}(U)$ has the required distribution. We can see this as follows:

If
$$X = F^{-1}(U)$$

then
$$\Pr(X \leq x) = \Pr(F^{-1}(U) \leq x)$$

and because $F(x)$ is the cumulative distribution function of a continuous random variable, $F(x)$ is a strictly monotonic increasing continuous function of x. This fact enables us to write

$$\Pr(F^{-1}(U) \leq x) = \Pr(U \leq F(x))$$

But, as U is a $U(0, 1)$ random variable,

$$\Pr(U \leq F(x)) = F(x) \qquad \text{(see Section 2.8)}$$

i.e.
$$\Pr(X \leq x) = F(x)$$

and so the X obtained by setting $X = F^{-1}(U)$ has the required distribution.

The above argument, which was given earlier in Equation (2.1), is perhaps best understood by considering a few examples. Figure 5.2 illustrates one cumulative distribution function for a truncated exponential random variable. We operate the rule $X = F^{-1}(U)$ by simply taking values of $U(0, 1)$ variates and projecting down on the x-axis as shown, using the graph of $y = F(x)$.

We shall now consider two further examples in more detail.

EXAMPLE 5.3
In the untruncated form, if X has an exponential density with parameter λ, then

$$f(x) = \lambda e^{-\lambda x} \qquad \text{for } x \geq 0, \lambda > 0,$$

and
$$F(x) = 1 - e^{-\lambda x}$$

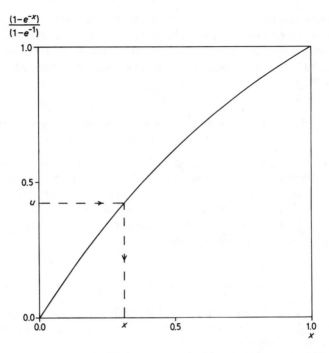

Figure 5.2   Operation of the table-look-up method for a continuous random variable. The curve has the equation $(1 - e^{-x})/(1 - e^{-1})$ and is the cumulative distribution function of a truncated exponential random variable, with probability density function, $e^{1-x}(e - 1)^{-1}$ for $0 \leq x \leq 1$.

To simulate X we set $X = F^{-1}(U)$, i.e., set

$$U = 1 - e^{-\lambda X}, \text{ and solve for } X.$$

This gives $$X = -\frac{1}{\lambda}\log_e(1 - U)$$

and for the same reasoning as in Example 5.1, we obtain the desired distribution for X from setting

$$X = -\frac{1}{\lambda}\log_e U \qquad (5.5)$$

A verification of this result is provided by the solution to Exercise 2.1, and this method is used in the IMSL routine GGEXN and the NAG routine GO5DBF (see Section A1.1).

As with Example 5.1, the result of (5.5) is deceptive in its simplicity, and it is no coincidence that the geometric and exponential distributions play similar rôles, the former in the discrete case and the latter in the continuous case, as discussed in Section 2.10, Exercise 2.23 and Exercise 5.12. It is unfortunately the case that it is often *not* simple to form $X = F^{-1}(U)$. The prime example of this occurs with the normal distribution, which has led to a variety of different approximations to both the normal cumulative distribution function and its inverse. We shall return to the subject of these approximations later in Section 5.7 and Exercise 5.9.

We conclude this section with an example of how this method can be used, in a 'table-look-up' fashion to simulate standard normal random variables.

EXAMPLE 5.4
If we take the same $U(0, 1)$ values as in Example 5.2 then we can use tables of the standard normal cumulative distribution function, $\Phi(x)$ to give the following realizations of an $N(0, 1)$ random variable X:

U	X (to two places of decimals)
0.0318	-1.85
0.4167	-0.21
0.4908	-0.02
0.2459	-0.69
0.3643	-0.35
0.8124	0.89
0.9673	1.84
0.1254	-1.15

Thus, for example, $0.8124 = \Phi(0.89)$, to the accuracy given.
Two points should be made here:

(a) Because of the symmetry of the $N(0, 1)$ distribution, the tables usually only give values of $x \geq 0$ and, correspondingly, values of $\Phi(x) \geq 0.5$, and so when $u < 0.5$ we have to employ the following approach which is easily verified to be correct:
We want x for which $u = \Phi(x)$.
If $u < 0.5$, then by the symmetry of the normal density, $x = -\Phi^{-1}(1-u)$.
(b) The accuracy of the numbers produced (in this case to two decimal places) depends on the accuracy of the tables, which also determines the degree of the truncation involved.

The 'table-look-up' method for continuous random variables is often called the *inversion* method, and a general algorithm is provided by the IMSL routine GGVCR (see Section A1.1). We shall use these terms interchangeably, though strictly they describe different ways of implementing the same basic method.

5.3   The rejection method for continuous random variables

Suppose we have a method for sprinkling points uniformly at random under any probability density function $f(x)$, and which may give rise to the pattern of points in Fig. 5.3. What is the probability that the abscissa, X say, of any one of these points lies in the range $\alpha \le X < \beta$, for any $\alpha < \beta$?

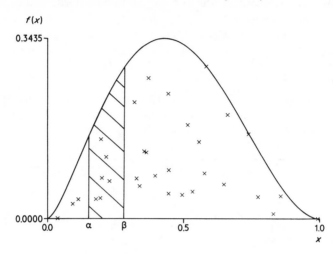

Figure 5.3   An illustration of points × uniformly and randomly distributed underneath the probability density function $f(x)$. For illustration we have used the B_e (2.5, 3) distribution.

The event, $\alpha \le X < \beta$ is equivalent to the point being in the shaded area shown in Fig. 5.3, and so, because of the assumed uniform distribution of the points, this event has probability

$$\frac{\text{area of shaded area}}{\text{total area under } f(x)}$$

This is

$$\int_\alpha^\beta f(x)\,\mathrm{d}x \bigg/ \int_{-\infty}^\infty f(x)\,\mathrm{d}x$$

i.e.

$$\int_\alpha^\beta f(x)\,\mathrm{d}x \ \text{ as } \ \int_{-\infty}^\infty f(x)\,\mathrm{d}x = 1,$$

since $f(x)$ is a probability density function.

   Thus we are saying that

$$\Pr(\alpha \le X < \beta) = \int_\alpha^\beta f(x)\,\mathrm{d}x, \qquad \text{for any } \alpha < \beta$$

where $f(x)$ is a probability density function, i.e. X has probability density function $f(x)$ (see Section 2.3).

Thus, for any probability density function $f(x)$, we can simulate random variables X from this density function as long as we have a method for uniformly and randomly sprinkling points under $f(x)$. Those who have read Section 4.2.2 will have already encountered a similar situation, the solution being in that case to enclose within a square the area to be sprinkled with points. It is a simple matter to distribute points uniformly at random over a square, and in Section 4.2.2, those points not within the area of interest were rejected. The same principle for any density function $f(x)$ results in a general rejection method, attributed to von Neumann (1951). While the rejection method (sometimes also called the 'acceptance–rejection' method) may be used for discrete random variables (see Fishman, 1979, for example), it is usually employed for continuous random variables, the case being investigated here.

If the probability density function $f(x)$ is non-zero over only a finite range, then it is easy to box it in, as shown in Fig. 5.4. Using $U(0, 1)$ random variables it is a simple matter to sprinkle points uniformly and randomly over the rectangle shown, simply by taking points with Cartesian co-ordinates $(\theta + (\alpha - \theta)U_1, \delta U_2)$, where U_1 and U_2 are independent $U(0, 1)$ random variables. Points landing above $f(x)$ are rejected, while for points landing below $f(x)$, we take $\theta + (\alpha - \theta)U_1$ as a realization of X.

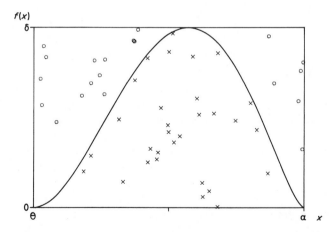

Figure 5.4   Simulating from the probability density function $f(x)$, which is non-zero over the finite range $[\theta, \alpha]$. The method used generates points (denoted by $\circ$ and $\times$) uniformly at random over the rectangle shown. Points denoted by $\circ$ are rejected (cf. Fig. 4.2), while the abscissae of the points $\times$ are accepted as realizations of the random variable with probability density function $f(x)$. For illustration we have used the $B_e(3, 2.5)$ distribution, for which $\theta = 0$, $\alpha = 1$, $\delta = 0.3435$.

As the area of the rectangle in Fig. 5.4 is $\delta(\alpha - \theta)$, and the area under the curve is unity, we see that the probability of accepting a point is $1/(\delta(\alpha - \theta))$, and so the smaller δ is, the larger is the probability of acceptance and, correspondingly, the more efficient the method. This is, of course, why a larger value of δ was not used in Fig. 5.4.

There are two snags with the above approach. A rectangle is, as we have seen, a convenient shape within which to simulate a random uniform spread of points, but it clearly cannot be used if the density $f(x)$ has an infinite range, as we can only simulate uniform random variables over a finite range. Furthermore, the probability of rejection could become quite large: if the density of Fig. 5.4 was replaced by a spiked density, for instance, such as would result from a Laplace distribution, truncated to have a finite range. In such a case the simplicity gained from distributing points uniformly over a rectangular region could be more than offset by the cost of frequent rejection.

Both of these snags can be overcome by using as the enveloping curve a suitable multiple of a *different* probability density function from $f(x)$, as we shall now see. Consider a p.d.f. $h(x)$, with the same range as $f(x)$, *but from which it is relatively easy to simulate*. It is then simple to obtain a uniform scatter of points under $h(x)$, by taking points (X, Y) such that X has density $h(x)$, while the conditional density of Y given $X = x$ is $U(0, h(x))$. For a uniform scatter of points, the conditional p.d.f. of Y clearly must be of this form, while the X co-ordinate must have the property that for any pair (α, β), with $\alpha < \beta$, $\Pr(\alpha \le X < \beta) \sim \int_\alpha^\beta h(x)\,dx$, i.e. X must have probability density function $h(x)$.

If it were possible to choose $h(x)$ to be of a roughly similar shape to $f(x)$ *and* then to envelop $f(x)$ by $h(x)$, we would obtain the desired scatter of points under $f(x)$ by first obtaining a scatter of points under $h(x)$ and then rejecting just those which were under $h(x)$ but not under $f(x)$. While it is often possible to choose an appropriate $h(x)$ to be of similar shape to $f(x)$, it is clearly not possible to envelop $f(x)$ by $h(x)$, so that, for all x, $f(x) \le h(x)$, since both $f(x)$ and $h(x)$ are density functions, and so $\int_{-\infty}^{\infty} f(x)\,dx = \int_{-\infty}^{\infty} h(x)\,dx = 1$. However, the solution to this last obstacle is easily obtained by, effectively, plotting $h(x)$ and the scatter of points obtained under $h(x)$ on stretchable paper, and then uniformly stretching the paper in a direction at right angles to the x-axis until $h(x) \ge f(x)$ for all x. Such stretching clearly does not change the uniformity of the scatter of the points. Mathematically this stretching is done, very simply, by taking as the conditional density of Y given $X = x$, $U(0, kh(x))$, where $k > 1$ is the stretching factor, and where X has probability density function $h(x)$.

Thus for suitable $h(x)$ and k, we have the following algorithm: if we write $g(x) = kh(x)$,

 (i) simulate $X = x$ from probability density function $h(x)$;
(ii) simulate Y to be $Ug(x)$, where U is an independent $U(0, 1)$ random variable;

(iii) accept $X = x$ as a realization of a random variable with probability density function $f(x)$ if and only if $Y < f(x)$.

The situation is illustrated in Fig. 5.5.

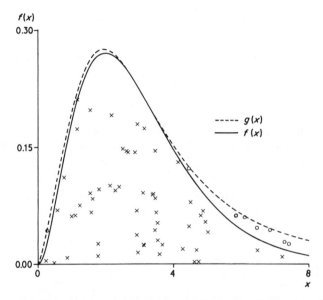

Figure 5.5   The points $\times$ and $\circ$ are uniformly and randomly distributed under the curve $g(x) = kh(x)$, where $k > 1$ is a constant, and $h(x)$ is a density from which it is easy to simulate. The points $\circ$ lie above the density function $f(x)$ and so are rejected. The points $\times$ are accepted and their abscissae are realizations of a random variable with probability density function $f(x)$. See Exercise 5.22 for an explanation of the p.d.f.'s used.

At first sight this algorithm seems unusual and confusing, since the test in (iii) concerns Y, but if the test is satisfied then it is $X = x$ which is accepted. However, in the light of the above discussion, we can now see that (iii) is just a component of testing whether a point constructed randomly and uniformly under $g(x)$ is also under $f(x)$.

The probability of rejection here is

$$\frac{\displaystyle\int_{-\infty}^{\infty} (g(x) - f(x))\,\mathrm{d}x}{\displaystyle\int_{-\infty}^{\infty} g(x)\,\mathrm{d}x} = 1 - \frac{1}{k}$$

reflecting the importance of small k, subject to $k > 1$.

We choose $h(x)$ with shape and convenience in mind. The next two examples provide two approaches for selecting k. The first example uses an exponential envelope to simulate normal random variables. An exponential envelope can only envelop half of the standard normal density, but it can envelop the 'half-normal' density, given by

$$f(x) = \sqrt{\left(\frac{2}{\pi}\right)}e^{-x^2/2} \quad \text{for } x \geq 0$$

i.e. $f(x) = 2\phi(x)$, for $x \geq 0$. If X has density function $f(x)$, then the random variable

$$\tilde{X} = \begin{cases} X \text{ with probability } \frac{1}{2} \\ -X \text{ with probability } \frac{1}{2} \end{cases}$$

clearly has the standard normal density $\phi(x)$ for $-\infty \leq x < \infty$. We shall therefore simulate from $\phi(x)$ by first simulating from $f(x)$, and then applying the above transformation, from $\tilde{X}$ to X.

EXAMPLE 5.5   *A rejection method for $N(0, 1)$ variables*

Here

$$f(x) = \sqrt{\left(\frac{2}{\pi}\right)}e^{-x^2/2} \quad \text{for } x \geq 0$$

and $$g(x) = ke^{-x} \qquad \text{for } x \geq 0$$

One way of choosing k is to consider the condition for equal roots arising from setting

$$ke^{-x} = \sqrt{\left(\frac{2}{\pi}\right)}e^{-x^2/2}$$

as the roots in x of this equation correspond to the intersection of $g(x)$ and $f(x)$. If this equation has no real roots, then k is too large. If the equation has two distinct roots, then k is too small. The case of two equal roots corresponds to the smallest possible value of k, and the two curves touch, as shown in Fig. 5.6.

Setting $k\sqrt{\left(\frac{\pi}{2}\right)} = e^{x-x^2/2}$

results in a quadratic equation in x:

$$x^2 - 2x + 2\log_e\left(k\sqrt{\left(\frac{\pi}{2}\right)}\right) = 0$$

which has equal roots if and only if

$$1 = 2\log_e\left(k\sqrt{\left(\frac{\pi}{2}\right)}\right)$$

i.e. $$k^2\frac{\pi}{2} = e$$

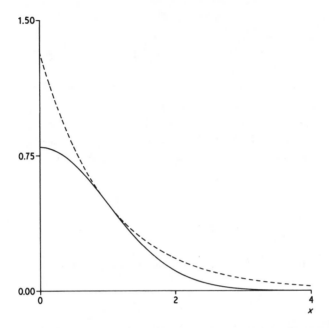

Figure 5.6   An illustration of the optimum choice of k. Here we illustrate the use of the exponential e^{-x} probability density function as the basis of an envelope for the half-normal density (solid line). The enveloping function is given by $\sqrt{\left(\dfrac{2e}{\pi}\right)}e^{-x}$ (dashed line), illustrated for $x \le 4$, as is $f(x)$.

i.e.
$$k = +\sqrt{\left(\frac{2e}{\pi}\right)} \approx 1.315\,489\,2$$

the equal roots occurring at $x = 1$, which is, in fact, also the point of inflexion for the half-normal density.

The algorithm therefore proceeds as follows:

(a) Simulate X from density function, e^{-x} for $x \ge 0$. We know, from Equation (5.5) above, that we can do this by setting $X = -\log_e U_1$, where U_1 is a $U(0, 1)$ random variable. An alternative approach is given in Exercises 5.33–5.35.

(b) If U_2 is an independent $U(0, 1)$ random variable, set $Y = kU_2e^{-X}$, i.e. $Y = kU_2U_1$.

(c) Accept X if and only if,
$$Y < \sqrt{\left(\frac{2}{\pi}\right)}e^{-X^2/2}$$

i.e.
$$kU_1U_2 < \sqrt{\left(\frac{2}{\pi}\right)}e^{-X^2/2}$$

i.e. $$U_1 U_2 < \exp(-(1 + X^2)/2),$$

since $$k = \sqrt{\left(\frac{2e}{\pi}\right)}.$$

Thus ultimately the algorithm does not involve k directly.

Finally, of course, we must convert the half-normal random variable X to the standard normal random variable $\tilde{X}$. While this last stage can always be done by selecting a new $U(0, 1)$ random variable, and then testing whether it is greater or less than $\frac{1}{2}$, we note that as Y is $U(0, g(X))$, then conditional on

$$Y < \sqrt{\left(\frac{2}{\pi}\right)} e^{-X^2/2},$$

Y has a $U\left(0, \sqrt{\left(\frac{2}{\pi}\right)} e^{-X^2/2}\right)$ distribution,

and the sign of $\tilde{X}$ can be decided by considering whether or not

$$Y < \frac{e^{-X^2/2}}{\sqrt{(2\pi)}}.$$

This is the same idea that was exploited in Exercise 4.2 and Section 4.4.1. A BASIC program for this algorithm is shown in Fig. 5.7.

The reason for using e^{-x} as the p.d.f. for the basis of the envelope here, rather than any other $\lambda e^{-\lambda x}$ p.d.f. can be found from a consideration of the probability of rejection, $1 - 1/k$, and we see in Exercise 5.21 that $\lambda = 1$ minimizes this rejection probability.

```
10    RANDOMIZE
20    INPUT M
30    REM PROGRAM TO SIMULATE M STANDARD NORMAL
40    REM RANDOM VARIABLES, USING A REJECTION METHOD
50    REM WITH A HALF-NORMAL PDF ENVELOPED BY A
60    REM MULTIPLE OF THE EXPONENTIAL PDF WITH
70    REM PARAMETER 1
80    FOR I = 1 TO M
90      LET U1 = RND
100     LET U2 = RND
110     LET X = -LOG(U1)
120     LET B = .5*EXP(-.5-X*X/2)
130     LET C = U1*U2
140     IF C < B THEN 170
150     IF C < 2*B THEN 190
160     GOTO 90
170     PRINT -X
180     GOTO 200
190     PRINT X
200   NEXT I
210   END
```

Figure 5.7 BASIC program for the rejection method illustrated in Fig. 5.6.

The exponential function provides a suitable envelope for the half-normal probability density function, as the rate at which e^{-x} tends to zero as $x \to \infty$ is less than the rate at which $e^{-x^2/2}$ tends to zero as $x \to \infty$.

A general way of finding k is to note that we want k to satisfy $kh(x) \geq f(x)$ for all x, and that we cannot have equality here for all x. k is therefore given by

$$k = \max_x \left(\frac{f(x)}{h(x)} \right)$$

if a finite maximum can be found, as then $kh(x) \geq f(x)$ for all x, with equality for at least one x.

A finite maximum will not result if $h(x)$ is unsuitable as a basis for an envelope of $f(x)$. For instance, we could have $h(x) = 0$ when $f(x) > 0$, or we might try setting $f(x) = e^{-x}$ and $h(x) = e^{-x^2/2}$. In this latter case,

$$\log \left(f(x)/h(x) \right) = \frac{x^2}{2} - x$$

which increases without bound as $x \to \infty$.

This approach should work, however, if a suitable $h(x)$ has been found. In this example we have

$$\frac{f(x)}{h(x)} = \sqrt{\left(\frac{2}{\pi} \right)} \frac{e^{-x^2/2}}{e^{-x}} = \sqrt{\left(\frac{2}{\pi} \right)} e^{x - x^2/2}$$

$$y = \log \left(f(x)/h(x) \right) = \log \left(\sqrt{\left(\frac{2}{\pi} \right)} \right) + x - \frac{x^2}{2}$$

$$\frac{dy}{dx} = 1 - x$$

$$\frac{d^2 y}{dx^2} = -1$$

Thus we maximize $f(x)/h(x)$ by setting $x = 1$, to give, as before, $k = \sqrt{\left(\frac{2e}{\pi} \right)}$.

In Section 4.3 we have already seen one way of simulating $\Gamma(n, \lambda)$ random variables, when n is a positive integer. The next example provides an alternative approach, for the case $n > 1$, using rejection, and an exponential envelope as in the last example. This approach may also be used when $n > 1$ is not integral.

Here we take $\lambda = 1$ for simplicity. If a random variable X results, then the new random variable $Y = X/\lambda$ will have a $\Gamma(n, \lambda)$ distribution, from the theory of Section 2.12 (see Exercise 2.2).

*EXAMPLE 5.6   *A rejection method for $\Gamma(n, 1)$ variables.*

Here
$$f(x) = \frac{x^{n-1} e^{-x}}{\Gamma(n)} \qquad \text{for } x \geq 0, \text{ and } n > 1$$

$$g(x) = ke^{-x/n}/n \qquad \text{for } x \geq 0.$$

As $n > 1$, then as $x \to \infty$, $f(x) \to 0$ faster than $g(x)$, implying that $g(x)$ is a suitable enveloping function for $f(x)$.

Let $y = f(x)/h(x)$. We seek k by maximizing y with respect to x.

$$\log_e y = (n-1)\log_e x - x + \frac{x}{n} + \log_e (n/\Gamma(n))$$

$$\frac{d}{dx}(\log_e y) = \frac{n-1}{x} - 1 + \frac{1}{n}$$

$$\frac{d^2}{dx^2}(\log_e y) = \frac{1-n}{x^2}$$

Thus, as $n > 1$, we maximize y when

$$\frac{n-1}{x} = 1 - \frac{1}{n}$$

i.e. when $x = n$, and so

$$k = n^n e^{1-n}/\Gamma(n)$$

It is now a simple matter to derive the following algorithm. Let U be a $U(0, 1)$ random variable, and let E be an independent exponential random variable with parameter n^{-1}. If

$$t(x) = \left(\frac{x}{n}\right)^{n-1} \exp\left[(1-n)\left(\frac{x}{n} - 1\right)\right] \qquad \text{for } x \geq 0$$

then conditional on $t(E) \geq U$, E has the required gamma p.d.f.

The above method, due originally to G. S. Fishman, is described by Atkinson and Pearce (1976). Of course, any density function $f(x)$ can be enveloped by a wide variety of alternative functions, and an alternative rejection method for simulating gamma random variables is given in Exercise 5.22.

For distributions over a finite range, an alternative approach is to envelop the distribution with a suitable polygon and then use the method of Hsuan (1979). A generalization of the rejection method is given in Exercise 5.29.

5.4   The composition method

Here again we encounter a general method suitable for discrete and continuous random variables. We shall begin our discussion of this method with an illustration from Abramowitz and Stegun (1965, p. 951).

For a binomial $B(5, 0.2)$ distribution we have the following probabilities, given to four places of decimals:

i	p_i
0	0.3277
1	0.4096
2	0.2048
3	0.0512
4	0.0064
5	0.0003

If we take p_0 as an example, we can write

$$p_0 = 0.3277 = 0.9 \times \frac{3}{9} + 0.07 \times \frac{2}{7} + 0.027 \times \frac{7}{27} + 0.003 \times \frac{7}{30}$$

and similarly,

$$p_1 = 0.4096 = 0.9 \times \frac{4}{9} + 0.07 \times \frac{0}{7} + 0.027 \times \frac{9}{27} + 0.003 \times \frac{6}{30}$$

$$p_2 = 0.2048 = 0.9 \times \frac{2}{9} + 0.07 \times \frac{0}{7} + 0.027 \times \frac{4}{27} + 0.003 \times \frac{8}{30}$$

and so on, so that in general,

$$p_i = 0.9 \, r_{i1} + 0.07 \, r_{i2} + 0.027 \, r_{i3} + 0.003 \, r_{i4} \qquad \text{for } 0 \le i \le 5 \qquad (5.6)$$

where $\{r_{i1}\}$, $\{r_{i2}\}$, $\{r_{i3}\}$ and $\{r_{i4}\}$ are all probability distributions over the same range, $0 \le i \le 5$.

We can see that in (5.6),

$$0.9 = 10^{-1} \times (\text{sum of digits in first decimal place of the } p_i)$$

$$0.07 = 10^{-2} \times (\text{sum of digits in second decimal place of the } p_i)$$

and so on

while, for example,

$$r_{01} = \frac{0.3}{0.9} = \frac{3}{9}$$

$$r_{11} = \frac{0.4}{0.9} = \frac{4}{9}$$

$$r_{21} = \frac{0.2}{0.9} = \frac{2}{9}$$

$$r_{02} = \frac{0.02}{0.07} = \frac{2}{7}$$

etc.

This explains the derivation of (5.6). We can now use (5.6) to simulate from the $\{p_i\}$ distribution as follows:

(i) Simulate a discrete random variable, R, say, according to the distribution:

j	$\Pr(R = j)$
1	0.9
2	0.07
3	0.027
4	0.003

(ii) If $R = j$, simulate from the $\{r_{ij}\}$ distribution for $1 \leq j \leq 4$. If the resulting random variable is denoted by X,

$$\Pr(X = i) = \sum_{j=1}^{4} \Pr(R = j)r_{ij} \qquad \text{(see for example } ABC, \text{ p. 85)}$$

i.e. $\Pr(X = i) = p_i$

i.e. X has the required binomial distribution.

Of course, a small amount of approximation has taken place here, as we have written the $\{p_i\}$ only to four places of decimals. Nevertheless, this approach may be used for any discrete distribution. While one has to simulate from two distributions $\{\Pr(R = j), 1 \leq j \leq 4\}$ and $\{r_{ij}\}$, most (97 %) of the time one is simulating from $\{r_{i1}\}$ and $\{r_{i2}\}$, and these component discrete distributions are of a very simple form. A disadvantage of this method is the need to store the component distributions. In (5.6) we have written the $\{p_i\}$ distribution as a *mixture*, or *composition*, of the $\{r_{ij}\}$ distributions; a further example of this kind is to be found in Exercise 5.42. We shall now consider the analogous procedure for continuous random variables.

It is not unusual to encounter probability density functions which are mixtures of other probability density functions, say

$$f(x) = \alpha f_1(x) + (1 - \alpha)f_2(x) \qquad 0 < \alpha < 1 \tag{5.7}$$

In psychology, for example, bimodal histograms of reaction times are sometimes encountered, which may reflect a tendency for subjects to behave in some standard fashion a proportion α of the time, producing reaction times with probability density function $f_1(x)$, say, but the remainder of the time, possibly due to a loss in concentration, to produce reaction times that tend to be longer than before, with probability density function $f_2(x)$, say. Cox (1966) provides further discussion of this example.

Another example is provided by human height histograms, which could be bimodal due to a mixture of different male and female height histograms. However, samples from such mixtures may not obviously reflect the mixture form of the underlying p.d.f., as is the case in the histogram of Fig. 5.8.

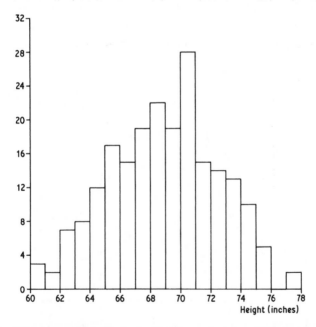

Figure 5.8   Histogram describing the heights of some undergraduates at the University of Kent (taken from Fuller and Lury, 1977, p. 14). For the 142 male undergraduates, heights range from 63 to 77 inches, with a modal height of 70 inches; for the 69 female undergraduates, heights range from 60 to 70 inches, with a modal height of 65 inches.

Indeed, we shall see that in many applications in simulation the mixture form of (5.7) is adopted solely for convenience, even for unimodal distributions such as the normal distribution. The convenience arises if α is fairly large and

$f_1(x)$ is a probability density function which is appreciably easier to simulate from than $f(x)$ itself. If we sample from $f_1(x)$ with probability α, and from $f_2(x)$ with probability $(1-\alpha)$, then because of the relationship of (5.7) we obtain a random variable, X, with probability density function $f(x)$. We see this simply as follows:

$$\Pr(X \leq x) = \alpha \Pr(X \leq x \mid \text{sample from } f_1(x)) + (1-\alpha)\Pr(X \leq x \mid$$
$$\text{sample from } f_2(x))$$

$$= \alpha \int_{-\infty}^{x} f_1(y)\,dy + (1-\alpha) \int_{-\infty}^{x} f_2(y)\,dy$$

$$= \int_{-\infty}^{x} f(y)\,dy \qquad \text{by (5.7)}$$

Let us now consider two examples which illustrate the use of the method.

EXAMPLE 5.7
Suppose we want to simulate random variables with the p.d.f. of Fig. 5.9.

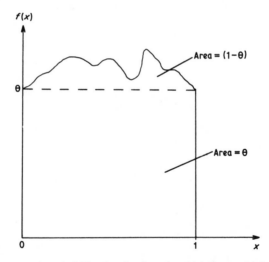

Figure 5.9   An unusual probability density function $f(x)$, from which we shall simulate using a composition, the first density, $f_1(x)$, of which is the $U(0, 1)$ density.

We note here that we can write

$$f(x) = \theta + f(x) - \theta \qquad \text{for any} \quad 0 \leq x \leq 1$$

i.e.,
$$f(x) = \theta \times 1 + (1-\theta)\left(\frac{f(x)-\theta}{1-\theta}\right) \tag{5.8}$$

As $\int_0^1 f(x)\,dx = 1$, then $\theta < 1$, and (5.8) is of the same form as (5.7). To simulate from $f(x)$, with probability θ we simply select a $U(0, 1)$ random variable, while with probability $(1-\theta)$ we simulate from the p.d.f.,

$$f_2(x) = \left(\frac{f(x) - \theta}{1 - \theta}\right)$$

which has subsumed the features of $f(x)$ which made it a difficult p.d.f. from which to simulate. We can simulate from both $f(x)$ and $f_2(x)$ using a rejection method, but the advantage of the composition of (5.8) is that one only simulates from $f_2(x)$ with probability $(1-\theta)$, and in the illustration of Fig. 5.9, $(1-\theta)$ is appreciably less than 0.5.

EXAMPLE 5.8
A random variable X with the simple beta probability density function

$$f(x) = 6x(1-x) \quad \text{for} \quad 0 \le x \le 1$$

can be simulated quite easily by either the table-look-up method, or the rejection method (see Exercise 5.18). Figure 5.10 shows how we may use a

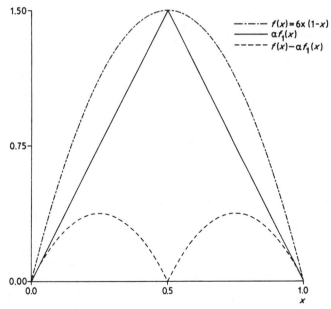

Figure 5.10   The beta density function, $f(x) = 6x(1-x)$ and the basis for a composition.

composition method. Here, $f_1(x)$ is simply the symmetric triangular density over (0,1), i.e.

$$f_1(x) = \begin{cases} 4x & \text{for} \quad 0 \le x \le 0.5 \\ 4(1-x) & \text{for} \quad 0.5 \le x \le 1 \end{cases}$$

From simple geometrical consideration we see that we must have $\alpha \le \frac{3}{4}$, and as we want α to be as large as possible, we take $\alpha = \frac{3}{4}$. The second density in the composition is then given by

$$f_2(x) = \frac{f(x) - \alpha f_1(x)}{1 - \alpha} = \begin{cases} 12x(1-2x) & \text{for} \quad 0 \le x \le 0.5 \\ 12(1-2x)(x-1) & \text{for} \quad 0.5 \le x \le 1 \end{cases}$$

We shall leave the reader to consider how we might simulate from $f_2(x)$.

In general, suppose we have a probability density function $f(x)$, and that $f_1(x)$ is a probability density function of roughly similar shape, but that it is appreciably easier to simulate from $f_1(x)$ than from $f(x)$. We shall see shortly why we want $f(x)$ and $f_1(x)$ to be of similar shape.

We can formally write, for any α in the range $0 < \alpha < 1$,

$$f(x) = \alpha f_1(x) + (1-\alpha)\left(\frac{f(x) - \alpha f_1(x)}{1-\alpha}\right)$$

and from the above discussion we see that we can simulate from $f(x)$ by simulating from $f_1(x)$ with probability α, and from $f_2(x) = (f(x) - \alpha f_1(x))/(1-\alpha)$ with probability $(1-\alpha)$. As $f_1(x)$ is chosen to be relatively easy to simulate from, we clearly want α to be as large as possible. The constraint on α is that for all x we must have $f(x) - \alpha f_1(x) \ge 0$, in order to ensure that $f_2(x)$ is also a probability density function (it is easy to see that its integral is unity). Now if

$$\alpha = \min_{x}\left(\frac{f(x)}{f_1(x)}\right)$$

and a positive, non-zero minimum can be found, then $\alpha \le f(x)/f_1(x)$, i.e. $f(x) - \alpha f_1(x) \ge 0$, as required, and there is at least one x for which $\alpha f_1(x) = f(x)$, so that a larger value of α cannot be found. This approach to finding α is, of course, analogous to the general approach given in the last section for finding k. Here we can consider α as shrinking $f_1(x)$ so that it just fits completely under $f(x)$, as we have seen in the last two examples. This of course explains why we seek an $f_1(x)$ to be of roughly similar shape to $f(x)$: the more similar in shape $f(x)$ and $f_1(x)$ are, then the larger the shrinking parameter α can be. For rejection, then, we envelop $f(x)$, but for composition it is $f(x)$ itself that plays the enveloping rôle. The method generalizes in a straightforward way, so that we could repeat the procedure for $f_2(x)$, and so on, ending ultimately with the mixture:

$$f(x) = \sum_{i=1}^{n+1} \alpha_i f_i(x)$$

in which

$$\sum_{i=1}^{n+1} \alpha_i = 1$$

and

$$\alpha_i > 0 \qquad \text{for} \quad 1 \le i \le n+1,$$

all the $f_i(x)$ are probability density functions, and

$$f_{n+1}(x) = \left(f(x) - \sum_{i=1}^{n} \alpha_i f_i(x) \right) \bigg/ \left(1 - \sum_{i=1}^{n} \alpha_i \right)$$

In choosing n, one has to counterbalance the difficulty of simulating from the final density function, $f_{n+1}(x)$, with the size of α_{n+1}, and the general desirability of keeping n small.

In the last two examples, all the p.d.f.'s considered had a finite range. As we shall see in the next section, it sometimes happens that $f(x)$ has an infinite range, but $f_1(x)$ has a finite range. In such a case, we have a range of x for which $f_1(x) = 0$, but $f_2(x) > 0$. In fact, as is shown in the next section, we can also have $f_2(x) = 0$ and $f_1(x) > 0$ for certain x.

*5.5 Combining methods for simulating normal random variables

In recent years much ingenuity has been devoted to devising composition methods for the standard normal probability density function. These approaches have also employed rejection, table-look-up and particular methods, and it is fascinating to see all of these different tools put to work on the one problem. As with the rejection method, many different compositions can be formed for any one probability density function, and here we shall just consider one, for the $N(0, 1)$ density. Due to Marsaglia and Bray (1964), the method gives rise to what has been termed their 'convenient' algorithm. Other methods are discussed in Exercises 5.36–5.38.

What many of the different methods proposed for the $N(0, 1)$ p.d.f. have in common, however, is the initial isolation of the *tails* of the normal density function, and the first composition usually taken is:

$$\frac{e^{-x^2/2}}{\sqrt{2\pi}} = \alpha\phi_1(x) + (1-\alpha)\phi_2(x) \tag{5.9}$$

in which

$$\phi_1(x) = \begin{cases} \dfrac{1}{\alpha} \dfrac{e^{-x^2/2}}{\sqrt{(2\pi)}} & \text{for} \quad -3 \le x \le 3 \\ 0 & \text{for} \quad |x| > 3 \end{cases}$$

and
$$\phi_2(x) = \begin{cases} \dfrac{1}{(1-\alpha)} \; \dfrac{e^{-x^2/2}}{\sqrt{(2\pi)}} & \text{for} \quad |x| > 3 \\[2ex] 0 & \text{for} \quad -3 \le x \le 3 \end{cases}$$

where

$$1 - \alpha = 2 \int_{-\infty}^{-3} \frac{e^{-x^2/2}}{\sqrt{(2\pi)}} \simeq 0.0027$$

to 4 places of decimals. Here, then, is an instance of the two component p.d.f.'s in (5.7) having different ranges. $|x| > 3$ is used to define the normal p.d.f. tails as 3 is suitably large and, as we shall see, ties in conveniently with the approaches adopted in what follows.

The composition of (5.9) means that most of the time we simulate from the *expanded* normal density, $\phi_1(x)$, over the finite range $|x| \le 3$, while with the very small probability $(1 - \alpha)$ we simulate from the p.d.f. $\phi_2(x)$, formed by expanding the tail areas from the standard normal p.d.f.

Let us consider $\phi_2(x)$ first of all. A random variable X with probability density function $\phi_2(x)$ is simply an $N(0, 1)$ random variable, conditioned to be $|X| \ge 3$. Such random variables result from the Box–Müller or Polar Marsaglia methods of Section 4.2 as follows: in the Box–Müller notation of Equation (4.1), if the exponential variable $-2 \log_e U_1 > 9$, then from the geometrical explanation of Section 4.2.1, there is a good chance that at least one of N_1 and N_2 is greater than 3 in modulus, as required. Certainly, if $-2 \log_e U_1 < 9$ then neither of N_1 and N_2 will be greater than 3, and so the standard approach of Section 4.2 towards constructing the conditioned normal variables that we require would be very wasteful. However, as is discussed in Exercise 5.24, $Y = 9 - 2 \log_e U_1$ is a random variable with the required exponential distribution, but *conditional on being greater than 9.* We can therefore simulate from the p.d.f. $\phi_2(x)$ by replacing $-2 \log_e U_1$ in Equation (4.1) by $(9 - 2 \log_e U_1)$, but only accepting a resulting N_1 or N_2 value if it is greater than 3 in modulus. Correspondingly, we can modify the Polar Marsaglia method by replacing $(-2 \log_e W)$ in Equation (4.2) by $(9 - 2 \log_e W)$, and proceeding in the same fashion.

There is more discussion of tail area simulation in the solution to Exercise 5.38. So far we have used the particular approaches of Section 4.2, and the table-look-up method to give exponential random variables of mean 2. Now we shall return to $\phi_1(x)$.

Figure 5.11 illustrates $\phi_1(x)$ and also the probability density function of the random variable

$$Y = 2(U_1 + U_2 + U_3 - 1.5) \qquad -3 \le Y \le 3$$

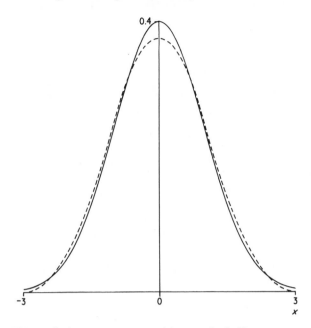

Figure 5.11   The preliminary to a composition method. Here we want to simulate from $\phi_1(x)$ (denoted by a solid line) the standard normal p.d.f. conditioned to the range $|x| \le 3$. It is proposed to use as the first p.d.f. in the composition, $f_1(x)$ (dashed line), the density of $Y = 2(U_1 + U_2 + U_3 - 1.5)$, where U_1, U_2 and U_3 are independent $U(0, 1)$ random variables.

in which U_1, U_2 and U_3 are independent $U(0, 1)$ random variables. (See Exercise 4.8). The two curves are of similar shape, and Y is clearly easy to simulate. We shall now, therefore, seek a composition for $\phi_1(x)$, with the first p.d.f. in the composition being $f_1(x)$, the probability density function of Y, given by (see Exercise 4.8)

$$f_1(x) = \begin{cases} (3 - x^2)/8 & -1 \le x \le 1 \\ (3 - |x|)^2/16 & 1 \le |x| \le 3 \\ 0 & |x| \le 3 \end{cases}$$

Using the approach outlined in the last section, we want to minimize $q(x) = \phi_1(x)/f_1(x)$ with respect to x varying over the range $|x| \le 3$. Because $f_1(x)$ is specified differently over different ranges for x, we shall deal with these different ranges separately.

First let us consider $0 \le x \le 1$. Here,

$$q(x) = \frac{8 \, e^{-x^2/2}}{\alpha(3 - x^2) \sqrt{(2\pi)}}$$

$$l(x) = \log q(x) = \text{constant} - \frac{x^2}{2} - \log(3 - x^2)$$

$$\frac{d}{dx} l(x) = -x + \frac{2x}{(3 - x^2)}$$

$$= 0 \text{ when } x = 0 \text{ and when } 3 - x^2 = 2, \text{ i.e. } x = 1$$

$$\frac{d^2 l(x)}{dx^2} = -1 + \frac{2}{(3 - x^2)} + \frac{4x^2}{(3 - x^2)^2}$$

i.e. $\frac{d^2 l(x)}{dx^2}$ is negative when $x = 0$, and positive when $x = 1$.

Next we shall consider the range $1 \leq x \leq 3$.

Here,
$$q(x) = \frac{16 e^{-x^2/2}}{\alpha(3 - x)^2 \sqrt{(2\pi)}}$$

$$l(x) = \log q(x) = \text{constant} - \frac{x^2}{2} - 2\log(3 - x)$$

$$\frac{d}{dx} l(x) = -x + \frac{2}{(3 - x)}$$

$$= 0 \qquad \text{when} \quad 3x - x^2 = 2$$

i.e. when $x = 1$ and when $x = 2$.

$$\frac{d^2}{dx^2} l(x) = -1 + \frac{2}{(3 - x)^2}$$

i.e. $d^2 l(x)/dx^2$ is negative when $x = 1$, and positive when $x = 2$, revealing a minimum to $q(x)$ when $x = 2$.

Thus for the case of $x \geq 0$, $q(x)$ has a maximum when $x = 0$, a saddle-point when $x = 1$, and a minimum when $x = 2$. We need not consider the case $x \leq 0$ separately because of the symmetry present, and so we can conclude that $q(x)$ has minima at $x = \pm 2$ in the range $|x| \leq 3$.

Hence if we write

$$\phi_1(x) = \alpha_1 f_1(x) + \alpha_2 f_2(x)$$

$$\alpha_1 = \frac{\phi_1(2)}{f_1(2)} = \frac{16 e^{-2}}{\alpha \sqrt{(2\pi)}}$$

and overall, from considering the compositions for $e^{-x^2}/\sqrt{(2\pi)}$ and $\phi_1(x)$, we simulate from $f_1(x)$ with probability

$$\alpha\alpha_1 = \frac{16 e^{-2}}{\sqrt{(2\pi)}} \approx 0.8638$$

Here we see a dramatic demonstration of the possible power of the composition method: over 86 per cent of the time we can expect to simulate an

$N(0, 1)$ random variable by simply taking a linear function of the sum of three independent $U(0, 1)$ random variables.

In fact there is still more of interest remaining in this example. With probability $\alpha\alpha_2$ we must simulate from the probability density function

$$f_2(x) = \left(\frac{\phi_1(x) - \alpha_1 f_1(x)}{1 - \alpha_1}\right) \qquad -3 \le x \le 3$$

i.e. with probability $\alpha\alpha_2 = \alpha(1 - \alpha_1) = (0.9973 - 0.8638) = 0.1335$.

Figure 5.12 presents a graph of $\phi_1(x) - \alpha_1 f_1(x)$, and the form of the graph suggests proceeding further with the composition for $\phi_1(x)$, by now using a triangular p.d.f. and setting

$$f_2(x) = \beta g(x) + (1 - \beta)h(x)$$

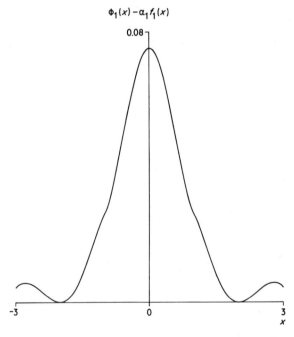

$$\phi_1(x) - \alpha_1 f_1(x)$$

Figure 5.12   The residual curve $\phi_1(x) - \alpha_1 f_1(x)$ following the composition method envisaged in Fig. 5.11.

where
$$g(x) = (6 - 4|x|)/9 \qquad \text{for } |x| \le 1.5$$
$$= 0 \qquad \text{for } |x| > 1.5$$

$g(x)$ is simply the probability density function of

$$Y = 1.5(U_1 + U_2 - 1)$$

where U_1 and U_2 are independent $U(0, 1)$ random variables (see Exercise 4.8). In this case, with the aim of determining β, $f_2(x)/g(x)$ cannot be minimized explicitly, but a numerical method such as Newton–Raphson readily provides us with $\beta \approx 0.8292$, the minimum occurring at $x = \pm 0.8739$. We thus simulate from $g(x)$ with probability $\alpha(1 - \alpha_1)\beta = 0.1107$, so that over 97 per cent of the time we use the two simple p.d.f. 's $f_1(x)$ and $g(x)$.

The three compositions that we have dealt with here can be written as one, to give

$$\frac{e^{-x^2/2}}{\sqrt{(2\pi)}} = 0.8638 \, f_1(x) + 0.1107 \, g(x) + 0.0027 \, t(x) + 0.0228 \, r(x)$$

$$\text{for} \quad -\infty \leq x \leq \infty \tag{5.10}$$

where $t(x)$ is the tail-area p.d.f. which we considered earlier, and $r(x)$ is the p.d.f. that remains for $|x| \leq 3$.

We simulate from the p.d.f. $r(x)$ only with probability 0.0228, and as $r(x)$ is of a fairly complicated form (shown in Fig. 5.18) we can simulate from it by means of simple rejection, using a rectangular enveloping region over the finite range $|x| \leq 3$, with, it can be shown, probability 0.53 of rejection (see Exercises 5.16 and 5.39).

The above derivation of (5.10) should not disguise the fact that (5.10) is a description of one way of dividing up the area under the $N(0, 1)$ probability density function, precisely as was done in a different case in Example 5.7. The end-result is shown in Fig. 5.13.

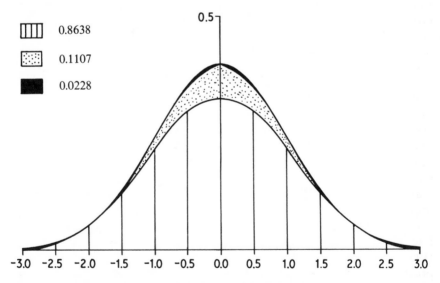

Figure 5.13   A representation of the composition given in Equation (5.10) for the range $|x| \leq 3$. The regions shown have the areas indicated above.

5.6 Discussion and further reading

The examples considered in this chapter form only a very small subset of the many interesting and complicated approaches that have been devised in recent years. For instance, having obtained a normal random variable, then one can even use the normal distribution itself as an enveloping distribution; see, for example, Ahrens and Dieter (1974) and Atkinson (1979). More examples are to be found in the exercises.

We have seen that uniform $U(0, 1)$ random variables are the building-blocks for the simulation of any other random variable. Complicated algorithms utilizing composition and rejection methods, and sometimes requiring the storage of a large number of constants, are designed for use on computers that work to high precision. These algorithms are often programmed in machine code and tend to be the most efficient. They are therefore most suitable if one is seeking to provide a computer with an efficient package of programs for simulating a variety of random numbers, which will be used frequently by a large number of individuals. Comparisons of different algorithms, using speed and efficiency, have been undertaken by a number of authors—see, for example, Atkinson and Pearce (1976), Kinderman and Ramage (1976) and Ripley (1983b). Appleton (1976) pointed out that certain methods can be programmed in the programming language APL, to take advantage of APL's vector-handling capabilities, and as an example he found the Box–Müller method to be 30 times faster than the Marsaglia and Bray 'convenient' method, when both were programmed in APL. This is partly due to the fact that APL programs are interpreted, rather than compiled, as are FORTRAN programs. Using FORTRAN, Atkinson and Pearce (1976) found Box–Müller to be roughly twice as slow as the convenient method. Distributed array processing is another factor which could influence the comparison of different algorithms (cf. Exercise 3.16).

Ripley (1983b) provides a list of relatively efficient simple algorithms for a variety of distributions. Most users of main-frame computers will be likely to use library subroutines such as those described in Section A1.1. Because of the time-lag before library subroutines are changed to accommodate new developments, these programs may not always be the most efficient. Each individual clearly has to experiment with the facilities available if it is suspected that long generation times of random variables could render a simulation impractical.

Of course, the simplest way for human beings to simulate random variables is to use tables of realizations of such random variables, such as those by Wold (1954), providing normal random variables, and those by Barnett (1965), which provide exponential random variables. In the absence of such tables, the table-look-up approach is also easily performed by hand if one has suitable tables of cumulative distribution functions, and only a small-scale simulation is envisaged. Some such tables can be found in Harter (1964), Lieberman and Owen (1961), Mardia and Zemroch (1978), Neave (1978), Odeh, *et al.* (1977), Williamson and Bretherton (1963) and Worsdale (1975).

Simulation of discrete random variables by the table-look-up method can be very time consuming. This occurs with the Poisson distribution, for example, if it has large mean, in which case the particular method for this distribution, described in the last chapter, will also be inefficient. There is further discussion of this point in Exercises 5.3 and 5.4. The range-dividing technique, discussed in Section 5.1, can be generalized by dividing the range into $d > 2$ parts, as in Neave (1972), who provides ALGOL programs for several discrete distributions. A faster search procedure is the optimum binary tree search described in Knuth (1968, p. 400). As can be seen from Section A1.1, the NAG library of computer programs simulates all discrete distributions by first of all establishing a reference vector of cumulative sums, and then performing an indexed search by means of the routine G05EYF. The IMSL routine for the table-look-up method for a general discrete distribution is GGDT (see Section A1.1).

The polar Marsaglia method of Section 4.2.2 shows that the ratio V_1 / V_2 of the co-ordinates of a point uniformly distributed over a disc of unit radius and centred on the origin has a Cauchy distribution (see Exercise 5.8). Kinderman and Monahan (1977) have generalized this result to provide a new general method for simulating random variables, viz., the *ratio* method—see Ripley (1983b) for illustrations of its use. A further new general method is the *alias* method for discrete random variables, described in Exercise 5.42.

In this chapter we have only considered univariate random variables, but table-look-up, rejection and composition methods may also be used for multivariate random variables. Kemp and Loukas (1978a, b) consider the table-look-up method for a bivariate Poisson distribution, and the table-look-up method for bivariate Poisson and normal distributions is discussed in Exercises 5.10 and 5.11. Best and Fisher (1979) use a rejection method on the circle, enveloping the von Mises distribution with a wrapped Cauchy distribution.

We shall conclude this chapter with some further discussion of methods for simulating normal random variables.

*5.7 Additional approaches for normal random variables

The table-look-up method for normal random variables is difficult to program for computers because of the intractable form of the standard normal cumulative distribution function, $\Phi(x)$, and its inverse, $\Phi^{-1}(x)$. Various authors have approached this problem by providing approximate methods—see, for example, Zelen and Severo (1966) and Wetherill (1965). Wetherill's approach employs the attractive idea that an efficient algorithm can result from using one approximation to $\Phi^{-1}(x)$ in the middle of the range for x, but another, more complicated algorithm in the tails, which would be used far less frequently. This idea is simply providing a composition method, the components of which are simulated using approximate table-look-up methods. Some other approaches are described below.

Because of the similar shapes of the normal and logistic probability density functions, it is natural to try to approximate the normal cumulative distribution function by the simple logistic cumulative distribution function. In order to obtain a good match over the middle of the range, the logistic cumulative distribution function that may be used is

$$F_1(x) = \left[1 + \exp\left(-2\sqrt{\left(\frac{2}{\pi}\right)}x\right)\right]^{-1} \qquad -\infty \le x \le \infty$$

as this curve has the same slope at $x = 0$ as does $\Phi(x)$. An alternative possibility which might be considered is the logistic cumulative distribution function,

$$F_2(x) = \left[1 + \exp\left(-\frac{\pi x}{\sqrt{3}}\right)\right]^{-1} \qquad -\infty \le x \le \infty$$

corresponding to a random variable with zero mean and unit variance. $F_1(x)$ and $F_2(x)$ are illustrated in Fig. 5.14, for $0 \le x \le 3$.

Table 5.1 is taken from Page (1977), who tries to improve a logistic approximation by adding an extra parameter, resulting in the cumulative distribution function

$$G(x) = \{1 + \exp[-2a_1 x(1 + a_2 x^2)]\}^{-1} \qquad -\infty \le x \le \infty$$

Note that as the coefficient of the new parameter, a_2, is x^3, and not x^2, which may have been considered a more natural choice, then we preserve the property $G(x) + G(-x) = 1$, and the corresponding probability density function is symmetric about $x = 0$.

If $a_1 = \sqrt{(2/\pi)}$, and a_2 is chosen by least squares, then a value of $a_2 = 0.044\,715$ is obtained. A slightly better approximation is obtained by allowing both a_1 and a_2 to be chosen by least squares, but the advantage of keeping $a_1 = \sqrt{(2/\pi)}$ is that if one wanted to approximate $\Phi(x)$ this way on a hand-calculator, only one constant needs to be remembered, most calculators having a 'π' key. To simulate approximate $N(0, 1)$ random variables we need $\tilde{x} = G^{-1}(U)$ (see Exercise 5.9). Some examples are given in Table 5.1.

Hamaker (1978) and Schmeiser (1979) provide further approximations that are suitable for computation on a hand-calculator, and more recent work is described in Bailey (1981) and Lew (1981).

Kinderman and Ramage (1976) use an even simpler p.d.f. for $f_1(x)$, the first p.d.f. in a composition for the standard normal density, than that resulting from the sum of three $U(0, 1)$ random variables. In their case, they used the p.d.f. of the sum of just two $U(0, 1)$ random variables, as illustrated in Fig. 5.15.

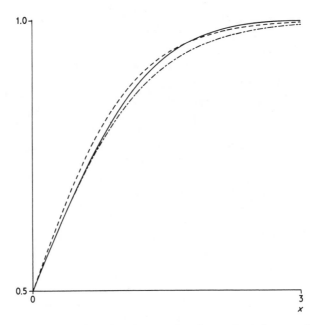

Figure 5.14 An illustration for $0 \le x \le 3$ of two logistic cumulative distribution functions, (—·—): $F_1(x) = \left[1 + \exp\left(-2\sqrt{\left(\dfrac{2}{\pi}\right)}x\right)\right]^{-1}$ and (———): $F_2(x) = \left[1 + \exp\left(-\dfrac{\pi x}{\sqrt{3}}\right)\right]^{-1}$, either of which may be used as a rough approximation to the normal cumulative distribution function, $\Phi(x)$, denoted by (———).

Table 5.1 Approximating the standard normal cumulative distribution function $\Phi(x)$. Two possible approximations are $F_1(x)$ and $G(x)$, explained in the text. $\tilde{x}$ results from inverting $G(x)$ (from Page, 1977).

x	$1 - \Phi(x)$	$1 - F_1(x)$	$1 - G(x)$	$\tilde{x}$
0	0.5	0.5	0.5	0
0.1	0.460 172 2	0.460 190 2	0.460 172 5	0.1
0.3	0.382 088 6	0.382 551 9	0.382 096 9	0.3
0.5	0.308 537 5	0.310 478 2	0.308 572 0	0.5001
1.0	0.158 655 3	0.168 573 8	0.158 808 0	1.0006
1.5	0.066 807 2	0.083 657 9	0.066 952 3	1.5011
2.0	0.022 750 1	0.039 485 4	0.022 701 2	1.9991
2.5	0.006 209 7	0.018 174 0	0.006 033 7	2.4901
3.0	0.001 349 9	0.008 266 0	0.001 212 5	2.9693
3.5	0.000 232 6	0.003 739 0	0.000 176 1	3.4332
4.0	0.000 031 7	0.001 687 1	0.000 017 6	3.8800

Figure 5.16 illustrates $\phi(x) - \alpha f_1(x)$ for $-3 \le x \le 3$, which may be simulated by means of rejection, the details of which are discussed in the

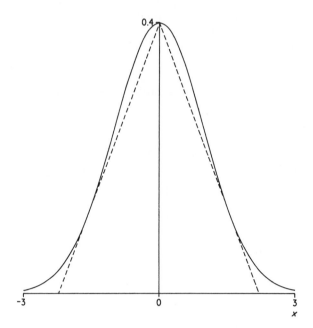

Figure 5.15 The standard normal density function $\phi(x)$ (———) over the range, $(-3, +3)$, and $\alpha f_1(x)$ (------), in the notation of the composition method of the Section 5.4. Here $f_1(x)$ is the probability density function of $\beta(U_1 + U_2 - 1)$, where U_1 and U_2 are independent $U(0, 1)$ random variables. α is chosen so that $\alpha f_1(0) = 1/\sqrt{(2\pi)}$, and β must be chosen to give the triangle illustrated here, i.e. the largest symmetric triangle with height $1/\sqrt{(2\pi)}$ which can be fitted under $\phi(x)$.

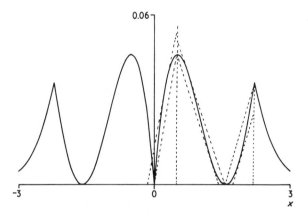

Figure 5.16 A graph of $\phi(x) - \alpha f_1(x)$ for $|x| \leq 3$, resulting from the curves of Fig. 5.15. The dotted lines relate to a particular approach used for the rejection method employed to simulate from the probability density function which is a positive multiple of this curve, as discussed in the solution to Exercise 5.38.

solution to Exercise 5.38. Also presented in Exercise 5.38 is the full algorithm of Kinderman and Ramage (1976).

5.8   Exercises and complements

(a) General

5.1   Use Equation (2.3) to show that when U is a $U(0, 1)$ random variable, then $1 - U$ is also a $U(0, 1)$ random variable.

5.2   When X has the half-normal p.d.f.

$$f_X(x) = \sqrt{\left(\frac{2}{\pi}\right)} e^{-x^2/2} \qquad \text{for} \qquad x \geq 0$$

show that $\tilde{X}$, defined by:

$$\begin{cases} \tilde{X} = X \text{ with probability } \tfrac{1}{2} \\ \tilde{X} = -X \text{ with probability } \tfrac{1}{2} \end{cases}$$

has the standard normal p.d.f.

5.3   Simulating Poisson random variables with large mean can be time consuming, whether one uses a particular approach, as in Chapter 4, or a general, table-look-up approach. Discuss one way of tackling this problem, in the context of the distribution of the sum of two independent random variables, each with Poisson distributions. See Exercise 2.8(b).

(b) Table-look-up methods

†**5.4**   Select a Poisson distribution with mode different from zero.
  (a) Simulate from this distribution using the table-look-up method.
  (b) Repeat (a), but employ a θ, as suggested in Section 5.1.
  (c) Repeat (a) but employ two such θ's, thus dividing the range into three parts.
  (d) Repeat (b) after having first ordered the probabilities in increasing order.
  Compare the efficiencies of these four approaches (cf. also Exercise 5.3, and Kemp, 1982).

†**5.5**   (a) Write a computer program to simulate a random variable, X, from

the triangular distribution defined by:

$$f(x) = \begin{cases} 0 \\ x \\ 2-x \\ 0 \end{cases} \qquad F(x) = \begin{cases} 0 & x < 0 \\ x^2/2 & 0 \le x \le 1 \\ 2x - (x^2/2) - 1 & 1 \le x \le 2 \\ 1 & x > 2 \end{cases}$$

using the inversion method. Here $f(x)$ is the probability density function of x, and $F(x)$ is the cumulative distribution function of x. This is the method used in the IMSL routine GGTRA (see Section A1.1).

(b) Compare the efficiency of this program with one which simulates such a random variable by simply summing two independent $U(0, 1)$ random variables.

5.6   Use the table-look-up method to simulate 10 random variables:

(a) from the binomial distribution $B(6, 1/3)$; and
(b) from the normal distribution $N(1, 2)$, using tables of the standard normal cumulative distribution function.

*5.7   Use the table-look-up method to simulate random variables with the simple beta probability density function

$$f(x) = 6x(1-x) \qquad \text{for } 0 \le x \le 1.$$

*5.8   (a) Explain how to simulate random variables from the Cauchy distribution, with probability density function,

$$f(x) = \frac{1}{\pi(1+x^2)} \qquad \text{for } -\infty \le x \le \infty$$

using the inversion method. An algorithm using this approach is provided by the IMSL routine, GGCAY (see Section A1.1).

(b) If N_1 and N_2 are independent standard normal random variables then, as we saw in Exercise 2.15(b) and Exercise 4.5(b), their ratio $C = N_1/N_2$ has the Cauchy probability density function of (a) above. Explain how this result may be deduced from (a) and an understanding of the Box–Müller method described in Section 4.2.1.

*5.9   The approximate approach for simulating $N(0, 1)$ random variables described in Section 5.7 involved setting $\tilde{x} = G^{-1}(u)$, where $G(x) = [1 + \exp\{-2a_1 x(1 + a_2 x^2)\}]^{-1}$. Solve for $\tilde{x}$.

*5.10   Discuss how you would use the table-look-up method for simulating from the bivariate Poisson distribution of Exercise 4.7.

*5.11 Discuss how you would use the inversion method for simulating from the bivariate normal distribution.

5.12 If X is a random variable with the exponential, $\lambda e^{-\lambda x}$ p.d.f., for $x \geq 0$, deduce the distribution of the integral part of X, viz., $Y = [X]$. Hence explain why, in (5.4), we obtain a geometric random variable by rounding up an exponential random variable.

†5.13 Use the inversion method to simulate from the following distributions:

(a) logistic: $f(x) = \dfrac{e^{-x}}{(1+e^{-x})^2}$ for $-\infty \leq x \leq \infty$.

Note that this method is implemented in the NAG routine: G05DCF (see Section A1.1).

(b) Weibull (see Exercise 2.3): $f(\omega) = \dfrac{\beta}{\gamma^\beta}\omega^{\beta-1}\exp\{-(\omega/\gamma)^\beta\}$ for

$0 \leq \omega < \infty,\ \beta > 0,\ \gamma > 0$.

Note that this method is implemented in the NAG routine G05DPF and the IMSL routine GGWIB (see Section A1.1).

(c) Pareto distribution:

$$\Pr(X \leq x) = 1 - \left(\frac{k}{x}\right)^a \qquad \text{for } a > 0,\ x \geq k > 0.$$

(d) Extreme-value distribution:

$$\Pr(X \leq x) = \exp\{-\exp((\xi - x)/\theta)\} \qquad \text{for } x \geq 0.$$

5.14 Provide a detailed algorithm for simulating from the logarithmic distribution of Exercise 4.22.

*5.15 Barnett (1980) presents the bivariate uniform p.d.f.:

$$f(u, v) = (1 - \alpha)[(2uv - u - v)\alpha + 1]\{\Psi(u, v)\}^{-3/2} \qquad (5.11)$$

where $\Psi(u, v) = (\alpha(u + v) - 1)^2 + 4\alpha(1 - \alpha)uv$ and $\alpha < 1,\ 0 \leq u, v \leq 1$. This probability density function is illustrated in Fig. 5.17 for the case $\alpha = -4$. It is constructed from a bivariate distribution of Plackett (1965), which is given implicitly by:

$$\frac{\{F(x, y)\{1 - F_X(x) - F_Y(y) + F(x, y)\}}{\{F_X(x) - F(x, y)\}\{F_Y(y) - F(x, y)\}} = 1 - \alpha$$

From Section 5.2 we can see that if we set $U = F_X(X)$ then U is $U(0, 1)$, and so is $V = F_Y(Y)$. This is an interesting reversal of the aim of Section 5.2, which is to progress from U to X. Verify that this substitution here results in the joint p.d.f. $f(u, v)$ of Equation (5.11). Derive further bivariate uniform distributions in this manner from the following bivariate distributions also presented by Barnett (1980):

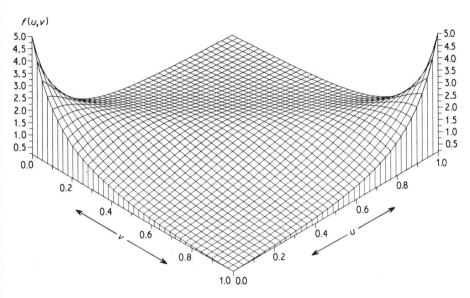

Figure 5.17   Isometric projection of the bivariate uniform density of Equation (5.11), from Morgan (1983).

(a) $F(x, y) = F_X(x) F_Y(y) \{ 1 - \alpha (1 - F_X(x))(1 - F_Y(y)) \}$
     for $|\alpha| < 1$.

(b) $f(x, y) = \{ (1 + \alpha x)(1 + \alpha y) - \alpha \} \exp(-x - y - \alpha x y)$
     for $0 < \alpha < 1$.

(this is a bivariate exponential distribution)

(c) $f(x, y) = \dfrac{1}{2\pi} (1 + x^2 + y^2)^{-3/2}$.

(this is a bivariate Cauchy distribution)

(c) Rejection methods

5.16   Figure 5.18 shows the probability density function $r(x)$ of Equation (5.10), resulting from the composition of (5.10). Explain how you would simulate from $r(x)$ using a rejection method.

5.17   To simulate from the probability density function given by

$$f(x) = \begin{cases} \dfrac{1}{\pi \sqrt{(1 - x^2)}}, & \text{for } -1 \le x \le 1 \\ 0 & \text{otherwise} \end{cases}$$

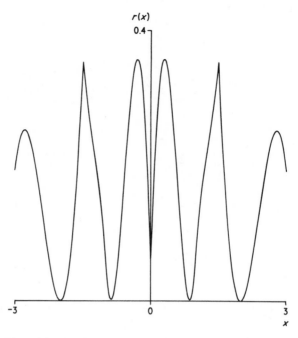

Figure 5.18   The p.d.f. $r(x)$ for $-3 \le x \le 3$, which is sampled with probability 0.0228 in the 'convenient' composition method of Equation (5.10).

we note that this is the probability density function of the random variable $X = \cos(\pi U)$, where U is a $U(0, 1)$ random variable. Use this result to devise a rejection method, based on the first quadrant of a circle and similar to the Polar Marsaglia method, for generating the required random variables.

5.18   Devise a rejection method, with an acceptance probability of not less than 8/9, for simulating random variables from the beta probability density function $f(x) = 6x(1 - x)$ for $0 \le x \le 1$.

†**5.19**   Describe how to simulate a random variable with the logistic probability density function, $e^{-x}(1 + e^{-x})^{-2}$ for $-\infty \le x \le \infty$, using a rejection method based on the exponential envelope, e^{-x} for $x \ge 0$.

5.20   Explain how to simulate normal random variables using a rejection method with an enveloping function based on a logistic p.d.f. Derive the probability of rejection (cf. Exercise 5.31).

†**5.21**   Repeat the approach adopted in Example 5.5 with $k\lambda e^{-\lambda x}$ for $x \ge 0$ as the enveloping function. Show that the probability of rejection is minimized if we take $\lambda = 1$, as in Example 5.5.

*5.22   Cheng (1977) presented a rejection method for simulating from the $\Gamma(\alpha, 1)$ distribution, where $\alpha > 1$. In his case,

$$h(x) = \lambda \mu x^{\lambda - 1} (\mu + x^{\lambda})^{-2} \qquad \text{for } x \geq 0$$

(a) Consider how you would simulate from the probability density function, $h(x)$.

(b) If $\mu = \alpha^{\lambda}$ and $\lambda = \sqrt{(2\alpha - 1)}$, determine the probability of rejection, and the mean number of variable selections until acceptance. The case $\alpha = 3$ is illustrated in Fig. 5.5, for $x < 8$.

*5.23   Compare the two rejection methods for simulating from a gamma distribution, given in Example 5.6 and Exercise 5.22.

*5.24   Explain how you would simulate a random variable that has an exponential distribution of mean 2, conditional on it being greater than 9 (cf. Exercise 5.26).

*5.25   Figure 5.16 presents a p.d.f. to be simulated from by means of a rejection method. Kinderman and Ramage (1976) used the method of triangles (see Marsaglia, MacLaren and Bray, 1964), which, in outline, is as follows.

 If a p.d.f. from which one wants to simulate can be sandwiched between two parallel lines, the X-value for the rejection method is simulated from an appropriate triangular distribution corresponding to the upper of the parallel lines. When the corresponding uniformly distributed Y value is less than the appropriate value on the lower of the parallel lines, then X is accepted, and it is not necessary to compute the formula for the curve. If, however, the Y value is greater than the appropriate value on the lower line then it is necessary to compute the formula for the curve in order to decide on rejection or acceptance.

 Discuss the objective of such an approach, and explain its use for the beta p.d.f. of Exercise 5.18 (cf. comments in the solution to Exercise 5.23).

*5.26   Marsaglia (1964) proposed the following method for simulating standard normal random variables X, conditional upon $X > a > 0$. Let U_1, U_2 be two independent $U(0, 1)$ random variables. Set

$$X = (a^2 - 2 \log_e U_1)^{1/2}.$$

Accept X as a realization of the required random variable if $U_2 X < a$. Otherwise, reject U_1 and U_2, and start again. Verify that X has the required distribution (cf. Exercise 5.24 and the comments of Section 5.5).

†5.27   If U_1 and U_2 are independent $U(0, 1)$ random variables, show that, conditional upon $(2U_1 - 1)^2 + U_2^2 \leq 1$, then $C = (2U_1 - 1)/U_2$ has the Cauchy distribution of Section 2.11. Note that this method is implemented in the NAG routine G05DFF (see Section A1.1).

5.28 If X_1 and X_2 are independent, identically distributed exponential random variables with mean unity, show that, conditional upon $(X_1 - 1)^2 < 2X_2$, then X_1 has a half-normal p.d.f., derived from an $N(0, 1)$ distribution. (This result is due to von Neumann—see Kahn, 1956, p. 39.)

*5.29 Suppose we have a probability density function $f(x)$ which can be written in the form:
$$f(x) = cg(x)r(x)$$
where $g(x)$ is also a p.d.f., $c > 0$ is a constant, and over the range of x, $0 \leq r(x) \leq m$, for some finite m. Show that we can simulate X from $f(x)$ as follows:

(i) Simulate X from $g(x)$
(ii) Accept X if $Um < r(X)$

  where U is an independent $U(0, 1)$ variable. Otherwise reject X and U and start again at (i).

What is the rejection probability? An example is provided by Butcher (1960), in which $f(x)$ is half-normal, and $g(x)$ is exponential. This generalization of the rejection method can give rise to efficient '*switching*' algorithms, in which the rôles played by $g(x)$ and $r(x)$ change for different parts of the x-range; see Atkinson and Whittaker (1976), and Atkinson (1979b).

(d) Composition methods

†5.30 Use the composition approach of Section 5.4, as applied in the example of Equation (5.6), to simulate random variables with the Poisson distribution of Example 5.2.

5.31 Explain why it is not possible to simulate normal random variables using a composition, the first element of which, $f_1(x)$, is a logistic density.

†5.32 A continuous random variable X has the 'wedge-shaped' probability density function, $f_1(x) = \alpha - \alpha^2 x/2$, for $0 \leq x \leq 2/\alpha$ and $\alpha > 0$.

(a) Explain how you would simulate X.
(b) It is desired to simulate from the exponential p.d.f. $f(x) = \lambda e^{-\lambda x}$ for $x \geq 0$ and $2\lambda > \alpha$, using a composition, the first p.d.f. of which is to be $f_1(x)$. Derive the shrinking factor for $f_1(x)$, and deduce that, by

suitable choice of α, the probability of simulating from $f_1(x)$ in the composition can be made as large as $2/e$.

5.33 (a) Show that the random variable X, with probability density function

$$f(x) = \frac{e^{m-x}}{(e-1)} \qquad \text{for } (m-1) < x \leq m, \text{ where } m \geq 1$$

is obtained simply by setting $X = m - Y$, where Y has probability density function

$$f(y) = \frac{e^y}{(e-1)} \qquad \text{for } 0 \leq y < 1$$

The cumulative distribution function for X when $m = 1$ is illustrated in Fig. 5.2.

(b) By expanding $f(y)$ as a power series in y, show that we can simulate from $f(y)$ by means of a composition, simulating from probability density function, $(i+1)y^i$ for $0 \leq y < 1$, with probability

$$\frac{1}{(i+1)!(e-1)} \qquad \text{for } i \geq 0.$$

5.34 (*continuation*) We note that

$$e^{-x} = \frac{(e-1)e^{-m} \times e^{m-x}}{(e-1)} \qquad \text{for any } m \geq 1$$

Explain, with reference to Fig. 5.19, how this result may be used as a basis for a composition method for simulating from the probability density function, e^{-x} for $x \geq 0$.

*5.35 (*continuation*) Explain the following algorithm, given by Marsaglia (1961), for simulating random variables from the exponential e^{-x} p.d.f.:

(i) Simulate a discrete random variable, I, from the distribution

$$\frac{1}{(i+1)!(e-1)} \qquad \text{for } i \geq 0$$

(ii) Set $W = \max(U_1, U_2, \ldots, U_{I+1})$,
where the $\{U_j\}$ are independent $U(0, 1)$ random variables.

(iii) Simulate a discrete random variable, M, from the distribution

$$(e-1)e^{-m} \qquad \text{for } m \geq 1$$

(iv) Set $X = M - W$.

*5.36 Consider how you would simulate standard normal random variables using a composition method, in which the first p.d.f. in the composition, $f_1(x)$, is of trapezoidal form. See Ahrens and Dieter (1972) for an algorithm based on this approach.

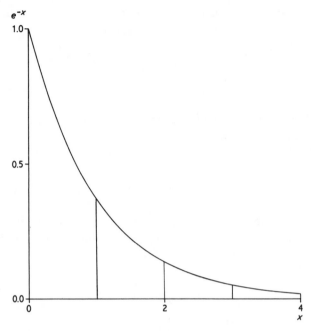

Figure 5.19　A breakdown of the p.d.f. e^{-x} into sections of area, $(e-1)e^{-m}$, for $m \geq 1$, for $x \leq 4$.

*5.37　Figure 5.20 illustrates a portion of the half-normal p.d.f. $2\phi(x)$, and $0.97f_1(x)$, in which $f_1(x)$ is a density composed of 97 rectangles, each of area 1/97. Figure 5.21 illustrates $2\phi(x) - 0.97f_1(x)$. Discuss how these curves may be used to simulate standard normal random variables. This method is due to Lenden–Hitchcock (1980) and is based on a method of Marsaglia, MacLaren and Bray (1964).

*5.38　Kinderman and Ramage (1976) produce the algorithm, given below, for their method discussed in Section 5.7. Explain how the method gives rise to this algorithm. (Note that $\xi = 2.216\,035\,867\,166\,471$ and $f(t) = \phi(t) - 0.180\,025\,191\,068\,563\,(\xi - |t|)$, for $|t| < \xi$. Here we preserve the high accuracy of constants given in the original source.)

Algorithm from Kinderman and Ramage (1976)

    1. Generate u. If $u < 0.884\,070\,402\,298\,758$, generate v and return $x = \xi \times (1.131\,131\,635\,444\,180\,u + v - 1)$.
    2. If $u < 0.973\,310\,954\,173\,898$, go to 4 below.
    3. Generate v, w. Set $t = \xi^2/2 - \log_e w$. If $v^2 t > \xi^2/2$, begin this step

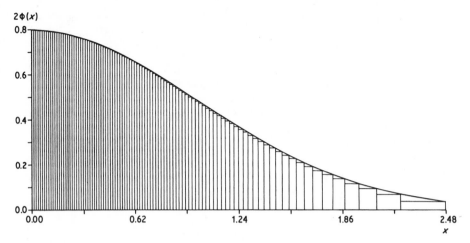

Figure 5.20   Part of the half-normal density,

$$2\phi(x) = \sqrt{\left(\frac{2}{\pi}\right)}\,e^{-x^2/2},$$

enveloping 97 rectangles, each of area 0.01.

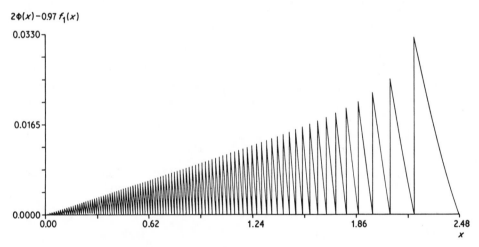

Figure 5.21   A graph of $2\phi(x)-0.97 f_1(x)$, from Fig. 5.20, in which $0.97 f_1(x)$ is the envelope of the rectangles shown in Fig. 5.20.

    again. Otherwise return $x = (2t)^{1/2}$ if $u < 0.986\,655\,477\,086\,949$ or
return $x = -(2t)^{1/2}$ if not.
4. If $u < 0.958\,720\,824\,790\,463$, go to 6 below.
5. Generate v, w. Set $z = v - w$ and $t = \xi - 0.630\,834\,801\,921\,960 \times$

$\min(v, w)$. If $\max(v, w) \leq 0.755\,591\,531\,667\,601$, go to 9. If $0.034\,240\,503\,750\,111|z| \leq f(t)$, go to 9. Otherwise, repeat this step.

6. If $u < 0.911\,312\,780\,388\,703$, go to 8.
7. Generate v, w. Set $z = v - w$ and

$$t = 0.479\,727\,404\,222\,441 + 1.105\,473\,661\,022\,070 \min(v, w).$$

If

$$\max(v, w) \leq 0.872\,834\,976\,671\,790$$

go to 9. If $0.049\,264\,496\,373\,128|z| \leq f(t)$ go to 9. Otherwise, repeat this step.

8. Generate v, w. Set $z = v - w$ and

$$t = 0.479\,727\,404\,222\,441 - 0.59550\,71380\,15940 \min(v, w).$$

If

$$\max(v, w) \leq 0.805\,577\,924\,423\,817$$

go to 9. If $0.053\,377\,549\,506\,886|z| \leq f(t)$, go to 9. Otherwise, repeat this step.

9. If $z < 0$, return $x = t$; otherwise return $x = -t$.

*5.39 For the residual p.d.f. $r(x)$ from the composition of Equation (5.10), show that the probability of rejection is 0.53 when we simulate from $r(x)$ using rejection and an enveloping rectangle.

(e) Additional methods

*5.40 Suppose W is $U(a, b)$, for some $b > a$, and suppose that for $a \leq x \leq b$, $0 \leq g(x) \leq 1$ for some function $g(x)$. Suppose N is the first integer ≥ 1 such that

$$g(W) \geq U_1 \geq U_2 \geq \ldots \geq U_{N-1} < U_N$$

where the $\{U_i\}$ are a sequence of independent, identically distributed $U(0, 1)$ variables. Thus $N = 1$, if and only if $g(W) < U_1$

$$N = 2, \text{ if and only if } g(W) \geq U_1 < U_2$$

etc.

Show that

$$\Pr(N = n | W = w) = \frac{g(w)^{n-1}}{(n-1)!} - \frac{g(w)^n}{n!} \qquad \text{for } n \geq 1$$

and deduce that

$$\Pr(N \text{ is odd} | W = w) = \exp(-g(w))$$

Finally, show that the conditional p.d.f. of W is given by

$$f_W(w \mid N \text{ is odd}) = \frac{\exp(-g(w))}{\displaystyle\int_a^b \exp(-g(w))\, dw} \tag{5.12}$$

*5.41   (*continuation*)   Use the results of the last exercise to provide a rejection method to simulate random variables with the p.d.f. of Equation (5.12) over the range (a, b). This is the basis of what is known as *Forsythe's method* (see Forsythe, 1972), which has been used for a variety of distributions (see Atkinson and Pearce, 1976). Of course, the requirement that $g(x) \leq 1$ is, by itself, very restrictive; however, this restriction can be overcome by dividing up the range of x into a number of intervals, and then first of all using a composition to determine the appropriate interval: if $\tilde{g}(x)$ is an increasing function of x, over the range $(0, \infty)$, say, then if the interval (q_i, q_{i+1}) is chosen by the first stage of the composition method, $\{\tilde{g}(q_i + x) - \tilde{g}(q_i)\}$ plays the rôle of $g(x)$ in the last exercise. The $\{q_i\}$ must be chosen so that

$$0 \leq \tilde{g}(q_i + x) - \tilde{g}(q_i) \leq 1 \qquad \text{for } 0 \leq x \leq (q_{i+1} - q_i)$$

One such choice of $\{q_i\}$ gives rise to what is called Brent's GRAND method for $N(0, 1)$ variables (see Brent, 1974). This is the method employed by the NAG routine, G05DDF (see Section A1.1). Further discussion and comparisons with other methods are given by Atkinson and Pearce (1976). One advantage of Forsythe's method is that it avoids time-consuming exponentiation.

*5.42   The random variable X takes the values 1, 2, 3, 4 with the following probabilities:

$$\begin{aligned}
\Pr(X = 1) &= \tfrac{1}{6} = \tfrac{1}{4}(\tfrac{2}{3} + 0 + 0 + 0) \\
\Pr(X = 2) &= \tfrac{1}{12} = \tfrac{1}{4}(0 + \tfrac{1}{3} + 0 + 0) \\
\Pr(X = 3) &= \tfrac{7}{12} = \tfrac{1}{4}(\tfrac{1}{3} + \tfrac{2}{3} + 1 + \tfrac{1}{3}) \\
\Pr(X = 4) &= \tfrac{1}{6} = \tfrac{1}{4}(0 + 0 + 0 + \tfrac{2}{3})
\end{aligned}$$

Thus, by analogy with Equation (5.6), we can write

$$\Pr(X = i) = \frac{1}{4} \sum_{j=1}^{4} r_{ij}$$

where the $\{r_{ij}, 1 \leq i \leq 4\}$ are all probability distributions, for each j, $1 \leq j \leq 4$. The difference as compared with Equation (5.6) is that now random variables with any of the four $\{r_{ij}, 1 \leq i \leq 4\}$ distributions take just one of at most two values, and the distributions in the composition have equal probability of being used. Show that any discrete random variable X over a finite range can be obtained by means of such a composition (Kronmal and Peterson, 1979). This composition results in the *alias* method, so called because if the $\{r_{ij}, 1 \le i \le 4\}$ distributions do not select the value $X = j$ then the 'alias' value for X is chosen by the $\{r_{ij}, 1 \le i \le 4\}$ distribution. For example, in the above illustration, with probability $\frac{1}{4}$ the component distribution, $\{r_{i2}, 1 \le i \le 4\}$ is selected, and then either $X = 2$, with probability $r_{22} = \frac{1}{3}$, or $X = 3$, the alias value, with probability $r_{23} = \frac{2}{3}$. For further discussion, see Peterson and Kronmal (1982). An attractive feature of this method is that it does not require more than two uniform random variables for each value of X. Can you suggest a way in which only one uniform random variable need be used? (See Kronmal and Peterson, 1979.) An algorithm for the alias method is provided by the IMSL routine GGDA (see Section A1.1).

6

TESTING RANDOM
NUMBERS

6.1 Introduction

The need for stringent testing of uniform random variables was emphasized in Chapter 3. When tables of random digits were first produced, tests were employed for uniform random digits. More recently, with the development of pseudo-random-number generators, the numbers to be tested are continuously distributed over the range (0, 1). In the latter case, tests for digits are frequently applied to the digit occupying the first decimal place, while in some cases of detailed testing other decimal places are also considered, as in Wichmann and Hill (1982a). An alternative approach, given by Cugini *et al.* (1980), is described in Section 6.3.

We have seen that congruential methods of random number generation are convenient and widely used, but that they can produce sequences of numbers with certain undesirable properties. For any particular application, the need is to determine what may be 'undesirable', so that random numbers should always be tested with an application in mind. This is often easier said than done, but we can see that it could entail testing not only uniform variables, but also variables of other distributions, obtained by methods such as those of the last two chapters. In Chapter 5 in particular, some of the algorithms given are very complicated, and in such cases testing is needed quite simply as a check that there have been no programming errors. In Chapter 4 we saw that particular properties of random variables and processes can be used to generate particular random variables. By the same token, similar properties may be used to test particular random variables, as we shall see in Section 6.6.

A room full of eternally typing monkeys will ultimately produce the plays of Shakespeare, and similarly, a large enough table of uniform random digits will, by the very nature of random digits, contain sections which, by themselves, will certainly fail tests for uniformity. This feature is noted in the table of Kendall and Babbington–Smith (1939a), which contains 100 000, digits. They tested their table as a whole, and also in parts, down to blocks of 1000 digits each. As

137

expected, some of these individual blocks failed certain tests, and a note was added to these blocks, to 'caution the reader from using them by themselves'.

EXAMPLE 6.1
As an illustration of this, let us consider the digits of Table 3.1. For the two halves of the table we obtain the following frequencies for single digits:

Digit	0	1	2	3	4	5	6	7	8	9	Totals
(a)	17	16	13	16	17	16	36	16	20	13	180
(b)	15	19	20	19	14	22	16	20	11	24	180
(a) + (b)	32	35	33	35	31	38	52	36	31	37	360

For the entire table, if the digits were random the expected number for each digit is $360/10 = 36$, and so the departures from 36 observed can be tested by the chi-square test of Section 2.14. Here no parameters have been estimated from the data, and so the number of degrees of freedom is 9. For the entire table we obtain $\chi_9^2 = 9.39$, which is not significant at the 5 % level. However, if we take part (a) of the table above, we find $\chi_9^2 = 22$, just significant at the 1 % level, for a one-tail significance test, or the 2 % level for a two-tail test. As we shall see later, two-tail tests are frequently used for testing random numbers.

In the context of pseudo-random numbers, we have already encountered this same point in Chapter 3, since congruential generators can be devised which have a low first-order serial correlation for their full cycle, but which result in much higher such correlations for fractions of the cycle (see Exercise 6.1). A property of a pseudo-random number generator for its entire cycle provides, effectively, a test of that generator, and a test of a kind that is not possible for physical random number generators. As well as serial correlations, the first- and second-order moments of Exercise 3.13 can be interpreted in this way. Such tests have come to be known as *theoretical* tests, and an elaborate such test is the spectral test of Coveyou and MacPherson (1967). Theoretical tests evaluate the generating mechanisms used, and do not make use of generated numbers. Knuth (1981, p. 89) states that all congruential generators that are thought to be good pass the spectral test, while those that are known to be bad fail it. Oakenfull (1979) and Knuth (1981, p. 102) provide the results of applying this test to a variety of congruential generators. Ultimately, however, we have to test the numbers produced by a generator in the context of their use, and this is done by a variety of *empirical* tests, which are the subject of this chapter. Atkinson (1980) describes when the spectral test is appropriate, and for a number of generators compares the results of theoretical and empirical tests. The same theoretical/empirical comparison is also made by Grafton (1981).

6.2 A variety of tests

When we are dealing with random variables such as Poisson or normal, we want to check that the generated values come from the distributions we think they do. In the case of Poisson variables this could involve checking that the differences between the bar-charts of Fig. 2.3, for example, are not significant, while for normal variables we would be comparing, for instance, the density function of Fig. 2.5(a) with the histogram of Fig. 2.5(b). Methods for making these comparisons will be considered later. In addition, we may well want to consider the serial dependence of the variables, as is done for instance by Barnett (1965) for exponential random variables.Obvious discrepancies can sometimes be spotted by inspection of a convenient graphical display, as can be done for the figures of Chapter 2, but ultimately significance tests must be applied. The scatter plot of Fig. 3.2 is 'obviously' non-random, but what can we say of the scatter of Fig. 3.3? The same question can be asked of the plot of Fig. 6.1, produced by the generator of Equation (3.1).

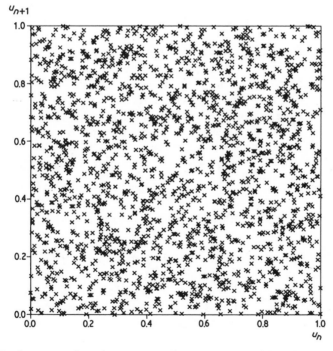

Figure 6.1 A scatter plot of u_{n+1} vs. u_n for a sequence of length 2000 from the generator of Equation (3.1).

Recently a sophisticated approach to judging the randomness of scatter plots has been provided by Ripley (1977, 1981), whose technique itself utilizes repeated simulations, and Atkinson (1977b) applied this technique to numbers resulting from the multiplicative congruential generator

$$x_{i+1} = 5x_i$$

discussed earlier in Equation (3.4).

In inspecting and testing scatter plots such as that of Fig. 6.1, we are implicitly considering how often different one- and two-dimensional intervals are represented. This corresponds to the basic frequency and serial tests which we shall soon describe. Apart from these, however, what other empirical tests should we apply? Thus, hard on the heels of the earlier problems of generator selection and, when appropriate, which method to use for transforming uniform random variables, is the problem of test selection. As mentioned earlier, the glib answer is that tests should be suggested by the use intended for the random variables, and this could result in the application of very specific tests, over and above those already applied to a basic source generator. Different producers of uniform random numbers have answered this question in different ways, and batteries of tests are to be found, for example, in Kendall and Babbington–Smith (1939), Tausky and Todd (1956), Craddock and Farmer (1971), Miller and Prentice (1968) and Wichmann and Hill (1982a).

It is important to realize that there is nothing magical or God-given about a particular set of tests. Clearly an infinity of tests is possible, and as we shall see, numbers which pass one test may fail another. Kendall and Babbington–Smith used just four tests, designed to check frequencies and various forms of sequential dependence, and this basic approach is that adopted by subsequent authors, though conventions have changed with time. We shall now describe certain standard tests for uniform random digits, and then see the results of applying these tests to sequences resulting from a variety of generators.

6.3 Certain tests for uniform random digits

When presented with tables of digits such as those of Tables 3.1, 3.2 and Exercise 3.10, the first reaction of most of us would be to count up the frequencies of occurrence of each digit and compare the observed frequencies with what we would expect for random digits. The statistical yardstick that is usually used in making this comparison is the chi-square test of Section 2.14, and we have already seen such examples of a *frequency* test in Example 6.1, and Exercise 2.24.

Deterministic sequences such as: ... 89012345678901 ... satisfy the frequency test (if a one-tail test is used – see later), but blatantly fail tests which consider the ordering of the elements in the sequence. The simplest such test, the *serial* test, takes a sequence of digits: $\ldots d_i, d_{i+1}, d_{i+2}, \ldots$, and from considering non-overlapping pairs of digits compares the observed matrix $\{O_{kl}\}$ with the expected $\{e_{kl}\}, 0 \le k, l \le 9$, in which digit l is observed to follow digit k, O_{kl} times, and e_{kl} is the corresponding number to be expected if we have a truly random sequence. If we have n non-overlapping pairs of digits, then $e_{kl} = n/100$, for $0 \le k, l \le 9$. The yardstick here is again a chi-square test, but this time on 99 degrees of freedom. An illustration of a serial test is given in Example 6.2.

Non-overlapping pairs of digits are taken so that the independence requirement of the chi-square test is preserved (see Section 2.14). If overlapping pairs are used, then a modified test, due to Good (1953), may be used. The IMSL routine GTPST, of Section A1.2 performs this test. It is an interesting footnote that Kendall and Babbington–Smith (1939a) used overlapping pairs and then incorrectly applied the standard chi-square test. Deterministic sequences such as that illustrated above do in fact produce *too good* an agreement with expectation in the frequency test, and this is indicated by significantly *small* values of the chi-square goodness-of-fit statistic. Consequently, chi-square tests of randomness are often two-tail tests, unlike customary chi-square tests in which only the upper tail is used as the critical region. An example of a sequence of digits that are too regular is provided by the first 2000 decimal digits of $e = 2.71828 \ldots$. Here the frequency test gives $\chi_9^2 = 1.06$, a value which is significant at the 0.2 % level, using a two-tail test. If the first 10 000 decimal digits of e are taken, then we obtain the satisfactory result: $\chi_9^2 = 8.61$; a more detailed breakdown can be found in Stoneham (1965), some of whose results are illustrated in Example 6.7 and Exercise 6.15.

Of course, as stated in Section 2.14, the chi-square test is an asymptotic test, and so is not appropriate if expected cell values are 'small'. The serial test generalizes to the consideration of triples, quadruples and so on, of digits, and the number of cells correspondingly increases geometrically. Thus, especially if one is considering non-overlapping n-tuples, care must be taken in tests of high-dimensional randomness to ensure that expected cell values are large enough for the chi-square test to be valid. An alternative test of randomness in high-dimensional space is the *collision* test described by Knuth (1981, pp. 68–70), and for which a FORTRAN program is given by Hopkins (1983b).

We can test random digits in a less routine way, by looking for patterns. One rudimentary way of doing this is provided by the *gap* test, which is as follows: select any digit, e.g. 7. We can now consider any sequence as consisting of 7's and 'not 7's', i.e., a binary sequence in which $\Pr(7) = 1/10$, and $\Pr(\text{not } 7) = 9/10$, if the sequence is random, and if successive digits are independent then the distribution of the number of digits between 7's is geometric (see

Section 2.6). Thus empirical and observed distributions of numbers of digits between 7's may be compared. For an illustration, see Example 6.2. Like the gap test, the 'coupon-collector' test is also based on a waiting-time, as it considers the number of digits until at least one of each of the digits 0–9 has appeared. This test treats all digits equally, and was first proposed by Greenwood (1955), who found that the test was satisfied by the first 2486 digits in the decimal expansion of $e = 2.71828\ldots$ and by the first 2035 digits in the decimal expansion of $\pi = 3.14159\ldots$; details of his test results can be found in Exercise 6.7.

A more obvious way of looking for patterns is provided by the *poker* test, which considers digits in sequences of length 5, and classifies the patterns according to the conventions of the game of poker: all different, two pairs, etc. Further discussion of the coupon-collector and poker tests is provided in Exercises 6.6, 6.7 and 6.15, and Example 6.7. The poker test may be performed by means of the IMSL routine–GTPOK (see Section A1.2).

Example 6.2 gives the results of applying the serial and gap tests to sequences produced by the random number generator of the Commodore PET microcomputer. This generator is not of a standard form, and will not be described here.

EXAMPLE 6.2   *The result of applying the serial and gap tests to the Commodore PET microcomputer random number generator*

(a) SERIAL TEST:

		Following value											Totals
		1	2	3	4	5	6	7	8	9	10	11	
	1	15	17	18	27	20	16	21	18	21	20	14	207
	2	30	24	20	18	25	13	18	24	27	25	17	241
	3	25	18	19	23	28	15	14	16	16	16	22	212
Preceding	4	14	24	23	14	22	16	17	16	18	19	19	202
value	5	24	16	16	15	15	23	17	21	24	23	18	212
	6	22	24	22	27	18	8	17	19	31	24	25	237
	7	26	22	21	15	19	24	13	20	19	19	17	215
	8	24	13	18	26	21	16	19	21	19	14	20	211
	9	14	17	24	22	18	18	17	15	18	18	21	202
	10	22	24	26	27	23	25	18	23	25	16	23	252
	11	22	13	24	25	26	18	21	18	25	14	23	229
Totals		238	212	231	239	235	192	192	211	243	208	219	2420

Here we obtain $\chi^2_{120} = 108.8$, which is clearly not significant, and so on the basis of this test we would not reject the hypothesis that the digits were uniform and random.

(b) GAP TEST

Gap size	Actual count	Expected count
0	36	25.90
1	36	23.31
2	23	20.98
3	20	18.88
4	17	16.99
5	6	15.29
6	15	13.76
7	10	12.39
8	11	11.15
9	10	10.03
≥ 10	75	90.31
Totals	259	258.99

Here $\chi^2_{10} = 19.924$, which is close to significance at the 5 % level (two-tail test), and one would want to repeat this test to see if other samples produced similar results.

Note that these and other test results presented later in this chapter were obtained using the suite of BASIC test programs of Cugini *et al.* (1980). Rather than work with digits, they divided the (0, 1) interval into 11 sections for the serial test, while for the gap test, gaps were recorded between numbers lying in the (0.03, 0.13) interval. Thus for the gap test,

$$25.90 = 259/10,$$

$$23.31 = 25.9 \times 0.9, \text{ etc.}$$

*6.4   Runs tests

A striking feature of a table of digits can be the occurrence of *runs* of the same digit. If such runs occur with greater frequency than one would expect for random digits then one might, for example, expect this feature to result in a significant departure from the geometric distribution of the gap test. One can, however, look at distributions of other types of runs, and this was done by Downham and Roberts (1967).

Runs tests are frequently applied to a sequence of $U(0, 1)$ variates. Here we shall just consider 'runs up'. To illustrate what is meant by a 'run up', consider the following sequence of numbers, given here to 3 decimal places:

$$(0.134 \ 0.279 \ 0.886) \ (0.197) \ (0.011 \ 0.923 \ 0.990) \ (0.876)$$

The 'runs up' are indicated in parentheses, so that here we have four such runs, of lengths 3, 1, 3, 1, respectively. We see that a 'run up' ends when the next item in the sequence is less than the preceding item, the next item then starting the next 'run-up'. Levene and Wolfowitz (1944) showed that in a random sequence of n $U(0, 1)$ variates, the expected number of 'runs up' of length $k \geq 1$, R_k, say, is given by:

$$\mathscr{E}[R_k] = \frac{(k^2 + k - 1)(n - k - 1)}{(k + 2)!} \qquad \text{for } 1 \leq k \leq n$$

(See also Knuth, 1981, pp. 65–68, for a derivation of this result.)

Typically, n is taken to be large, so that

$$\mathscr{E}[R_k] \approx \frac{(k^2 + k - 1)}{(k + 2)!} n \qquad \text{for } k \ll n$$

Clearly, for fixed n, $E[R_k]$ decreases as $k \to n$, and it is usual to consider the joint distribution of $(R_1, R_2, \ldots, R_j, S_j)$, for some $j > 1$, where $S_j = \sum_{k=j+1}^{n} R_k$; $j = 5$ is frequently adopted. Successive run lengths are not independent, and so a standard chi-square test for comparing observed and expected numbers of runs is inappropriate. The test-statistic used (see Levene and Wolfowitz, 1944) is

$$U = \frac{1}{n} \sum_{i=1}^{6} \sum_{j=1}^{6} (X_i - \mathscr{E}[X_i])(X_j - \mathscr{E}[X_j]) a_{ij} \qquad (6.1)$$

in which $X_k = R_k$ for $1 \leq k \leq 5$, and $X_6 = S_6$,

the $\{a_{ij}\}$ form the inverse of the variance–covariance matrix of the $\{X_k\}$, and for large n are given by:

$$\mathbf{A} \approx \begin{Bmatrix} 4529.4 & 9\,044.9 & 13\,568 & 18\,091 & 22\,615 & 27\,892 \\ & 18\,097 & 27\,139 & 36\,187 & 45\,234 & 55\,789 \\ & & 40\,721 & 54\,281 & 67\,852 & 83\,685 \\ & & & 72\,414 & 90\,470 & 111\,580 \\ & & & & 113\,262 & 139\,476 \\ & & & & & 172\,860 \end{Bmatrix}$$

the lower half of this matrix being obtained from symmetry. The exact expression is given by Knuth (1981, p. 68). U is referred to chi-square tables on 6 (not 5) degrees of freedom. As with the usual chi-square test, an asymptotic approximation is being made when this test is used, and Knuth recommends taking $n \geq 4000$. An illustration of the outcome of applying this test is given in the following example.

EXAMPLE 6.3   *The result of applying the 'runs up' test to a sequence of length* $n = 5000$ *from the generator* $(131, 0; 2^{35})^\dagger$

---

† Note that for convenience we shall henceforth use the notation: $(a, b; m)$ for the congruential generator of Equation (3.2).

Run length (k)	R_k	$\mathscr{E}[R_k]$
1	824	833.34
2	1074	1041.66
3	440	458.33
4	113	131.94
5	42	28.77
≥ 6	7	5.95

$\chi_6^2 = 18.10$, significant at the 2 % level, using a two-tail test. Thus here the test rejects the hypothesis that the variables are random and uniform.

Before the work of Levene and Wolfowitz (1944), runs tests were incorrectly used, incorporating the standard chi-square approach. Unfortunately the algorithm by Downham (1970) omitted the $\{a_{ij}\}$ terms of Equation (6.1). That this omission could possibly result in erroneous conclusions is demonstrated by Grafton (1981), who provides a brief comparison between the correct runs test and the spectral test. Grafton (1981) provides a FORTRAN algorithm which tests 'runs down' as well as 'runs up', though the two tests are not independent. See also Section A1.2 for the IMSL routines GTRN and GTRTN. Accounts of the power of runs tests vary, and are clouded by incorrect uses of the tests. Kennedy and Gentle (1980, pp. 171–173) provide the theory for the case of runs up and down.

6.5   Repeating empirical tests

One might expect a poor generator to fail empirical tests, but a failure of an empirical test need not necessarily indicate a poor generator. Conversely, a poor generator can pass empirical tests, and both of these instances are illustrated in the following two examples.

EXAMPLE 6.4
The frequency test was applied to the (781, 387; 1000) generator, starting the sequence with 1. The full cycle was divided into 20 consecutive groups of 50 numbers each. For any group the frequency test was satisfied, but the 20 chi-square statistics took just one of the three values, 10.0, 8.8, 7.2.

EXAMPLE 6.5
The PET generator produced the borderline 5 % significance result of Example 6.2(b) under the gap test. Nine subsequent gap tests produced the insignificant statistics of:

$$9.49, \ 14.88, \ 6.50, \ 13.73, \ 7.80, \ 7.80, \ 4.36, \ 8.12, \ 7.80$$

A similar 'unlucky start' is found with the frequency test applied to the decimal digits of e (Stoneham, 1965).

These difficulties can sometimes be resolved by repeating an empirical test, producing in effect a more stringent test. Chi-square values from repeating tests can be interpreted in a number of ways: a simple graphical representation can be obtained by probability (also called Q–Q) plots (see for example, Chernoff and Lieberman, 1956; Gerson, 1975; and Kimball, 1960), in which a sample of size n from some distribution (chi-square in our case) is ordered and plotted against the expected values of the order statistics. The expected order statistics for chi-square distributions are provided by Wilk *et al.* (1962), and two illustrations are provided by the following two examples.

EXAMPLE 6.6
The RANDU generator, $(65\,539, 0; 2^{31})$, resulted in the probability plot shown in Fig. 6.2 for 30 applications of the 'runs up' test, each applied to a sequence of 5000 numbers.

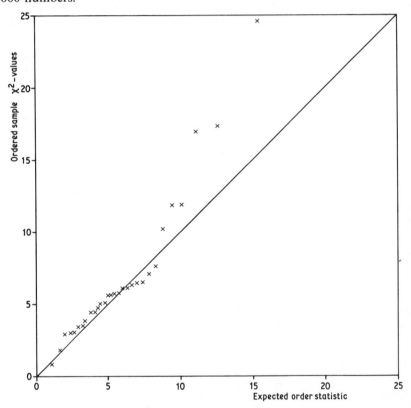

Figure 6.2   A probability plot of 30 test-statistics resulting from the 'runs up' test applied to the RANDU generator. The ordered sample is plotted against the expected order statistics for a sample of size 30 from a χ^2_6 distribution.

EXAMPLE 6.7

Stoneham (1965) made a study of the first 60 000 decimal digits of *e*. The results of 12 applications of the poker test are illustrated in Fig. 6.3, each test being applied to a block of 5000 consecutive digits. Some of the detail is presented in Exercise 6.15.

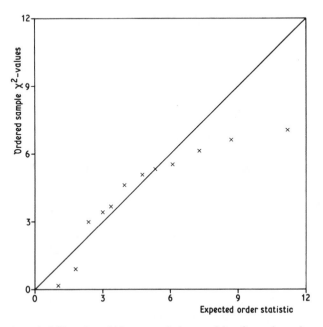

Figure 6.3   A probability plot of 12 test-statistics resulting from the poker test applied to decimal digits of *e*. The ordered sample is plotted against the expected order statistics for a sample of size 12 from a χ_5^2 distribution.

Note, however, that Wilk *et al.* (1962) remark that it is difficult to interpret such plots for fewer than 20 points, reaching their conclusion after applying the plots to simulated data.

Whether or not these plots indicate significant departures from the appropriate chi-square distribution can also be judged by means of a further chi-square test, if the sample size permits, as demonstrated in the next example.

EXAMPLE 6.8

The Cugini *et al.* (1980) program for the frequency test applies the test to 1050 numbers in the (0, 1) range, categorizing them according to 21 equal-length intervals. The test is then repeated 60 times, and the resulting chi-square statistics are themselves categorized according to the percentile range into which the values fall. Applying this test to the PET generator produced the following result:

% range	Actual count	Expected count
0–1	4	0.6
1–5	0	2.4
5–10	2	3
10–25	10	9
25–50	18	15
50–75	15	15
75–90	7	9
90–95	2	3
95–99	2	2.4
99–100	0	0.6
	60	60

Combining the first three and the last three rows allows us to perform a chi-square test, now on 5 degrees of freedom, to this table. We obtain the value of $\chi_5^2 = 1.82$, which is not significant at the 10 % level, and so these numbers pass this empirical test.

An alternative approach is to use the Kolmogorov–Smirnov test, which is described by Hoel (1954, p. 345) and Knuth, 1981 (pp. 45–52) who provides some interesting comparisons of power between the chi-square and Kolmogorov–Smirnov tests. Categorization of the individual chi-square values is unnecessary for the Kolmogorov–Smirnov test, and when applied to the sample of size 12 illustrated in Fig. 6.3, the test does not reveal a significant departure from the expected χ_5^2 distribution at the 5 % level. While the same is true for the sample of size 30 illustrated in Fig. 6.2, in that case the result is significant at the 6 % level.

6.6 Tests of non-uniform random variables

We have already seen, in the last example, an illustration of the chi-square test being used to test a non-uniform (in this case also chi-square) random variable. The same approach may be used, with suitable combining of categories when necessary, for any distribution; see Exercise 6.19 for an illustration. The Kolmogorov–Smirnov test may also be used for any distribution, and, as above, does not require categorization. We shall now briefly consider some particular tests for non-uniform random variables.

6.6.1   Normal variables

Wold (1954) obtained standard normal random variables, to 2 decimal places, by transforming the digits of Kendall and Babbington–Smith (1939); the digits were initially grouped to correspond to $U(0, 1)$ variates and then the table-look-up method of Section 5.2 was used (see Example 5.4). Except for normal variables in the tails of the distribution, for the two-place accuracy needed, just four-decimal place accuracy was necessary for the $U(0, 1)$ variates. The resulting table had 25 000 items, which were tested as a whole, as well as in groups of 500 and 5000. Four tests were employed:

(a) The numbers in a group were summed, and the result referred to the appropriate normal distribution (see Exercise 2.8(a)).
(b) The squares of the numbers in a group were summed, and the result referred to the appropriate chi-square distribution (see Exercise 2.5). As the group sizes are ≥ 500, we can use the approximation that if X has a χ_v^2 distribution, $(\sqrt{(2X)} - \sqrt{(2v - 1)})$ is approximately $N(0, 1)$ (see Exercise 6.24).
(c) From the solution to Exercise 6.4 we see that if R is the range from a random sample of size n from an $N(0, 1)$ distribution, then

$$\Pr(R \leq r) = n \int_{-\infty}^{\infty} (\Phi(x + r) - \Phi(x))^{n-1} \phi(x)\, dx \qquad (6.2)$$

and thus the ranges of such samples of size n can be obtained and compared with what one would expect, using a chi-square test. The distribution of (6.2) is tabulated in Pearson and Hartley (1970, 178–183).
(d) A runs test was applied to the runs of signs only of the sequence of numbers.

As with the Kendall and Babbington–Smith (1939a) tables, a note was appended to each set of numbers that failed any test.

Other tests for normality are discussed by Pearson *et al.* (1977) and Wetherill *et al.* (1984, chapter 8). One of these tests, by Shapiro and Wilk (1965), tests for departures from linearity in the appropriate probability plot.

*6.6.2   Multivariate normal variables

If (X_1, X_2) has the bivariate normal density function of Section 2.15, then the derived univariate statistic,

$$D^2 = \frac{1}{(1 - \rho^2)} \left\{ \left(\frac{X_1 - \mu_1}{\sigma_1}\right)^2 - 2\rho \frac{(X_1 - \mu_1)(X_2 - \mu_2)}{\sigma_1 \sigma_2} + \left(\frac{X_2 - \mu_2}{\sigma_2}\right)^2 \right\} \qquad (6.3)$$

has a χ_2^2 distribution (i.e. exponential of mean 2)—see Exercise 6.5. Healy (1968a) proposed using sample values of D^2 and comparing them with the chi-square distribution they would have if (X_1, X_2) is indeed bivariate normal. Once again a graphical examination can be made with the aid of a probability plot, the expected order statistics in a sample of size n from a χ_2^2 distribution being:

$$\frac{2}{n}, \left\{\frac{2}{n} + \frac{2}{(n-1)}\right\}, \left\{\frac{2}{n} + \frac{2}{(n-1)} + \frac{2}{(n-2)}\right\}, \ldots, \left\{\frac{2}{n} + \frac{2}{(n-1)} + \ldots + \frac{2}{1}\right\}$$

(see Cox and Lewis, 1966, p. 27). In practice the parameters of (6.3) have to be estimated from the data, and there is discussion of this in Barnett and Lewis (1978, pp. 212–215 and 226). This approach can also be extended for general multivariate normal distributions. For additional tests see Mardia (1980), Gnanadesikan (1977, p. 161) and Royston (1983).

6.6.3 Exponential and Poisson variables

A random sample from any exponential density may be illustrated by using the order-statistics of the last section after a preliminary scaling (see Exercise 2.2). Tests for exponential random variables were used by Barnett (1965), who, in contrast to Wold, transformed pseudo-random variables, using a multiplicative congruential generator with $m = 2^{27}$, and the transformation of Equation (5.5). Five tests were then applied to the resulting numbers, including an extension of the test of Cox (1955) for detecting the presence of first-order serial correlation in a sequence of exponential variables. Barnett (1965) also generated and tested χ_1^2 variates by squaring $N(0, 1)$ variables derived by the Box–Müller method of Section 4.2.1. In connection with some of his tests, Barnett was confident that only the right-hand tail of the chi-square distribution need be used for the test critical region.

The mean and variance of Poisson random variables are equal, and the *index of dispersion* test makes use of this result to provide a particular test for the Poisson distribution. If $(x_1, \ldots, x_n)$ is a random sample from a Poisson distribution of parameter λ, then

$$\frac{\sum\limits_{i=1}^{n} (x_i - \bar{x})^2}{\bar{x}}$$

is, approximately, a realization of a χ_{n-1}^2 random variable, where

$$\bar{x} = \frac{\sum\limits_{i=1}^{n} x_i}{n}$$

See for example *ABC*, p. 314.

6.7   Discussion

The very first tabulation, by Tippett (1927), of random digits did not include an account of any systematic testing. By contrast, the testing of random variables has now become a standard procedure, and a description of a variety of computerized algorithms which may be used is given in Section A1.2. A suite of test programs such as that of Cugini *et al.* (1980) indicates the kind of compromise that may be reached in the choice of a suitable subset of empirical tests.

The need to match tests of numbers to the intended application for those numbers is graphically illustrated by the insignificant result of the Kolmogorov–Smirnov test of Example 6.6. The RANDU generator that is tested here has very poor properties when one considers successive triples of numbers, as explained by Exercise 3.25, yet the generator does not fail at the 5 % level the repeated runs test of Example 6.6.

The RANDU generator failed the extension of the serial test to three dimensions when this test was applied by Dieter and Ahrens (1974, p. A8): each time the test was applied the resulting chi-square test statistics were roughly 100 standard deviations from the expected chi-square mean for the test. However, only one (a poker test) of the many other empirical tests applied indicated that the generator had poor properties. Caution is clearly the key word. The possible problems with pseudo-random numbers are evident, and true random numbers could be biased in unexpected ways. For instance, Kendall and Babbington–Smith (1938) selected digits from the London telephone directory and found appreciably fewer 5's and 9's than one would expect (see Exercise 6.8). They attributed this to the high acoustic confusion between five and nine (airline pilots use 'fife' and 'niner' respectively), and telephone engineers selecting numbers to try to reduce this effect (for related work, see Morgan *et al.*, 1973, and Exercise 9.10).

Neave (1973) showed that when certain pseudo-random variables were transformed by the Box–Müller transformation of Section 4.2.1, the resulting variables displayed unusual characteristics. For instance, observed frequencies in the intervals $(-\infty, -3.3)$ and $(3.6, \infty)$ were zero, compared with expectations (for 10^6 generated values) of 483 and 159 respectively. However, it has been pointed out subsequently (see, e.g., Golder and Settle, 1976) that this effect is mainly due to the $(131, 0; 2^{35})$ generator used, which was considered earlier in Example 6.3 (see also Exercise 6.25). Atkinson (1980) in fact uses the Box–Müller transformation combined with a test of normal random variables as a test of the underlying generator.

6.8   Exercises and complements

6.1   Compare the bounds on the first-order serial correlation given by Equation (3.3) for a mixed congruential generator with empirical first-order serial correlations obtained for a sequence of length 1000, for the following generators:

$$(781, 387; 10^3)$$
$$(6941, 2433; 10^4)$$
$$(5^{17}, 0; 2^{42})$$

Note that an additional test of random numbers is provided by comparing empirical serial correlations with their expectation for a random sequence (see, e.g., Cugini *et al.*, 1980).

†**6.2**   Perform the index-of-dispersion test for the Poisson distribution using the following sample statistics obtained using the PET generator and the program of Fig. 4.4

λ	n	$\bar{x}$	s^2
5	500	4.960	4.804
2	500	1.856	1.679
1	500	1.060	1.140
1	500	0.974	1.076
1	500	1.018	1.080
0.5	500	0.438	0.463

Here $s^2 = \sum_{i=1}^{n} (x_i - \bar{x})^2 / (n - 1)$.

†**6.3**   Two dice were thrown 216 times, and the number of sixes at each throw were:

No. of sixes	0	1	2	Total
Frequency	130	76	10	216

Test the hypothesis that the probability of a six is $p = 1/6$.

Explain how this test would be modified if the hypothesis to be tested is that the distribution is binomial with the parameter p unknown. (Based on an Oxford A-level question, 1978).

6.4   Verify the formula for the distribution function of Equation (6.2), for the range of a random sample of size n from an $N(0, 1)$ distribution.

***6.5**   Use the formula of Equation (2.4) to verify that the random variable of Equation (6.3) has a χ_2^2 distribution.

*6.6 (*The coupon-collector problem applied to the digits* 0–9) The probability that the full set of digits is obtained for the first time at the jth digit of a sequence is given by:

$$\Pr(j) = 10^{1-j} \sum_{v=1}^{10} (-1)^{v+1} \binom{9}{v-1} (10-v)^{j-1} \qquad \text{for } j \geq 10.$$

Two different ways of proving this result are suggested below.

(i) If the number of digits until the first occurrence of a complete set is denoted by S, then (verify) we can write

$$S = 1 + \sum_{i=2}^{10} X_i$$

where X_i has the geometric distribution of Section 2.6, with $p = ((11-i)/10)$, $2 \leq i \leq 10$.

   Show that the probability generating function (see Section 2.16) of S is given by:

$$G(z) = \frac{9! \, z^{10}}{\prod_{i=1}^{9} (10 - iz)}$$

which, further, may be written as:

$$G(z) = \frac{9z^{10}}{10^8} \sum_{i=1}^{9} (-1)^{9-i} \frac{i^8}{(10 - iz)} \binom{8}{i-1}$$

Finally, by expanding this expression as a power series in z, verify the distributional form of S given above.

(ii) An alternative approach uses the theory of occupancy problems, in which r balls are thrown at random into n cells, with $r \geq n$. In our case, each digit corresponds to a ball and each *type* of digit $(0, 1, \ldots,$ etc.) is a cell.

   If $u(r, n) = \Pr$(no cell is empty when r balls are thrown at random into n cells),

   then we see that

$$u(r, 10) = \Pr(S \leq r),$$

and so $\Pr(S = r) = u(r, 10) - u(r-1, 10)$.
Use this approach, coupled with the fact that

$$\Pr(\text{no cell is empty}) = 1 - \Pr(\text{at least one is empty})$$

to obtain $\Pr(S = r)$.

   NOTE (Feller, 1957, p. 59) that the median of the distribution of S is 27; $\Pr(S > 50) > 0.05$; $\Pr(S > 75) \approx 0.0037$. Note further that $u(r, n)$ may be used to solve the 'birthday problem': $n = 365$,

r = number of people in a room; if e.g., $r = 1900$, Pr(no day is not represented as a birthday) ≈ 0.135.

†6.7   Greenwood (1955) obtained the following results from the coupon-collector test applied to the first 2486 digits in the decimal expansion of $e = 2.71828 \ldots$ and the first 2035 digits in the decimal expansion of $\pi = 3.14159 \ldots$ (reproduced by permission of the American Mathematical Society)

Number of digits to	π		e	
the full collection	Observed	Expected	Observed	Expected
10–19	13	11.604	12	14.202
20–23	13	11.720	11	14.344
24–27	9	11.491	14	14.064
28–32	5	11.480	15	14.050
33–39	13	10.195	17	12.477
40+	14	10.510	13	12.863
X^2 values:	6.436		2.826	

Verify the expected values given above and discuss the non-significance of the result for e in relation to the failure of the frequency test by these digits (see Metropolis *et al.*, 1950).

6.8   (i) Kendall and Babbington–Smith (1938) obtained the following distribution of 10 000 digits from the London telephone directory:

Digit	0	1	2	3	4	5	6	7	8	9	Total
Frequency	1026	1107	997	966	1075	933	1107	972	964	853	10 000

Verify that the frequency test results in $\chi_9^2 = 58.582$.

(ii) Fisher and Yates (1948) obtained the following distribution of digits obtained from suitably reading tables of logarithms:

Digit	0	1	2	3	4	5	6	7	8	9	Total
Frequency	1493	1441	1461	1552	1494	1454	1613	1491	1482	1519	15 000

Verify that the frequency test results in $\chi_9^2 = 15.63$, and discuss their decision to remove at random 50 of the 6's, and then replace them

with other digits, chosen at random. For additional discussion, see
Kendall and Babbington–Smith (1939b).

6.9   Comment on the following test statistics resulting from applying the
'runs up' test to sequences of 5000 numbers from the generators
indicated:

	Sequence		
Generator	1	2	3
$(131, 0; 2^{35})$	9.89	2.70	18.10
$(65539, 0; 2^{31})$	13.16	5.59	12.70
$(23, 0; 10^{8} + 1)$	6.83	14.62	11.90
$(3025, 0; 67\,108\,864)$	10.87	3.53	3.75
PET	5.16	2.26	8.20
The generator of Equation (3.1)	5.03	4.23	6.90

6.10   Verify that for the sequence of numbers from the $(781, 387; 10^{3})$
generator there are no runs up of length greater than 4, and discuss this
result.

6.11   Apply tests of this chapter to the digits of Tables 3.1 and 3.2, and of
Exercise 3.10. For many years the established decimal expansion for π
was that of William Shanks, computed over a 20-year period to 707
decimal places. It was noted that 7 appeared only 51 times. In 1945 it
was noticed that Shanks made an error on the 528th decimal, and all
subsequent decimals are wrong. In the correct series the frequency of 7's
is as one would expect (Gardner, 1966, p. 91).

6.12   Consider how you might construct a sequence of numbers which pass
the frequency test, but which fail the serial, gap, poker and coupon-
collector tests.

6.13   A test which is sometimes used (see, e.g., Cugini *et al.*, 1980) is the
'maximum-of-t' test. Here numbers are taken in disjoint groups of size t,
and the largest number is recorded for each group. The resulting
maxima are then compared with the expected distribution.

$$\text{If } M = \max(U_1, U_2, \ldots, U_n)$$

where the $\{U_i\}$ are independent, identically distributed $U(0, 1)$ random
variables, show that M has the density function

$$f_M(x) = nx^{n-1} \qquad \text{for } 0 \leq x \leq 1.$$

What is the distribution of M^n?

†6.14 A further test is the permutation test, in which, again, numbers are taken in groups of size t. Here the ordering of the numbers is recorded, and the empirical distribution of the orderings compared with expectation, which allots a probability of $1/t!$ for independent uniformly distributed numbers. (The possibility of tied values is not considered.) The following results were obtained for the PET generator and the case $t = 4$.

Permutation	Number of cases	Permutation	Number of cases
1	9	13	11
2	17	14	12
3	16	15	10
4	9	16	7
5	6	17	9
6	6	18	13
7	10	19	9
8	12	20	12
9	10	21	7
10	10	22	8
11	8	23	10
12	11	24	8

Assess the significance of these results using a chi-square test.

*6.15 (Stoneham, 1965)   The detail of six of the poker test results presented in Example 6.7 and Fig. 6.3 is given below:

Block	Hands					
	All different	One pair	Two pairs	One triple	One triple and one pair	4 or 5 of the same kind
1	316	506	98	70	5	5
2	307	499	108	80	6	0
3	317	503	90	72	13	5
4	299	511	114	58	12	6
5	299	498	99	84	18	2
6	307	503	111	67	8	4
Theoretical frequencies for random digits	302.4	504	108	72	9	4.6

Verify the theoretical frequencies and the resulting chi-square values:

Block	1	2	3	4	5	6
χ^2_5	3.42	6.62	5.53	4.61	5.32	7.05

6.16   Invent a test of your own for uniform random numbers.

6.17   Investigate and test the random number generators that are available to you. This can be quite revealing. Miller (1977a, 1977b) and Bremner (1981) have revealed errors in Texas hand-calculator multiplicative congruential generators. Furthermore, Bremner has pointed out that the RND function available in the University of Kent implementation of BASIC is (3025, 0; 67 108 864), and not (3125, 0; 67 108 864), as intended, and for which test results were available! (See Pike and Hill, 1965). Nevertheless, the (3025, 0; 67 108 864) generator passes the empirical tests of Cugini *et al.* (1980). (See also Exercise 6.27.)

6.18   (Cooper, 1976)   The Box–Müller method involves computing the functions, log, square-root, sin and cos, for each pair of normal random variables generated. If (as in Barnett, 1965) the aim is to simulate χ^2_1 variables, show how the number of functions computed can be reduced.

6.19   Use a chi-square test to compare the p.d.f. and histogram of Fig. 2.5. The frequencies illustrated by the histogram are:

$$2, 4, 8, 18, 19, 12, 14, 14, 5, 2, 2$$

Repat this approach for other appropriate figures from Chapter 2, reading the frequencies from the histograms/bar-charts.

6.20   Wold defined the *P*-value for each test as the two-tail probability of being as, or more, extreme as the resulting value of the test-statistic. Thus, for example, the sum of the first 500 numbers was $S = 159.97$, $\Phi(159.97/\sqrt{5000}) = 0.9882$, and $P = 2(1 - 0.9882) = 0.0237$. In addition to the tests already described, he wrote:

'For each type of test, the distribution of *P*-values obtained from the 50 page sets has been compared with the expected distribution, which is rectangular over the interval (0, 1). On the whole, the agreement with the expected distribution is good. The deviations have been tested by the χ^2 method, grouping the distribution in 10 equal intervals. The *P*-values obtained for the 4 tests are 49.4, 13.7, 29.0 and 91.1 % respectively. The agreement was also tested by the method of Kolmogoroff, mentioned above, a method not involving grouping, with the results $P = 15.5, 42.6, 26.6$ and 98.9%'.

Discuss his approach and conclusions (cf. Section 6.5).

6.21 The 30 test statistics illustrated in Fig. 6.2 are given below:

0.84	1.82	2.92	3.01	3.06	3.43	3.51	3.84	4.43	4.45
4.74	5.02	5.09	5.61	5.64	5.73	5.77	6.11	6.12	6.33
6.47	6.52	7.09	7.62	10.20	11.84	11.88	16.93	17.32	24.59

Use a chi-square test to assess whether these values come from a χ^2_6 distribution. The Kolmogorov–Smirnov test of Example 6.6, applied to these data, was made with the aid of the NAG FORTRAN routine GO1BCF, which evaluates the right-hand tail areas, assuming these values form a random sample from χ^2_6; and GO8CAF, using the option: null $= 1$, which performs a Kolmogorov–Smirnov test of whether these tail areas are uniform (cf. Exercise 6.20). Tail areas for chi-square densities are also given in Pearson and Hartley (1972, p. 160). For computational formulae see Kennedy and Gentle (1980, Section 5.7).

6.22 Use the results of Exercise 2.8 to construct particular tests for exponential and Poisson variables. How might you make use of the result of Exercise 4.17?

*6.23 Use the results of Exercise 4.14 to simulate bivariate normal random variables, and test them using the approach of Section 6.6.?

*6.24 In Section 6.6.1 we used the result:
If X has a χ^2_ν distribution, for large ν, then
$N = \sqrt{(2X)} - \sqrt{(2\nu - 1)}$ is approximately $N(0, 1)$.
Why is this? (Note that the results of Section 2.12 are exact, while here we are seeking an approximate relationship.)

*6.25 Neave (1973) combined a multiplicative congruential generator with only the sine form of the Box–Müller transformation (see Section 4.2.1), obtaining

$$N = (-2 \log_e U_1)^{1/2} \sin (2\pi U_2)$$

in which $x_2 = ax_1 \pmod{m}$,

and $\qquad\qquad U_1 = x_1/m; \; U_2 = x_2/m.$

Show that we can write N in the form:

$$N = (-2 \log_e U)^{1/2} \sin (2\pi \, aU)$$

Chay *et al.* (1975) suggested using the $\{U_i\}$ from the multiplicative congruential generator in the opposite order to that above, resulting in:

$$N = (-2 \log_e U_2)^{1/2} \sin (2\pi U_1)$$

Show that $x_1 = a^* x_2 \pmod{m}$
where $aa^* = 1 \pmod{m}$

so that the 'Chay interchange' is equivalent to changing the multiplier in the generator, and keeping to the original sequence. Kronmal (1964) applied the Box–Müller transformation to pseudo-random numbers from two mixed congruential generators, one for U_1 and another for U_2, and found that the resulting numbers passed a variety of tests.

6.26   Write computer programs to perform the tests considered in this chapter.

†**6.27**   Conduct empirical tests of the $(25\,173,\ 13\,849;\ 2^{16})$ generator. T. Hopkins has pointed out that this generator, proposed by Grogono (1980), performs badly on the spectral test. The choice of multiplier here appears to be particularly unfortunate, as literally hundreds of alternative multipliers give rise to a much better result on the spectral test. Consider, for example, $(13\,453,\ 13\,849;\ 2^{16})$.

6.28   What will the result be if the frequency test is applied to the entire cycle of a full-period mixed congruential generator? What are the implications of this result?

*Eeoe*6.29   Given a sequence of pseudo-random numbers, how would you test whether or not a cycle was present?

7

VARIANCE REDUCTION
AND INTEGRAL ESTIMATION

7.1 Introduction

In the preceding chapters we have seen how to generate uniform random variables, and we have considered ways of transforming these to produce other common random variables. Having tested our random variables, we are well prepared for using them in a simulation exercise. However, before pressing on in a bull-at-a-gate fashion it is worth while first of all considering whether the efficiency of the approach to be adopted could be increased. Andrews (1976) writes:

'In a recent Monte Carlo study of a regression problem the computing cost was about £250. The cost of generating the required 160 000 Gaussian [normal] deviates was 50p, a negligible amount relative to the total cost. I have found that variance reduction methods often apply. As these affect sample size they affect the remaining £249.50. Modest gains in efficiency result in large savings; very efficient methods can often be found.'

Thus variance reduction is a way of improving value for money, and it can result in much greater savings than those involved in just changing from one algorithm to another for generating variates. As we shall see, there are many different variance-reduction techniques, and a ready way of illustrating these techniques arises in the context of integral estimation using random numbers. The above quotation used the term 'Monte Carlo'; this is now frequently employed as a more evocative synonym for simulation when random variables are employed. 'Monte Carlo' frequently also has an implied connotation of some variance-reduction method having been used. (See, for example, Cox and Smith, 1961, p. 128; Gross and Harris, 1974, p. 383; and Schruben and Margolin, 1978, for related discussion.) It is an item of folklore that this term was introduced as a code-word for secret simulation work in connection with the atomic bomb during the Second World War (Rubinstein, 1981, p. 11).

The basic idea of variance reduction is contained in the following example.

160

EXAMPLE 7.1 *Buffon's cross*

The Buffon needle experiment has already been described in Exercise 1.2. If a thin needle of length l is thrown at random on to an infinite horizontal table with parallel lines a distance $d \geq l$ apart, then the probability that the needle will cross a line is given by $2l/\pi d$. This probability may be estimated by the proportion of crossings in an experiment consisting of a number of successive throws of a needle, and knowledge of l and d then enables us to estimate π. From the data of Exercise 1.2, we see that π is not very precisely estimated in this way (cf. the precision of Exercise 3.10), even for as many as 960 throws of the needle. Soldiers recovering from wounds sustained during the American Civil War had the time, and apparently also the interest (Hammersley and Handscomb, 1964, p. 7), for multiple repeats of the needle experiment, but present-day experimenters are unlikely to be so patient.

One way to speed the process up is to throw more than one needle each time, and then the picking up of the needles is facilitated if the needles are joined together. In its simplest form, this is accomplished by fusing two needles of equal length at right-angles at their centres, to form a cross. If Z denotes the total number of lines intersected from a single throw of the cross, we can write $Z = X + Y$, where X and Y separately denote the number of crossings of each of the two needles. The distribution of X, and equivalently Y, is unaffected by the presence of the other needle, and so $\mathscr{E}[X] = \mathscr{E}[Y] = 2l/(\pi d)$, and $\mathscr{E}[Z] = 4l/(\pi d)$. The best approach is to take $l = d$ (see Exercise 1.2), and let us, in this case set $\theta = 2/\pi$. X and Y are simple binomial random variables and so (see Table 2.1), $\mathrm{Var}(X) = \mathrm{Var}(Y) = \theta(1 - \theta)$.

It can be shown that the distribution of Z is given by:

$$\mathrm{Pr}(Z = 0) = 1 - \frac{2\sqrt{2}}{\pi}; \; \mathrm{Pr}(Z = 1) = 4(\sqrt{2} - 1)/\pi;$$

$$\mathrm{Pr}(Z = 2) = 4(1 - 1/\sqrt{2})/\pi$$

This enables us to evaluate $\mathrm{Var}(Z) = \mathrm{Var}(X) + \mathrm{Var}(Y) + 2\,\mathrm{Cov}(X, Y)$, yielding, ultimately,

$$\mathrm{Cov}(X, Y) = 2(\pi(2 - \sqrt{2}) - 2)/\pi^2 \approx -0.0324$$

reflecting the fact that the needles are fixed together, and X and Y are not independent: $\mathrm{Corr}(X, Y) = -0.14$.

In the original Buffon experiment, $\mathrm{Var}(\hat{\theta}) \approx 0.2313/n$, where n denotes the number of throws of the needle. In the case of the cross, from one throw, $\hat{\theta} = \frac{1}{2}(X + Y)$, and so, because of the term $\frac{1}{2}$ we immediately have a reduction in $\mathrm{Var}(\hat{\theta})$, for

$$\mathrm{Var}(\hat{\theta}) = \tfrac{1}{4}(\mathrm{Var}(X) + \mathrm{Var}(Y) + 2\,\mathrm{Cov}(X, Y)) = \tfrac{1}{2}\mathrm{Var}(X) + \tfrac{1}{2}\mathrm{Cov}(X, Y)$$

and $\mathrm{Cov}(X, Y) < 0$

Thus fixing the two needles together has a utility over and above the added ease of collecting the needles. For n throws of the cross, $\text{Var}(\hat{\theta}) \approx 0.0995/n$. So we see that using a cross, rather than a single needle, is a variance-reduction technique; it results in greater precision, i.e. an estimator of smaller variance. Against this gain must be offset the labour of correctly fixing the needles to form a cross (though this is more easily done by etching a cross on a clear perspex disc), and the computation of the new theory. These losses occur once only, and would clearly be worth while if a very large experiment were envisaged. There is no reason why further gains should not be obtained from the use of more than two fused needles, and Kendall and Moran (1963, p. 72) provide the result for the case of a star shape; see Hammersley and Morton (1956) for details. In the case of a star, further additional labour (small for a cross) is involved in counting the number of crossed lines. The converse to changing the needle is changing the grid, and Perlman and Wichura (1975) provide the theory for the case of square and triangular grids. Further discussion and elaboration are to be found in Mosteller (1965, pp. 86–88) and Ramaley (1969), as well as Exercises 7.1–7.4.

The extended example above illustrates the basic features of a variance-reduction method, and we shall encounter these features again in the next section. Of course the above example is artificial in that we already know π, which permits a simple evaluation of the variance reduction achieved.

A fundamental aspect of the above example, and others to follow, is the estimation of a parameter θ by an estimator $\hat{\theta}$, with

$$\mathscr{E}[\hat{\theta}] = \theta \qquad \text{and} \qquad \text{Var}(\hat{\theta}) \propto n^{-1}.$$

In both the needle and the cross cases, $\hat{\theta}$ is proportional to a sum of random variables, and hence for large n (and typically n is large in such experiments), central limit theorems apply, so that $\hat{\theta} \approx N(\theta, \kappa/n)$, for appropriate κ. Thus as well as simply producing the estimate $\hat{\theta}$, we can also obtain approximate confidence intervals for θ; for example, a 95% confidence interval is $(\hat{\theta} \pm 1.96\,\kappa^{1/2}/n^{1/2})$, when the normal approximation is valid. The width of this interval is $\propto n^{-1/2}$, so that, for instance, to halve an interval width one has to quadruple the number of observations. Because of this feature it is clearly desirable to employ a variance-reduction technique that results in as small a value for κ as possible.

7.2   Integral estimation

A definite integral, such as $\Phi(x)$, which cannot be explicitly evaluated, can be obtained by a variety of numerical methods. Some of these are described by Conte and de Boor (1972), and algorithms are available for programmable hand-calculators, as well as within computer subroutine libraries such as

NAG. For numerical evaluation of integrals in a small number of dimensions one would therefore be unlikely to use simulation. However, simulation methods can be viable for high-dimensional integration, say in the dimensional range 6–12 (Davis and Rabinowitz, 1975, p. 314). In this section we refer solely to simple one-dimensional integrals, as they provide a convenient vehicle for illustrating some basic methods of variance reduction. It is in any case interesting to see how random numbers may be used to evaluate deterministic integrals. In the following we shall again consider estimation of π, but now through the representation:

$$\frac{\pi}{4} = \int_0^1 \sqrt{(1-x^2)}\,dx \qquad (7.1)$$

each side of (7.1) being the area of a quadrant of a circle of radius unity.

7.2.1 Hit-or-miss Monte Carlo

The integral of (7.1) is the area of a quadrant of the circle, radius 1 and centre 0. If that quadrant is enclosed by a unit square, and points thrown independently at random on to the square, then the proportion R/n of n points thrown that land within the quadrant can be used as an estimate of the probability, $\pi/4$, of any point landing within the quadrant; see Fig. 7.1(a). Thus $4R/n$ can be used as an estimate of π. Now R is a random variable with a $B(n, \pi/4)$ distribution, with $\mathrm{Var}(R) = n\frac{\pi}{4}(1-\frac{\pi}{4})$ so that $\mathrm{Var}(4R/n) = 16\,\mathrm{Var}(R)/n^2 = \pi(4-\pi)/n \approx 2.697/n$.

7.2.2 Crude Monte Carlo

As $\dfrac{\pi}{4} = \displaystyle\int_0^1 \sqrt{(1-x^2)}\,1\,dx$, we can write

$\pi/4 = \mathscr{E}[\sqrt{(1-U^2)}]$, where U is a $U(0, 1)$ random variable, and so if we take a random sample, $U_1, U_2, \ldots, U_n$, we can estimate $\pi/4$ by:

$$I = \sum_{i=1}^n \sqrt{(1-U_i^2)}/n$$

This approach is termed 'crude' Monte Carlo.

Clearly, $\mathrm{Var}(I) = n^{-1}\,\mathrm{Var}(\sqrt{(1-U^2)})$

$$= n^{-1}\left(\int_0^1 (1-x^2)\,dx - (\int_0^1 \sqrt{(1-x^2)}\,dx)^2\right)$$

$$= n^{-1}\left(\frac{2}{3} - \frac{\pi^2}{16}\right)$$

$$\approx 0.498/n$$

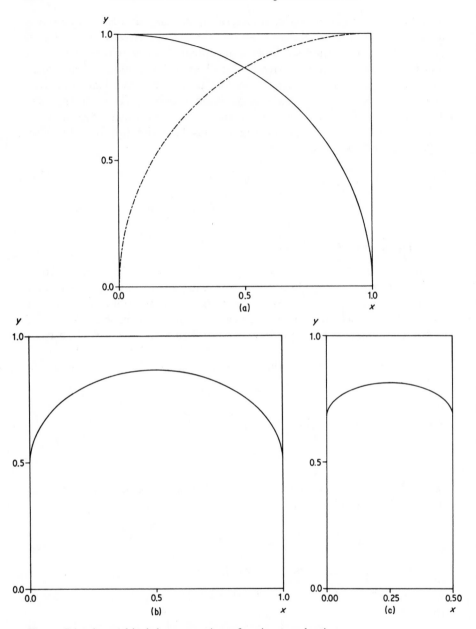

Figure 7.1   A graphical demonstration of variance reduction

(a) ——$y = \sqrt{(1 - x^2)}$; — · — · $y = \sqrt{(1 - (1 - x)^2)}$

(b) $y = \frac{1}{2}\{\sqrt{(1 - x^2)} + \sqrt{(1 - (1 - x)^2)}\}$

(c) $y = \frac{1}{4}\{\sqrt{(1 - x^2)} + \sqrt{(1 - (1 - x)^2)} + \sqrt{(1 - (\frac{1}{2} - x)^2)} + \sqrt{(1 - (\frac{1}{2} + x)^2)}\}$

The area under the curves of (a) and (b) is $\pi/4$; the area under the curve of (c) is $\pi/8$.

To construct a point at random within a unit square, we only need to take a point with Cartesian co-ordinates (U_1, U_2), where U_1 and U_2 are independent $U(0, 1)$ variables. In order to compare the 'hit-or-miss' and crude Monte Carlo approaches to estimating π we can take $2n$ $U(0, 1)$ variables in each case, resulting in the respective variances of estimators of π:

$$2.697/n \qquad \text{and} \qquad 0.0498 \times 16/2n = 0.398/n$$

So we see that the variance in the hit-or-miss case is roughly seven times larger than in the crude case, indicating that crude Monte Carlo is much more efficient.

7.2.3 Using an antithetic variate

Let
$$H = \tfrac{1}{2}\{ \sqrt{(1 - U^2)} + \sqrt{(1 - (1 - U)^2)}\},$$

where U is $U(0, 1)$. $(1 - U)$ is the 'antithetic' variate to U. As both U and $(1 - U)$ are $U(0, 1)$ random variables,

$\mathscr{E}[H] = \pi/4$, but now

$$\begin{aligned}
\text{Var}(H) &= \tfrac{1}{4}\{\text{Var}(\sqrt{(1 - U^2)}) + \text{Var}(\sqrt{(1 - (1 - U)^2)}) \\
&\quad + 2\,\text{Cov}(\sqrt{(1 - U^2)}, \sqrt{(1 - (1 - U)^2)})\} \\
&= \tfrac{1}{2}\{\text{Var}(\sqrt{(1 - U^2)}) + \text{Cov}(\sqrt{(1 - U^2)}, \sqrt{(1 - (1 - U)^2)})\} \\
&= \frac{1}{2}\left(\frac{2}{3} - \frac{\pi^2}{16}\right) + \frac{1}{2}\mathscr{E}\left[\sqrt{((U + 1)U(U - 1)(U - 2))} - \frac{\pi^2}{16}\right]
\end{aligned}$$

It can be shown that

$$\mathscr{E}[\sqrt{((U + 1)U(U - 1)(U - 2))}] = \frac{\pi}{4}\left\{\frac{71}{96} - 6\sum_{k=2}^{\infty} \frac{(2k - 3)!\,(2k - 1)!\,(12)^{-2k}}{(k - 2)!\,(k - 1)!\,k!\,(k + 1)!}\right\}$$

$$\approx 0.5806$$

leading to:
$$\text{Var}(H) = 0.0052.$$

Thus if $2n$ $U(0, 1)$ variates were used to estimate π using this antithetic approach, the resulting estimator would have variance $0.042/n$. The crude estimator variance is just over nine times larger than this, while the hit-or-miss estimator variance is roughly 64 times larger. In real terms this means that to obtain the same precision using the hit-or-miss and antithetic approaches, we need 64 times as many uniform variates in the hit-or-miss approach. Of course there may be losses in the different types of arithmetic involved between these two different approaches. We commented on this aspect in our comparison of the Buffon needle and cross, and we can now see that the second needle of the cross produced an antithetic variate, resulting in the negative correlation between X and Y in Section 7.1. The idea of using antithetic variates was formally introduced by Hammersley and Morton (1956), who explained the idea through the example of Buffon's needle.

*7.2.4 Reducing variability

The reduction in variability obtained by the use of an antithetic variate as above is simply seen from a comparison of the curves of Fig. 7.1(a) and 7.1(b): the ranges of the y-values are, respectively, 1, ($\sqrt{3}-1)/2$, while the areas under the curves are each $\pi/4$. We can clearly reduce variability even further by using the curve of Fig. 7.1(c), enabling us to estimate π using:

$$H = \{ \sqrt{(1 - U^2)} + \sqrt{(1 - (1 - U)^2)} + \sqrt{(1 - (\tfrac{1}{2} - U)^2)} + \sqrt{(1 - (\tfrac{1}{2} + U)^2)}\} \tag{7.2}$$

where now U is $U(0, 0.5)$. This process can be continued without end, rather like the testing of random numbers. Again a compromise has to be reached, in this case between variance reduction and increase in computation. For more discussion of this approach, see Morton (1957) and Shreider (1964, p. 53).

The variability in $y = \sqrt{(1 - x^2)}$ can be reduced in a number of additional ways. For instance, we can write

$$y = \{1 - x^2\} + \{ \sqrt{(1 - x^2)} - (1 - x^2)\} \qquad \text{for } 0 \le x \le 1 \tag{7.3}$$

as suggested in Simulation I (1976, p. 41). In (7.3), of course, both the components of y can be integrated explicitly, but if one knows how to integrate $\{1 - x^2\}$ and not $\sqrt{(1 - x^2)}$, then the decomposition of (7.3) replaces the variability of $\sqrt{(1 - x^2)}$ by the smaller variability of $\{ \sqrt{(1 - x^2)} - (1 - x^2)\}$. A decomposition of y can also be obtained without introducing a new function, simply by splitting up the range of x, and evaluating the integral as the sum of the integrals over the separate parts of the x-range. This is called *stratified sampling*, and is familiar to students of sampling theory (see Barnett, 1974, p. 78). For illustration, suppose the function to be integrated is $y = f(x)$, over the range $(0, 1)$, (see Fig. 7.2).

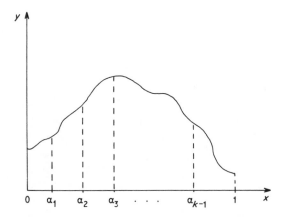

Figure 7.2   Stratified sampling for Monte Carlo integral estimation.

We shall break the range of integration into k pieces, of length $(\alpha_j - \alpha_{j-1})$, for $1 \le j \le k$, with $0 = \alpha_0 < \alpha_1 < \ldots < \alpha_k = 1$. Clearly, the variability of y within each piece is less than the variability of y over the full range. Estimation of the sub-integrals may be done in each case by, for example, crude Monte Carlo. If we use n_j $U(0, 1)$ variates for the jth interval, then we can estimate $\theta = \int_0^1 f(x)\,dx$ by:

$$\hat{\theta} = \sum_{j=1}^{k} \sum_{i=1}^{n_j} \frac{(\alpha_j - \alpha_{j-1})}{n_j} f(\alpha_{j-1} + (\alpha_j - \alpha_{j-1})U_{ij})$$

in which the U_{ij} are independent $U(0, 1)$ random variables. Thus

$$Y_{ij} = \alpha_{j-1} + (\alpha_j - \alpha_{j-1})U_{ij} \quad \text{is} \quad U(\alpha_{j-1}, \alpha_j)$$

as required for crude Monte Carlo estimation within the jth interval. The terms $(\alpha_j - \alpha_{j-1})$ are weights which are needed to ensure that $\hat{\theta}$ is unbiased.

We see that

$$\mathscr{E}[\hat{\theta}] = \sum_{j=1}^{k} \sum_{i=1}^{n_j} \frac{(\alpha_j - \alpha_{j-1})}{n_j} \int_0^1 f(\alpha_{j-1} + (\alpha_j - \alpha_{j-1})x)\,dx$$

$$= \sum_{j=1}^{k} \left(\frac{\alpha_j - \alpha_{j-1}}{n_j} \right) \int_{\alpha_{j-1}}^{\alpha_j} \frac{f(x)\,dx}{(\alpha_j - \alpha_{j-1})} \left(\sum_{i=1}^{n_j} 1 \right)$$

$$= \sum_{j=1}^{k} \int_{\alpha_{j-1}}^{\alpha_j} f(x)\,dx = \int_0^1 f(x)\,dx = \theta$$

In order to examine the precision of this approach we need the variance of $\hat{\theta}$:

$$\text{Var}(\hat{\theta}) = \sum_{j=1}^{k} \sum_{i=1}^{n_j} \frac{(\alpha_j - \alpha_{j-1})^2}{n_j^2} \text{Var}(f(\alpha_{j-1} + (\alpha_j - \alpha_{j-1})U_{ij}))$$

which leads readily to:

$$\text{Var}(\hat{\theta}) = \sum_{j=1}^{k} \frac{1}{n_j} \left\{ (\alpha_j - \alpha_{j-1}) \int_{\alpha_{j-1}}^{\alpha_j} f^2(x)\,dx - \left(\int_{\alpha_{j-1}}^{\alpha_j} f(x)\,dx \right)^2 \right\} \quad (7.4)$$

In using stratified sampling, one has to choose k, $\{\alpha_j\}$, $\{n_j\}$. Increasing k, the number of pieces, or strata, is likely to increase precision, but results in more arithmetic labour, and as before, a compromise is usually reached. For given k and $\{\alpha_j\}$, one can try to choose the $\{n_j\}$ to minimize the variance of (7.4).

We can write

$$\text{Var}(\hat{\theta}) = \sum_{j=1}^{k} \frac{a_j}{n_j}$$

say, and we want to choose the $\{n_j\}$ to minimize $\text{Var}(\hat{\theta})$, subject to a restriction such as $\Sigma_{j=1}^{k} n_j = n$, for fixed n. We can incorporate the constraint by introducing a Lagrange multiplier, λ, and then minimizing the Lagrangian:

$$L = \sum_{j=1}^{k} \frac{a_i}{n_j} - \lambda \left(n - \sum_{j=1}^{k} n_j \right)$$

Stationary values are obtained by setting

$$\frac{\partial L}{\partial n_j} = 0 \qquad \text{for } 1 \leq j \leq k,$$

i.e. $a_j = \lambda n_j^2$, for $1 \leq j \leq k$.

Standard theory (see Exercise 7.6) verifies that this stationary value is indeed a minimum, obtained by selecting the

$$n_j \propto a_j^{1/2} = (\alpha_j - \alpha_{j-1}) \sqrt{\{\text{Var}(f(\alpha_{j-1} + (\alpha_j - \alpha_{j-1})U_{ij})\}} \qquad (7.5)$$

Unfortunately, as we can see from (7.4), a_j involves the very integral we are seeking, and so is unknown. However, the message of (7.5) is clear, suggesting that the larger strata, and strata with more variable function values, should receive relatively more variates, as one would expect. Thus one could, as a rough rule of thumb, choose the $\{\alpha_j\}$ to correspond, as closely as possible, to parts of the curve with a constant range of y-values, and then allot the $\{n_j\}$ in proportion to $\{(\alpha_j - \alpha_{j-1})\}$. Alternatively, one could conduct a preliminary experiment to estimate the unknown variances in (7.5), and then use those estimates in deciding upon the $\{n_j\}$ for a full, subsequent investigation.

*7.2.5 Importance sampling

In stratified sampling, proportions n_j/n of n $U(0, 1)$ variables are transformed to the range (α_{j-1}, α_j). This is not dissimilar from selecting the U_{ij} variates which go to form $\hat{\theta}$ from the composition density function:

$$\Psi(x) = \sum_{j=1}^{k} \left(\frac{n_j}{n} \right) \eta_j(x) \qquad (7.6)$$

in which $\eta_j(x) = \begin{cases} 1/(\alpha_j - \alpha_{j-1}) & \text{for } \alpha_{j-1} \leq x < \alpha_j \\ 0 & \text{otherwise} \end{cases}$ for $1 \leq j \leq k$.

The continuous analogue of (7.6) is found in *importance sampling*, so called because, as with (7.6), $f(x)$ is evaluated at the important parts of the range more frequently than otherwise. Continuing with the above illustration, let us further suppose $f(x) > 0$ for $0 \leq x \leq 1$, and also that $g(x)$ is a probability density function over this range.

$$\theta = \int_0^1 f(x)\,dx = \int_0^1 \frac{f(x)}{g(x)} g(x)\,dx$$
$$= \mathscr{E}[f(X)/g(X)], \text{ when } X \text{ has probability density function } g(x).$$

If $(X_1, \ldots, X_n)$ constitutes a random sample from $g(x)$, then we can estimate θ by:

$$\hat{\theta} = \frac{1}{n} \sum_{i=1}^{n} \frac{f(X_i)}{g(X_i)}$$

which has variance given by

$$\text{Var}(\hat{\theta}) = \frac{1}{n}\left\{ \int_0^1 \frac{f^2(x)}{g(x)}\,dx - \theta^2 \right\}$$

$$= 0 \text{ if } g(x) = \theta^{-1}f(x) \qquad (7.7)$$

which would clearly be a good form to adopt for $g(x)$. Unfortunately, this entails knowledge of the unknown θ before $g(x)$ can be specified. However, if $g(x)$ is of roughly the same shape as $f(x)$ then (7.7) will hold approximately, and so we would expect the $\hat{\theta}$ that results to have small variance.

EXAMPLE 7.2 *An illustration of importance sampling*

Evaluation of $\Phi(x) = \int_{-\infty}^{x} \frac{e^{-y^2/2}}{\sqrt{(2\pi)}}\,dy = \int_{-\infty}^{x} \phi(y)\,dy$

We have seen already in Example 5.4 how this distribution function may be used in simulation, and we know that it is not possible to evaluate the integral explicitly. A density curve of similar shape to $\phi(y)$ is the logistic:

$$f(y) = \frac{\pi \exp(-\pi y/\sqrt{3})}{\sqrt{3}(1 + \exp(-\pi y/\sqrt{3}))^2}$$

with mean 0 and variance 1, already encountered in Section 5.7.

$$\Phi(x) = \int_{-\infty}^{x} \frac{k\phi(y)}{f(y)}\frac{f(y)}{k}\,dy$$

where k is chosen so that $f(y)k^{-1}$ is a density function over the range $(-\infty, x)$.

Thus $f(y)k^{-1} = \dfrac{\pi \exp(-\pi y/\sqrt{3})(1 + \exp(-\pi x/\sqrt{3}))}{\sqrt{3}(1 + \exp(-\pi y/\sqrt{3}))^2}$ (7.8)

and if Y is a random variable with the density function of (7.8) then,

$$\Phi(x) = (1 + \exp(-\pi x/\sqrt{3}))^{-1} \mathscr{E}[\phi(Y)/f(Y)]$$

We can therefore estimate $\Phi(x)$ by:

$$\hat{\theta} = \frac{1}{n}(1 + \exp(-\pi x/\sqrt{3}))^{-1} \sum_{i=1}^{n} \phi(Y_i)/f(Y_i)$$

where $\{Y_i, 1 \le i \le n\}$ is a random sample from the density of Equation (7.8), conveniently simulated by means of the inversion method of Section 5.2 as follows. If U is a $U(0, 1)$ random variable,

set $U = F(Y) = (1 + \exp(-\pi x/\sqrt{3}))/(1 + \exp(-\pi Y/\sqrt{3}))$

then we seek $Y = F^{-1}(U)$, resulting in:

$$Y = -\frac{\sqrt{3}}{\pi} \log_e \{(1 + \exp(-\pi x/\sqrt{3}))U^{-1} - 1\}$$

A BASIC program to evaluate $\Phi(x)$ in this way is given in Fig. 7.3 for a selection of x-values, and results from using this program are shown in Table 7.1.

```
10    RANDOMIZE
20    REM PROGRAM TO CALCULATE PHI(X)
30    REM USING IMPORTANCE SAMPLING
40    INPUT N
50    LET P1 = 1.813799364
60    LET X = -2.5
70    FOR K = 1 TO 4
80    LET X = X+.5
90    LET S = 0
100   FOR I = 1 TO N
110   LET R = (1+EXP(-X*P1))/RND
120   LET Y = -(LOG(R-1))/P1
130   LET P2 = (EXP(-Y*Y/2))/2.506628275
140   LET Q = (P2*(1+EXP(-Y*P1))^2)*EXP(Y*P1)
150   LET S = S+Q
160   NEXT I
170   LET S = S/N
180   LET S = S/(P1*(1+EXP(-X*P1)))
190   PRINT X,S,N
200   NEXT K
210   END
```

Figure 7.3   A BASIC program to evaluate $\Phi(x)$ using importance sampling. Note that $\pi/\sqrt{3} \approx 1.813\,799\,364$, and $\sqrt{(2\pi)} \approx 2.506\,628\,275$.

An application of importance sampling in queueing theory is provided by Evans *et al.* (1965).

Table 7.1

| | Estimated $\Phi(x)$ | | | Actual $\Phi(x)$ (from |
| | | | | tables) to 4 d.p. |
x	$n = 100$	$n = 1000$	$n = 5000$	
-2.0	0.0222	0.0229	0.0229	0.0227
-1.5	0.0652	0.0681	0.0666	0.0668
-1.0	0.1592	0.1599	0.1584	0.1587
-0.5	0.3082	0.3090	0.3081	0.3085

We shall not here consider $\text{Var}(\hat{\theta})$ for this example, but see the solution to Exercise 7.7.

7.3 Further variance-reduction ideas

7.3.1 Control variates

With antithetic variates, negative correlation was used to reduce variance. One can also use positive (or negative) correlation with some additional, *control* variate. As with stratified sampling, comparisons can be made here with elements of sampling theory.

Suppose X is being used to estimate a parameter θ, and $\mathscr{E}[X] = \theta$. If Z is a random variable with known expectation μ, then for any positive constant c, we can write

$$Y = X - c(Z - \mu)$$

Thus Y, like X, is an unbiased estimator of θ, as $\mathscr{E}[Y] = \mathscr{E}[X] = \theta$. Whether or not Y is a better estimator of θ than X depends on the relationship between X and Z. Now $\mathrm{Var}(Y) = \mathrm{Var}(X) + c^2 \mathrm{Var}(Z) - 2c \mathrm{Cov}(X, Z)$, and so if $\mathrm{Cov}(X, Z) > c \mathrm{Var}(Z)/2$, then $\mathrm{Var}(Y)$ will be less than $\mathrm{Var}(X)$, indicating that Y is the better estimator. The maximum variance reduction is obtained when $c = \mathrm{Cov}(X, Z)/\mathrm{Var}(Z)$, and while $\mathrm{Cov}(X, Z)$ (and possibly also $\mathrm{Var}(Z)$) may not be known, it could be estimated by means of a pilot investigation. However, many investigators have simply taken $c = \pm 1$ as appropriate.

We shall see an example of the use of a control variate in Section 9.4.2. More than one control variate may be used, and a variety of different approaches have been employed to obtain the desired correlation between the variate of interest, X, and the control variate, Z. See, for example, Law and Kelton (1982, Section 11.4). Improvements in the use of control variates are considered by Cheng and Feast (1980). A recent application is provided by Rothery (1982), in the context of estimating the power of a non-parametric test, and an illustration from queueing theory is given in the following example.

EXAMPLE 7.3 (*Barnett, 1965, p. XVII*): *Machine interference*
A mechanic services n machines which break down from time to time. We suppose that machines break down independently of one another, and that for any machine, breakdowns are events in a Poisson process of rate λ. We suppose also that the time taken to repair any machine is a constant, μ. The interference arises if queues of broken-down machines form. This process can be solved analytically, but it was presented by Barnett as an illustration of the use of a control variate. It is clearly simple to simulate the process, and to estimate the 'machine availability', S, over a time period of length t, by,

$$\hat{S} = \frac{\text{total cumulative running time for all machines}}{nt}$$

As the control variate, Barnett used the estimate, $\hat{L}$, of $1/\lambda$, given by

$$\hat{L} = \mu \times \frac{\text{total cumulative running time for all machines}}{\text{total cumulative repair time for all machines}}$$

It was estimated empirically that the correlation between $\hat{S}$ and $\hat{L}$ was $\approx$ $+0.95$ for a variety of values of n, t and the product $\lambda\mu$. Thus S was estimated by

$$\hat{S}_1 = \hat{S} - c\left(\hat{L} - \frac{1}{\lambda}\right)$$

and c was chosen as indicated above for maximum variance reduction, using estimates of second-order moments obtained from a pilot study. It was estimated that $\text{Var}(\hat{S})/\text{Var}(\hat{S}_1) \approx 9.87$. (Further discussion of this example is given in Exercises 7.22 and 7.23.)

The standard approach for estimating $\theta = \mathscr{E}[X]$ is by forming

$$\hat{\theta} = \frac{\sum\limits_{i=1}^{n} X_i}{n}$$

where $\{X_i, 1 \leq i \leq n\}$ forms a random sample from the distribution of X. This is completely analogous to the averaging approaches used in the previous examples in integral estimation, which is to be expected, since here θ is a mean value which, for continuous random variables, can be written as an integral, and vice versa. As was pointed out in Section 7.2, integrals of low dimensionality are probably best evaluated by a numerical method which does not involve simulation. However, while one can certainly think of the estimation of a mean of a random variable in terms of evaluating an integral, in this case the integrand is itself almost certainly going to be a function of the (unknown) mean, and so simulation methods are then appropriate.

When a model is to be simulated under different conditions, and comparisons made between the different simulations, then the variation between the simulations can be reduced by using *common random numbers* in the different simulations. This is a very popular method of variance reduction and, as with many uses of control variates, it relies on an induced positive correlation for its effect. We shall return to the use of common random numbers in Chapter 9. A good example, involving the comparison of alternative queueing mechanisms, is given by Law and Kelton (1982, p. 352).

*7.3.2   Using simple conditioning

The principle here is best illustrated by means of an example.

EXAMPLE 7.4 (*Simon*, 1976)

Suppose we want to estimate the mean value θ of a random variable X, which has the beta $B_e(W, W^2 + 1)$ distribution, where W itself is a random variable, with a Poisson distribution of known mean, η. The obvious approach is to simulate n X-values and simply average them. However, this involves simulation from Poisson and beta distributions, and an alternative approach is as follows.

We know, from Section 2.11, that

$$\mathscr{E}[X|W = w] = w/(w^2 + w + 1)$$

Furthermore,
$$\sum_{w=0}^{\infty} \mathscr{E}[X|W = w] \frac{e^{-\eta}\eta^w}{w!} = \theta$$

(here we are using a property of conditional expectation—see Grimmett and Stirzaker, 1982, p. 44).

So we may estimate θ by

$$\hat{\theta} = \frac{1}{n} \sum_{i=1}^{n} w_i/(w_i^2 + w_i + 1)$$

where the $\{w_i, 1 \le i \le n\}$ form a random sample from the Poisson distribution, parameter μ; a procedure which does not, in fact, involve simulating X. Thus this approach certainly saves labour. Discussion of how the variance of the above estimator may be further reduced is given in Exercise 7.20. In a different context, Lavenberg and Welch (1979) use conditioning to reduce variance in a particular queueing network, and their example is reproduced by Law and Kelton (1982, p. 364).

7.3.3 The M/M/1 queue

It is interesting to see how variance-reduction techniques that have been clearly expressed for simple procedures, such as the evaluation of one-dimensional integrals, may be employed in more complicated investigations. We shall here consider the M/M/1 queue. This model of a simple queue has already been encountered in Exercise 2.27, which also provided a BASIC program for the simulation of the queue.

We are often interested in the average customer waiting-time in a queue. The waiting time of the nth customer, from arrival at the queue until departure, W_n, may be very simply expressed as:

$$\left. \begin{array}{ll} W_n = (W_{n-1} - I_n + S_n) & \text{if} \quad W_{n-1} \ge I_n \\ W_n = S_n & \text{if} \quad W_{n-1} < I_n \end{array} \right\} \tag{7.9}$$

where S_n is the service-time of customer n, and

I_n is the time between the arrival of the nth and $(n-1)$th customers, for $n \ge 2$.

Note here that we take $W_1 = S_1$, i.e., the first customer arrives to find an empty queue. Figure 7.4 provides a BASIC program for simulating this queue, for which the service and inter-arrival times are both exponential, with respective parameters $\mu = 1$, $\lambda = 0.6$. We see that the average waiting-time is computed for 200 customers. The process is then repeated 100 times so as to provide an estimate of the variance of the average waiting time. The program of Fig. 7.4 provides a much simpler way of estimating average waiting-time than direct use of the program of Exercise 2.27, and we shall return to this point in Chapter 8.

```
10  REM BASIC PROGRAM TO ESTIMATE THE AVERAGE
20  WAITING TIME OF THE FIRST 200 CUSTOMERS
30  REM AT AN M/M/1 QUEUE, STARTING EMPTY, USING (7.9)
40  LET L=.6
50  LET M=1
60  RANDOMIZE
70  LET T1=0
80  LET T2=0
90  FOR J=1 TO 100
100   LET S2=0
110   LET W=0
120   FOR I=1 TO 200
130    LET U=RND
140    LET S=(-LOG(U))/M
150    LET U=RND
160    LET T=(-LOG(U))/L
170    IF W<T THEN 200
180    LET W=W+S-T
190    GOTO 210
200    LET W=S
210    LET S2=S2+W
220   NEXT I
230   LET T1=T1+(S2/200)
240   LET T2=T2+(S2/200)^2
250  NEXT J
260  LET V=(T2-(T1*T1)/100)/99
270  PRINT "VARIANCE OF AVERAGE WAITING TIME = ",V
280  PRINT "MEAN = ",T1/100
290  END
```

Figure 7.4   A BASIC program to estimate the average waiting-time of the first 200 customers at an M/M/1 queue, starting empty. The procedure is repeated 100 times. Note that in lines 140 and 160 the method of Equation (5.5) is used.

This is an example where an antithetic-variate approach could prove useful. Figure 7.5 provides another BASIC program for simulating this queue. In this case we duplicate each block of 200 customers, and in the duplicate block each original U is replaced by $(1 - U)$, with the result that long service times are replaced by short services times, and vice versa, and similarly also for inter-arrival times. Each block average therefore still estimates the same average waiting time, but the two duplicate block averages might now be expected to have a negative correlation. Table 7.2 illustrates the results of the start of a run of the program of Fig. 7.5, and we can see here the anticipated relationship developing between the two sets of W_n values. Proofs that variance reduction will occur when antithetic variates are used in this, and more general, queueing

```
10   REM ILLUSTRATION OF VARIANCE REDUCTION USING
20   REM ANTITHETIC VARIATES IN AN M/M/1 QUEUE
30   DIM R(400)
40   LET L = .6
50   LET M = 1
60   LET I1 = 0
70   RANDOMIZE
80   LET U1 = 0
90   LET U2 = 0
100  LET T1 = 0
110  LET T2 = 0
120  LET N = 50
130  FOR J = 1 TO N
140   FOR I = 1 TO 400
150    LET R(I) = RND
160   NEXT I
170  LET S2 = 0
180  LET W = 0
190  LET K = 0
200  FOR I2 = 1 TO 200
210   LET K = K+1
220   LET U = R(K)
230   LET S = (-LOG(U))/M
240   LET K = K+1
250   LET U = R(K)
260   LET T = (-LOG(U))/L
270   IF W < T THEN 300
280   LET W=W+S-T
290   GOTO 310
300   LET W = S
310   LET S2 = S2+W
320  NEXT I2
330  IF I1 = 1 THEN 360
340  LET S5 = S2/200
350  GOTO 400
360  LET S5 = (S5+S2/200)/2
370  LET T1 = T1+S5
380  LET T2 = T2+S5*S5
390  GOTO 460
400  FOR I = 1 TO 400
410   LET R(I) = 1-R(I)
420  NEXT I
430  REM THIS FORMS THE ANTITHETIC VARIATES
440  LET I1 = 1
450  GOTO 170
460  LET I1 = 0
470  NEXT J
480  LET V = (T2-T1*T1/N)/(N-1)
490  PRINT "VARIANCE OF AVERAGE WAITING TIME=",V
500  PRINT "ESTIMATE OF MEAN WAITING TIME =",T1/N
510  END
```

Figure 7.5 A BASIC program to estimate the average waiting-time of the first 200 customers in an M/M/1 queue, starting empty. The procedure is based upon Equation (7.9) and uses antithetic variates, as explained in the text.

models are provided by Mitchell (1971) and others (see Kleijnen, 1974, p. 190), who also provide empirical investigations, as do Law and Kelton (1982, p. 356).

A variety of results from running the programs of Figs 7.4 and 7.5 are given in Table 7.3. We can see, by considering the results from different runs, that the estimate of efficiency gain can vary appreciably, but in all comparisons there is a gain in efficiency. Use of equations (7.9) does in fact contravene a basic rule for variance reduction, already encountered in (7.3) (see also Exercises 7.8 and

Table 7.2   An illustration of the use of Equation (7.9) to compute waiting times in an M/M/1 queue, and the effect of replacing service (S_n) and inter-arrival times (T_n) by their antithetic counterparts.

		Main block				Antithetic block		
n	S_n	T_n	W_n		n	S_n	T_n	W_n
1	1.35	—	1.35		1	0.30	—	0.30
2	0.20	0.40	1.15		2	1.70	1.34	1.70
3	0.75	1.89	0.75		3	0.64	0.22	2.12
4	0.17	0.36	0.56		4	1.88	1.42	2.58
5	0.43	0.97	0.43		5	1.05	0.60	3.03
6	1.83	0.39	1.87		6	0.18	1.34	1.23

7.20), as we shall now explain. We can write

$$W_n = Q_n + S_n \qquad \text{for } n \geq 1$$

where Q_n is the time spent by the nth customer queueing before being served. Q_n and S_n are independent, and $\mathrm{Var}(Q_n) < \mathrm{Var}(W_n)$. Therefore in order to estimate $\mathscr{E}[W_n]$ it is more efficient to estimate $\mathscr{E}[Q_n]$ and then add on the known $\mathscr{E}[S_n] = 1/\mu$. This can be seen from a comparison of Tables 7.3(a) and (b). This comparison also suggests, however, that the use of (7.9) combined with an antithetic-variate approach can increase efficiency, relative to the use of (7.10) below combined with antithetic variates. Note that

$$Q_n = \max(Q_{n-1} + S_n - I_n, 0) \tag{7.10}$$

Table 7.3   (a) Sample variance of the estimator of the mean waiting-time of the first 200 customers in an M/M/1 queue, starting empty and with $\mu = 1$. 100 replications were used in each case, with 50 matched pairs when antithetic variates were employed. In this case the waiting-times were simulated including the service-times, i.e. using Equation (7.9)

λ		No variance reduction	Using antithetic variates
	Run		
	1	0.1338	0.0760
0.5	2	0.1426	0.0736
	3	0.2174	0.0561
		0.1646	0.0686
	1	0.3102	0.1338
0.6	2	0.4010	0.1851
	3	0.5233	0.1676
		0.4115	0.1622

(contd.)

Table 7.3   (*contd.*)

λ		No variance reduction	Using antithetic variates
0.7	1 2 3	2.3254 5.4315 1.0977	0.2434 0.4734 0.3401
		2.9515	0.3523
0.8	1 2 3	2.7757 6.6679 6.7081	1.2451 1.6539 2.1598
		5.3839	1.6863

Corresponding average waiting-times of the first 200 customers

λ		No variance reduction	Using antithetic variates	Theoretical value in equilibrium (see Exercise 7.24)
	Run			
0.5	1 2 3	1.889 2.035 2.030	1.998 2.020 1.984	2.0
		1.985	2.001	
0.6	1 2 3	2.513 2.467 2.533	2.464 2.408 2.543	2.5
		2.504	2.472	
0.7	1 2 3	3.320 3.497 3.051	3.041 3.199 3.241	3.33
		3.289	3.160	
0.8	1 2 3	4.244 4.416 5.152	4.392 4.404 4.451	5.0
		4.604	4.416	

(b) The following results are obtained by simulating the waiting-times without the service-times, i.e. using Equation (7.10). First of all we give the sample variances, as in (a).

λ		No variance reduction	Using antithetic variates
	Run		
	1	0.1314	0.0927
0.5	2	0.1547	0.0902
	3	0.1225	0.0466
		0.1362	0.0765
	1	0.3228	0.2487
0.6	2	0.6141	0.1521
	3	0.5180	0.1182
		0.4850	0.1730
	1	0.8762	0.3201
0.7	2	1.1808	0.5234
	3	0.9866	0.6251
		1.0145	0.4895
	1	6.4527	1.3440
0.8	2	3.2618	2.0066
	3	5.4324	1.6008
		5.049	1.6703

Corresponding average waiting-times of the first 200 customers in an M/M/1 queue, starting empty and with $\mu = 1$, as above. Values are obtained by computing the average waiting-time, excluding service, and then adding on the known mean service-time.

λ		No variance reduction	Using antithetic variates	Theoretical values in equilibrium (see Exercise 7.24)
	Run			
	1	2.028	2.019	
0.5	2	2.000	2.034	2.0
	3	1.936	1.989	
		1.988	2.014	
	1	2.354	2.533	
0.6	2	2.550	2.418	2.5
	3	2.559	2.401	
		2.488	2.451	

		No variance reduction	Using antithetic variates	Theoretical values in equilibrium (see Exercise 7.24)
0.7	1	3.129	3.300	
	2	3.134	3.165	3.33
	3	3.198	3.282	
		3.154	3.249	
0.8	1	4.594	4.444	
	2	4.447	4.445	5.0
	3	4.576	4.531	
		4.539	4.473	

As one might expect, whether we are using Equations (7.9) or (7.10), the amount of variance reduction achieved depends on the relationship between λ and μ: if λ is appreciably smaller than μ, then the queue will frequently be empty, reducing the negative correlation. The values of λ and μ also affect the rate at which a steady-state system is reached (for the case $\lambda < \mu$ – see Exercise 7.24). Barnett's (1965) tables of exponential random variables provide values of $-\log_e(1-U)$ as well as $-\log_e U$, with just such antithetic investigation in mind (see Example 7.22).

An alternative approach to antithetic variance reduction in simple queues was applied by Page (1965), who used the following idea. Suppose we are simulating an M/M/1 queue, constructing service and inter-arrival times respectively from:

$$S = -\frac{1}{\mu}\log_e(U_1)$$

$$T = -\frac{1}{\lambda}\log_e(U_2)$$

for independent $U(0, 1)$ variables U_1 and U_2.
A duplicate run can be made with

$$\tilde{S} = -\frac{1}{\mu}\log_e(U_2)$$

$$\tilde{T} = -\frac{1}{\lambda}\log_e(U_1)$$

In this case U_1 values giving rise to large service times in the original run will be translated into large waiting times in the duplicate run, and vice versa. Page showed that

$$\text{Corr}\,((S-T),\,(\tilde{S}-\tilde{T})) = -2\rho/(1+\rho^2) \qquad \text{where } \rho = \lambda/\mu.$$

*7.3.4 A simple random walk example

In random walks we are interested in the distribution of the position of a particle which moves along a line according to probability rules. A simple example results when the particle moves between absorbing barriers at 0 and a, a being a positive integer, according to the rules specified in Fig. 7.6. Exercise 1.4 provided an example of a random walk with a reflecting barrier at 0.

The particle position can be used to describe features of more complicated processes such as the population size of a colony of bacteria; the particular example of Fig. 7.6 is often called the 'gambler's ruin' problem, as the particle position can be taken as the capital of one of two gamblers, with combined capital of a units. In the game played by the gamblers, money changes hands in single units according to the probabilities p and q, and the game ends when one of the gamblers loses all his/her capital, corresponding to the particle reaching one of the barriers. Various features of this walk are of interest, such as the values $\{d_k,\ 1 \le k \le a-1\}$, where d_k is the average number of steps to termination of this walk, when the walk starts at k.

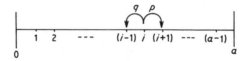

Figure 7.6   Illustration of a simple random walk. When the particle is at i (i an integer in the range $1 \le i \le a-1$), then, independently of the past, it moves to $(i+1)$ with probability p, and to $(i-1)$ with probability $q = (1-p)$. Once either of the barriers at 0 and a is reached, then the walk terminates.

It can be shown that the $\{d_k\}$ satisfy:

$$d_k = 1 + pd_{k+1} + qd_{k-1} \qquad \text{for } 1 \le k \le a-1$$
$$d_0 = d_a = 0 \tag{7.11}$$

(see, e.g., Bailey, 1964, p. 27, and Exercise 7.30). While these equations have an explicit solution, given in Exercise 7.31, the $\{d_k\}$ can be estimated by simulation (cf. Exercise 3.29). Thus one can select a value of k, and simulate n walks, starting at k, running until absorption. d_k may then be estimated by the sample mean time to absorption. A very simple variance-reduction idea which may be used here is outlined by Barnett (1962a). If a walk starting at k passes through some point *j* at a later stage, say after *r* steps, then if it takes *n* steps for the original walk to end, we also have, from this walk, an example of a walk taking $(n - r)$ steps to absorption, starting from *j*. Thus a single walk starting from any *k* can provide information on mean times to absorption for walks starting from points other than *k*. Clearly by this method the mean-time estimators for different values of *k* will not be independent. For discussion of this see Barnett (1962a), who used this approach for a two-dimensional random walk without an explicit solution; and Morgan and Robertson (1980), who considered an intractable one-dimensional problem (see Exercise 7.32).

7.4 Discussion

The aim of this chapter has been to underline the importance of variance reduction, and to introduce some of the methods that are used. Further methods and illustrations will be encountered later. Quite apart from its importance, variance reduction is attractive because of the extra information that can often be squeezed out of single random variables, this process frequently requiring a 'flash of insight', in Barnett's (1976) words. Because of the dramatic gain in value that can result, variance-reduction techniques are sometimes also termed 'swindling' (see Simon, 1976, and Schruben and Margolin, 1978, p. 524). Of course, this idea of obtaining as much value as possible from a single random variable has already been encountered, for instance in Exercise 4.2, and Example 5.5. We can also note the analogies between the rejection method of Section 5.3, and importance sampling and hit-or-miss Monte Carlo, as well as between the composition method of Section 5.4, and stratified sampling and the method of extracting an easier integral, in Section 7.2.4.

Much more detail of variance reduction is provided by Law and Kelton (1982, chapter 11), and Kleijnen (1974, chapter 3). Frequently it is necessary to run a pilot study in order to assess the possible value of a variance-reduction technique, for in complicated systems theoretical justification for employing such a technique is usually not possible. Indeed, it is unfortunately the case that in some applications variances have inadvertently been increased from using a 'reduction' approach. Cheng (1982) points out the importance of *high* negative correlation when antithetic variates are used, and suggests a modified procedure which has been applied successfully to a variety of models.

We saw in Section 7.1 that in many applications simulation estimators of parameters are expressed as sums of independent random variables. In these and related cases it is a simple matter to estimate variances of estimators, which are vital for the interpretation of estimates. In Section 7.2.2 the variance of the estimator involved the parameter that was being estimated. In that case, for illustration, the known parameter value was used in calculating the variance, while in practice the variance itself would only be an estimate. Knowledge of the answer (π) has undoubtedly also affected the reporting of experiments involving Buffon's needle, as well as the decision of when to stop. As Mantel (1953) points out, Lazzerini's experiment, conducted in 1901, produced $\hat{\pi} = 3.141\,592\,9$ after 3408 throws, but ending the experiment one throw sooner or later inevitably loses half the decimal place accuracy. When one is estimating waiting-times in queues, one is averaging *dependent* random variables, which complicates variance estimation. A number of possible approaches for such cases are described and compared by Moran (1975), and we shall return to this topic in Chapter 8.

Additional reductions in variance may be obtained by the judicious combination of different methods (see Exercise 7.23 and Schruben and Margolin, 1978), though here again one must proceed with caution, as Kleijnen (1974, Section III.8) has shown. 'Proceed with care' is therefore clearly the watchword for variance reduction, as it was with the use of pseudo-random numbers. However, as with the use of pseudo-random numbers, the benefits from using an appropriate variance-reduction technique can be substantial. Finally, we may note that common random variables and antithetic variates are variance-reduction techniques of *general* applicability, in so far as the same approach is adopted, whatever the problem. In contrast, methods such as importance sampling, stratified sampling, and the use of control variates all have to be individually tailored to particular problems.

7.5   Exercises and complements

(a) On Buffon's needle

7.1   What is the mean number of lines crossed if $l > d$? For discussion of this case, see Mosteller (1965, p. 88) and Mantel (1953).

†7.2   (Gani, 1980)   Suppose the centre of the needle lands at a distance x from a line, and that the needle makes an angle θ with the direction of the lines. Map out the sample space for (x, θ), and by identifying the subset of the sample space corresponding to the needle crossing a line, show that, for $l \leq d$, Pr(needle crosses a line) $= 2l/\pi d$. When $l > d$, show that this probability must be corrected by the amount

$$\frac{2}{\pi} \cos^{-1}(d/l) - \frac{2l}{\pi d}\sqrt{(1 - d^2/l^2)}.$$

*7.3   (Perlman and Wichura, 1975)   The case of a single needle thrown on to a double grid. Here we have grids, A and B, say, each of parallel lines a distance d apart, the grids being at right angles to each other. This problem was originally studied by Laplace. Let $r = d/l$, and let p_{AB} = Pr(needle crosses an A-line and a B-line), $p_{A\bar{B}}$ = Pr(needle crosses an

A-line but not a B-line), etc. Show that

$$p_{AB} = \frac{r^2}{\pi}$$

$$p_{A\bar{B}} = p_{\bar{A}B} = \frac{2r}{\pi} - \frac{r^2}{\pi}$$

$$p_{\bar{A}\bar{B}} = 1 - \frac{4r}{\pi} + \frac{r^2}{\pi}$$

Separate evaluation of these probabilities may be used to estimate π, in different ways, and which one to use is an intriguing question. If in n throws of the needle there are $n_{\bar{A}\bar{B}}$, $n_{\bar{A}B}$, $n_{A\bar{B}}$ and n_{AB} throws in the four possible categories, then (verify)

$$\frac{n_{AB}}{n} \text{ is an estimator of } \frac{r^2}{\pi}$$

$$\frac{n_{A\bar{B}} + n_{\bar{A}B} + 2n_{AB}}{n} \text{ is an estimator of } \frac{4r}{\pi}$$

$$\frac{n - n_{\bar{A}\bar{B}}}{n} \text{ is an estimator of } \frac{4r - r^2}{\pi}$$

If $r = 1$, Perlman and Wichura show that the variances of the resulting estimators are, respectively, $5.63/n$, $1.76/n$, $0.466/n$.

†**7.4**   (Continuation)   The following data were collected from class experiments by E. E. Bassett:

	n	$n_{\bar{A}\bar{B}}$	$n_{\bar{A}B}$	$n_{A\bar{B}}$	n_{AB}
experiment 1	400	16	112	125	147
experiment 2	990	64	315	304	307

Use these data and the estimators of the last exercise to provide a variety of estimates of π.

Further data are provided by Kahan (1961), who describes practical problems such as the blunting of the needle with use, and Gnedenko (1976, pp. 36–39), who also considers the throwing of a convex contour. Historical background is found in Holgate (1981), who conjectures on how Buffon obtained his solution. Mantel (1953) obtains an estimator of π from the estimation of a variance, rather than a mean.

7.5   (Holgate, 1981)   Another problem studied by Buffon was the 'Jeu du franc-carreau': a circular coin of radius b is thrown on to a horizontal square grid of side $2a$. Show that if $b/a = (1 - 2^{-0.5})$ then the coin is as likely as not to land totally within a square.

(b) Integral estimation

*7.6   Show that the stationary value of the Lagrangian

$$L = \sum_{j=1}^{k} \frac{a_j}{n_j} - \lambda\left(n - \sum_{j=1}^{k} n_j\right)$$

given by: $n_j \propto a_j^{1/2}$ is a minimum.

†7.7   When the computer program of Fig. 7.3 is run for values of $x = 0.5, 1.0,$ 1.5 and 2.0, the following values result:

	\multicolumn{3}{c}{Estimated $\Phi(x)$}	$\Phi(x)$
x	$n = 100$	$n = 1000$	$n = 5000$	to 4 d.p.
0.5	0.6999	0.6905	0.6898	0.6915
1.0	0.8477	0.8405	0.8423	0.8413
1.5	0.9253	0.9386	0.9337	0.9332
2.0	0.9867	0.9822	0.9761	0.9773

We obtain better accuracy with the results of Table 7.1. Use the argument of Section 7.2.4 to explain why we might expect this.

7.8   Hammersley and Handscomb (1964, p. 51) define the relative efficiency of two Monte Carlo methods for estimating a parameter θ as follows: The efficiency of method 2 relative to method 1 is:

$$(n_1 \sigma_1^2)/(n_2 \sigma_2^2)$$

where method i takes n_i units of time, and has variance σ_i^2, $i = 1, 2$. Write BASIC programs to estimate the integral:

$$I = \int_0^1 e^{-x^2} dx$$

by hit-or-miss, crude and antithetic variate Monte Carlo methods, and compare the efficiencies of these three methods by using a timing facility. Suggest, and investigate, a simple preliminary variance-reduction procedure. Investigations of variance reduction when

$I = \int_0^1 g(x)dx$ are given by Rubinstein (1981, pp. 135–138).

7.9   Write a BASIC program to estimate the integral of Exercise 7.8 using a stratification of four equal pieces, and 40 sample points. How should you distribute the sample points?

7.10   Explain how the use of stratification and antithetic variates may be combined in integral estimation.

*7.11   Verify that $\int_0^1 \sqrt{\{(x+1)x(x-1)(x-2)\}}\,dx \approx 0.5806$.

7.12   Given the two strata, $(0, \sqrt{3}/2)$, $(\sqrt{3}/2, 1)$, for evaluating $\int_0^1 \sqrt{(1-x^2)}dx$, how would you allot the sampling points?

†7.13   In crude Monte Carlo estimation of $\int_0^1 \sqrt{(1-x^2)}dx$, how large must n be in order that a 95 % confidence interval for the resulting estimator of π has width v? Evaluate such an n for $v = 0.01, 0.1, 0.5$.

*7.14   Daley (1974) discusses the computation of integrals of bivariate and trivariate normal density functions. Describe variance-reduction techniques which may be employed in the evaluation of such integrals using simulation. For related discussion, see Simulation I (1976, Section 13.8.3) and Davis and Rabinowitz (1975, Section 5.9).

7.15   Repeat Exercise 7.8, using the pseudo-random number generator of Equation (3.1).

(c) General variance reduction

7.16   Show that the maximum variance reduction in Section 7.3.1 is obtained when $c = \mathrm{Cov}(X, Z)/\mathrm{Var}(Z)$.

7.17   (Kleijnen, 1974, p. 254)   Suppose one is generating pseudo-random uniform variables from the $(a, 0; m)$ generator, with seed x_0. Show that the corresponding antithetic variates result from using $(m - x_0)$ as the seed.

7.18   Suppose one wants to estimate $\theta = \mathrm{Var}(X)$, and $X = U + V$, when U, V are independent random variables, and $\mathrm{Var}(U)$ is known. Clearly here a simulation should be done to estimate $\mathrm{Var}(V) < \mathrm{Var}(X)$. Use this result to estimate $\mathrm{Var}(M)$, where M is the median of a sample of size n from a $N(0, 1)$ distribution. You may assume (see Simon, 1976) that $\bar{X}$ and $(M - \bar{X})$ are independent, where $\bar{X}$ denotes the sample mean.

7.19   (*Continuation*)   Conduct an experiment to estimate the extent of the variance reduction in Exercise 7.18.

7.20   (Simon, 1976)

(i) Verify that $\left(\dfrac{W}{W^2+W+1}\right) = \dfrac{1}{(W+1)} - \dfrac{1}{(W+1)(W^2+W+1)}$.

(ii) If W has the Poisson distribution of Example 7.4, show that $E[1/(W+1)] = (1 - e^{-\eta})/\eta$.

(iii) Show that

$$\tilde{\theta} = \left(\frac{1-e^{-\eta}}{\eta}\right) - \frac{1}{n}\sum_{i=1}^{n}\{(W_i+1)(W_i^2+W_i+1)\}^{-1}$$

is an unbiased estimator of θ, and that $\mathrm{Var}(\tilde{\theta}) < \mathrm{Var}(\hat{\theta})$.

(d) Queues

*7.21 Investigate the use of the mean service-time as a control variate for the mean waiting-time in an $M/M/1$ queue.

7.22 For any simulation giving rise to the estimate $\hat{S}$ in the machine-interference study of Example 7.3 we can construct an antithetic run, replacing each $U(0, 1)$ variate U in the first simulation by $(1 - U)$ in the second. If we denote the second estimator of S by $\hat{S}'$, then a further estimator of S is:

$$\hat{S}_2 = \tfrac{1}{2}(\hat{S} + \hat{S}').$$

Barnett (1965) found empirically that the correlation between $\hat{S}$ and $\hat{S}'$ was ≈ -0.64, a high negative value, as one might expect. Estimate the efficiency gained (see Exercise 7.8) from using this antithetic-variate approach (cf. Fig. 7.5).

7.23 (*Continuation*) Barnett (1965) considered the further estimator of S:

$$\hat{S}_3 = \tfrac{1}{2}\{(\hat{S} + \hat{S}') - k(\hat{L} + \hat{L}' - 2/\lambda)\}$$

where $\hat{L}'$ is the estimator of $1/\lambda$ from the antithetic run. Discuss this approach, which combines the uses of control and antithetic variates. Show how k should be chosen to maximize the efficiency gain, and compare the resulting gain in efficiency with that obtained from using control variates and antithetic variates separately.

7.24 In an $M/M/1$ queue, when $\lambda < \mu$, then after a period since the start of the queue, the queue is said to be 'in equilibrium', or to have reached the 'steady state'. The distribution of this period depends on λ, μ and the initial queue size. In equilibrium the queue size Q has the geometric distribution

$$\Pr(Q = k) = \rho(1-\rho)^k \qquad \text{for } k \geq 0$$

where $\rho = \lambda/\mu$, and is called the 'traffic intensity'. Use this result to show that the customer waiting-time in equilibrium (including service time) has the exponential density: $(\mu - \lambda)e^{(\lambda - \mu)x}$, and hence check the theoretical equilibrium mean values of Table 7.3. Further, comment on the disparities between the values obtained by simulation and the theoretical equilibrium mean values. For related discussion, see Rubinstein (1981, p. 213) and Law and Kelton (1982, p. 283).

Unfortunately, it is the cases when ρ is near 1, for $\rho < 1$, that are often of practical importance, but also the most difficult to investigate using simulation.

7.25   Exponential distributions are, as we have seen, often used to model inter-arrival and service-times in queues. Miss A. Kenward obtained the data given below during a third-year undergraduate project at the University of Kent. Illustrate these data by means of a histogram, and use a chi-square test to assess whether, in this case, the assumption of exponential distributions is satisfactory (cf. Exercise 2.26).

The following data were collected from the sub-post office in Ashford, Kent, between 9.00 a.m. and 1.00 p.m. on a Saturday in December, 1981.

Inter-arrivals

Time in seconds	0–10	10–20	20–30	30–40	40–50	50–60	60–70	70–80	80–90	90–100
No. of arrivals	179	108	79	37	32	21	10	13	8	4

Time in Seconds	100–110	110–120	120–130	130–140	140–150	150–160	160–170
No. of arrivals	5	1	3	2	1	0	2

Service times

Time in minutes	0–0.5	0.5–1	1–1.5	1.5–2	2–2.5	2.5–3	3–3.5	3.5–4
No. of customers	63	32	21	10	7	6	0	2

Time in minutes	4–4.5	4.5–5	5–7	7–7.5	7.5–8
No. of customers	0	1	0	1	1

*7.26   (Gaver and Thompson, 1973, p. 594)   Sometimes service in a queue takes a variety of forms, performed sequentially. For example, if one has two types of service: payment for goods (taking time T_1), followed by packing of goods (taking time T_2), then the service time $S = T_1 + T_2$. It is an interesting exercise to estimate $\mathscr{E}[S]$ by simulation, using antithetic variates. In an obvious notation, this would result in:

$$\hat{S} = \frac{1}{2n} \sum_{i=1}^{n} (T_{1i} + T_{2i} + T'_{1i} + T'_{2i}),$$

in which T'_{ji} is an antithetic variate to $T_{ji}, j = 1, 2, 1 \le i \le n$. If T_1 and T_2 are independent, exponential variables, with density e^{-x}, then show that the usual approach of taking, for example, $T'_{1i} = \log_e (1 - U)$, where $T_{1i} = \log_e U$, results in:

$$\text{Var}(\hat{S}) = \frac{1}{n}\left\{1 + \int_0^1 \log_e x \log_e(1-x)\,dx\right\} = \frac{1}{n}\left(2 - \frac{\pi}{6}\right)^2 \approx 0.36/n.$$

Compare this value with that which results from a usual averaging procedure. Of course in this simple example the distribution of S is known to be $\Gamma(2, 1)$ (see Section 2.10). However, we have here the simplest example of a *network*, and for a discussion of more complicated networks see Gaver and Thompson (1973, p. 595), Rubinstein (1981, pp. 151–153) and Kelly (1979).

7.27 Investigate further the findings of Table 7.3 by means of a more extensive simulation. Validate your conclusions by also using the generator of Equation (3.1).

7.28 Ashcroft (1950) provides an explicit solution to the machine-interference problem with constant service-time, while Cox and Smith (1961, pp. 91–109) and Feller (1957, pp. 416–420) provide the theory and extensions for the case of service-times with an exponential distribution. Discuss how you would simulate such a model. Bunday and Mack (1973) consider the complication of a mechanic who patrols the machines in a particular order.

7.29 (Page, 1965)   In the simulation of Fig. 7.5, let D_n and D'_n be defined by:

$$D_n = S_n - I_n$$
$$D'_n = S'_n - I'_n$$

Show that Corr $(D, D') = -0.645$ (cf. Exercise 7.26).

(e) Gambler's ruin

7.30 From a consideration of the first step taken by the particle in the gambler's ruin problem of Section 7.3.4, verify the relationships of Equation (7.11).

*7.31 Show that the solution of Equation (7.11) is given by:

(i) the case $p \neq q$:

$$d_k = \frac{k}{(q-p)} - \frac{a}{(q-p)}\frac{(1-(q/p)^k)}{(1-(q/p)^a)}   0 \leq k \leq a.$$

(ii) the case $p = q = \frac{1}{2}$:

$$d_k = k(a-k)   0 \leq k \leq a.$$

*7.32   Write a BASIC program to simulate the gambler's ruin problem of Section 7.3.4, employing the variance-reduction technique of that section, and compare estimated values of $\{d_k\}$ with the theoretical values given in the last exercise.

8

MODEL CONSTRUCTION
AND ANALYSIS

As discussed in Chapter 1, there are many uses of simulation in statistics, and a variety of these will be considered in Chapter 9. In this chapter we shall concentrate on the problems that arise when we use simulation to investigate directly the behaviour of models.

8.1 Basic elements of model simulation

8.1.1 Flow diagrams and book-keeping

In Table 7.2 we used Equation (7.9) to calculate the waiting-time of customers in an $M/M/1$ queue. An alternative approach is simply to chart the occurrence of the different events in time, and from this representation deduce the desired waiting-times. This is done for the main block simulation of Table 7.2 in Fig. 8.1.

In order to construct the time sequence of Fig. 8.1 from the service and inter-arrival times we have used the approach that is now standard in many simulations of this kind. This involves using a 'master-clock' which records the current time, and which changes not by constant units, as in Exercise 1.4, for example, but by the time until the next event. The time until the next event is obtained by reference to an 'event-list' which provides the possible events which may occur next, and which also designates when they will occur. Thus in the illustration of Fig. 8.1, when the master-clock time is 0.40 the second customer has just arrived, and the event list has two events: the arrival of customer 3 (after a further time of 1.89) and the departure of customer 1 (after a further time of 0.95). This queue operates on a 'first-in-first-out' policy (FIFO), so that at this stage the event list need not contain the departure of customer 2 as customer 1 is still being served. Customer 1 departs before customer 3 arrives and the master-clock time becomes 1.35, and so on.

Customer	Service time	Time to arrival of next customer
1	1.35	0.40
2	0.20	1.89
3	0.75	0.36
4	0.17	0.97
5	0.43	0.39

The sequence of events in time is then as shown below: Ai (Di) denotes the arrival (departure) of customer i, $i \geq 1$:

Type of event	Time from start of simulation	Queue size	Total waiting times
A1	0	1	
A2	0.40	2	
D1	1.35	1	1.35
D2	1.55	0	1.15
A3	2.29	1	
A4	2.65	2	
D3	3.04	1	0.75
D4	3.21	0	0.56
A5	3.62	1	
A6	4.01	2	
D5	4.05	1	0.43
⋮	⋮	⋮	⋮

Figure 8.1   Details of the main block simulation of an M/M/1 queue, from Table 7.2, the first customer arriving at time 0 to find an empty queue and an idle server.

This procedure is usually described by means of a 'flow-diagram', which elaborates on the basic cycle:

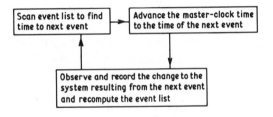

Naturally there have to be starting configurations and stopping conditions, as well as 'book-keeping' procedures for calculating statistics of interest, such as the waiting times in Fig. 8.1. In the simulation of continuous-time Markov processes, an alternative approach may be used which is described in Section 8.3, and which was employed in the program of Exercise 2.27. Another alternative approach is to advance the clock by fixed units of time, each time checking for changes to the system, but this approach is not regarded as generally useful (Law and Kelton, 1982, p. 51).

8.1.2 Validation

In Table 7.3 we were interested in the effect of using antithetic variates on the sample variance of an estimator of mean waiting-time in an M/M/1 queue. In addition we computed average waiting-times and compared them with the known theoretical values for an M/M/1 queue in equilibrium. Had the comparisons been poor, say out by a factor of 10, then we would have seriously questioned the validity of our model and/or programming, and sought a possible error. This is a simple example of *validation*, which is akin to the work of Chapter 6 on testing random numbers. Thus in validation one is checking that the model is working as intended. Further discussion is given in Naylor and Finger (1967) and Law and Kelton (1982, chapter 10). When a model is simulated via a computer program, then validation is very much like the debugging of any computer program. However, detailed checking of the simulation output of a model against observed data from the real-life system that is being modelled, can provide challenging time-series problems. One approach, adopted by Hsu and Hunter (1977) is to fit standard Box–Jenkins models to each output series, and then test for differences between the parameters of the two fitted models. A more complicated approach examines the similarities between the serial correlations in both series by comparing their spectra (see Naylor, 1971, p. 247).

8.2 Simulation languages

Not only do model simulations have features such as master-clocks and event-lists in common, but any model simulation is likely to require random variables from various distributions, as well as book-keeping routines for calculating and printing summary statistics. Simulation languages have been developed which automatically provide these required facilities for computer simulation. This allows much faster programming than if such facilities have to be programmed laboriously in languages such as BASIC or FORTRAN.

One of the most popular simulation languages is GPSS (which stands for General Purpose Simulation System), described, for example, by O'Donovan (1979). In order to program a simple FIFO queue in this language it suffices to describe the experience of a single customer passing through the queue. Distributions for inter-arrival and service-times can be selected by means of a simple index and the entire simulation can then be effected by a dozen or so instructions. Additionally, such languages can perform certain checks for errors. A defect of GPSS is that it can only simulate discrete random variables, causing one to have to reverse the common procedure of approximating discrete random variables by continuous random variables, if the latter are required for a simulation.

Other simulation languages allow more detail to be provided by the user, as in the FORTRAN-based GASP IV, and the ALGOL-basd SIMULA. Law and Kelton (1982) present a FORTRAN-based language, SIMLIB, that is similar to but simpler than GASP IV, and it is instructional to consider their listings of the FORTRAN routines which the language uses. As they point out, there are interesting problems to be solved when one considers the relative efficiencies of different ways of sorting through event-lists to determine the next event to occur.

The advantage of greater detail is that it is easier to incorporate features such as variance reduction, and also to provide additional analyses of the results. On the other hand, the disadvantage of greater detail is that the programming takes longer, and no doubt GPSS is popular on account of its extreme simplicity. There is also, associated with GPSS instructions, a symbolic representation, which is an aid to the use and description of the language. Tocher (1965) provided a review of simulation languages, while a more up-to-date comparison is to be found in Law and Kelton (1982, chapter 3), who also give programs for simulating the M/M/1 queue in FORTRAN, GASP IV, GPSS, and a further commonly used simulation language, SIMSCRIPT II.5. A further useful reference is Fishman (1978).

8.3 Markov models

8.3.1 Simulating Markov processes in continuous time

The Poisson process of Section 2.7 is one of the simplest Markov, or 'memoryless', processes. To take a process in time as an illustration, the probability rules determining the future development of a Markov process depend only upon the current state of the system, and are independent of past history. Generally speaking, Markov models tend to be easier to analyse than non-Markov models, and for this reason are popular models for real-life processes. In Section 9.1.2 we consider the simulation of a Markov chain, for which the time variable is discrete; here we shall consider Markov processes with a continuous time variable.

Kendall (1950) pointed out that we may simulate one such process, the linear birth-and-death process, in a different way from that of Section 8.1.1, though that approach may also of course be used. In the linear birth-and-death process the variable of interest is a random variable, $N(t)$, which denotes the number of live individuals in a population at time $t \geq 0$, and with $N(0) = n_0$, for some known constant n_0. Independently of all others, each individual divides into two identical 'daughter' individuals with birth-rate $\lambda > 0$, and similarly each individual has death-rate $\mu > 0$. For a full description, see Bailey (1964, p. 91). Thus at time t the event list of Section 8.1.1 contains $N(t)$ events (either births or deaths) as well as their times of occurrence. A simpler way of proceeding is to simulate the time to the next event as a random variable T with the exponential probability density function given by:

$$f_T(x) = (\lambda + \mu) N(t) \exp\left[-(\lambda + \mu) N(t) x\right] \qquad \text{for } x \geq 0$$

and then once the time to the next event has been chosen, select whether a birth or a death has occurred with respective probabilities $\lambda/(\lambda + \mu)$ and $\mu/(\lambda + \mu)$. The justification for this procedure comes in part from realizing that when there are $N(t)$ individuals alive at time t, then for each there is a Poisson process of parameter λ until a birth, and a Poisson process of parameter μ until a death, resulting in a Poisson process of rate $(\lambda + \mu)$ until one of these events, so that overall there is a Poisson process of rate $(\lambda + \mu) N(t)$ until the next event for the entire population. It is this Poisson process which results in the exponential density above. Conditional upon an event occurring, the probabilities of it being a birth or a death, given above, are intuitively correct (for full detail see Kendall, 1950). In this simulation the time to the next event is a single random variable from a single distribution, and it is not read from an event list. In this case there is a different kind of list which contains just two possibilities, i.e. a birth or a death, and now associated with these are relative probabilities of occurrence. If we now return to the BASIC program of Exercise 2.27 we see that this approach is used there for simulating an M/M/1 queue. This approach may also be used for quite general Markov processes in continuous time (see Grimmett and Stirzaker, 1982, pp. 151–152).

One feature of the M/M/1 queue is the different qualitative behaviour between a queue for which $\lambda \geq \mu$, and a queue for which $\lambda < \mu$, where λ and μ in that case refer to the arrival and service rates for the queue, respectively (see Exercise 2.27). A similar result holds for the linear birth-and-death process, simulations of which are given in Fig. 8.2.

In the case of the birth-and-death process, where λ and μ now refer respectively to the birth and death rates, if $\lambda \leq \mu$ then with probability 1 the population ultimately dies out, and then, unlike the case of queues, which start up again with the arrival of the next customer, in the absence of immigration that is the end of that population. However, if $\lambda > \mu$ then the probability of extinction of the population is $(\mu/\lambda)^{n_0}$. Simulations such as that of Fig. 8.2 are useful in demonstrating these two qualitatively different types of behaviour. We also find such differences in more complex Markov models such as models

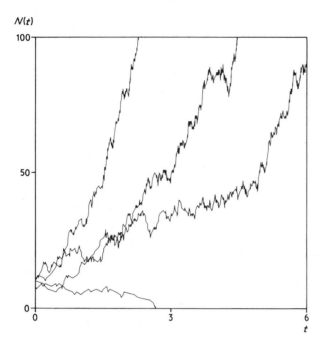

Figure 8.2   Four simulations of the linear birth-and-death process, for $\lambda = 1.5$, $\mu = 1$ and $n_0 = 10$.

for the spread of epidemics in a population (see e.g. Kendall, 1965), and also when we investigate far more complicated models for the development of populations.

The linear birth-and-death process could be used as a model for the development of only the simplest of populations, such as bacteria, though even here the model may be too simple (Kendall, 1949). McArthur *et al.* (1976) simulated a much more complicated model, for the development of a monogamous human population, with death and fertility rates varying with age, with rules for deciding on who was to marry whom, with (or without) incest taboos, and so forth. One set of their results is illustrated in Fig. 8.3. Population F is heading for extinction (in fact all of its 6 members after 135 years were male) while populations A, B and C appear to be developing successfully, and the future for populations D and E is unclear at this stage.

Quite often, in examples of this kind, simulation is used to investigate the qualitative behaviour of a model. As a further example, consider the elaboration of the linear birth-and-death process which occurs when some of the individuals carry an inanimate marker, which may not reproduce. Thus these markers are only affected by the death rate μ, if we suppose that when a cell divides the marker moves without loss into one of the daughter cells.

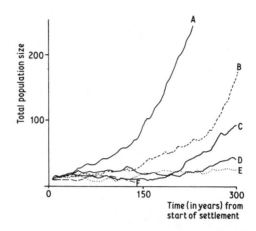

Figure 8.3   Taken from McArthur *et al.* (1976), this figure charts the development of six simulated human populations, under the same conditions of mortality, mating and fertility. Each population started from three married couples, with wives aged 21, 22 and 23 years and respective husbands aged 2 years older.

Meynell (1959) described experiments on cells, in which the marker was either a particular gene, or a bacteriophage. The aim, in his experiments, was to try and use the development of the population of markers to assist in the separate estimation of the cell birth and death rates, λ and μ, respectively. A mathematical analysis of the corresponding birth-and-death process model is given in Morgan (1974). If $N_1(t)$ and $N_0(t)$ denote the numbers of marked and unmarked cells, respectively, at time t, then

$$\mathscr{E}[N_1(t)] = N_1(0)e^{-\mu t}$$

$$\mathscr{E}[N_1(t) + N_0(t)] = (N_1(0) + N_0(0))e^{(\lambda - \mu)t}$$

and
$$\mathscr{E}[N_1(t)]/\mathscr{E}[N_1(t) + N_0(t)] \propto e^{-\lambda t}$$

an observation which prompted Meynell's experiments. It is therefore of interest to see how the ratio $N_1(t)/(N_1(t) + N_0(t))$ develops with time, and this is readily done by simulations such as those of Fig. 8.4.

8.3.2   Complications

Markov processes are named after the Russian probabilist, A. E. Markov, and the 'M' of an M/M/1 queue stands for 'Markov'. As mentioned above, Markov processes may be far too simple even for models of lifetimes of bacteria, as they are also for the lifetimes of higher organisms such as nematodes (see Schuh and Tweedie, 1979; Read and Ashford, 1968). A further complication, that rates may vary with time, is the subject of the paper by Lewis and Shedler (1976),

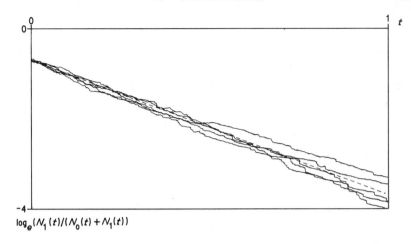

Figure 8.4   Simulations of a marked birth-and-death process. $N_1(t)$ denotes the number of marked cells at time t and $N_0(t)$ denotes the number of unmarked cells at time t. $N_1(0) = N_0(0) = 50$, $\lambda = 3$ and $\mu = 1$. It was of interest to compare the simulations with the dashed line, which is a graph of $\log_e (\mathscr{E}[N_1(t)]/\mathscr{E}[N_0(t) + N_1(t)])$ vs. t. Here, and in related simulations, the time t and the parameters λ and μ are measured in the same units.

who discuss the simulation of a non-homogeneous Poisson process; the IMSL library routine GGNPP of Section A1.1 allows one to simulate such a process. McArthur *et al.* (1976) emphasize that an unrealistic feature of their simulation study is that it assumes that the probability rules do not change with time. When lifetimes can be seen as a succession of stages in sequence, then each stage may be described by a Markov process, though the entire lifetime will no longer be Markov. For the case of equal rates of progression through each stage, the resulting lifetime will have a gamma distribution (see Section 2.10). Because service-times in queues can be thought to consist of a number of different stages, the gamma distribution is often used in modelling queues. This is the case in the example which now follows, and which illustrates a further complication in the simulation of queues.

EXAMPLE 8.1   *Simulating a doctor's surgery*
During the 1977 Christmas vacation, Miss P. Carlisle, a third-year mathematics undergraduate at the University of Kent, collected data on consultation and waiting times from the surgery of a medical practice in a small provincial town in Northern Brittany. From eight separate sessions, overall mean consultation and waiting times were, respectively, 21.3 minutes and 43.3 minutes. The surgery did not operate an appointment system, and it was decided to investigate the effect of such a system on waiting times by means of a simulation model. It was assumed that consultation times for different patients

were independent, and that inter-arrival times were independent, according to an appointment system with superimposed early/late values simulated from an empirical distribution in Bevan and Draper (1967). Patients were seen in the order in which they arrived and patient non-attendance was taken as 10% (see Gilchrist, 1976; Blanco White and Pike, 1964). For some simulations, consultation times were simulated from an empirical distribution (using a histogram to give the c.d.f.), while in others they were simulated from a fitted $\Gamma(9.13, 0.43)$ distribution, using the algorithm of Cheng (see Exercise 5.22). Simulation allows one to experiment with different length appointment systems, and some results are given in Fig. 8.5.

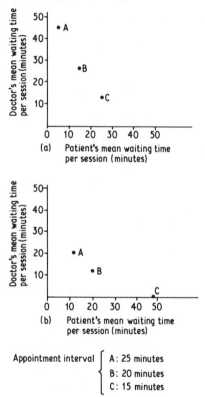

Figure 8.5 (from Carlisle, 1978) Mean waiting times from simulating a doctor's surgery with an appointment system: (a) simulating from the empirical distribution of consultation times, and (b) simulating from a gamma distribution fitted to the consultation times (see Exercise 8.17).

These results are qualitatively as one would expect, but simulation allows one to gauge effects in a quantitative manner. As described by Bevan and Draper (1967), appointment systems can be difficult to establish, and studies of this kind are needed so that the gains can be fully appreciated. This study was only tentative, provided no estimates of error (the subject of the next section) and did not employ any variance-reduction technique. Comparisons between the use of different appointment intervals could be improved by the use of common random numbers (see Section 7.3.1). More detailed studies of this kind are described by Bailey (1952) and Jackson *et al.* (1964), relating to British surgeries, and related studies can be found in Duncan and Curnow (1978), *ABC*, p. 159, and Gilchrist (1976) (cf. also Exercises 1.6 and 1.7).

As can be seen from Nelsen and Williams (1968), the analytic approach to queues with delayed appointment systems is mathematically complex, and so simulation here can play an important rôle. Furthermore, analytic solution of queues is often directed at equilibrium modes, while in the case of the above surgery, over the eight sessions of observation, the number of individuals attending varied from 6 to 13, reflecting a quite different situation from that investigated for the M/M/1 queue in Section 7.3.3, in which average waiting-times were computed for groups of size 100 customers. In practice we shall often want to investigate distinctly non-equilibrium conditions for models, and even for the simple M/M/1 queue mathematical solution is then difficult. However, simple models have a particular utility in preparing us for different possible modes of behaviour, as can be appreciated from a comparison of Figs 8.2 and 8.3.

*8.4   Measurement of error

A frequent end-product of simulation modelling is the point estimate of a population parameter, such as a mean waiting-time, but point estimates are of little use unless they are accompanied by measures of their accuracy. This is a subject that we touched upon in Chapter 7, and here we consider in greater detail the approach to be adopted when, as is frequently the case, the point estimate is a function of a sum of *dependent* random variables.

Let us suppose that we observe a process in equilibrium, such as the M/M/1 queue for $\lambda < \mu$, that we take observations $X_1, \ldots, X_n$ on some random variable, and we want to estimate $E[X_i] = \theta$, $1 \le i \le n$. The usual estimator would be:

$$\bar{X} = \frac{\sum\limits_{i=1}^{n} X_i}{n}$$

Let us now suppose that $\mathrm{Var}(X_i) = \sigma^2$, $1 \le i \le n$, and $\mathrm{Corr}\,(X_i, X_{i+s}) = \rho_s$, dependent only upon the separation of X_i and X_{i+s}, and not upon i. Then it can be shown (see Exercise 8.8) that

$$\mathrm{Var}\,(\bar{X}) = \frac{\sigma^2}{n}\left\{1 + 2\sum_{s=1}^{n-1}\left(1 - \frac{s}{n}\right)\rho_s\right\} \tag{8.1}$$

Ignoring the effect of the $\{\rho_s\}$ can result in too small an estimate of $\text{Var}(\bar{X})$; for instance, Conway (1963) reported a factor of 8, found from a particular empirical study. Law and Kelton (1982, p. 146) point out that the three main simulation languages, GPSS, GASP IV and SIMSCRIPT II.5, make the serious mistake of estimating $\text{Var}(\bar{X})$ by σ^2/n alone. In practice one does not know σ^2 or $\{\rho_s\}$ and so they have to be estimated from the data. The simplistic approach of a method of moments estimation of each ρ_s (see Exercise 8.15) is not in general to be recommended, as seen from Law and Kelton (1982, p. 147).

Moran (1975) has compared a variety of approaches. One way to proceed, suggested by Hannan (1957) is to estimate $\text{Var}(\bar{X})$ by:

$$\hat{V}_1 = \frac{n}{(n-k)(n-k+1)} \left\{ \sum_{j=-(k-1)}^{k-1} \left(1 - \frac{|j|}{n}\right)(c_j - \bar{X}^2) \right\}$$

where
$$c_j = (n-j)^{-1} \sum_{s=1}^{n-j} X_s X_{s+j} \quad \text{for } 0 \le j \le k-1 \tag{8.2}$$

and k is chosen large enough so that

$$\rho_j \approx 0 \qquad \text{for } j \ge k$$

As shown by Moran (1975), the bias in $\hat{V}_1$ involves only terms incorporating ρ_j for $j \ge k$. Jowett (1955) has considered the variance of $\hat{V}_1$. We see from (8.2) that computation of $\hat{V}_1$ could involve the calculation of a large number of cross-product terms.

An alternative approach, suggested by Blackman and Tukey (1958, p. 136), and which we have in fact already used in Section 7.3.3, is to divide the sequence of $X_1, \ldots, X_n$ into blocks, and to compute the sample mean of each block. The idea here is that for reasonable size blocks, the block means, which will each estimate the parameter of interest, θ, will be effectively independent and then we can estimate $\text{Var}(\bar{X})$ by:

$$\hat{V}_2 = \left(\sum_{i=1}^{b} \bar{X}_i^2 - b\bar{X}^2 \right) / (b(b-1)) \tag{8.3}$$

where b denotes the number of blocks, and $\bar{X}_i$ is the block mean of the ith block, $1 \le i \le b$. Moran (1975) considers the case $\rho_s = \rho^s$ and demonstrates that in this case it is possible that appreciable bias can result from the use of (8.3). We shall examine this case in the following example.

EXAMPLE 8.2   (*Moran*, 1975)
An investigation of the bias resulting from using block means as a device for estimating the variance of an estimator which is a function of dependent random variables.

Suppose $n = bk$, that $\sigma^2 = 1$,

and $$\rho_s = \rho^s \quad \text{for } -1 < \rho < +1$$

for $1 \le s \le n$, while for $\hat{V}_2$ we assume

additionally that $\rho_s = 0$ for $s > k$.
Here,

$$\text{Var}(\bar{X}) = n^{-1}\left(1 + \frac{2\rho}{1-\rho} - \frac{2\rho}{(1-\rho)^2}(1-\rho^n)\right)$$

and $$\mathscr{E}[\hat{V}_2] = \frac{1}{(b-1)}\left\{\frac{1}{k}\left(1 + \frac{2\rho}{(1-\rho)} - \frac{2\rho}{(1-\rho)^2}(1-\rho^k)\right)\right.$$

$$\left. - \frac{1}{n}\left(1 + \frac{2\rho}{(1-\rho)} - \frac{2\rho}{(1-\rho)^2}(1-\rho^n)\right)\right\}$$

Whence, the bias in $\hat{V}_2$ is

$$\mathscr{E}[\hat{V}_2] - \text{Var}(\bar{X}) = \frac{2\rho}{k(b-1)(1-\rho)^2}\left\{\frac{(1-\rho^n)}{n} - \frac{(1-\rho^k)}{k}\right\}$$

Examples are given in Table 8.1.

Table 8.1   Values of the bias in $\hat{V}_2$, given by Equation (8.3)

(a) for $n = 1000$, $k = 50$, $b = 20$

ρ	0.2	0.4	0.5	0.6	0.8	0.95	0.99
Bias	−0.000 012 5	−0.000 044 4	−0.000 08	−0.000 15	−0.000 80	−0.013 97	−0.143 8

(b) for $n = 100$, $k = 5$, $b = 20$

ρ	0.2	0.4	0.5	0.6	0.8	0.95	0.99
Bias	−0.001 25	−0.004 40	−0.007 74	−0.013 8	−0.052 4	−0.282 4	−0.721 6

The values for $\rho = 0.5, 0.95$ and 0.99 were given by Moran (1975); however, we now see that for smaller values of ρ, which might be expected in practice (see Daley, 1968), the bias is far less serious, especially if, as is likely, experiments of size approximating (a), rather than (b), are employed.

In practice a difficulty lies in the choice of b, and further empirical results are given by Law and Kelton (1982, p. 301).

8.5   Experimental design

Gross and Harris (1974, p. 425) describe the simulation of a toll booth, where the question of interest was whether or not to change a manned booth to an automatic booth, which could only be used by car-drivers with the correct change. From observations on the inter-arrival times of cars, the times for service at a manned booth, and also a reasonable guess at the service time at an automatic booth (including the possibility of drivers sometimes dropping their change) there remained, effectively, just the one parameter to vary, viz. the ratio of manned : automatic booths; even then the constraints of the problem meant that only a small number of possibilities could be tried.

We have seen from Example 8.1 that a similar situation arises when one investigates the optimum time-interval for patient arrival at a waiting-room (such as a doctor's surgery) with an appointment system.

However, more complicated systems certainly arise, in which one is interested in the effect of changing a large number of different factors. In ecology, for instance, quite complex models have been proposed for the behaviour of animals such as mice and sheep, and these models can contain as many as 30 parameters. In such cases one needs to simulate the model a large number of different times, each corresponding to a different combination of parameter values. A simple instance of this is given in the next example, from the field of medicine.

EXAMPLE 8.3

Schruben and Margolin (1978) present the results of a pilot study, carried out to investigate the provision of beds in a special hospital unit which was to be set up in a community hospital for patients entering with heart problems. On entry to the unit, patients were assessed and passed either to an intensive-care ward, or a coronary-care ward. Whichever ward a patient entered, after a period of time the patient was either transferred to an intermediate-care ward, prior to release from the hospital, or released from the hospital, possibly due to death. It was thought that the hospital could cope with numbers of beds in each ward of roughly the following magnitudes:

$$
\left.
\begin{array}{lr}
\text{intensive care} & \text{14 beds} \\
\text{coronary care} & \text{5 beds} \\
\text{intermediate care} & \text{16 beds}
\end{array}
\right\}
\tag{8.4}
$$

and presumably these numbers also reflected the hospital experience and skills in dealing with patients with heart problems. Such experience also enabled the system to be modelled as follows:

(a) A proportion 0.2 of patients leave the hospital via intensive care, without spending time in the intermediate care beds.
(b) A proportion 0.55 of patients pass through intensive and intermediate care wards.
(c) A proportion 0.2 of patients pass through coronary and intermediate care wards.
(d) The final proportion, 0.05 of patients leave via the coronary care wards, without spending time in the intermediate care beds.

Stays in all wards were taken to have a log-normal distribution (see Exercise 4.21) with the following parameters (time is measured in days):

	μ	σ
Intensive care	3.4	3.5
Coronary care	3.8	1.6
Intermediate care for intensive care patients	15.0	7.0
Intermediate care for coronary care patients	17.0	3.0

Arrivals at the unit were taken to be Poisson with parameter 3.3. We thus have here a more complicated example of a network of queues than in Exercise 7.26. The problems arise when queues begin to form, for instance of individuals who have completed their stay in coronary care and are due for a period in intermediate care, yet find that no intermediate-care beds are available. The effect of such congestion can be a lack of free beds in the intensive-care and coronary-care wards for new patients, and the response Y, say, of the model was the number of patients per month who were not admitted to the unit because of a lack of suitable beds. Without simulation of the system, no one knew the effect on the response of the distribution of beds between wards.

In the simulation it was therefore decided to consider small variations in the bed distribution of (8.4). The number of beds in each ward was taken as a separate factor, and each factor was taken at one of the two levels: the value of (8.4) ± 1, resulting in 8 different, separate simulations. Each simulation was run for a simulated time of 50 months, after a 10-month initialization period (cf. Exercise 8.16). The results obtained are given in Table 8.2.

Table 8.2

Simulation experiment	Number of intensive care beds	Number of coronary care beds	Number of intermediate care beds	Mean response $\bar{Y}$	Estimated* variance of $\bar{Y}$
1	13	4	15	54.0	1.75
2	13	4	17	47.9	1.59
3	13	6	15	50.2	1.81
4	13	6	17	44.4	1.65
5	15	4	15	55.2	1.35
6	15	4	17	48.9	1.76
7	15	6	15	50.5	1.26
8	15	6	17	44.1	1.62

* These values were obtained by means of a blocking approach, as described in the previous section.

Comparison of the results of experiments 1 and 8 reveals what one might expect, viz., an increase in numbers of beds in all wards reduces the mean response. However, more detailed comparisons are more revealing, and we see that the change in the number of intensive-care beds does not seem to have much effect on the response. The information in Table 8.2 can itself be modelled and suggestions such as this can then be formally tested within the model structure. When this was done by Schruben and Margolin, they concluded with the model:

$$\mathcal{E}[\bar{Y}] = 51.1 - 2.1\,(X_1 - 5) - 3.1\,(X_2 - 16) \tag{8.5}$$

where X_1 and X_2 respectively denote the number of beds in the coronary-care and intermediate-care wards. In fact the model of (8.5) arose from a slightly different set of experiments from those of Table 8.2, and we shall explain the difference shortly.

This example was only a pilot study, but although (8.5) only holds for the range of factor levels considered, it provides an indication for further investigation. In order to try and reduce variability, common random numbers (see Section 7.3.1) were used for all of the experiments in Table 8.2. A novel feature of the study by Scruben and Margolin (1978) was the repeat of half of the experiments of Table 8.2, using antithetic random number streams, rather than uniformly common random number streams (see Exercise 8.13 for the results using the antithetic simulations). Both of these variance-reduction techniques were shown to perform well, compared with the approach of using no variance reduction, and simply employing different streams of random numbers for each experiment. (See Schruben and Margolin, 1978, for further discussion and for a comparison between the two different variance-reduction techniques.) Duncan and Curnow (1978) discuss the general problem of bed allocations in hospitals.

Consultant statisticians are accustomed to advising scientists on how to design their experiments. In the context of model simulation statisticians also adopt the rôle of experimenters, and this can unfortunately lead to a neglect of advice that would be given to other experimenters. The experimental design of Example 8.3 was that of a 2^3 factorial design, and the IMSL routine AFACN (see Section A1.4) provides one example of a computerized algorithm for the general analysis of a 2^k factorial design. When k is appreciably larger than 3, the number of separate simulation experiments may become prohibitively large for a full factorial design, and in such cases alternative designs such as fractional replicates may be used. For example, the experiments of Exercise 8.13 form a 2^2 fractional replicate of the experiments in Example 8.3. Standard analysis of standard designs allows one to investigate how the response variable for the model is affected by the factor levels adopted in the design, and the conclusion of the analysis may be described by a simple equation such as that of (8.5), for the ranges of factor levels considered. The exploratory approach of varying factor levels with the intention of trying to *optimize* the response is termed 'response-surface methodology', and is described by Myers (1971); we shall return to this topic in Section 9.1.3.

8.6   Discussion

We have seen from this chapter that model simulations may range from elementary explorations of the qualitative predictions of a model, to elaborate explorations of a realistic model described by a large number of parameters, all of which may be varied. Whatever the simulation, the approach would make use of the material in Chapters 3–7, simulating random variables from a variety of distributions and if possible using variance-reduction techniques.

An interesting feature of the use of common random numbers for variance reduction, and of the use of antithetic variates, is that they preclude using rejection methods for the generation of derived random variables unless the uniform random variables used in one simulation run are stored before being used in a comparison run. However, Atkinson and Pearce (1976) show how Forsythe's method, a rejection method described in Exercise 5.41, may be used to produce approximately antithetic variates. Thus, for example, if such variance-reduction methods were likely to be used, and normal random variables were required, one might well prefer to use the Box–Müller method, rather than the Polar Marsaglia method, even though the latter may be faster, as the former does not involve rejection (see Section 4.2). For an explanation of how to use common random numbers in GPSS, see O'Donovan (1979, chapter 8).

The analysis of the results of simulation experiments is a subject in itself, and in Sections 8.4 and 8.5 we have only touched briefly upon certain aspects of this analysis. For further discussion see Hunter and Naylor (1970) and Kleijnen (1975, 1977). Schruben and Margolin (1978) used a simulation language for the simulation described in Example 8.3. Many large-scale model simulations are made with the aid of simulation languages, but as we have seen, these may not always provide the required analyses.

The topic of this chapter has been the simulation of models, and model simulation also occurs in Chapter 9. However, as we shall see, in that chapter the main interest is often in the evaluation of techniques, rather than in the models themselves.

8.7   Exercises and complements

8.1   Write a BASIC program to simulate an M/M/1 queue by the approach of Section 8.1.1, and compare the complexity of this program with that already given in Exercise 2.27. Modify your program to simulate $M/D/1$ and $M/E_2/1$ queues, in which service-times are, respectively, constant (D stands for 'deterministic') and $\Gamma(2, \mu)$ (E stands for 'Erlangian', a synonym for $\Gamma(n, \mu)$ distributions in which n is a positive integer – named after A. K. Erlang).

8.2 Modify the program of the last exercise to admit 'discouragement', i.e. let the arrival parameter be a decreasing function of n; specifically, let $\lambda = \lambda/(n+1)$. It can be shown theoretically that such a queue with $\lambda = 2$ and $\mu = 1$ has the same equilibrium mean size as the $M/M/1$ queue with parameters $\lambda = 2$ and $\mu = 3$. Verify this by simulation, and compare the variability in queue size for the two different queues. cf. Exercise 2.28.

†8.3 Repeat the simulation approach of Fig. 8.1, but using the antithetic block data from Table 7.2.

8.4 If the birth-rate λ in the linear birth-and-death process is set to zero, we obtain a linear death process. Show that for this process, if the death-rate $\mu = 1$ then the time to extinction of the population starting with n individuals, is given by the random variables Y and Z of Exercise 2.11.

8.5 Simulate a linear birth-and-death process, and consider how to incorporate the complexity of marked individuals as in the simulation of Fig. 8.4.

8.6 A further bivariate birth-and-death process is presented in Cox and Miller (1965, p. 189). In this model individuals are either male or female, and only females may reproduce. Any female thus has two birth-rates, λ_1 (for female offspring) and λ_2 (for male offspring). Death-rates for males and females are, respectively, μ_1 and μ_2. It can be shown analytically that if $\lambda_1 > \mu_1$, then as time $t \to \infty$, the ratio of the expected number of females to the expected number of males is $(\lambda_1 - \mu_1 + \mu_2)/\lambda_2$. Discuss this model, and how you would investigate this result using simulation.

*8.7 The Galton–Watson branching process is described in Cox and Miller (1965, p. 102). In this process a population of identical individuals is considered in successive generations, and any individual in any generation independently gives rise to a family of i descendants in the next generation with distribution $\{p_i, i \geq 0\}$. It can be shown analytically that if this distribution has probability generating function, $G(z) = \Sigma_{i=0}^{\infty} z^i p_i$, and mean $\mu = \Sigma_{i=0}^{\infty} i p_i$, then the probability of ultimate extinction of the population is given by the smaller root, η, of the equation $x = G(x)$. Furthermore, if $\mu \leq 1$ then $\eta = 1$, but if $\mu > 1$ then $\eta < 1$. From a study of white American males in 1920, Lotka (1931) suggested the following form for $G(x)$:

$$G(x) = (0.482 - 0.041\,x)/(1 - 0.559\,x)$$

In this example the individuals in the population are white males, and the Galton–Watson branching process may be used in a rudimentary fashion to determine the probability of survival of a family surname. Use simulation to investigate the probability of extinction for the population with this $G(x)$.

*8.8   Prove the variance formula of Equation (8.1).

8.9   Use the early/late data of the histogram of Fig. 8.6 to repeat the simulation of a doctor's waiting room, described in Example 8.1, based on the fitted gamma distribution. Use common random numbers to compare the effect of different appointment intervals and discuss your findings (cf. Exercise 1.6).

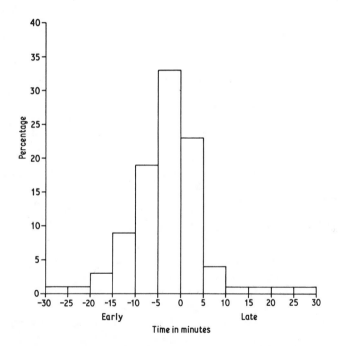

Figure 8.6   Histogram showing the distribution of patients' arrival times, relative to appointment times, based on data collected by Bevan and Draper (1967).

8.10   Devise a rejection method for simulating exponential random variables and then use this method to repeat the antithetic variate investigations of Table 7.3. Comment upon your results.

†**8.11** Produce a flow diagram for the toll-booth simulation mentioned in Section 8.5. Discuss how you might validate results arising from simulating the toll-booth process.

8.12 Give a flow diagram for the simulation reported in Example 8.3. In connection with this example, comment upon the proportion of individuals turned away.

*8.13 The antithetic simulations of Schruben and Margolin (1978) resulted in the following data:

Simulation experiment	Number of intensive care beds	Number of coronary care beds	Number of intermediate care beds	Mean response $\bar{Y}$	Estimated variance of $\bar{Y}$
9	13	4	17	51.3	1.73
10	13	6	15	53.9	1.49
11	15	4	15	58.5	1.83
12	15	6	17	47.8	1.56

Use the analysis of variance to verify the fitted model of Equation (8.5), resulting from these data, together with the data from experiments 1, 4, 6, 7 in Table 8.2.

8.14 (Moran, 1975)   Verify the form given for $\text{Var}(\bar{X})$ in Example 8.2.

*8.15 Use your program from Exercise 8.1 to simulate an M/M/1 queue and estimate the mean delay experienced by a customer when the queue is in equilibrium. Use Equation (8.1) and a method-of-moments estimation of ρ_s to estimate the variance of your estimator of mean delay. Daley (1968) provides the theoretical values of ρ_s which may be used to gauge the performance of Equation (8.1).

8.16 Exercises such as Exercise 8.15 presuppose that one is able to judge when a queue has reached equilibrium. Schruben and Margolin (1978), for example, ran their simulation, described in Example 8.3, for a simulated time of 50 months, but first ran it for an initialization period of 10 months. Consider how you would approach this problem. For further discussion, see Gross and Harris (1974, p. 419).

8.17 Discuss the differences between Figs 8.5 (a) and (b), and comment on the relative merits of simulating from an empirical distribution and a fitted theoretical distribution.

9

FURTHER EXAMPLES
AND APPLICATIONS

Statisticians are likely to employ simulation in a wide variety of situations. Some of these uses of simulation were mentioned briefly in Section 1.5, and in this chapter we shall consider these examples in greater detail, together with additional applications.

9.1 Simulation in teaching

We have already encountered examples of the use of simulation in teaching, in the comparison of the histograms of such as those of Figs 2.8–2.11, in the pebble-sampling experiment of Exercise 3.5, and in the Markov process simulations of Figs 8.2 and 8.4. Further illustrations are considered here.

9.1.1 Simulation in MINITAB

The MINITAB computer package is widely used for a range of elementary statistical analyses, described by Ryan, Joiner and Ryan (1976). This in itself gives rise to an invaluable teaching aid. However, a bonus provided by MINITAB is that it incorporates within the package a number of simulation facilities, described in chapter 3 of Ryan et al. (1976). Thus, for example, the student is encouraged to simulate convolutions such as that of Exercise 4.8, in addition to using imagination and mathematical analysis (Ryan et al., 1976, p. 80). In MINITAB it is possible to simulate directly from the $U(0, 1)$, $N(\mu, \sigma^2)$, Poisson and binomial distributions, as well as simulate uniform random integers and any finite-range discrete distribution. Instructions for simulating from exponential and Cauchy distributions are given in Fig. 9.1. Note that in MINITAB it is possible to operate directly on column vectors of data, such as C1, and 'LOGE OF C1, PUT IN C2' will take the natural logarithms of each element of the column C1, and enter them in the same order in column vector C2.

```
NOPRINT
URANDOM 50 OBSERVATIONS, PUT IN C1
LOGE OF C1, PUT IN C2
DIVIDE C2 BY -2, PUT IN C3
PRINT C3


NOPRINT
NRANDOM 50 OBSERVATIONS, WITH MU=0.0, SIGMA=1.0, PUT IN C1
NRANDOM 50 OBSERVATIONS, WITH MU=0.0, SIGMA=1.0, PUT IN C2
DIVIDE C1 BY C2, PUT IN C3
PRINT C3
```

Figure 9.1   MINITAB programs for (a) simulating 50 variables with an exponential distribution, parameter $\lambda = 2$, using the method of Equation (5.5); (b) simulating 50 variables with a standard Cauchy distribution, using the method of Exercise 2.15(b). The 'NOPRINT' commands ensure that the individual uniform and normal variates are not printed following their generation.

Such a simulation facility, combined with a simple histogram command, 'HISTO C3', for example, enables a ready investigation of central limit theorems (Section 2.9) and a consideration of how rate of approach to normality of the distribution of a sum of random variables as the number of terms in the sum increases, can depend on the form of the distribution of the component variables comprising the sum. Further examples of this use of MINITAB are given in Exercise 9.1.

MINITAB contains a number of programs for carrying out statistical tests. A graphical teaching aid results when such programs are combined with in-built simulation mechanisms, as can be seen from the MINITAB programs of Example 9.1. The statistical terminology of this example is given in *ABC*, chapters 16 and 17, for example.

EXAMPLE 9.1   *Investigation of* (a) *confidence-intervals and* (b) *type I error, using MINITAB.*

```
NOPRINT
 STORE
 NRANDOM 10 OBSERVATIONS, WITH MU=69, SIGMA=3, PUT IN C1
 ZINTERVAL 90 PERCENT CONFIDENCE, SIGMA=3, DATA IN C1
 END
EXECUTE 30 TIMES


NOPRINT
 STORE
 NRANDOM 10 OBSERVATIONS, WITH MU=0.0, SIGMA=1.0, PUT IN C1
 ZTEST MU=0, SIGMA=1, DATA IN C1
 END
EXECUTE 30 TIMES
```

The example of (a) is suggested by Ryan *et al.* (1976, p. 122), the normal distribution adopted being suggested as a possible distribution for male adult human heights (cf. Fig. 5.8). The 'EXECUTE' statement executes the commands between 'STORE' and 'END' the number of times specified after the EXECUTE statement. Thus the program (a) produces 30, 90 % confidence intervals for μ, assuming $\sigma = 3$, each interval being based on a separate random sample of size 10 from an $N(69, 9)$ distribution. The theory predicts that roughly 90 % of these intervals will contain the value $\mu = 69$, and students can now compare expectation with reality.

Program (b) proceeds in a similar way, producing 30 two-sided significance tests of the null hypothesis that $\mu = 0$, each test at the 5 % significance level. When, as here, the null hypothesis is true, then the null hypothesis will be rejected (type I error) roughly 5 % of the time, and again it is possible to verify this using the simulations. Other computer packages, such as S, mentioned in Section A1.4, and BMDP also contain simulation facilities.

9.1.2 Simulating a finite Markov chain

Feller (1957, p. 347) and Bailey (1964, pp. 53–56) consider mathematical analyses of a 'brother–sister' mating problem, arising in genetics. In stochastic process terms the model considered is a 6-state Markov chain (see Grimmett and Stirzaker, 1982, chapter 6), with states E_1 and E_5 absorbing, i.e., once entered they are never left, like the random-walk barriers of Section 7.3.4. The transition-matrix is given below and the BASIC simulation program is given in Fig. 9.2.

Probability transition matrix for the 'brother–sister' mating problem

		Following state					
		E_1	E_2	E_3	E_4	E_5	E_6
	E_1	1	0	0	0	0	0
	E_2	0.25	0.5	0.25	0	0	0
Preceding	E_3	0.0625	0.25	0.25	0.25	0.0625	0.125
state	E_4	0	0	0.25	0.5	0.25	0
	E_5	0	0	0	0	1	0
	E_6	0	0	1	0	0	0

If we denote this matrix as $\{p_{ij}, 1 \leq i, j \leq 6\}$ then

$$p_{ij} = \Pr(\text{next state} = j \,|\, \text{last state} = i)$$

Thus each row of the matrix constitutes a probability distribution over the integers 1–6, and as we can see from the following BASIC program, movement to the next state can be determined by, say, the table-look-up method applied to the distribution determined by the current state.

Results from a single run of the above program, which simulates the chain 10

```
10    RANDOMIZE
20    DIM A(6,6),B(6)
30    REM SIMULATION OF A SIMPLE 6-STATE
40    REM MARKOV CHAIN,STARTING IN STATE N
50    REM INPUT TRANSITION MATRIX
60    FOR I = 1 TO 6
70     FOR J = 1 TO 6
80      READ A(I,J)
90     NEXT J
100   NEXT I
110   LET K = 1
120   REM CALCULATE CUMULATIVE SUMS OF PROBABILITIES FOR
130   REM EACH ROW, FOR TABLE-LOOK-UP SIMULATION
140   FOR I = 1 TO 6
150    FOR J = 2 TO 6
160     LET A(I,J) = A(I,J)+A(I,J-1)
170    NEXT J
180   NEXT I
190   FOR K1 = 1 TO 10
200    PRINT
210    LET N = 3
220    PRINT "    1    2    3";
230    PRINT "    4    5    6"
240    FOR I = 1 TO 6
250     LET B(I) = 0
260    NEXT I
270    LET B(N) = 1
280    FOR I = 1 TO 6
290     IF B(I) = 0 THEN 320
300      PRINT "   •";
310      GOTO 340
320      PRINT "    ";
330    NEXT I
340    IF N = 1 THEN 430
350    IF N = 5 THEN 430
360    LET U = RND
370    FOR I = 1 TO 6
380     IF U < A(N,I) THEN 400
390    NEXT I
400    LET N = I
410    PRINT
420    GOTO 240
430   NEXT K1
440   DATA 1,0,0,0,0,0
450   DATA 0.25,0.5,0.25,0,0,0
460   DATA 0.0625,0.25,0.25,0.25,0.0625,0.125
470   DATA 0,0,0.25,0.5,0.25,0
480   DATA 0,0,0,0,1,0
490   DATA 0,0,1,0,0,0
500   END
```

Figure 9.2   BASIC simulation program for 6-state Markov chain.

times, each time starting in state 3, are shown below:

$3 \to 1$

$3 \to 1$

$3 \to 3 \to 2 \to 2 \to 3 \to 4 \to 5$

$3 \to 4 \to 4 \to 4 \to 5$

$3 \to 6 \to 3 \to 2 \to 2 \to 1$

$3 \to 4 \to 4 \to 3 \to 3 \to 4 \to 4 \to 5$

$3 \to 5$

$3 \to 5$

$3 \to 5$

$3 \to 2 \to 2 \to 3 \to 4 \to 5$

Thus we can estimate Pr(end in absorbing state 1|start in state 3) by 0.3 in this case. Further discussion of this example is given in Exercises 9.2 and 9.3; the latter also provides the theoretical solution to the probabilities of ultimately ending in state 1 or state 5.

9.1.3   Teaching games

Simulation can play an important rôle in teaching games. One illustration is provided by Farlie and Keen (1967), where the technique being taught is an optimization procedure. A variety of further examples are provided by Mead and Stern (1973), one being a capture–recapture study for the estimation of a population size already known to the simulator, and we shall return to this topic in the next section. In Section 8.5 we mentioned the exploratory side of experimental analysis and design, and this is a feature which is well suited to teaching by means of a game. In the game the teacher establishes a known model for response to factor levels, together with conventions for possible plot and block effects; the aim of the student is to try to optimize the response using experimental designs within special constraints. Mead and Freeman (1973) describe such a game for a fictitious crop yield, which was expressed as a known function of the levels of six nutrients, together with an additive plot effect. Also described are the performances of a number of students, some of whom used rotatable response surface designs. Clearly a similar approach could be adopted for a situation such as that of Example 8.3, replacing Equation (8.5) by a more complicated model outside the region of the parameter space considered in Example 8.3. White *et al.* (1982) make extensive use of simulation examples in the teaching of capture–recapture methodology, one of the topics of the next section.

9.2   Estimation and robustness in ecology

International agreements to limit the exploitation of certain species such as whales are made in the light of estimates of the sizes of mobile animal populations. Naturally, such estimates result from sampling of some kind, as in transect sampling, where an observer traverses an area of habitat and records the individuals seen (see Burnham *et al.*, 1980).

In the case of birds, observations can be taken during migration (see Darby, 1984), or by the censusing of particular areas, as occurs in the Christmas Bird Census of North America, and the Common Bird Census of the United Kingdom (see Mountford, 1982; Upton and Lampitt, 1981). The approach we shall consider here is one of repeated sampling of an area, in which captured individuals are marked in some way prior to release. For full discussion of this topic, see Ralph and Scott (1981), and Morgan and North (1984).

Suppose the population size to be estimated is *n* and one captures and marks *m* of these and then releases them. The distribution of marked animals in any subsequent sample is hypergeometric (see, e.g. *ABC*, p. 147) if one assumes that the marking process does not affect recaptures. Typically, *n* will be very much larger than *m*, and the hypergeometric distribution can be approximated by a more tractable binomial or Poisson distribution. Suppose that in a further sample of size $\tilde{n}$ the number of marked individuals is *M*, then a natural estimate of *n* is

$$\hat{n} = \frac{m\tilde{n}}{M}$$

Attributed to Petersen (1896), this is the maximum-likelihood estimate of *n* under the various parametric models that may be proposed (see Cormack, 1968). Using a normal approximation to the distribution of *M*,

$$M \approx N\left(\tilde{n}\frac{m}{n}, \tilde{n}\frac{m}{n}\left(1 - \frac{m}{n} \right) \right)$$

enables us to write down approximate confidence intervals for *n* if we approximate m/n by $M/\tilde{n}$. For example, if $n = 10\,000$, $m = 500$ and $\tilde{n} = 500$, respective 95 % and 90 % confidence intervals for *n*, given $M = 26$, are: (6 997, 15 366) and (7 317, 14 018). We explain below why we have taken $M = 26$ to obtain these intervals. It is disappointing that the width of the confidence intervals is of the same order of magnitude as *n* itself. This conclusion follows also from simulation of this capture–marking–recapture experiment, conducted by Clements (1978) and illustrated in Fig. 9.3. The value of $M = 26$ chosen for illustration above was the modal value of the appropriate set of simulations.

More precise estimates of *n* result if the capture–recapture cycle is repeated many times (see Exercise 9.4). The modes of analysis that may then be used are described by Cormack (1973), and Bishop and Sheppard (1973) compared the performance of different approaches by means of a simulation study: the simulated population was sampled either 10 or 20 times (at equal time-intervals), the probability of survival from sample to sample was taken to be 0.5 or 0.9, the population size *n* was taken as either 200, 1000 or 3000, and the proportion of the population sampled was taken as either 0.05, 0.09, or 0.12. In all there were therefore 36 separate simulations. Simulation is clearly a most useful procedure in such investigations as it enables us to try out an estimation technique in a situation where all the parameters are known.

Models such as these rely on assumptions such as 'equal catchability', i.e. marked individuals returned to the population are assumed to be no more or less likely to be caught when the next sample is taken. Evidence exists that this assumption may well be violated (Lack, 1965, regularly found a particular robin in his trapping nets) and simulation may then be used to investigate the *robustness* of a methodology with respect to departures from assumptions. A simple illustration is provided by the data of Table 9.1, taken from Bishop and Bradley (1972). Here the population size of interest was the number of taxi-

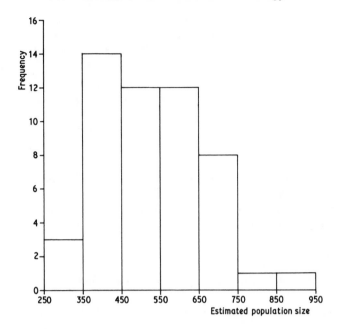

Figure 9.3   (From Clements, 1978) Results of a simulation study of the Petersen estimate $\hat{n} = m\tilde{n}/M$. Here the known value of $n = 500$; $m = 100$, $\tilde{n} = 50$, and 50 simulations were taken.

cabs in Liverpool (known at that time to be 300). Only 62 of these taxis operated regularly from Lime Street Station, and so repeated samples, taken between 14.30 and 15.30 hrs at that station were inappropriately analysed by a model which assumed equal catchability.

Table 9.1   Estimates of the known number (300) of taxi-cabs in Liverpool (source: Bishop and Bradley, 1972), using the capture–recapture model of Jolly (1965). Data were collected from two different sampling points: Lime Street, and Lime Street Station.

Day of sampling	Lime Street	Lime Street Station
Wednesday	208.4	97.1
Thursday	336.3	98.0
Friday	352.8	130.4
Saturday	286.9	178.8
Sunday	213.9	267.2
Monday	230.7	90.5
Tuesday	240.7	59.6
Mean	267.1	131.6

A computer simulation study is reported by Buckland (1982): 13 successive samples were taken on a cohort of 500 animals, each of which was given the probability 0.95 of surviving between samples. It was shown that significantly large biases arose in the estimation of the probability of survival if the simulated animals were given different catchabilities, not accounted for by the model used. An elementary theoretical approach to unequal catchability is to be found in Exercise 4.19, where we can consider the $\Gamma(n, \theta)$ distribution for the Poisson parameter λ as describing unequal catchability. If this model holds then numbers of individuals caught will be described by a negative-binomial distribution, rather than a Poisson distribution in the case of equal catchability. For far more detailed theoretical studies, see Burnham and Overton (1978) and Cormack (1966).

Here we have been considering robustness in the context of particular ecological models. A more standard study of robustness, using simulation, is the work of Andrews *et al.* (1972).

9.3 Simulation in multivariate analysis

9.3.1 Linear discriminant analysis

Discrimination and classification are natural human activities—for instance, we classify individuals as men or women, and we can also devise rules for discriminating between them. In statistics the traditional way of discriminating between two populations is to use the linear discriminant function: $d = \mathbf{x}'\boldsymbol{\alpha}$, in which $\boldsymbol{\alpha}$ is a previously derived vector of coefficients and $\mathbf{x}$ is a vector of measures taken on an individual. If $d > c$, for some additional derived constant c, then the individual is assigned to population 1, but if $d \leq c$, the individual is assigned to population 2. The construction of $\boldsymbol{\alpha}$ and c from $\mathbf{x}$ values on already classified individuals is described by Lachenbruch (1975, p. 9) and other texts on multivariate statistics. Suppose we have vector means $\bar{\mathbf{x}}_1$ and $\bar{\mathbf{x}}_2$ from two samples, one from each population, and let $\mathbf{S}$ denote the pooled sample variance–covariance matrix, estimated from both of the samples. The linear discriminant function is then (with an additional term)

$$d = \{\mathbf{x} - \tfrac{1}{2}(\bar{\mathbf{x}}_1 + \bar{\mathbf{x}}_2)\}' \mathbf{S}^{-1} (\bar{\mathbf{x}}_1 - \bar{\mathbf{x}}_2)$$

If it is equally likely that the individual with measures $\mathbf{x}$ comes from either of the two populations, and if the costs of the two possible incorrect decisions can be taken as equal, then the rule which minimizes the expected cost of misclassification is that the individual is classified into population 1 or 2 according as $d > 0$ or $d \leq 0$, respectively. (For an example of the data that may arise, see Exercise 9.5.) Of course, in classifying further individuals one may incorrectly assign individuals; this occurs for one of the women in Exercise 9.5, if one applies the discrimination rule back to the data that produced it. The likelihood of misclassifying individuals should depend upon the amount of separation between the two populations, and indeed, $\Phi(-\hat{\delta}/2)$ is used as an estimate of the probability of misclassification, where $\hat{\delta}$ is the sample Mahalanobis distance between the two samples, given by:

$$\hat{\delta}^2 = (\bar{\mathbf{x}}_1 - \bar{\mathbf{x}}_2)' \mathbf{S}^{-1} (\bar{\mathbf{x}}_1 - \bar{\mathbf{x}}_2)$$

Unfortunately, the sampling distribution of $\Phi(-\hat{\delta}/2)$ is unknown, and Dunn and Varady (1966) have investigated it using simulation.

Let $\tilde{p}$ denote the true probability of correct classification when the linear discriminant function is used, let the cost of misclassifying an individual be the same, from whichever population the individual comes, and suppose that individuals are equally likely to come from either population. $\tilde{p}$ is estimated by $\Phi(\hat{\delta}/2)$, and in a simulation $\tilde{p}$ is also calculated explicitly from the known parameters of the simulation. Dunn and Varady (1966) estimated percentiles of the sampling distribution of $\{\tilde{p}/\Phi(\hat{\delta}/2)\}$, using 1000 pairs of random samples, each pair being of n variables of dimension k. Table 9.2 gives their estimates of α such that $\Pr(\tilde{p}/\Phi(\hat{\delta}/2) \leq \alpha) = \theta$, for a range of θ values. The two populations were taken with population means separated by a distance δ, and each population was multivariate normal, with the identity matrix as the variance–covariance matrix. Standard normal random variables were obtained from mixed congruential generators followed by the Box–Müller transformation, an approach employed and tested by Kronmal (1964) (see Exercise 6.25).

Dunn and Varady (1966) also use their results to obtain confidence intervals for $\tilde{p}$, given $\Phi(\hat{\delta}/2)$. Clearly these results are a function of the particular simulation parameters employed, but they allow one to appreciate the influence of values of n and k in a known context. One might query the number of simulations used in this study, and the use of three decimal places in presenting the results of Table 9.2, without any discussion of accuracy of the estimates of the percentiles, and we shall discuss these matters in Section 9.4.2. A device for improving efficiency is suggested in Exercise 9.6.

We shall return to this use of simulation in Section 9.4. For further discussion, and alternative discriminatory techniques, see Lachenbruch and Goldstein (1979) and Titterington *et al.* (1981).

9.3.2 Principal component analysis

The measures taken on individuals are unlikely to be independent, and a useful and commonly used multivariate technique consists of finding the 'principal components', $\mathbf{y} = \mathbf{A}\mathbf{x}$, for which the elements of $\mathbf{y}$ are uncorrelated, and have progressively smaller variances (see, for example, Morrison, 1976, p. 267). A difficult but important question concerns the number of principal components that will suffice to provide a reasonable description of the original data set, and

Table 9.2 (Source: Dunn and Varady, 1966 reproduced with permission from The Biometric Society) An illustration of using simulation to determine the sampling distribution of $\{\tilde{p}/\Phi(\hat{\delta}/2)\}$ in linear discriminant analysis. The table gives estimates, from 1000 simulations in each case, of α, such that $\Pr(\tilde{p} \leq \alpha\Phi(\hat{\delta}/2)) = \theta$, for given values of θ, δ, k and n, explained in the text.

		δ					
		2			6		
n	k	$\theta = 0.05$	$\theta = 0.50$	$\theta = 0.95$	$\theta = 0.05$	$\theta = 0.50$	$\theta = 0.95$
25	2	0.911	0.984	1.080	0.998	0.999	1.006
	10	0.825	0.899	0.972	0.991	0.997	1.000
500	2	0.982	0.999	1.019	0.999	1.000	1.001
	10	0.975	0.995	1.013	0.999	1.000	1.000

Jeffers (1967) provides a 'rule-of-thumb', resulting from experience. Simulation provides a useful way of augmenting experience, and Jolliffe (1972a) derived such rules by analysing simulated data sets of known structure. He suggested retaining those principal components corresponding to eigenvalues greater than ≈ 0.7, when the components follow from finding the eigenvalues and eigenvectors of the sample correlation matrix of the original variables.

Principal components retained may be many fewer in number than the original variables, but each principal component is still a linear combination of the original variables, and a further question is whether one might dispense with some of the original variables without the loss of too much information. Jolliffe (1972a) considered this problem for a number of structured data sets, one of which involved the measurement of variables $\{X_i, 1 \leq i \leq 6\}$ on each individual, where $X_i = N_i$, for $i = 1$, 2, 3, $X_4 = N_1 + 0.5N_4$, $X_5 = N_2 + 0.7N_5$, $X_6 = N_2 + N_6$, and the N_i, $1 \leq i \leq 6$, were independent standard normal random variables, constructed by means of the generalized rejection method of Butcher (1960) (see Exercises 5.29 and 9.7). Further discussion of this example is given in Exercise 9.8. Applying rejection methods to real data sets, Jolliffe (1972b) found that the pictures produced by principal component analyses did not differ appreciably if a substantial fraction of the original variables were removed in an appropriate manner.

9.3.3 Non-metric multidimensional scaling

A useful feature of a principal component analysis can be an *ordination* of individuals, i.e. a scatter-plot in which proximity of individuals is related to their similarity. Ordination is the aim of techniques such as non-metric multidimensional scaling, described by Mardia *et al.* (1979, p. 413), Gordon (1981, Section 5.3) and Kruskal and Wish (1978), with additional illustrations found in Morgan (1981) and Morgan, Woodhead and Webster (1976). Non-metric multidimensional scaling tries to position points representing individuals in some r-dimensional space so that the more similar individuals are then the closer together are the corresponding points. 'Similarity' here can be gauged from the distance separating individuals in the original n ($> r$) dimensional space resulting if n variables are measured on each individual, but it is often calculated in other ways. The performance of non-metric multidimensional scaling in r dimensions can be judged by a measure called the 'stress', which is usually expressed as a percentage. Large stress values might suggest that a larger value of r is necessary in order to obtain a reasonable representation of the similarities between the individuals. Kruskal (1964) gave guidelines as to what are acceptably small stress values, but his guidelines were not calibrated with respect to the number of individuals concerned, viz. $0\% =$ perfect fit, $2.5\% =$ excellent fit, $5\% =$ good fit, $10\% =$ fair fit, 20% = poor fit. Intuitively, one would expect such a calibration of stress to change with respect to the number of individuals involved, and by the end of the 1960s simulation studies of the distribution of stress were undertaken by Stenson and Knoll (1969), Klahr (1969) and Wagenaar and Padmos (1971). The results of Table 9.3 are taken from Klahr (1969).

Table 9.3 (From Klahr, 1969) Stress values (given as percentages) when non-metric multidimensional scaling is applied to uniform random similarities between n individuals. For $n = 8$, 100 separate runs were used, while for $n = 12$ and $n = 16$, 50 separate runs were used. Each of the $\binom{n}{2}$ similarity values was obtained by selecting an independent $U(0, 1)$ random variable.

Number of individuals		Number of dimensions (r)		
(n)		2	3	4
8	Average	16.0	6.5	1.6
	standard deviation	3.4	2.7	1.8
12	Average	24..0	14.4	8.8
	standard deviation	1.7	1.6	1.6
16	Average	27.9	18.5	13.0
	standard deviation	1.4	1.0	1.1

The pattern of Table 9.3 thus provides a quantification of one's intuition in connection with this technique: 'Good' stress values were often obtained for 8 individuals in 3 dimensions, but for $n > 10$ no 'good' stress values were obtained when the similarities were randomly generated.

Further details are provided in the source papers. While Arabie (1973) points out a possible flaw in the scaling procedure used, the value of the basic approach is quite clear.

9.3.4 Cluster analysis

A logical preliminary to the discrimination discussed in Section 9.3.1 is classification (also called cluster analysis), described by Everitt (1979) as 'probably one of the oldest scientific pursuits undertaken by man'. Even so, it is certainly true that the area of classification can be something of a minefield. This is because there are many different methods of cluster analysis, as well as possibly different algorithms for any particular method. In addition, for any one method a variety of different 'stopping-rules' may be proposed, which suggest the appropriate number of clusters to describe the data (see Everitt, 1980, Section 3.4). Unfortunately these rules may suggest structure when none is present, and alternatively they may miss known structure. For illustration, see Day (1969), and Everitt (1980, p. 74). An obvious way of assessing and comparing different methods is by simulating populations of known structure, and then applying the methods to such populations. The following example is taken from Gardner (1978).

EXAMPLE 9.2
Figure 9.4 illustrates three random samples, each of size 70, one from each of three bivariate normal distributions, with respective means $(-2, 0)$, $(2, 0)$, $(0, 3.46)$ and unit variance–covariance matrix. The population means of these samples therefore lie at the corners of an equilateral triangle with side equal to 4 standard deviations. Also shown on Fig. 9.4 is the result of the three-cluster solution of a cluster analysis using the method of Beale (1969), an algorithm for which is provided by Sparks (1973, 1975). We can see that the cluster analysis has captured most of the known structure in the simulated data, with only 9 of the 210 points misclassified. Calinski and Harabasz (1974) propose a measure, C, for indicating an appropriate number of clusters for describing a set of data, and it is interesting to evaluate the performance of such a measure on simulated data. For Beale's method and the data of Fig. 9.4, we find:

Number of clusters sought (k)	C
6	239
5	254
4	279
3	341
2	133

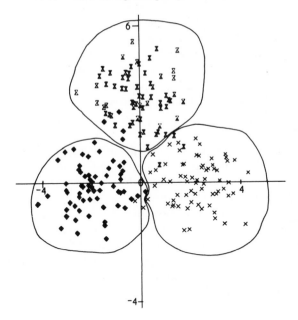

Figure 9.4 (From Gardner, 1978) Three bivariate normal samples together with a three-cluster solution using the method of Beale (1969). Objects within each contour form a cluster, and the different symbols indicate the different sample memberships.

A local maximum in the graph of C vs. k is an indication of a good description of the data for that value of k, and we see that for these data this measure is performing as anticipated.

Many more examples of the uses of simulation in classification are referenced by Gordon (1981), and further discussion of Example 9.2, as well as additional illustrations, are found in Exercise 9.9.

9.4 Estimating sampling distributions

The use of simulation to investigate sampling distributions arising in multivariate analysis was encountered in the last section, and in Section 1.5 we encountered the early example of this use by W. S. Gossett. Two further examples are given here.

9.4.1 The chi-square goodness-of-fit test

This test was described in Section 2.14, and employed in Chapter 6. As emphasized already, the chi-square test is only an approximate test when expected values are 'small' (say less than 5), and one way to examine how and when the approximation breaks down is by means of simulation. An illustration is provided by Hay (1967), who considered both the October mean pressure at the point 63°N, 20°W in the Atlantic, and a measure of subsequent winter temperature in Central England, for 45 specified years (see Table 9.4). In this case the chi-square goodness-of-fit statistic is computed as explained in *ABC*, p. 277, resulting in a value of $X^2 = 20.2$ with 8 degrees of freedom. This value is just significant at the 1 % level using the standard chi-square test, but in Table 9.4, under the assumption of independent row and column classifications, *all* the expected cell values are less than 5, and so one would question the validity of the standard chi-square test, while at the same time drawing some comfort from the extreme significance of the result obtained. Craddock and Flood (1970) used simulation to show that the standard chi-square test is in fact conservative for this application. For each of a variety of table sizes, and grand totals of table entries, they simulated 10 000 values of X^2 with independent row and column cassifications and uniform marginal distributions, and drew up histograms of the resulting sampling distributions. The relevant estimated percentiles for the above application are given in Table 9.5.

Table 9.4   Relationship between monthly mean pressure in October at a location 63°N, 20°W and a measure of subsequent winter temperature in Central England for 45 years. (From Hay, 1967). The figures in the body of the table are frequencies of years.

October mean pressure (terciles)	Winter temperature (quintiles)					Total
	1	2	3	4	5	
Upper	8	3	1	2	2	16
Middle	1	2	2	4	5	14
Lower	0	4	6	3	2	15
Total	9	9	9	9	9	45

Table 9.5   Value of z such that $p = \Pr(X \le z)$. (From Craddock and Flood, 1970)

Grand total of (5 × 3) table	100 p									
	1	2	5	10	50	90	95	98	99	99.9
40	1.84	2.25	2.98	3.80	7.7	13.3	15.2	17.5	19.2	24.9
50	1.80	2.19	2.91	3.70	7.6	13.3	15.3	17.7	19.4	25.2
∞	1.65	2.03	2.73	3.49	7.3	13.4	15.5	18.2	20.1	26.1

Estimates of error are also given by Craddock and Flood (1970), who suggest that, apart from the 99 and 99.9 percentile values, which are subject to larger errors, the estimated percentiles should differ from the true values by less than 0.2 (see Kendall and Stuart, 1969, p. 236). We shall return to error estimation in sampling distributions in Section 9.4.2.

It is interesting to note that Craddock and Flood (1970) were unfortunate in their choice of random number generator, discarding a discredited generator of Kuehn (1961) and replacing it by:

$$x_{n+1} = (kx_n + x_{n-1}) \mod (2^{48})$$

where $k = 5^{20}$, which, at a later stage, was found to fail the runs up-and-down test of Downham (1969), though this test has itself since been shown to be incorrect (see Chapter 6)! We shall continue discussion of this application in Exercise 9.13. Note that no variance-reduction method was used, as is true also in a more recent study by Cox and Plackett (1980), who use simulation to evaluate the distribution of X^2 for small samples in three-way as well as two-way tables. Hutchinson (1979) provides a bibliography of related research. By contrast, the first use of control variates in simulation was by Fieller and Hartley (1954), whose approach we shall describe in the next section.

*9.4.2 Precision and the use of control variables

As we have seen, if one has a random variable of unknown distribution then one approach is to simulate the variable many times and then draw a histogram to visualize the resulting sample. This procedure is illustrated by Fieller and Hartley (1954) for the range of a sample of size 5 from an $N(0, 1)$ distribution. In this case of course the distribution of the range is given by Equation (6.2), and Fieller and Hartley were able to produce the following table:

	Value of the range						
	0–1	1–2	2–3	3–4	4–5	5–∞	Total
Observed values	56.0	330.5	414.0	163.0	33.5	3.0	1000
Expected values	45.0	336.6	407.5	173.2	34.0	3.7	1000

If p_i is the probability that the range is in the ith category above, for $1 \leq i \leq 6$, then we can estimate p_i by

$$\hat{p}_i = \frac{n_i}{n} \quad \text{and} \quad \text{Var}(\hat{p}_i) \approx n_i(n - n_i)/n^3$$

when n_i of the n values are found to lie in the ith category, $1 \leq i \leq 6$. Thus we obtain the estimated distribution given in Table 9.6.

Table 9.6

	Value of the range					
	0–1	1–2	2–3	3–4	4–5	5–∞
Estimated probability of range being in this interval	0.056	0.331	0.414	0.163	0.034	0.003
$n \times$ variance of estimated probability	0.053	0.221	0.243	0.136	0.032	0.003

The error for any interval is $(N_i - np_i)$, and from Table 9.6 we see that while the variance of the error is greatest for $p_i \approx 0.5$, the variance of $(\hat{p}_i/p_i)$ is greatest for $p_i \approx 0$ or 1. In deciding upon the value of n, it is sensible therefore to consider the *relative* error,

$$X = \left(\frac{N_i - np_i}{np_i} \right)$$

for which

$$\mathscr{E}[X] = 0 \quad \text{and} \quad \text{Var}(X) = (1 - p_i)/(n_i p_i)$$

Suppose we want to choose n so that

$$\Pr\left\{ \left| \frac{N_i - np_i}{np_i} \right| \leq \delta \right\} = \theta \tag{9.1}$$

for some small $\delta > 0$ and $1 \geq \theta \geq 0$. N_i has a $B(n, p_i)$ distribution, and making a normal approximation to the binomial distribution and employing a continuity correction (see, e.g., *ABC*, p. 227) we can approximate to (9.1) by

$$2\Phi\left(\frac{\delta - \frac{1}{2}}{\sqrt{(q_i/np_i)}} \right) = \theta \quad \text{where} \quad q_i = (1 - p_i)$$

i.e.,

$$n = \left\{ \frac{\Phi^{-1}(\theta/2)}{\delta - \frac{1}{2}} \right\}^2 \left(\frac{q_i}{p_i} \right) \tag{9.2}$$

We see from (9.2) that if we want to choose n so that (9.1) is satisfied for all i, then the size of n is determined by $\min_i \{p_i\}$. Thus we may obtain initial estimates of the $\{p_i\}$ by means of a pilot simulation, and then select n according to (9.2) and the smallest value of $\hat{p}_i$, to correspond to predetermined values of θ and δ. There was no discussion of precision in the studies reported in Tables 9.2 and 9.3, for which the choice of the number of simulation runs taken was not explained.

As we have seen in Chapter 7, for a given precision the size of a simulation experiment can be reduced by using a variance-reduction technique, and we shall now consider the use of a control variate for estimating the sampling distribution of the range described above. The idea of using a control variate is the same as in Section 7.3.1, namely, we want to estimate the distribution of X using simulated random variables, and we introduce a correlated random variable Y of known distribution. We use the same random variables to estimate the distribution of Y and hope to increase the precision for X by making use of the correlation between X and Y and the known distribution of Y. The details which follow come from Fieller and Hartley (1954) and while they are qualitatively different from the analysis of Section 7.3.1, they utilize the same underlying idea. Y will be taken as the sample corrected sum-of-squares, $(n-1)s^2$, and X will be the sample range. Each simulated sample of size 5 from an $N(0, 1)$ distribution therefore gives rise to a value of X and a value of Y. In this particular example we would certainly expect a high positive correlation between X and Y. If n such random samples are taken in all, and if the range of X is divided into n_x categories, while the range of Y is divided into n_y categories, then

$$n = \sum_{i=1}^{n_x} \sum_{j=1}^{n_y} n_{ij}$$

where n_{ij} samples give rise to an X value in the ith X category and a Y value in the jth Y category. For $n = 1000$, Fieller and Hartley (1954) presented the data reproduced in Table 9.7.

Table 9.7   An illustration of the use of a control variate in the investigation of a sampling distribution, (From Fieller and Hartley, 1954).

Categories for the range, X	Categories for the corrected sum-of-squares, Y						Totals
	0–0.71	0.71–2.75	2.75–4.88	4.88–9.49	9.49–14.86	14.86–∞	
0–1	49.5	6.5					56
1–2	5.0	305.0	20.5				330.5
2–3		39.0	306.0	69.0			414
3–4			2.5	142.0	18.5		163
4–5				10.0	19.5	4.0	33.5
5–∞						3.0	3.0
Totals	54.5	350.5	329.0	221.0	38.0	7.0	1000

Let $p_{ij} = \Pr(X \in i\text{th } X\text{-category and } Y \in j\text{th } Y\text{-category})$.

Thus $\sum_i \sum_j p_{ij} = 1$.

For simplicity, let us also use the summation notation:

$$n_{i.} = \sum_j n_{ij}, \qquad p_{.j} = \sum_i p_{ij}, \text{ etc.}$$

We want to estimate $\{p_i\}$. We can introduce the effect of the control variate by writing:

$$p_{i.} = \sum_j p_{ij} = \sum_j \left(\frac{p_{ij}}{p_{.j}} p_{.j} \right)$$

and substituting the known values for $\{p_{.j}\}$ while estimating $(p_{ij}/p_{.j})$ from the data. Thus we can form:

$$\hat{p}_{i.} = \sum_j \left(\frac{n_{ij}}{n_{.j}} \right) p_{.j} \qquad \text{as long as } n_{.j} > 0.$$

If $n_{.j} = 0$ for some j it is suggested that $(n_{ij}/n_{.j})$ be replaced by $(n_{i.}/n)$, but we are unlikely to have $n_{.j} = 0$ in a well-designed sampling experiment (see below). It is shown by Fieller and Hartley (1954) that

$$\mathscr{E}[\hat{p}_{i.}] = p_{i.} - \sum_j (1 - p_{.j})^{n-1} (p_{ij} - p_{i.}p_{.j})$$

and

$$\mathrm{Var}(\hat{p}_{i.}) \approx p_{i.}(1 - p_{i.})/n - \sum_j (p_{ij} - p_{i.}p_{.j})^2/(np_{.j})$$

and they also present the exact formula for $\mathrm{Var}(\hat{p}_{i.})$. We see that if X and Y are independent, $p_{ij} = p_{i.}p_{.j}$, and so the use of the variate Y accomplishes nothing. However, for $p_{ij} \neq p_{i.}p_{.j}$, the variance of $(\hat{p}_{i.})$ is reduced from its value when Y is not employed. The price to pay for this variance reduction is the bias in $\hat{p}_{i.}$, but this will be small as long as the (known) $p_{.j}$ are not too small (Fieller and Hartley suggest choosing the intervals for Y such that $np_{.j} \geq 8$).

For the data of Table 9.7 we obtain the values:

							Total
$n\hat{p}_{i.}$	51.8	327.9	396.0	184.8	37.3	2.1	999.9

and the comparison of the variances is shown in Table 9.8.

Table 9.8   Evaluating the effect of the use of the control variate in Table 9.7. Here $\hat{p}_{i.} = (n_{i.}/n)$.

Categories for the range, X	Estimated variance		$\dfrac{\text{Var}(\tilde{p}_{i.})}{\text{Var}(\hat{p}_{i.})}$
	$\text{Var}(\tilde{p}_{i.})$	$\text{Var}(\hat{p}_{i.})$	
0–1	0.053	0.011	4.8
1–2	0.221	0.063	3.5
2–3	0.243	0.104	2.3
3–4	0.136	0.063	2.2
4–5	0.032	0.021	1.6
5–∞	0.003	0.002	1.5

Tocher (1975, p. 92) presents the results of a simulation study with $n = 819$, for which X is a sample median and Y is a sample mean. Further discussion, involving the choice of the number of categories to use, is given by Fieller and Hartley (1954).

*9.5   Additional topics

In this section we take a brief look at two relatively new developments in statistics, each of which involves simulation.

9.5.1   Monte Carlo testing

In Section 9.4.1 simulation was used to estimate the sampling distribution of the X^2 statistic for a particular type of two-way table, and this enabled us to judge the significance of one X^2 value for such a table. This example reveals two features: when estimating distributions using simulation in this way one is frequently interested in the tails of the distribution, and we also saw that the tails are estimated with least precision, with much of the simulation effort being of little relevance to the tails. An alternative approach was suggested by Barnard (1963), resulting in what is now called Monte Carlo testing. We shall illustrate what is involved here by reference to an example considered by Besag and Diggle (1977), which is relevant to questions posed in Chapter 6 regarding the randomness or otherwise of two-dimensional scatters of points.

One approach to scatters of points in two dimensions is to compute the distance from each point to its nearest neighbour and then form the sum, s_1 say, of these nearest-neighbour distances. For regular patterns of points, small nearest-neighbour distances will not occur, and so s_1 will tend to be large, while for patterns of points that appear to be clustered, as in Fig. 9.4, s_1 will tend to be small. For a given scatter of n points we can compute s_1 and then we need to be able to assess its significance. One approach is to simulate many uniformly random scatters of k points, for the area in question, possibly using the rejection approach of Hsuan (1979). For each such random scatter s_1 can be calculated and ultimately the sampling distribution can be estimated, using ideas such as those of Section 9.4.2. An alternative approach involves $(n-1)$ such simulations, for some n, resulting in the sums of nearest-neighbour distances $\{s_2, s_3, \ldots, s_n\}$. The Monte Carlo test approach is simply to form the ordered sample $s_{(1)}, s_{(2)}, \ldots, s_{(n)}$; the significance level of s_1 is then obtained from the rank of s_1 among the n order-statistics. Investigations of the power of such a test have been carried out by Hope (1968) and Marriott (1979). It is of interest to apply a Monte Carlo test to the problem of Craddock and Flood (1970) considered earlier, and that is done in the following example.

EXAMPLE 9.3
Suppose we want to perform a one-tail significance test of size α on some given null hypothesis. In general we choose an integer n and carry out $(n-1)$ simulations under the given null hypothesis. The value of the test statistic is calculated for each simulation and also for the observed data. In addition, we choose an integer m such that $m/n = \alpha$, and the null hypothesis is rejected if the value of the test statistic for the data is among the m largest of the n values of the statistic. Marriot recommends that we ensure $m \geq 5$. Two-way tables of the same dimension as Table 9.4 were simulated with column totals of 9 and with each row given probability 1/3 of occupation by each datum. For $\alpha = 1\%$ the values of $m = 5$, $n = 500$ were chosen, and for three separate runs the following values for the rank, r, of the value $X^2 = 20.2$ were obtained: $r = 12$, $r = 14, r = 12$, each resulting in a significant result at the 3 % level, in reasonable agreement with the finding of Craddock and Flood (1970) from Table 9.5, though the discrepancy merits further investigation.

9.5.2   The bootstrap

Associated with the parameter estimates of Section 9.2 were estimates of error, obtained from the curvature of the likelihood surface and making use of an asymptotic approximation to a sampling distribution. Suppose that in a general problem we have a random sample of data $(x_1, \ldots, x_n)$ which is used to estimate a parameter θ, resulting in the estimate $\hat{\theta}$. As discussed in Section 8.4, $\hat{\theta}$ by itself is of little use without some indication of its precision, which is a property of the sampling distribution of $\hat{\theta}$. The usual approach is to make an asymptotic approximation, but this could be misleading if confidence intervals are required, and repeated sampling to produce further sets of real data may be impractical. The idea of the bootstrap is to sample repeatedly, uniformly at random, and with replacement, from the one sample of real data $(x_1, \ldots, x_n)$.

Successive bootstrap samples will therefore contain the elements $x_1, \ldots, x_n$, but some may be repeated several times for any sample, and then others will be missing. For each bootstrap sample the estimate $\hat{\theta}$ of θ may be calculated, enabling a form of sampling distribution to be estimated, which may then be used to gauge the precision of $\hat{\theta}$, as well as a wealth of further detail. Due to Efron (1979a), this technique is well illustrated in Efron (1979b), with further discussion in Efron (1981). The bootstrap relies upon multiple simulations of a discrete uniform random variable, which could be done by means of the NAG routines GO5DYF or GO5EBF, or the IMSL routine GGUD, as well as within MINITAB (see Appendix 1). Except in cases when theoretical analysis is possible (see Reid, 1981, for example), the bootstrap procedure would have to utilize a computer, and is therefore a truly modern statistical technique, as implied by the title of Efron's (1979b) paper.

9.6   Discussion and further reading

The examples we have considered here reveal the power and versatility of simulation methods. At the same time we can see the limitations of some of these methods – as when variance–covariance matrices are chosen arbitrarily to be the identity matrix, for example.

An alternative approach to the simulation of spatial patterns of points is given by Ripley (1979). Forests provide real-life examples of spatial arrangements, in which the points are trees; when foresters want to estimate the density of forests they may use quadrat sampling, which for long and narrow 'belt' quadrats, or transects, is analogous to the transect method mentioned in Section 9.2, which was investigated using simulation by Gates (1969). Quadrat sampling can be extremely time-consuming and laborious, and recently there has been much interest in the use of methods which involve measuring distances from points to trees and from trees to trees; see Ripley (1981, p. 131) for an introduction. Here again the various methods proposed have been evaluated using simulation; see Diggle (1975) and Mawson (1968).

Simulation continues to be used in multivariate analysis, as can be seen from the work of Krzanowski (1982) who used simulation to help in the comparison of different sets of principal components. Just as simulated data are useful for demonstrating and investigating methods of cluster analysis, such data may be similarly used in discriminant analysis, for example, and are also used in the evaluation of 'imputation' procedures for estimating missing values from sample surveys. Principal components and discriminant analysis use linear combinations of the original variables, which may be thought to be somewhat restrictive (for this reason, Cormack, 1971, favoured non-metric multidimensional scaling). In discriminant analysis an alternative approach is to employ a quadratic discriminant function, and Marks and Dunn (1974) have used simulation to compare the performance of linear and quadratic discriminant functions.

In this chapter we have concentrated on examples of the use of simulation in a small number of areas, and even here it has been necessary to select only a very small subset of the possible examples. There are many more applications of simulation. Rubinstein (1981, chapters 5–7) describes some of these, showing, for instance, that the solution to certain integral equations is given in terms of the mean time to absorption in particular Markov chains with an absorbing state. In this application an aspect of a completely deterministic problem is estimated by simulation of the related Markov chain (cf. Sections 9.1.2 and 7.3.4). There are also important uses of simulation in statistical physics.

9.7   Exercises and complements

(Some of the following are best appreciated if one has a prior knowledge of MINITAB and multivariate analysis.)

†9.1   Discuss the following MINITAB program for simulating chi-square random variables on 10 degrees of freedom.

```
NOPRINT
  STORE
  NRANDOM 10 OBSERVATIONS, WITH MU=0.0, SIGMA=1.0, PUT IN C1
  LET C2=C1*C1
  SUM C2, PUT IN K1
  JOIN K1 TO C3, PUT IN C3
  END
EXECUTE 100 TIMES
HISTO C3
```

Suggest how you might write MINITAB programs to simulate logistic and beta random variables.

9.2   Modify the program of Fig. 9.2 to produce estimates of probabilities of absorption, and suggest how you might employ the variance-reduction approach of Section 7.3.4.

*9.3   Feller (1957, p. 395) provides the following analytical solution to the brother/sister mating problem. Starting in the jth state, for $j = 2, 3, 4, 6$, the probability of being absorbed in the first state after n steps is:

$$p_{j2}^{(n-1)}/4 + p_{j3}^{(n-1)}/16$$

and the corresponding probability for the fifth state is:

$$p_{j3}^{(n-1)}/16 + p_{j4}^{(n-1)}/4$$

where
$$p_{jk}^{(n-1)} = \sum_{r=1}^{4} \frac{\theta_{jk}^{(r)}}{s_r^{n-1}}$$

$$s_1 = 2, \ s_2 = 4, \ s_3 = \sqrt{5} - 1 \quad \text{and} \quad s_4 = -(\sqrt{5} + 1),$$

$$\theta_{jk}^{(r)} = c_r x_j^{(r)} \ y_k^{(r)}$$

where $c_1 = \frac{1}{2}$, $c_2 = 1/5$, $c_3 = (1 + \sqrt{5})^2/40$, $c_4 = (\sqrt{5} - 1)^2/40$,

and $\qquad\qquad \{x_j^{(1)}\} = (1, 0, -1, 0) = \{y_k^{(1)}\}$

$$\{x_j^{(2)}\} = (1, -0, 1, -4)$$
$$\{x_j^{(3)}\} = (1, (\sqrt{5} - 1), 1, (\sqrt{5} - 1)^2)$$
$$\{x_j^{(4)}\} = (1, -1 - \sqrt{5}, 1, (1 + \sqrt{5})^2),$$
$$\{y_k^{(2)}\} = (1, -1, 1, -0.5)$$
$$\{y_k^{(3)}\} = (1, \sqrt{5} - 1, 1, (\sqrt{5} - 1)^2/8)$$
$$\{y_k^{(4)}\} = (1, -1 - \sqrt{5}, 1, (1 + \sqrt{5})^2/8).$$

Use these expressions to calculate the probabilities of ultimate absorption in states 1 and 5, and compare your results with the estimates of Exercise 9.2. Comment on the relative merits of the simulation and analytic approaches to this problem.

*9.4 Alternative approaches to the estimation of the size of mobile animal populations include multiple marking, following a succession of recaptures (see, e.g., Darroch, 1958), and inverse sampling, in which following marking, the second-stage sampling continues until a prescribed number of marked individuals are caught. Identify the distribution of the number of animals that will be taken at the second-stage sampling. Find the mean of this distribution, and suggest a new estimator of the population size (see Chapman, 1952).

*9.5 The following data provide measurements, in inches, on members of a final-year undergraduate statistics class at the University of Kent.

			Men			
Chest	Waist	Wrist	Hand	Head	Height	Forearm
37.5	31.0	5.5	8.0	23.0	67.0	18.0
34.5	30.0	6.0	8.5	22.5	69.0	18.0
41.0	32.0	7.0	8.0	24.0	73.0	19.5
39.0	34.0	6.5	8.5	23.0	75.0	20.0
35.0	29.0	6.4	8.0	22.5	66.0	16.0
34.0	28.0	6.2	8.1	19.7	66.0	17.3
34.0	29.0	6.5	7.5	22.0	66.0	18.0
36.0	31.0	7.0	8.0	23.0	73.0	18.0
40.0	36.0	7.5	8.5	25.0	74.0	19.0
38.0	30.0	7.5	9.0	22.0	67.0	17.0
38.0	33.0	6.5	8.5	22.3	70.0	19.2
34.5	30.0	6.0	7.8	21.8	68.0	19.3
36.0	29.0	6.0	8.0	22.0	70.5	18.0

Women

Chest	Waist	Wrist	Hand	Head	Height	Forearm
35.0	25.0	6.0	7.5	22.0	63.0	16.0
35.0	25.0	5.8	7.3	21.5	67.5	18.0
34.0	26.0	6.0	7.0	21.0	62.8	17.0
36.0	27.0	6.0	7.0	23.0	68.0	17.5
34.0	26.0	6.0	7.0	21.0	65.0	18.0
35.0	26.0	5.0	7.0	22.0	65.0	16.0
36.0	30.0	6.5	7.5	24.0	70.0	17.0
34.0	24.0	6.5	7.0	22.0	63.0	16.5
36.0	26.0	6.5	7.5	22.0	67.5	16.5
30.0	23.0	6.0	7.0	21.0	59.0	14.5
32.0	24.0	5.8	7.0	22.5	60.0	15.0
35.5	28.0	6.0	7.0	22.0	69.5	17.5
34.0	22.0	5.0	6.0	22.0	66.0	15.0
36.0	27.0	6.0	7.0	22.0	66.5	17.5
38.0	29.0	6.0	7.5	22.0	64.5	17.5

(i) Compute the principal components, using the correlation matrix, for men and women separately. Can you suggest variables which may be omitted?

(ii) Calculate the linear discriminant function and discuss the probability of misclassification using the linear discriminant function for these data.

*9.6 Discuss the use of $\hat{\hat{\delta}}$ as a control variate in the sampling investigation summarized in Table 9.2.

9.7 As mentioned in Section 9.3.2, Jolliffe (1972b) used a method of Butcher (1960) for simulating standard normal random variables. The method employed the generalized rejection method of Exercise 5.29 for the half-normal density, writing

$$\sqrt{\left(\frac{2}{\pi}\right)} e^{-x^2/2} = cg(x)r(x) \qquad \text{for } x \geq 0$$

in which

$$g(x) = \lambda e^{-\lambda x}$$

and

$$r(x) = \exp(\lambda x - x^2/2)$$

Deduce c, and show that the resulting algorithm is that already considered in Example 5.5.

9.8 For the structured data set of Section 9.3.2, if ρ_{ij} denotes the correlation between X_i and X_j, show that $\rho_{14} = 0.894$, $\rho_{25} = 0.819$, $\rho_{26} = 0.707$, and $\rho_{56} = 0.579$. If it is decided to reject three out of the six variables, explain why the best subsets of variables to retain are: $\{X_1, X_2, X_3\}$ and

$\{X_2, X_3, X_4\}$, while the next best ('good') subsets are: $\{X_1, X_3, X_5\}$, $\{X_1, X_3, X_6\}$, $\{X_3, X_4, X_5\}$, $\{X_3, X_4, X_6\}$. Two of the variable-rejection rules considered by Jolliffe (1972a, b) were: B2—reject one variable to correspond to each of the last three principal components; and B4—retain one variable to correspond to each of the first three principal components. For B4 it was found that for 11 out of 96 simulated data sets, the best subsets were retained, while for B2 it was found that the subsets retained were always good, but never best.

†**9.9**   Discuss the results given below, from Gardner (1978). Two further populations, A and B, were simulated as in Example 9.2, with the differences that for A the three group means were ($-1.75, 0$), ($1.75, 0$), ($0, 3.03$), and for B there was no group structure, all the values being obtained from a bivariate normal distribution with mean ($0, 0$). Samples of size 60 were taken from the populations, and the results from two of the samples from A are given below for the 3-cluster solution produced by Beale's (1969) method:

	Cluster no.	Cluster size	Cluster	centre	No. of mis-classifications
	1	15	2.3	0.0	1
1st sample	2	27	−1.7	0.2	4
from A	3	18	−0.2	3.2	1
	1	20	1.7	0.5	3
2nd sample	2	17	−2.2	0.3	1
from A	3	23	−0.2	2.9	2

In addition, the values of the C measure for 4 samples of size 60 are as follows:

No. of clusters	Population of Example 9.2	Population A, 1st sample	Population A, 2nd sample	Population B
6	69.6	60.1	68.9	48.2
5	78.5	64.6	72.8	48.6
4	91.3	69.5	73.5	49.4
3	101.0	77.5	66.7	47.3
2	38.2	39.0	31.1	39.1

9.10   Morgan, Chambers and Morton (1973) applied non-metric multidimensional scaling to measures of similarity between pairs of the digits

1–9. 'Similarity' here was derived from experimental results summarizing how human subjects acoustically confused the digits. A two-dimensional representation is given in Fig. 9.5 in which the six largest similarities are also linked. Comment on the stress value of 6.44%, in the light of the results of Table 9.3 (cf. the discussion of digit confusion in Section 6.7).

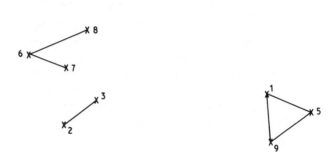

Figure 9.5　Two-dimensional non-metric multidimensional scaling representation of the acoustic confusion of digits.

9.11　Use the result of Exercise 4.14 to simulate bivariate normal samples as in Exercise 9.9, but with a different dispersion matrix.

9.12　Everitt (1980, chapter 4) simulated a bivariate sample of size 50, 25 elements from the bivariate normal:

$$N\left(\begin{pmatrix}0\\0\end{pmatrix}, \begin{pmatrix}16.0, 1.5\\1.5, 0.25\end{pmatrix}\right)$$

and the remaining 25 from the bivariate normal:

$$N\left(\begin{pmatrix}4\\4\end{pmatrix}, \begin{pmatrix}16.0, & 1.5\\1.5, & 0.25\end{pmatrix}\right)$$

Using single-link cluster analysis he obtained the following 5-cluster solution (numbers 1–25 refer to the first population, and 26–50 to the second):

Cluster	Objects
1	1–13, 15, 16, 18–25
2	26–41, 43–50
3	14
4	17
5	42

Comment on these results. Single-link cluster analysis is described by Everitt (1980, p. 9). He simulated his population using the Box–Müller method followed by the transformation of Exercise 4.14.

†9.13   Combine categories in Table 9.4 in such a way as to enable a standard chi-square test to be justified, and compare the results of this test with that for which there is no pooling. Suggest a possible control variate for use in the sampling distribution investigation of Section 9.4.1.

9.14   Repeat the study of Section 9.4.2 for a different pair of correlated variables.

*9.15   Write a computer program to provide a Monte Carlo test of whether the pattern of points in Fig. 9.4, or a similar pattern, generated as in Exercise 9.11, is random.

9.16   Investigate further the discrepancy observed in Example 9.3. If a BASIC program is written for the Monte Carlo test, an ordering subroutine can be obtained from the variety provided by Cooke, Craven and Clarke (1982, chapter 3).

9.17   Comment upon the relationship between the estimated probability and the variance of the estimated probability in Table 9.6.

9.18   Consider how you would use MINITAB to simulate Gossett's simulation mentioned in Section 1.5.

9.19   How would you use MINITAB to investigate central limit theorems? (cf. Mead and Stern, 1973, p. 194).

*9.20   Any hierarchical method of cluster analysis, such as single-link analysis mentioned in Exercise 9.12, can be represented by a set of 'ultrametric' distances between the objects (see Johnson, 1967). Consider how you might compare such ultrametric distances with the original between-object similarities, and suggest how simulation may be used to investigate the behaviour of your measure of comparison (see Gower and Banfield, 1975; Jenkinson, 1973).

APPENDIX 1

COMPUTER ALGORITHMS FOR GENERATION, TESTING AND DESIGN

The main aim of this section is to document some of the more convenient computer algorithms for the generation and testing of pseudo-random numbers. We shall also indicate the availability of programs for simple experimental designs.

A useful starting point in a search for such algorithms is the GAMS (1981) index, which indicates the IMSL and NAG libraries as the main sources of the computer algorithms of interest to us. IMSL is a commercial library containing about 500 FORTRAN subroutines. The library is available from IMSL, Inc., Sixth Floor, NBC Building, 7500 Bellaire Boulevard, Houston, Texas 77036, USA. NAG is a commercial library containing about 600 subroutines in FORTRAN and ALGOL, and NAG is available from the Numerical Algorithms Group Ltd, Mayfield House, 256 Banbury Road, Oxford OX2 7DE, UK, and 1131 Warren Avenue, Downer's Grove, Illinois 60515, USA. Other sources of algorithms exist, and we shall discuss some of these later, but first we shall describe what is available in NAG and IMSL. It is important to realize that this appendix is not intended as a reference manual for these libraries, and it will be necessary to consult the manuals for detailed running instructions. The information below is obtained from Edition 8 of the IMSL library, and mark 10 of the NAG library, and we only list routines which are user-callable. For both libraries, names of routines are strings of letters and numbers, as seen below.

A1.1   Generation routines available in NAG and IMSL

Beta

IMSL:GGBTR

In the notation of Section 2.11, the method used depends on the size of α and β:

(i) $\alpha < 1$ and $\beta < 1$, the method of Jöhnk (1964) is used.

(ii) $\alpha < 1$ and $\beta > 1$, or $\alpha > 1$ and $\beta < 1$, the method of Atkinson (1979b) is used.

(iii) $\alpha > 1$ and $\beta > 1$ and fewer than 4 variates required, then the algorithm BB of Cheng (1978) is used.

(iv) $\alpha > 1$ and $\beta > 1$ and ≥ 4 variates required, the algorithm B4PE of Schmeiser and Babu (1980) is used.

(v) $\alpha = 1$ or $\beta = 1$ then the inversion method is used (cf. Exercise 5.33(b)).

NAG: GO5DLF

Uses the method of Exercise 2.14, obtaining the component gamma variates from GO5DGF.

GO5DMF

Generates a variate from a beta distribution of the *second* kind, defined as (and generated by) X/Y in the notation of Exercise 2.14.

Binomial

IMSL: GGBN

Simulates $B(n, p)$ variates. If $n < 35$, a simple counting procedure is employed with re-use of uniform variates to improve efficiency. If $n \geq 35$, the method of Relles (1972) (see Exercise 4.17 and Section 4.4.1) is used.

NAG: GO5DZF

Simulates a $B(1, p)$ variate, with responses 'True' and 'False'.

GO5EDF

Uses the table-look-up method; used in conjunction with GO5EYF.

Cauchy

IMSL: GGCAY

Uses the inversion method of Exercise 5.8.

NAG: GO5DFF

Uses the rejection method of Exercise 5.27.

Chi-square

IMSL:GGCHS

Uses the method of Section 4.3, involving convolutions of exponentials, and possibly the further sum of a squared $N(0, 1)$ variate if necessary.

NAG:GO5DHF

Uses GO5DGF for gamma variates.

Exponential

IMSL:GGEXN

Uses the inversion method of Section 4.3.

GGEXT

Simulates from a mixture of two exponential densities.

NAG:GO5DBF

Also uses the method of Section 4.3.

F

NAG:GO5DKF

Sets $F = (nY)/(mZ)$, where Y, Z are independent χ^2_m and χ^2_n variables, respectively, obtained as $\Gamma(m/2, 2)$ and $\Gamma(n/2, 2)$, from GO5DGF, described below.

Gamma

IMSL:GGAMR

Simulates $\Gamma(a, 1)$ variates, the method used depending on the value of a. If $a < 1$ and $a \neq 0.5$, a rejection method described in Ahrens and Dieter (1974) is used. If $a = 0.5$, squared and halved standard normal variates are used. If $a = 1$, exponential deviates are used, while if $a > 1$ a ten-region rejection procedure of Schmeiser and Lal (1980) is employed.

NAG : GO5DGF

Simulates $\Gamma(n, \lambda)$ where $2n$ is a non-negative integer, using the approach of Section 4.3.

General distributions

IMSL : GGDA

Simulates from any given discrete distribution using the alias method of Exercise 5.42.

GGDT

Simulates any given discrete distribution using the table-look-up method, using bisection in the early stages.

GGVCR

Simulates any continuous random variable, using the inversion method. The user may provide only partial information on the distribution function, and the method of Akima (1970) is then used to interpolate, using piecewise cubics. The method, together with descriptions of accuracy, is described in Guerra *et al.* (1972).

NAG : GO5EXF

Used in conjunction with GO5EYF, for any given discrete distribution. This routine forms the cumulative sums needed for the table-look-up method.

GO5EYF

Performs an indexed search through tables of cumulative sums of probabilities.

Geometric

IMSL : GGEOT

Simulates a geometric random variable using the definition of such a variable in Section 2.6, i.e. $U(0, 1)$ variates are chosen sequentially until one is less than p, the geometric success probability.

Hypergeometric

IMSL: GGHPR

Simulates hypergeometric random variables using the definition of the hypergeometric distribution (see for example *ABC*, p. 147).

NAG: G05EFF

Employs the table-look-up method. Used in conjunction with G05EYF.

Pseudo-random integers

IMSL: GGUD

Generates n uniform random integers over a specified range, using GGUBFS and truncation.

NAG: G05DYF

Generates a uniform random integer over a specified range, using G05CAF and truncation.

G05EBF

The table-look-up approach for uniform random integers; used in conjunction with G05EYF.

Logistic

NAG: G05DCF

Uses the table-look-up method (see Exercise 5.13(a)).

Log-normal (See Exercise 4.21)

IMSL: GGNLG

Returns the value $\exp(\alpha + \beta N)$, where N is an $N(\mu, \sigma^2)$ random variable obtained from GGNPM.

NAG: G05DEF

Returns the value $\exp(N)$, where N is a suitable normal variate obtained from G05DDF.

Multivariate

IMSL:GGMTN

Simulates multinomial random variables. In the notation of Section 2.15, first a $B(n, p_1)$ distribution is simulated, and then successive conditional binomial simulations are used, simulating X_j from
$B(n - x_1 - x_2 \ldots - x_{j-1}, p_j/(1 - p_1 - p_2 - \ldots - p_{j-1}))$, for
$2 \leq j \leq m$, with $X_i = x_i$, for $1 \leq i \leq j - 1$.

GGNSM

Simulates multivariate normal random variables with zero means when the dispersion matrix, Σ, is specified. The subroutine LUDECP provides the required triangular factorization of $\Sigma = \mathbf{A}\mathbf{A}'$ (see the solution to Exercise 4.14).

GGSPH

Simulates points uniformly at random from the surface of the unit sphere in 3 or 4 dimensions (see Marsaglia, 1972a). In three dimensions, for example, if V_1 and V_2 are independent and $U(-1, 1)$, then conditional upon $S = V_1^2 + V_2^2 < 1$, set $Z_1 = 2V_1 \sqrt{(1 - S)}$, $Z_2 = 2V_2 \sqrt{(1 - S)}$ and $Z_3 = 1 - 2S$. The $\{Z_i\}$ are then the Cartesian co-ordinates of the required points.

ZSRCH

Generates points in n-dimensional space, for use with nonlinear optimization routines that require starting points. Such a procedure is needed for nonmetric multidimensional scaling (see Section 9.3.3).

NAG:GO5EAF

Sets up a reference vector which is used in conjunction with GO5EZF to generate multivariate normal random variables. The same approach is used as with GGNSM.

Negative binomial (cf. Exercise 4.19)

IMSL:GGBNR

Employs the table-look-up method.

NAG : GO5EEF

Employs the table-look-up method. Used in conjunction with GO5EYF.

Normal

IMSL : GGNML

Uses the inversion method, the inverse normal c.d.f. function being provided by the IMSL routine MDNRIS: n variates are generated.

GGNPM

Uses the Polar Marsaglia method of Section 4.2.2.

GGNQF

The version of GGNML that is used if only a single $N(0, 1)$ variate is required at each call.

NAG : GO5DDF

Uses Brent's (1974) GRAND algorithm, which employs Forsythe's method (see Exercise 5.41).

Order statistics

IMSL : GGNO

Generates a set of order statistics from the normal distribution. Initially GGUO is used, and then the ordered uniform variates are transformed, using the inversion method as in GGNML.

GGUO

Generates a set of order statistics from the $U(0, 1)$ distribution. If a full set is required, spacings between successive order statistics are obtained as exponential variables. If only a subset is required, a beta variable is used for one of the order statistics and then the others are generated conditionally. For general discussion of the generation of order statistics, see Gerontidis and Smith (1982).

Permutations

IMSL : GGPER

Generates a random permutation of the integers 1 to k, using an exchange sort procedure, described by Cooke *et al.* (1982, p. 15).

NAG : GO5EHF

Method used is only possible for $k < 20$.

Poisson

IMSL : GGNPP

Simulates a non-homogeneous Poisson process (see Lewis and Shedler, 1976; and Section 8.3.2).

GGPON

For use when the Poisson parameter λ changes from call to call, as may occur, e.g., if one was using the Poisson and gamma distributions to simulate a negative-binomial random variable (see also Kemp, 1982; and Exercise 4.18). Uses the method of Fig. 4.4 for $\lambda \le 50$. For $\lambda > 50$ a normal approximation is used.

GGPOS

For use when the Poisson parameter does not change often. For $\lambda \le 50$ the standard table-look-up method is used. For $\lambda > 50$ a normal approximation is employed.

NAG : GO5ECF

Calculates probabilities for use with GO5EYF and the table-look-up method.

Sampling from a finite population without replacement

IMSL : GGSRS

Given a population of size n and a sample of size $m < n$, this routine selects this sample sequentially without replacement, so that when $j < m$ items have been selected, the ith object is selected with probability $(m-j)/(n+1-i)$ (see

Bebbington, 1975, for an explanation, and McLeod and Bellhouse, 1983, for a new approach).

Stable distribution

IMSL : GGSTA

Simulates from a range of stable distributions indexed by two parameters, using the method of Chambers *et al.* (1976). Possible distributions include the Cauchy and the normal.

t

NAG : GO5DJF

Calculates $W = Y \sqrt{(n/Z)}$, in which Y has a $N(0, 1)$ distribution, provided by GO5DDF, and Z has an independent χ_n^2 distribution, obtained as $\Gamma(n/2, 2)$ from GO5DGF.

Time series

NAG : GO5EGF

Sets up a reference vector for an auto-regressive moving-average time series model with normally distributed errors. The series is initialized to a stationary position using the method of Tunnicliffe–Wilson (1979).

GO5EWF

Simulates the next term from an auto-regressive moving-average time series model, using the reference vector set up in GO5EGF.

Triangular

IMSL : GGTRA

Simulates random variables from the symmetric triangular density of Exercise 5.5, but over the range (0, 1). The inversion method is used, as in Exercise 5.5(a).

Uniform

IMSL : GGUBFS

Uses the multiplicative congruential generator $(7^5, 0; 2^{31} - 1)$. Note that for simplicity, $U(0, 1)$ pseudo-random deviates are obtained from normalizing by 2^{31}, and not $2^{31} - 1$. The seed may be specified by the user.

GGUBS

Generates n pseudo-random variates using GGUBFS.

GGUBT

Uses the multiplicative congruential generator $(397\,204\,094, 0; 2^{31} - 1)$. The algorithm is slower than GGUBFS, due to portable coding. The seed may be specified by the user.

GGUW

Employs 'shuffling', with $g = 128$ (see Section 3.6) applied to GGUBFS.

NAG : GO5CAF

Uses the multiplicative congruential generator $(13^{13}, 0; 2^{59})$ with initial value, $x_0 = 123\,456\,789 \times (2^{32} + 1)$ unless GO5CBF or GO5CCF is also used (see Best and Winstanley, 1978, for comments).

GO5CBF(I)

Sets the seed for GO5CAF as $x_0 = 2I + 1$.

GO5CCF

Generates a random seed for GO5CAF, using the real-time clock.

GO5CFF

Records the current value in a sequence from GO5CAF. The sequence can then be continued at a later stage by calling the routine GO5CGF.

GO5CGF

Restarts a sequence from GO5CAF, following a previous call to GO5CFF.

GO5DAF

Simply transforms a $U(0, 1)$ pseudo-random variable from GO5CAF, to give a $U(a, b)$ variate.

Weibull

IMSL:GGWIB

Uses the inversion method of Exercise 5.13 (b).

NAG:GO5DPF

Also uses the inversion method of Exercise 5.13 (b).

A1.2   Testing routines available in NAG and IMSL

IMSL:GFIT

Chi-square goodness-of-fit test.

GTCN

Used prior to a chi-square goodness-of-fit test, to determine suitable number of class-intervals (see Mann and Wald, 1942).

GTDDU

Computes the squared distance between successive pairs of pseudo-random $U(0, 1)$ variates and then forms an appropriate tally which is used in GTD2T.

GTD2T

Performs a test of the tallies produced by GTDDU (see Gruenberger and Mark, 1951).

GTNOR

Uses a chi-square goodness-of-fit test for $N(0, 1)$ variates.

GTPBC

Performs a count of the number of zero bits in a specified subset of a real word in binary form; tests are then applied to the resulting counts.

GTPKP

Used in the preparation of expected values for the poker test; this routine evaluates $B(n, 0.5)$ probabilities.

GTPL

Produces tallies for use in the poker test, performed by GTPOK.

GTPOK

Performs the poker test, using statistics computed by GTPL.

GTPR

Produces a tally of pairs (or lagged pairs) in a sequence of pseudo-random numbers; used in GTPST.

GTPST

Performs a test on pairs produced by GTPR.

GTRTN

Computes numbers of runs up and down in a sequence of variates.

GTRN

Tests the runs produced by GTRTN. Note that this test does *not* use the appropriate covariance terms (cf. Equation (6.1)) (see Kennedy and Gentle, 1980, pp. 171–173).

GTTRT

Produces a tally of triplets in a sequence of pseudo-random numbers; used in GTTT.

GTTT

Tests the randomness of triplets produced by a pseudo-random number generator, and previously tallied by GTTRT.

NKS1

Performs the Kolmogorov–Smirnov one-sample test, with the user supplying the theoretical distribution via a FORTRAN subroutine.

NKS2

Performs the Kolmogorov–Smirnov two-sample test.

NAG : GO8CAF

Performs the Kolmogorov–Smirnov one-sample test. The parameter 'null' provides alternative forms for the theoretical distribution. These are: $U(a, b)$, $N(\mu, \sigma^2)$, Poisson and exponential. This routine was used in connection with Exercise 6.21.

A1.3   Features of the NAG and IMSL algorithms

Because of inertia, library subroutines are unlikely to keep pace with the latest developments in the generation of random variables; while some of the routines of Section A1.1 are remarkably up to date, using methods of Chapter 5, some are clearly less so. Comparisons of efficiency for certain algorithms for beta, gamma and normal variables are given by Atkinson and Pearce (1976). It is interesting to contrast the different emphases used in these two different libraries. Thus IMSL places much importance on the testing of random variables, as can be seen from the last section. An attractive feature of certain of the routine descriptions is that they specify the empirical tests satisfied by the generated numbers, and provide the results of such tests. A number of the IMSL tests proceed in two stages, the first stage forming a tally which is then used by the second stage. A similar two-stage procedure is employed in NAG for simulating discrete random variables. The table-look-up method is employed through the GO5EYF routine which performs an indexed search through tables of cumulative sums of probabilities, set up individually by a separate routine for each distribution.

A1.4   Further algorithms

Published computer algorithms for simulation can be found in the journals: *Communications of the Association for Computing Machinery, Computer Bulletin, Computer Physics Communications, Computer Journal, Computing, The Journal of the Association for Computing Machinery*, and *Applied Statistics*. All of the algorithms in *Applied Statistics* up to the end of 1981 are indexed in *Applied Statistics* (1981), **30** (3), 358–373, while simulation algorithms from the other journals mentioned here are indexed in *Collected Algorithms* from ACM. *Applied Statistics* algorithms which complement those of Section A1.1 are:

Ripley (1979), on simulating spatial patterns (cf. Section 9.5.1);

Patefield (1981), on simulating random $r \times c$ tables with given row and column totals (cf. Section 9.4.1);

Smith and Hocking (1972), on simulating Wishart variates (cf. Exercise 4.15);

Wichmann and Hill (1982b) provide a FORTRAN program for their portable random generator, which combines the multiplicative congruential generators (171, 0; 30 269), (172, 0; 30 307) and (170, 0; 30 323) in the way explained in Exercise 3.18.

Golder (1976a, b) provides the spectral test, mentioned in Section 6.1.

This last test has recently been revised and a FORTRAN program for the new algorithm is provided by Hopkins (1983a). Hopkins (1983b) provides a FORTRAN algorithm for the collision test mentioned in Section 6.3. As mentioned in Chapter 6, the package of Cugini *et al.* (1980) provides a range of empirical test programs, written in BASIC.

Computerized algorithms for the analysis of experiments can be found in standard statistical packages such as GENSTAT, BMDP and SPSS. In addition, the IMSL routine AFACN provides an analysis for a full factorial experiment (see Section 8.5).

Other statistical packages provide simulation facilities. Some of those to be found in MINITAB have already been described, and we have seen how these may be extended by the use of the MINITAB language (see Exercise 9.1). The S language, developed at Bell Laboratories, also provides a wide range of simulation possibilities, and the opportunity arises within BMDP3D for using simulation for simple tests of robustness. See: C17, 'Quick and Dirty Monte Carlo'—p. 865 of the 1976 (not 1981) BMDP handbook.

Simulation programs in BASIC can also be found in Cooke *et al.* (1982, chapters 7 and 9) and in Wetherill (1982, Appendix 1).

APPENDIX 2

TABLES OF UNIFORM RANDOM DIGITS, AND EXPONENTIAL AND NORMAL VARIATES

These tables provide pseudo-random, rather than random, variates, but they should be adequate for small-scale investigations.

Table A2.1   Pseudo-random digits

Each digit was obtained from the first decimal place of pseudo-random $U(0, 1)$ variables resulting from the NAG routine GO5CAF. The digits have been grouped as shown for convenience of reading. The digits may be used to provide $U(0, 1)$ variates to whatever accuracy is required, by placing a decimal point in front of groups of digits. Thus for 5 decimal places we would obtain, from the first row of the table: 0.36166, 0.15217, etc.

```
36166  15217  88906  60493  36211  02862  68789  35346  83423  38001
48734  61061  82801  32055  99587  51156  61919  85682  16253  06162
29474  76062  40096  88802  52678  92156  61784  75192  39087  90198
24632  13950  13723  02027  29179  28792  75928  03507  74295  27971
84842  84503  07919  72887  77041  57728  05468  08203  97672  00856

63337  02467  90923  50023  69684  01854  68186  17018  31268  51312
61917  09485  90672  72283  97043  20066  92073  93723  60124  67424
88162  29658  70156  63238  55560  00192  09480  91738  99359  04009
19864  59125  89677  42774  93979  39397  78970  42590  49189  26010
57823  07638  75417  25906  85532  80853  47920  32719  79086  76277

95883  03872  88357  22660  41639  51747  20188  33676  37997  58689
86817  63621  51718  85194  56953  49026  53298  05380  23302  72846
75071  05936  63958  80809  09052  43912  31379  24941  82897  55358
19385  05924  17643  37034  81099  86478  66570  32685  81290  47747
50653  95687  02929  88847  59817  97697  52342  77772  41516  21306

62108  71354  38481  64582  68355  98234  89441  50133  33179  46922
38756  20124  18911  68285  40299  69862  57529  00433  57503  13604
86402  23441  04471  92961  50458  12385  24719  54217  17412  56950
21521  24909  18469  94693  58424  50233  87267  77388  66093  36902
77782  26088  39561  99008  29704  73404  42854  72034  35340  65979
```

```
36999  40918  12940  42293  81239  70291  36004  13710  74061  82940
03841  08765  62184  81399  07316  64641  94691  65074  61898  04083
00807  78726  59416  63247  45718  88664  85898  13795  87046  85866
02917  82605  09600  29041  81189  18604  19172  93031  05855  69612
64601  93871  48040  57314  34586  32937  16346  31772  30045  14411

56595  13487  12824  01773  73622  45794  65307  27776  66889  20934
16032  78673  69922  10028  83325  45572  77482  15638  25912  65162
14456  95942  95841  77315  60149  09003  24361  99812  55686  26936
67222  14182  76751  05780  02212  58651  77991  62466  96086  90989
10247  27376  83657  45033  35590  09304  23567  22613  76589  37302

21422  63373  21711  38058  75287  34529  69725  93297  65289  93188
82388  94499  77254  24158  55167  26300  93631  15096  33918  18141
72684  89170  04762  90070  98818  03719  65486  42392  45248  15136
28699  19170  73753  23241  68546  25212  02970  16920  45402  02343
71021  32190  86137  89319  80906  99572  41952  72071  42417  91892

83980  22152  44200  73564  83758  25394  64887  07239  94281  16310
75389  30623  64567  55612  45316  88660  67815  73256  32369  29460
62047  58958  59585  76603  86878  21102  16943  11897  90282  52079
75247  99939  86152  58634  35428  30692  14171  85562  77202  25123
13137  81688  37850  72674  83095  32217  75930  41517  99201  61852

96445  95977  07199  22046  11044  36433  02260  26243  32479  88557
13460  82626  45375  01319  55155  49979  47690  96011  43922  02048
50803  46870  12341  45281  52413  65600  72523  82302  69144  72124
10865  30925  15863  84504  64695  92529  00309  63797  65493  48420
02274  91320  96809  78475  72529  79329  61466  46642  99924  53647

18542  64085  02190  75560  31551  75421  65491  74278  54110  24016
81494  88532  44757  52378  26088  46991  83883  02387  84262  07839
55678  51663  00409  51342  30827  34336  46256  29043  71083  65480
18828  06215  71236  38371  82643  12269  85808  80589  52286  94826
71579  84319  57755  52609  00554  58061  63462  45746  31033  34576

92243  23481  71940  43545  68539  71537  86147  99940  53937  07833
47839  75693  93902  88982  35549  59077  16083  44916  02950  24889
11381  70221  12843  72933  31494  01026  33125  93650  73428  56666
43407  45189  23534  91948  16877  50161  56625  66547  82253  64260
21605  99899  13588  22150  95336  50900  33526  97706  19356  22391

71671  20240  28871  14852  81695  92408  68110  43747  44635  98671
68905  40211  21246  11299  29360  37829  32326  15975  17952  70767
90950  72574  79462  16444  76097  89310  72939  47577  71549  17270
24440  50126  62637  99076  65063  54117  89122  00806  96591  44700
77885  12787  57930  48813  36474  93627  85568  44294  86627  56531

18957  98793  64674  03885  09056  02257  50615  30698  27609  20132
33692  50936  18026  14555  14991  92073  34368  98843  68804  69235
86728  19108  69750  35394  87272  32514  69272  71707  77148  51250
99061  08714  57491  55037  60722  17467  21814  82460  77249  11848
14932  41756  57749  77093  29400  64792  21228  55738  31278  34764

84678  98353  15743  07920  45989  04466  38259  07986  44432  12614
00911  64580  95310  62810  42932  54119  08522  03883  38260  30253
93849  32655  56899  37396  39997  75403  31062  95681  39325  89946
81791  87778  46219  27048  61745  72576  73076  43531  08671  19885
93070  82481  36492  76023  61538  46864  41831  14911  20292  16995

94374  47133  71545  13578  80441  49114  99380  87654  15887  55035
46498  96367  09824  04516  24591  18289  72452  03971  77274  71729
73585  61883  00121  94029  65071  07975  99806  92880  88828  62018
24062  65158  44854  58600  91996  67859  34081  99534  64688  38482
82957  29903  57694  06955  92314  88307  94859  30320  50689  84158
```

(contd.)

Table A2.1 (contd.)

58116	77716	50045	86995	75517	66352	52300	55676	52761	20817
43105	70000	74113	92189	07764	73417	37965	98864	74921	71121
48642	29111	60116	96879	19642	70697	48955	94622	03892	98759
57806	93529	78862	30676	75643	23121	42213	85314	07570	23608
69391	03367	32175	76497	73839	97718	25877	24022	22924	28968
80830	78624	85966	95513	90641	21868	87699	00841	77126	30649
30771	33366	52247	67479	90247	46755	60993	49041	63318	69732
67996	79930	84564	70705	57831	20618	86605	94286	90522	52203
31856	84574	93822	15991	08263	01539	87868	49187	56313	67487
70795	88068	55503	75541	85589	70521	00714	94507	81677	63946
10220	57484	30795	06334	03875	52013	30290	04465	50883	81493
69375	77419	22914	76692	06106	71121	58998	89485	14909	07367
26587	58120	81086	28041	98928	80003	67575	93899	92768	21919
19494	51024	11371	27222	60832	54625	55611	74639	01664	47290
47191	40142	35681	85465	88982	92657	22040	64126	50301	98272
93090	51159	41773	20704	89490	23192	45756	10748	22563	01230
09344	48492	90624	04082	26159	79189	65601	97914	97136	25494
02312	84666	63794	94848	53765	80070	73313	74398	10206	92343
68661	01193	11425	42226	95654	60885	29321	89929	07239	39568
41379	50393	65240	89433	51113	86395	78271	71631	81527	73933
67084	13061	22106	11072	26118	97197	34766	47556	71441	40461
31358	26123	18891	76728	34605	12075	46392	04324	28985	99182
00767	05805	76506	28204	79129	46979	98472	63428	03755	53764
29777	38405	57199	09465	10119	73633	12131	16098	41159	34470
63711	36059	27403	19035	97961	79169	29648	36408	62714	85352
75928	21317	70128	73744	17432	81679	27578	69882	99225	76110
33986	07916	77054	49838	53563	23448	38564	55787	45299	67177
00097	14652	78153	41081	56950	80972	83711	32490	42143	97116
43287	72228	88220	54054	69210	69473	99599	61404	15400	40170
25488	90700	20844	30501	53500	60058	10689	11854	93029	26010
49015	81077	73106	84215	59146	74846	47773	92276	27200	31617
07679	61012	85057	21966	96622	44041	34440	88431	93782	34566
13983	76295	46221	90843	30436	94432	50801	05014	57670	35168
25092	44402	37977	89575	84117	87202	57610	11668	52681	12062
54144	47384	51232	84698	14973	87909	78415	88714	65015	58385
94659	48073	92404	40755	34113	04453	34126	71323	03381	21602
81582	98238	04333	76282	45360	92503	14556	45931	77687	31809
02828	99064	35230	35302	48969	20868	61852	48547	53949	61402
68395	72433	73080	63041	20266	37790	34964	90692	00908	41666
47855	78200	05131	36223	09609	01310	59600	59347	93379	09306
45200	59476	58331	58983	04752	42041	04034	86689	79104	10358
01918	91945	76546	29141	76238	83569	53021	17257	82111	82930
61540	28614	16849	03525	76376	00034	19491	08069	79091	29986
21784	77463	00629	59748	31781	72763	35753	06669	25366	58467
76309	28173	66598	78866	18983	55363	51882	53612	18179	90367
64868	96259	94895	31664	72553	46067	79129	41317	43249	94019
54222	89245	18609	81706	89002	10318	43117	56097	16846	46140
00670	64654	56324	63234	20928	88497	86712	71340	19249	46024
89738	90350	88020	14263	68442	65743	08167	97330	71354	76866
78724	47107	75096	75954	55046	05828	12946	08225	68254	19754
17253	08757	72241	73101	48783	75679	08801	67848	69759	60701
01695	03347	76198	21391	45264	54639	94383	29324	61533	69315
18179	57886	72056	10430	10098	34826	01970	42570	29802	97836
93284	26630	00436	89979	18825	59778	80234	18065	92327	38938
89674	64578	56497	00294	35061	66174	56981	70068	37099	92404

70758	58852	44908	12902	80367	20544	55501	61804	25625	53932
22158	02610	03366	21233	70854	47126	67408	41123	46509	15397
73754	19177	90691	47458	46097	91095	77467	16462	18557	93615
24481	96281	13584	33060	64401	43071	32464	40241	07161	80686
91059	04211	83450	30025	72293	74025	73148	45185	63359	83061
27916	02822	96352	09633	58924	75279	99967	17782	58272	73690
92001	18234	97267	12545	80438	75775	87711	63518	22947	87339
70714	33659	47341	51517	65875	84552	18719	87924	92700	81501
70980	02613	90570	91853	35434	01132	03392	56199	22093	15606
99346	17545	97398	39715	56819	04725	97472	60312	52951	23979
81870	76792	41803	58485	23160	92630	56568	62125	26546	66474
37833	26472	60993	61249	21816	51996	86750	52771	19772	38716
19729	57397	70534	51836	25213	26797	09106	47203	28940	99930
82538	67161	66611	82661	11668	10649	18831	70045	47794	83740
82615	42264	16995	56773	31603	07879	23724	62811	34053	20721

Table A2.2   Pseudo-random standard normal variates, i.e. with zero mean and unit variance, resulting from the NAG routine G05DDF. If a random variable N has an $N(0, 1)$ distribution, then we may obtain a random variable X with an $N(\mu, \sigma^2)$ distribution by setting $X = \mu + \sigma N$.

-0.3817	1.7309	0.1020	0.1781	-2.1976	0.2756	0.8807	-0.0815	0.9479	0.8028
-1.4627	-0.0787	0.9825	-0.9018	0.1992	1.2393	-0.8749	-0.6252	-0.1124	0.5100
0.3425	-2.5194	0.7917	-0.6435	-0.8305	0.5140	1.0899	1.1004	-0.3905	-0.0751
0.2134	0.2637	0.0484	0.4427	0.7435	0.1946	1.1059	-0.7671	1.6726	-2.3685
1.9877	-0.3445	0.2755	0.3305	-1.2995	0.5384	0.8722	-1.0991	0.5924	-1.0997
-0.8110	1.5247	0.6916	-0.2860	-0.0646	0.9836	-0.8654	0.2936	-0.6669	-2.7949
1.2422	3.2607	1.8091	-0.1199	-0.9951	-1.2879	1.2767	-0.5209	-0.4755	-0.4369
1.7016	-0.8645	-0.5770	0.6983	0.9749	1.9517	0.5771	-0.8968	1.0238	-1.8262
-2.2935	-0.2379	0.4626	-0.5964	-0.8931	0.7290	-1.7184	-0.8512	-1.0430	-0.5693
0.7397	0.0578	-0.8564	0.0379	1.1788	-1.5122	-1.2791	0.3306	0.9229	0.5169
0.2863	-0.2170	0.7516	-0.6194	0.7882	-1.8403	1.1957	-0.0081	-0.3016	0.2164
-0.9733	-1.2938	-1.1274	-0.8729	0.6601	-1.2445	1.2535	0.1542	0.3109	-0.2190
1.0996	-0.4117	-0.3625	-0.9987	-1.0937	0.6487	2.0155	0.2707	1.5944	-0.9749
-0.0864	-0.9278	-0.3483	-1.1420	0.0262	1.7378	-0.0644	1.4343	0.9960	0.3269
0.2443	0.5603	1.9258	0.5536	-1.0949	0.6032	-1.4501	0.4264	-0.6752	0.0338
-0.4326	-1.0698	-1.6900	0.3524	0.5440	-0.4991	-0.2465	2.2318	-1.3405	-0.5392
1.6935	1.1556	0.2844	0.3268	0.1115	-0.8292	1.0769	-1.4394	0.4225	-0.1589
0.2875	0.1739	0.3121	-0.4311	0.3326	-0.0997	0.1986	0.1609	0.4360	1.2305
0.3443	1.5596	0.3378	-1.7689	2.3660	-0.5133	-0.6426	0.3270	-0.6778	0.7603
0.5787	0.4870	0.4032	0.6988	-0.3843	0.2841	1.5372	-0.6841	0.7115	-0.7106
0.7738	0.5707	-0.0395	0.3851	-0.1621	-1.9569	1.6740	0.8431	1.2604	-0.6320
0.9049	-1.3804	0.7392	0.6756	-0.4579	-1.3956	0.3706	-1.0403	0.8779	-0.4107
-0.4386	1.7108	-0.4503	-0.6238	0.5754	1.6640	0.0940	0.2241	-0.4473	0.6381
0.7867	-2.5127	-0.3218	2.3873	-0.6108	1.1309	1.8718	0.8949	0.0993	-0.4548
-1.2941	-0.0209	-0.6836	2.0827	0.6614	1.9945	-0.7858	0.0739	-2.1420	0.4447
-0.2455	0.0987	-2.5699	0.0936	-0.1062	0.3717	1.0177	0.3336	-0.6937	-1.7130
-1.4142	0.9285	-0.8045	-0.3002	-1.1250	1.2432	-1.5822	-0.2261	0.9079	-0.4599
-0.6906	-0.8680	0.1212	-1.0759	0.3097	0.0314	-0.2841	-1.5271	-0.7704	0.1679
-1.3015	-0.1933	-1.0627	-0.4595	-1.2822	-1.4311	1.7182	-0.1883	0.4226	-0.2443
-1.0915	0.3373	-0.8352	-0.7284	-0.8288	0.2353	-0.0307	1.3551	0.6767	1.1028

(*contd.*)

Table A2.2 (contd.)

-0.7447	-0.8772	0.0127	0.7654	-1.7962	-0.6207	-1.6598	-0.3508	-0.0217	-0.8522
-0.1038	-0.5699	-0.0175	0.8108	0.1930	0.4096	-0.3671	-0.3834	-0.1277	-0.1389
-0.1214	0.0438	-0.2015	0.5480	0.4599	-0.0736	0.0520	1.5282	0.3362	1.1986
0.6330	-0.1110	0.2029	0.2977	-0.2666	-0.2293	0.0287	0.3880	-0.5429	1.4054
0.5491	-0.4730	0.1286	0.2657	0.0265	-1.2633	-1.8893	0.4530	0.7296	0.7892
0.2180	0.2324	-0.1088	-0.1978	-0.4070	0.3857	1.3599	0.0081	-0.1153	-0.7821
0.1982	-1.0065	0.2581	1.0381	0.7913	-0.7412	1.1229	0.6890	-1.3086	-0.9308
-0.7594	1.4248	0.3888	0.0650	-2.2091	-1.2866	1.1597	-2.1434	-0.5391	1.0803
-1.6500	-0.5480	-0.4918	1.2628	-0.3704	0.5616	-0.2671	0.0672	-0.2623	-0.9198
1.5428	-0.1753	-0.5112	-0.6321	-0.0249	0.0315	-1.2220	0.8575	1.2922	-0.6323
1.4909	-0.5305	0.4103	1.7201	0.7324	-1.1984	0.0065	-0.2614	1.8041	-0.6362
-1.0791	-0.6958	-0.1386	-2.1384	-1.4571	0.1021	-0.6138	-1.1380	-0.7032	0.5105
-1.4939	0.4900	0.8626	-2.1653	-1.8681	-0.0380	-1.0705	-0.3707	0.8859	-0.9370
0.2543	-0.2685	0.9255	-0.2155	0.9099	-1.2442	-0.8209	0.7985	0.5065	1.1831
-0.1023	-0.0806	-0.2510	0.3311	0.3626	-0.9315	-1.6741	0.0759	-0.2548	0.1467
0.3196	-1.1455	-0.0005	-1.2296	1.3916	-0.9621	-0.4561	-0.7797	0.1499	0.9646
-1.4982	-0.4769	0.5693	-0.0364	-0.0107	0.2022	-0.9352	-0.5443	-0.7543	0.1192
0.1937	0.2299	-0.1068	0.2201	-0.3486	0.0246	-1.1463	1.1911	0.2864	0.2987
0.8412	0.7450	0.3409	-0.1437	-1.3503	-0.6071	0.4920	-0.6524	-0.1416	-0.6042
0.5355	0.3768	0.3261	-0.8482	1.1350	0.0969	0.4290	0.9840	0.0393	-0.1375
1.1153	0.4419	-1.1267	0.9633	1.7656	0.9445	0.1617	-1.1596	-3.3947	-0.0885
-1.4848	-0.8722	-0.5615	0.9176	-1.1794	-1.4326	1.9189	-0.4979	-0.1419	-1.6472
-0.3317	-2.0402	1.9038	-0.4714	1.2864	-0.2311	-1.5293	1.0073	-2.4238	0.2967
-0.1971	-2.4073	1.9169	-2.1454	-0.4002	0.4891	-0.8868	0.4606	-0.8800	-1.3264
-0.6606	0.6312	0.1003	-0.6720	0.8703	0.9839	-0.3304	0.1415	-0.1812	-0.5166
-1.2673	0.0231	1.2253	1.1969	-2.0353	-0.0101	-0.6852	-0.0361	-0.8582	1.9879
0.4946	0.2027	-1.3595	-0.6540	-0.2104	0.0999	1.5437	-0.1192	-0.0978	-1.1926
0.0113	0.5291	0.6703	-0.4035	0.4518	0.5733	-0.6426	-0.2681	0.9294	-0.5792
-0.7008	0.4030	0.9852	-0.4456	0.1074	-2.1391	1.2920	-1.2210	0.1845	-0.4000
0.3877	0.4916	-0.0256	0.7662	-0.9510	-1.2123	0.7984	0.4067	0.2765	0.0120
0.2176	-0.0410	1.1940	0.5138	0.0490	0.2083	-0.3644	-0.4395	0.6796	-2.1861
2.4568	1.1273	-1.8763	0.9266	-0.5361	0.4281	1.0042	-1.3309	0.7990	0.2737
-0.5189	-0.7305	0.1622	-0.3210	2.6591	-1.2294	-1.3681	-0.1728	-1.6764	-1.5392
0.4605	-1.5739	-2.0301	0.6998	-0.7805	-0.0328	-0.5359	1.3519	1.0376	0.5738
-0.2112	0.2712	-0.5583	-0.6672	-1.7556	1.5365	-0.9355	-2.1280	-0.5481	-0.4610
0.5879	0.3855	0.3576	0.6718	0.2949	0.8121	-1.1718	0.1422	-0.7861	1.3364
1.2061	-1.1127	1.1804	0.6032	-0.5295	1.3583	0.7758	0.3685	1.2931	0.4544
0.0002	0.4946	-0.6920	0.4599	2.2684	0.3157	1.2824	0.4818	0.0673	0.4632
-0.2253	0.3013	-0.2282	-0.4879	-0.8050	-0.8271	-0.3503	-0.8143	0.1021	-0.3118
0.5471	0.9078	0.2249	0.2213	-1.1060	0.4849	-0.7031	-0.0051	-2.0247	0.0957
0.1652	-0.2295	-0.1413	-0.2856	-0.1498	-0.2851	-0.4935	0.8605	0.7028	1.7356
-0.6826	0.1470	0.1734	0.1971	0.8672	0.6629	1.6643	0.5277	-0.2714	-0.1074
-0.1796	0.2998	1.2826	-1.2854	1.4688	-1.1570	-2.1032	-0.8779	0.8949	1.1306
-1.1163	0.3578	2.0840	-0.3008	-0.1733	0.6392	-2.5278	-0.6316	0.1519	0.7871
1.7477	0.6345	0.5110	-0.6359	0.4579	0.7412	-0.5617	0.5451	-1.1775	0.2286
1.0788	0.4764	-1.6086	0.4507	0.7641	1.0745	1.0140	0.6907	-0.5315	0.7717
-1.2137	0.7048	-0.3749	0.4213	-0.0404	-1.5166	-0.2283	0.1414	-1.2080	-0.5067
0.6453	0.5468	-0.0953	0.8074	-0.4192	1.2891	0.4142	-0.3544	0.2319	0.0946
0.6607	-0.6891	-0.1132	0.8950	0.0689	1.1429	0.0596	0.2354	-1.1305	0.0949
0.5430	-0.7035	0.0226	0.7450	-0.1272	0.1311	0.5547	0.1567	-0.4982	0.0276

0.3460	0.7198	-1.2223	1.2202	0.0050	-0.2626	-0.5388	-1.1806	0.2583	1.6386
-0.2589	-0.2162	1.1548	-2.6792	-1.1973	-0.5551	-0.1702	-1.5690	-1.6533	2.3711
-1.1040	0.0462	-1.3020	-0.0301	-1.0134	2.4415	0.3828	0.7004	-0.1860	-1.3155
0.9261	0.3282	1.1284	0.9049	1.1271	1.1463	-2.1408	-0.4295	-1.2538	1.0178
2.0035	0.7039	0.1641	-1.1972	-1.2732	-1.2802	-0.5247	-1.5819	0.4825	-2.5208
0.8375	-0.5823	-0.3391	0.7842	-0.2856	1.1753	-0.1701	0.2960	-0.1099	-2.2621
-0.2829	1.1759	0.6676	-0.4985	-0.1122	0.2444	-1.3184	-0.5538	2.1025	-0.9317
-0.1922	-0.0924	1.0498	1.2507	0.7364	-0.6094	0.3139	-0.8567	-2.3104	1.0240
0.2171	0.6905	0.2018	-0.9239	-0.4136	-0.1313	0.0318	0.2032	1.4773	0.7604
-1.3191	-1.6618	-0.0480	0.0743	0.2409	0.9758	-0.0492	-1.6672	-1.0366	-0.7924
0.5560	0.1231	1.1472	-2.0845	0.7485	-0.5174	-0.3905	0.0276	0.8491	1.1972
-1.5706	0.2128	0.7111	0.3511	-0.0728	-0.2712	-0.8729	0.2279	-0.1911	0.6673
-0.9456	0.6403	1.7273	0.6093	-1.9654	-0.5398	-0.0139	-0.5650	0.1421	0.5615
-0.1165	-1.5958	-0.9846	-0.2547	1.0011	1.1833	-1.0947	0.1415	-0.9593	-1.1241
-0.3379	0.6091	-0.1162	0.0141	0.5700	0.2966	0.0885	-1.3566	-0.0854	0.3437
0.7381	-1.2642	0.5075	0.4373	-1.2345	-0.4483	-0.5594	-0.4709	0.4174	0.2098
-0.8579	0.0646	0.9651	1.2438	0.4470	-0.9279	0.3644	0.3957	-0.6866	0.0309
-1.5837	0.3119	0.6497	-0.2339	0.4062	-0.2241	2.1615	1.5279	0.7134	0.9049
-0.3588	-1.9944	-0.6753	-0.3621	0.4242	-0.2129	-0.5683	0.0644	0.3488	-0.2284
0.6035	-0.2967	-0.8854	0.0490	-1.8515	0.4484	-1.1920	0.7712	-0.0391	2.4790
0.1209	0.9153	-0.6865	-0.2160	0.1633	-0.4249	-0.0635	0.4999	-1.4120	-0.6085
-0.0868	-0.0022	0.7671	-0.7399	1.0856	0.2305	-0.9537	-1.8478	-0.8942	-1.0270
-2.0612	0.6422	0.1293	-0.5036	0.8353	0.7489	0.0347	-1.0474	1.7749	-0.6057
0.6965	0.0510	-0.1628	-0.6597	-0.3132	-1.5787	2.1354	1.2608	1.4539	0.4451
-0.4552	0.5669	-0.6006	1.7116	0.3294	0.0586	-1.5486	0.5342	-0.0617	-0.5186
0.8342	0.2675	-0.6018	-0.6515	-0.4376	-0.8055	-0.3741	-1.1115	-0.3490	-0.5677
0.3083	0.4172	-0.7705	-0.5042	0.3185	1.0291	-0.9206	1.4008	1.3564	0.4660
-0.9023	-0.0349	-1.7964	-1.1040	0.7958	0.0920	0.6723	-0.9108	-0.8027	-1.1455
-0.0739	-1.0777	-0.9581	0.5148	-1.3367	0.5243	-1.0246	-0.3141	0.9019	-1.0278
0.9581	1.8532	-1.6620	1.4843	-1.2946	0.4273	-0.0871	0.2727	0.0847	-0.2205
-1.1082	0.0403	-0.9305	1.0516	0.2338	0.9036	-0.3409	0.2501	0.0306	0.7370
-0.3651	0.9010	0.1728	0.3939	0.3449	0.7790	2.3268	0.6927	-0.2406	-0.9163
-0.0878	1.4724	-0.5195	-0.1867	0.4123	-0.6673	-0.0279	-0.4646	0.5088	0.0256
1.3949	-0.4008	-0.2809	-0.2315	0.9563	0.1550	-1.3613	-2.3085	-0.1026	1.9934
-0.9507	-0.8173	-0.6481	-0.5517	0.3518	0.3320	-1.1557	0.6753	-2.0882	-0.5401
-0.2975	1.0719	0.2245	-0.6437	-0.5597	0.7993	0.9188	0.3066	-1.8867	0.3124
0.3503	-1.1069	-1.1431	-0.0498	-0.7034	-0.4407	-1.7980	0.2801	0.4698	1.1033
1.4696	-1.8542	0.3778	1.4505	-1.8632	-1.3904	0.6922	1.7818	-0.2424	-0.6377
-0.6435	-1.1988	0.8109	-0.0664	1.1235	-0.1914	-0.2877	-1.7789	-2.2507	-0.0532
0.9780	-2.4088	-0.9229	0.8373	1.0870	-1.0420	0.4932	-2.4639	-0.5129	-2.3132
-0.7672	-1.2232	-0.1937	0.1592	0.1617	-1.5393	0.9292	-1.1157	-0.1974	0.0498
-0.4779	0.0708	-0.3298	-0.3235	0.8747	-1.4498	2.7573	-1.2530	-0.3147	-1.2958
-2.8830	-1.3078	1.0889	0.0017	-1.0147	-1.1387	1.1743	0.3187	0.0707	0.0575
0.1218	0.0250	0.7535	-1.3361	0.2371	0.8131	0.7527	1.3997	1.4066	-0.8182
-0.5065	-0.4000	-0.9144	1.8379	0.6187	1.5437	-0.7353	0.4858	-0.3588	-0.4198
0.7468	0.8258	0.0800	-0.1586	-1.3957	-0.9638	-0.3441	-0.1508	0.8359	-1.6932
-0.6942	-0.0367	-1.4847	-0.4460	-0.2375	-0.6544	0.0349	0.2408	1.6872	0.4398
-0.1597	-0.8739	1.6248	-0.5259	1.0277	1.1312	1.4566	0.5325	-1.1900	0.4827
-0.6789	-0.5154	-2.3406	0.6902	1.1743	0.8016	0.1503	0.1682	0.2502	0.7707
0.1046	-1.9092	2.1039	-0.6266	-0.8288	-0.3829	-1.0186	-0.7065	-2.1309	0.3020

(contd.)

Table A2.2 (contd.)

-0.0203	-1.0896	-0.0822	-0.0484	-0.8855	-0.4400	0.4770	0.5804	0.3339	1.0071
-0.9952	0.3641	0.2855	1.0495	1.0266	0.2986	-0.1652	-1.1661	-2.0677	1.7818
-1.0542	-0.2909	-0.2539	1.1788	0.5722	0.2289	-0.8464	-0.0999	0.0906	0.0099
1.2289	-0.7314	-0.9808	0.7221	-2.2937	0.8828	-1.3563	-1.3484	1.4520	0.6500
1.1963	0.9358	-0.7070	-0.3253	0.8227	0.1428	-1.2687	0.1695	0.7293	-0.9262
2.4309	-2.1713	0.0093	1.1039	1.6404	0.7972	-0.1515	1.0183	0.4111	-1.0596
-0.1958	0.3087	0.2570	0.6028	-1.0269	-0.3193	1.5805	-0.0009	1.9738	1.1816
1.5457	-0.5316	0.0106	2.0898	-0.1731	-0.0113	0.3493	0.7890	-2.1588	-0.3330
0.2267	0.0184	-1.9819	-1.1457	-1.0812	0.5891	1.2352	-0.6801	-0.1868	0.0982
-0.2612	-1.4163	-0.8742	-0.2906	-0.7184	1.7338	-1.1651	0.3753	-1.0519	1.1569
-0.8688	-0.8280	-1.2184	-1.4862	1.1289	0.0146	-0.4793	-0.6081	1.5993	-0.8757
-0.6955	1.1250	1.0204	1.0408	-0.9633	-1.0514	-1.3220	-0.1590	1.6191	0.4754
-0.4519	-0.6287	0.6040	1.3683	0.0742	-0.0605	0.1352	-1.4517	0.8944	0.9633
-0.6602	0.6240	0.2646	-0.3819	-0.0084	1.9886	1.1334	0.1009	0.0250	-1.0182
-0.1344	-0.1182	-1.1789	1.2170	1.5683	-1.4281	-0.1510	0.5672	0.0847	-0.0274
1.2640	0.0703	1.0076	-0.3489	1.2867	-1.7079	-0.0700	-1.2367	-0.3906	-1.6084
-1.9019	-0.8625	0.5933	-0.0961	-0.8162	0.3065	0.4936	0.3531	-0.0541	-1.2877
1.2089	0.3296	-0.3806	0.6904	1.3698	1.1400	-0.3950	0.1592	1.2722	-0.6745
0.8973	1.1281	0.4162	-0.1413	0.1180	0.2408	-1.9449	-0.6814	-0.2933	-0.3693
0.8424	-0.1854	-1.0996	2.1092	1.6036	-0.2453	1.3750	-0.2262	0.3836	-0.1857
-0.1649	0.6406	0.2668	0.7307	0.2373	0.6176	-0.0584	-1.1246	1.0290	-1.0958
-1.5278	1.5875	-0.2620	2.2898	0.2245	-0.4571	0.1227	-0.2653	-0.7171	0.8032
0.2816	1.0437	0.0577	0.7579	-0.4012	-0.9804	-1.9196	0.0878	0.2968	-0.3963
-0.2777	1.1807	1.0090	1.1751	-0.3217	-0.7736	0.8179	-0.1823	-0.6675	0.8259
-0.7149	-1.0806	-0.9845	1.4536	1.1983	0.4366	1.3928	0.4531	1.3993	-2.8221
-0.6375	1.8113	0.0048	1.1811	-0.8531	-1.9459	1.0539	-0.6002	-0.0357	-0.4037
-0.6547	-0.6046	0.5760	-0.9200	-0.5664	1.4467	0.0387	-0.4900	0.2345	2.0943
-0.3402	-0.7560	1.3698	0.3334	-0.8621	-0.0175	0.3165	0.8733	0.2367	0.5469
1.7028	0.8892	0.1971	-0.7230	-0.6838	-0.3847	1.3806	-1.8337	-0.9801	-1.0126
-1.5093	-0.5261	-1.7557	-1.5625	1.5003	-0.2731	0.2979	-1.7462	-0.5448	2.1122
-0.9819	-0.0530	-1.6216	-0.3596	1.4475	-0.1451	0.8106	2.0099	-0.9352	-0.1305
-0.6849	0.5243	0.2511	-0.5422	-0.0144	0.1959	0.1298	0.2764	-0.5023	-0.3567
0.7862	0.1847	2.2407	-0.6112	1.0421	-0.5862	-0.0981	0.9954	1.1684	0.5852
-1.0386	1.4752	-0.6907	1.6814	1.6319	0.4324	0.0041	0.3231	0.8942	-0.6281
-0.0153	0.4514	-0.9402	1.7857	-0.8629	-0.5054	0.1679	-0.4131	-0.5420	-0.6098
0.5000	-0.9703	-0.4820	-0.0260	-0.0864	1.9893	-0.9936	-0.1658	0.1289	-0.8644
0.3434	-0.0508	0.8022	-0.1013	-1.5885	-0.2033	1.1547	-1.3096	-1.3238	-1.8837
1.6728	0.7494	0.0275	0.5911	0.1651	-1.2728	-0.7027	0.3364	0.4084	3.1030
-0.1636	0.5416	0.1107	-0.1274	0.5649	0.4839	1.4732	0.8408	-0.2285	0.1433
0.9941	-1.4626	-0.6725	-0.2028	0.2368	0.1373	-0.3076	-3.1763	-0.7637	-0.4866
1.0448	1.2089	3.3747	0.1110	0.6652	-2.1108	0.7605	0.0861	1.1947	-1.3761
0.4274	1.0991	0.0780	-0.6661	0.7646	0.4445	-0.9208	1.2264	-0.5873	-0.1421
0.9986	-1.0404	-0.0846	-2.4126	0.5175	1.9468	1.7186	0.4114	-1.1982	2.0531
3.0454	-0.5958	0.0123	-0.4782	-0.6522	0.7619	0.9982	-0.2226	2.2064	0.0642
-0.3035	1.1293	-0.4138	-1.6477	1.6793	0.2961	2.1618	0.2845	1.2680	-0.5597
-0.1821	-0.2756	1.8769	-0.2809	-0.9521	-0.3845	-0.0187	1.6745	1.2674	-1.9071
-1.9166	1.4852	-1.3189	0.4355	-0.1369	0.3874	-0.7681	-0.8912	0.3885	-1.0778
-0.2656	1.3838	-0.0729	-0.2775	0.7002	-0.0637	-0.1728	1.1653	-0.2935	-1.1375
-1.4067	-0.9086	0.1491	-0.4140	-1.6494	-0.4681	-0.2469	0.3487	0.6849	-1.1233
-2.0422	-0.6236	0.2436	-0.8592	-0.8945	-0.3383	-0.4052	0.1950	0.5988	-1.0363

-0.3704	1.5122	1.0163	0.7257	-1.6742	-0.7219	-0.8545	-0.7573	1.0481	-1.9817
2.3111	0.0824	-1.1191	0.3736	0.0403	-0.7051	0.1438	0.1740	0.1905	-0.2504
-1.2928	1.6848	-0.4283	2.2530	-2.2737	-0.8996	-0.8064	-0.9003	1.5859	1.6073
-0.0680	0.2739	1.7214	0.6838	1.1518	0.5303	-1.4901	3.0725	1.5888	0.0467
0.6079	2.9847	-1.2016	-0.9288	-0.6611	1.0695	0.1654	1.3661	0.3479	0.9196
0.2902	-0.0783	-0.5685	1.7392	-0.8415	-0.2260	1.1166	-1.4446	1.0936	0.4532
1.4997	-0.5749	-1.7208	1.4801	1.1530	1.3533	0.0609	0.4370	0.6746	-0.1997
0.1881	1.2962	-1.2836	-0.4941	-0.0138	-0.3771	-1.5323	-0.1888	-0.0857	-0.0968
-1.0821	-0.8844	0.1861	0.9863	-0.9521	0.6014	0.5074	0.0574	-0.6358	-0.0886
-0.7167	-0.5368	-0.6665	-0.5636	-0.0041	0.1911	-0.8054	0.1000	1.7903	-0.1474
0.3655	-0.2901	2.6336	-1.0341	-0.0606	-0.5633	1.0543	-1.9596	-0.0259	0.5221
0.5883	0.5970	-0.6511	-0.2552	0.2872	-0.1955	0.1831	-2.0900	0.9477	0.1326
-0.1159	0.4299	0.9338	-0.4526	1.4113	0.2747	1.7919	-0.7276	-0.1354	-0.3181
0.8231	-0.5494	0.7583	-1.1222	-1.3870	-0.6589	-0.5949	1.6004	-0.2463	-0.7921
-0.5248	0.4338	0.3673	-1.9040	0.8940	-1.4962	₁0.0825	2.4494	-0.2543	0.0029
-0.4950	1.0338	-1.9280	-0.8219	-0.8489	-0.7477	-0.3854	-0.8685	-0.3945	1.1731
-0.5616	-0.1437	-3.1074	0.8348	0.5676	-0.5008	-0.6084	0.8369	-1.1583	-0.9287
-2.4219	0.2816	-0.8823	-0.5755	-0.3019	0.8035	0.9731	1.5311	0.5951	0.8272
-1.2578	-0.4639	-1.0632	-1.6230	1.0035	-0.1920	-0.4635	-0.3706	-2.3201	-0.1589
0.5826	-0.5626	0.3296	0.2349	0.1979	-0.8413	-1.3038	0.1725	-1.0930	-0.6337

Table A2.3   Pseudo-random exponential variates with mean unity, resulting from setting $E = -\log_e U$, where the U values result from the NAG routine GO5CAF. If we set $X = \mu E$ then X will have the exponential p.d.f.,

$$f_X(x) = \frac{e^{-x/\mu}}{\mu} \quad \text{for } x \geq 0.$$

1.8572	3.4592	1.0743	3.6763	0.0957	0.8504	0.4698	0.1111	1.0375	1.7342
0.5251	0.4830	0.2855	1.6446	0.2093	0.0736	2.0515	0.8111	0.3192	0.0668
3.2347	0.1517	0.2225	0.7250	1.0593	1.5821	0.4814	0.5294	2.4075	0.1771
0.1458	0.5038	2.1378	0.5180	0.2783	5.0171	0.2582	1.8211	1.4403	0.1366
0.8193	1.7958	0.9195	0.1614	1.6518	3.2883	0.7644	5.6159	1.5269	0.9151
0.9144	0.5225	1.0022	0.0600	0.9241	0.1664	1.1494	1.3339	2.4800	0.6779
0.5722	0.2910	0.3653	3.0089	0.8273	2.9518	0.3656	1.4026	0.4348	0.6532
0.1169	0.1251	0.5084	0.5678	1.2514	0.4883	0.8996	0.9814	0.2377	0.2227
0.2150	1.1252	0.0134	0.9149	0.2773	0.4019	0.4207	0.0516	0.7958	0.4141
0.9534	0.2030	0.2731	0.5001	0.0300	1.2154	0.1014	0.6190	0.1772	0.5083
0.4205	0.3205	1.3315	1.7900	0.9252	1.2341	2.6929	2.7975	0.0547	0.4471
1.0300	1.6961	1.7440	0.0587	0.1239	0.6041	0.3901	0.5108	1.2378	0.2289
0.5837	0,1656	5.1307	1.2200	0.6698	0.0185	0.1918	0.9945	0.2738	0.1020
0.1853	0.5972	1.3244	0.5484	1.1069	1.5228	0.4044	1.1261	0.0252	0.3613
0.0885	0.2271	2.8642	0.4575	1.1305	0.3324	1.7452	0.9543	0.3530	0.6410
0.1712	1.5262	2.9107	0.0398	0.3061	0.5607	2.9709	0.1740	0.6842	1.3186
0.5914	2.3353	0.9906	0.1750	1.6806	2.7451	0.1197	2.6395	0.[174	2.1589
0.3200	0.3388	0.6579	0.9450	0.3465	0.4968	0.4769	1.1113	0.0247	0.2832
1.6550	0.6806	0.1719	0.3168	0.6605	2.8238	0.0597	0.3388	1.4157	1.5625
1.5747	0.7014	0.2086	0.0838	0.3878	0.3077	0.9526	0.0637	1.1553	0.5379

(*contd.*)

Table A2.3 (contd.)

0.4185	1.2300	0.8475	0.9256	0.3846	0.6269	1.1691	0.6065	3.7019	2.8047
1.2737	1.4079	0.3099	0.5729	0.2212	1.6878	0.4891	0.4054	0.8321	1.0964
1.1325	1.6126	2.4728	0.9263	0.5674	0.4978	0.9920	1.4322	0.5031	0.5459
2.2052	0.6565	1.0535	2.3828	0.4405	2.0129	0.4054	0.4980	1.2940	0.0433
0.0701	1.0057	0.0953	2.1642	0.1510	1.6404	1.6011	0.9120	0.6383	0.1711
0.1714	3.4097	1.9012	0.0833	1.2162	2.9613	0.1718	0.9876	0.0804	1.3418
0.1110	2.6290	0.1302	0.1271	2.2356	0.6415	2.7836	1.8770	0.3003	0.0654
2.8029	0.2799	0.1435	0.5791	1.2004	2.0409	0.5535	0.8929	0.3591	0.7373
2.3754	0.5073	0.7327	0.4555	2.5153	0.7893	1.8496	1.6322	1.0184	0.4702
0.0041	0.9784	0.1867	2.6548	0.5169	0.5486	0.7794	0.0859	1.0186	0.2387
2.8148	0.7120	0.1445	0.6907	1.1812	1.7549	0.0111	1.2587	0.8341	0.3339
0.0751	1.5227	0.0236	1.9825	0.1325	1.1399	0.3508	0.4137	0.6494	1.8879
0.1474	1.3915	2.7638	0.9405	1.8763	0.4607	2.4377	2.2926	0.0510	2.5343
0.0466	0.4257	2.3732	1.8579	3.4050	0.0480	2.1816	2.0132	0.2805	0.0573
0.0083	4.4788	0.9013	0.4316	0.8931	0.0744	1.6461	1.0149	0.2268	1.4858
0.3649	0.7327	0.6490	0.1008	0.1406	0.4946	0.2665	0.1807	0.2858	0.6055
1.0391	3.6150	0.8064	0.2634	0.0073	0.2048	2.8955	0.1219	2.3012	0.5159
0.5666	0.5058	0.5475	3.3509	1.9163	1.2568	3.3307	0.8367	0.0436	0.2315
6.2686	0.0182	1.3739	0.4793	2.2085	0.0251	0.9262	1.2752	2.6874	0.4613
0.2839	0.0818	2.2989	0.3025	0.2859	1.3330	0.7476	2.5947	2.1751	0.2562
0.7553	0.1958	0.6482	3.1597	0.7049	0.4731	0.2981	0.9889	0.4722	0.3513
0.5294	0.0283	0.1890	1.8211	1.0868	0.8588	0.4446	0.9306	2.0225	1.1114
0.2212	0.0344	0.0365	1.1418	0.3201	2.2238	0.0531	0.0442	0.1343	2.1465
1.3380	0.5349	0.1021	1.8462	1.7937	1.0681	0.1962	0.4485	0.2526	3.5403
0.7430	0.1428	0.9713	0.4437	0.3263	0.1411	3.5229	1.8129	0.9101	1.5802
0.5901	1.8289	0.4033	0.1098	2.2648	2.0006	1.4166	0.1625	2.6244	0.3690
1.5884	0.2744	0.7705	0.6616	1.3312	0.2171	0.4255	0.2302	1.6358	2.4900
0.5440	0.2728	0.6584	2.8711	0.3936	0.3822	2.6415	0.4134	0.4690	2.0769
1.0460	1.0403	0.1330	0.1673	2.1810	0.0238	0.4644	1.7693	0.7595	0.2740
1.7797	0.0356	2.3025	0.7837	0.2083	1.8412	0.0175	0.4230	0.9848	0.9767
0.0223	0.6693	0.8409	0.6856	1.8164	6.6056	1.7954	0.9148	2.0384	0.2238
0.3709	0.4719	3.7242	0.2236	1.4922	0.3131	0.2265	0.1156	2.7813	2.9407
0.7348	1.0829	2.8747	0.0057	0.2925	0.5631	0.8403	1.5795	1.1995	2.0519
0.2760	1.9640	0.8123	0.6752	0.7699	0.0591	1.7510	2.3686	1.2948	0.2045
0.0221	1.2288	0.0772	0.1937	1.5765	0.3312	0.5948	2.7803	0.9025	0.1465
0.5576	2.9216	0.7447	0.8725	0.9607	0.0060	0.0655	0.1110	0.4979	0.8917
0.3385	0.5854	1.9777	0.9338	0.4273	0.5296	0.4373	0.6929	0.1340	0.9231
0.2356	3.6138	0.2794	0.6032	4.4243	0.8540	0.2926	2.1885	0.3949	0.1318
0.5869	2.8752	0.3111	0.0049	0.7986	2.6738	0.4131	2.7132	0.5547	3.4642
0.1087	1.8766	3.7290	1.2497	0.4815	1.3412	0.1428	0.6227	1.0827	0.0069
1.7004	0.3988	3.7033	0.9512	0.6112	4.6257	0.3721	0.1397	4.5403	0.1618
4.1954	3.4717	0.3393	0.2818	0.6557	0.5678	1.4490	0.4012	2.0571	0.3174
3.1273	0.2123	3.2346	0.4919	0.4834	1.2086	2.3945	0.2160	0.7635	0.2176
3.3386	0.5137	2.8209	2.1816	0.1608	0.2346	1.0335	0.1672	0.7071	1.0351
0.4727	0.3168	0.3568	0.1580	2.1702	0.2393	0.2997	5.6316	0.6097	0.4229
1.0465	0.3252	7.1367	0.4669	0.4372	2.2244	0.5163	0.2632	0.4837	0.7102
0.8417	1.5129	2.9798	0.9627	2.2240	1.8782	1.8176	2.5195	0.1047	1.1865
0.4068	0.5325	1.4654	0.2384	0.2246	0.6089	0.1324	1.4154	1.5960	0.3706
0.1895	2.3601	0.3702	0.9370	1.3506	1.7658	0.2896	0.1202	0.4319	0.4869
0.6706	0.7942	0.4774	0.1276	3.9749	1.2749	1.8331	1.3803	0.9938	2.1865

0.2262	2.8267	0.1052	0.2362	1.7041	0.2734	0.4391	2.8258	0.6541	1.7958
3.2953	0.1135	0.1731	0.1587	0.2409	0.6744	0.5500	0.5134	0.8409	1.8484
0.6126	0.3490	0.4303	0.5224	1.8155	0.3143	1.6027	3.3937	0.1878	1.2754
0.3283	0.3141	1.8492	0.8709	0.3674	1.8772	2.1114	2.1362	1.3396	0.1503
1.0728	0.5917	0.2623	0.8331	0.0419	0.2985	0.8854	2.3474	1.3154	0.0851
0.1265	0.1137	1.0678	3.5661	0.5626	1.4893	0.8538	0.0404	0.3173	1.5016
2.0436	0.3795	1.2001	0.7641	0.7103	0.0398	1.5690	0.6858	0.4605	0.1801
2.5115	0.6820	2.0325	0.5374	0.2298	0.2618	0.9126	0.4163	1.8925	1.0032
0.3692	0.0972	1.1210	0.7469	0.5936	0.3059	2.1837	0.6422	1.9291	0.0381
4.1908	1.9557	0.9553	1.2938	0.2662	0.8478	2.6194	0.3188	0.3720	0.4913
1.7484	0.4912	1.4788	0.3386	0.1233	0.4302	2.5079	0.6642	0.4174	0.9704
0.8150	0.5232	0.2354	0.6321	0.8006	0.0826	0.8072	0.4663	0.7566	0.2717
0.8636	1.4733	0.3092	0.0898	0.2649	2.6429	0.1031	0.7353	0.0149	0.2657
0.5870	1.7315	0.3341	0.1121	0.2112	2.0134	1.2306	0.9863	3.4826	1.9196
0.6496	1.4335	0.3021	1.3425	0.0060	0.7569	0.0021	2.9950	2.2585	1.5043
0.5196	0.1848	0.8745	1.6484	0.1868	0.1593	0.0461	1.1629	1.1766	0.9316
1.4844	0.8120	0.7313	1.8527	0.6266	0.3292	0.2989	0.3474	0.1179	0.8653
0.5861	1.7172	1.3469	0.4608	0.5784	0.5551	0.1093	0.0462	0.8539	0.6057
0.4101	1.4035	2.1997	1.1978	0.3406	0.4611	1.5759	0.3700	0.7084	1.1784
0.5488	1.0422	1.0366	0.6334	0.1717	0.3166	0.1696	0.0643	0.2461	2.4901
0.1236	0.5649	0.5670	0.2612	0.5140	0.2137	0.8628	0.0793	1.3068	0.2376
0.6278	0.6140	0.6958	0.0740	0.0807	0.2824	0.9182	0.1132	0.1802	0.0479
0.5319	0.7430	0.7235	0.3241	1.2398	0.2892	2.5901	0.2241	0.0082	2.3213
0.5642	1.4134	0.1912	2.8957	1.8297	1.1298	1.2301	1.9680	0.3615	0.4673
2.3539	0.0207	1.5929	0.0600	1.8815	1.7101	0.0981	0.1547	1.9830	0.2007
1.9547	1.0600	1.6758	0.6544	0.1895	0.3309	4.3876	3.2156	0.5301	0.5060
2.2706	0.1829	0.5427	3.0325	0.9998	0.0633	0.0367	1.2302	1.5948	0.5088
0.4744	0.2108	0.9395	2.1109	0.8330	1.0813	0.0515	1.1784	1.8747	1.4503
0.7436	0.1659	0.3701	0.3925	1.1715	1.1459	0.3943	1.9392	3.6668	0.7452
4.2728	0.5936	1.6973	1.4687	2.7646	2.4024	0.6428	0.2875	1.6920	0.0018
0.0957	0.1430	0.2623	0.0217	0.2225	1.5412	3.3999	0.0035	0.9985	0.5645
0.9998	0.2178	0.0795	0.7800	1.3048	0.8586	0.3137	0.3108	0.0850	2.6314
0.5791	1.1190	0.3662	2.0195	0.2196	1.1779	2.4838	0.1701	0.0454	0.9289
3.1114	0.0944	3.9372	3.1869	0.8534	0.8903	0.0448	1.2545	0.8944	5.1982
1.0353	0.6959	1.8216	0.1180	1.0511	0.3947	0.4461	0.2814	1.4110	1.7693
1.9735	0.1029	0.4293	0.7401	0.8278	1.3691	0.4680	1.1357	3.6728	0.3248
0.1142	0.1513	0.8532	0.7742	0.7439	0.4131	0.7486	1.8577	0.2026	0.2623
2.5522	1.0665	0.0970	0.6111	2.3773	3.3147	0.9087	4.3171	1.2309	1.1514
0.1614	1.0738	0.0914	0.1081	1.1710	0.0784	0.3549	2.2707	1.8493	1.3236
0.0816	1.1954	0.6541	0.0907	0.4436	2.3404	1.4163	1.4363	0.7996	3.2709
1.2960	1.3858	0.5328	1.6771	1.0142	1.7405	0.8354	2.2597	0.2152	0.1080
0.2821	0.1510	1.0522	1.1828	0.1748	0.0983	1.7560	0.2089	0.7639	1.8624
1.5536	0.5723	1.7027	0.5789	0.3960	0.4795	0.8997	0.1382	0.0579	1.4898
0.4561	0.3852	0.7866	0.2477	0.3633	0.8102	0.0418	1.1972	0.2539	0.2425
0.7597	2.0661	3.5845	1.1805	0.0867	0.6872	0.4028	0.4954	2.3350	0.0021
1.5397	0.4409	0.2984	0.6214	0.9617	0.4515	0.1254	2.3894	1.6741	0.6003
2.5211	0.0564	2.3985	1.5490	1.2817	0.4700	0.2423	1.9756	1.0043	1.0038
0.3574	0.9720	0.6930	0.3806	0.0809	0.6515	0.8627	2.3706	3.3770	1.3854
1.0960	0.1730	0.6055	0.9324	0.2532	0.0956	0.1002	1.4651	0.0865	0.5684
1.4266	0.2631	0.4505	0.5244	0.2732	0.0784	0.0839	0.1109	0.5341	0.1828

(contd.)

Table A2.3 (contd.)

0.0020	1.5506	0.1412	1.3845	1.0397	0.3017	1.8554	0.4013	1.2064	0.5483
0.8056	0.8211	4.3835	0.7733	2.6632	0.5644	1.0729	1.6415	1.0340	0.9698
2.3716	0.7611	0.2588	0.7346	0.0218	2.3108	0.2370	0.2574	3.5895	1.7510
0.8240	0.0189	2.9614	0.9967	3.6090	3.9467	0.9286	1.0109	1.5118	0.3809
0.0273	1.3570	1.2247	1.4077	0.2343	0.9078	1.0065	0.7209	0.9502	0.7829
0.2267	0.0247	0.3038	0.2400	0.2400	0.7751	2.6584	1.3811	0.8652	1.3459
0.5598	0.1680	0.8589	0.4105	1.0735	2.7258	0.6697	1.8197	0.9354	2.8150
2.0794	0.0497	0.0814	0.2887	0.2973	0.9280	0.3838	1.5025	0.0980	1.2042
0.1951	0.5824	0.0336	3.2740	1.0851	0.3307	1.2155	0.9797	1.6466	2.8243
0.0868	0.1754	0.9027	0.7336	0.1776	0.2053	2.1264	0.1210	6.2721	1.0742
0.7221	0.1250	1.2740	5.3879	0.4954	0.9579	0.5828	2.5294	1.9390	1.8485
0.0204	0.0423	0.0501	0.4905	0.2521	1.1115	4.3949	2.2700	0.3386	0.8384
0.8472	3.1575	2.8907	0.4862	0.3371	1.4351	0.5208	0.3927	1.6363	0.4697
0.9920	0.9528	1.0351	0.6715	0.1306	0.5560	0.4613	0.0649	0.0304	0.9037
0.3503	0.4055	0.2853	0.7234	0.4398	3.6355	0.6676	0.5404	2.4822	0.1294
1.0599	1.0631	1.0945	0.3827	0.0704	0.2922	0.4514	0.1156	1.2851	0.2050
0.2141	0.6771	0.4807	0.7462	0.8734	1.2791	1.2755	3.4165	0.6051	0.2491
0.1566	1.4162	0.7977	1.4290	0.2720	1.4363	0.2324	2.2534	1.2951	1.6273
0.4487	1.0669	2.4955	0.0868	1.0839	0.0637	0.1374	0.5649	0.6746	0.3855
0.3972	0.3519	0.2145	0.5495	0.0113	0.3364	4.7241	0.2439	3.8726	0.1497
5.5211	1.9647	0.4753	0.2350	0.1076	0.6676	1.4092	0.8066	0.4367	0.1922
1.1163	0.1145	2.2686	0.3810	0.8976	1.6584	0.6624	1.0529	0.8873	2.4664
1.3387	0.1553	0.2712	0.8206	0.1716	0.4042	0.6780	0.0934	0.4501	1.0862
1.6541	0.9755	0.1364	0.1942	1.1767	0.2006	0.1681	1.0499	0.2130	0.0662
0.2194	3.8295	1.7801	0.7952	0.0226	0.2201	2.2478	0.2974	0.6153	0.1645
0.8518	2.1984	0.4149	0.9990	1.3600	0.1521	0.9890	4.0700	1.2248	0.0405
0.6637	2.1369	2.9399	0.4496	1.4639	1.1258	0.6278	1.7280	0.0522	1.8045
2.0584	0.5447	0.3175	0.1708	1.5913	1.2572	0.0060	1.7244	2.9564	1.1627
0.3382	2.5738	0.5194	0.0429	1.6915	0.2239	0.1227	0.1405	0.8199	1.6706
2.0646	0.0767	0.3615	0.6232	0.1053	5.5540	1.6225	0.0757	2.7629	0.2297
0.1164	0.9222	0.8609	0.4352	0.2178	0.0403	2.2668	1.7038	2.2501	3.7455
0.4037	0.5324	2.2477	0.5693	0.6259	1.6627	0.7400	1.3183	0.2363	0.6818
1.2446	0.5208	0.1301	0.7437	1.2288	0.0627	1.2719	0.6786	0.3839	0.0121
0.1328	0.0227	2.1513	0.3881	0.4592	0.1841	3.0301	0.4392	0.0048	1.3458
0.2047	0.3365	0.3302	0.5236	0.0138	0.1869	2.8403	0.5162	0.8853	0.4528
0.7170	0.2295	0.0153	0.5862	0.6261	0.0188	1.1097	0.9166	0.1952	0.2451
1.2079	1.4569	0.2900	0.6899	1.2776	0.1906	1.1876	0.9581	0.2968	0.8214
0.2446	0.3464	0.7005	3.2946	1.3201	0.2762	0.2149	0.0184	2.1297	1.7916
0.0430	5.8433	0.0135	1.8503	0.8965	0.8207	0.3920	1.4928	0.0278	0.0293
3.0440	0.0599	0.2378	0.6456	0.7093	0.5758	0.5150	0.7358	0.9877	0.9797
0.9938	0.2199	3.7546	0.5525	0.7431	0.7999	0.3385	1.0272	1.8057	1.7799
1.8878	0.8093	2.0029	2.6923	0.2624	0.3475	5.4931	3.2558	0.9715	1.3978
0.0627	0.9066	0.4246	0.9075	0.9085	1.7473	2.2634	1.1941	0.6448	0.2894
2.7327	0.2536	1.0699	0.3799	0.1146	0.7085	0.2637	1.8162	0.0179	0.4552
1.2805	0.3679	0.4568	1.1520	0.6807	0.5302	0.6891	0.8711	0.1018	4.9238
0.4077	0.6182	2.7196	0.0453	0.0994	1.1690	0.1256	0.6729	1.5018	1.3534
1.6524	0.0578	0.9401	0.4844	1.7384	0.4012	0.0896	0.2522	0.7065	0.7156
3.0129	0.5579	0.4581	1.0525	0.2073	0.1652	0.3050	0.2480	0.8802	0.2568
2.5483	0.2625	0.3315	1.0044	2.2402	0.6137	0.1468	0.2303	0.3303	3.8178
0.9173	0.3956	1.3693	4.4865	0.3643	1.8122	0.1770	0.0925	1.3591	2.4642

2.2040	0.6658	2.3207	0.1306	0.5352	2.0896	0.3820	0.7302	0.0935	2.1066
0.7918	2.3115	1.0047	0.0639	1.5118	0.4024	0.0268	2.3029	1.0775	0.0326
0.5436	3.1605	0.1936	1.2563	0.1289	0.0199	0.0419	0.4355	1.7336	0.3374
0.7121	1.1316	0.2907	0.2430	0.2678	1.0157	0.1725	5.8671	0.9236	0.7738
0.9205	0.4097	0.8732	0.7612	1.2238	0.0666	1.1736	0.5410	0.2813	0.3833
0.5685	1.8259	0.1638	0.6100	1.3283	0.2501	0.3153	1.1460	0.4860	0.3625
0.8819	1.3797	0.1310	1.6174	3.8816	1.1261	4.3691	0.5803	0.1833	1.1955
0.7753	3.6829	1.5551	0.5724	0.8455	0.4765	0.2525	2.3464	0.3067	1.3141
0.0397	0.9742	0.3634	0.6421	0.9689	1.3898	2.1468	0.2756	0.0560	0.3212
0.2504	1.2670	0.1704	0.6064	1.0659	0.8960	0.1546	0.0858	1.1147	4.2771
1.7969	0.6427	2.0124	0.6597	0.1216	1.6819	1.6728	3.2960	0.1293	1.6918
0.0224	1.0260	0.5243	0.1674	0.9266	0.2525	0.3436	0.1084	0.5761	0.3710
1.4744	0.6261	0.0737	0.5459	0.0148	0.8895	0.8107	2.7103	0.0233	1.5065
2.9960	1.1638	0.8562	0.9402	0.4523	1.6983	2.5498	0.3047	0.4901	0.1664
0.0432	0.1314	1.8979	0.0683	0.3236	0.2117	0.1061	0.8164	2.9188	2.2930
0.0121	0.1066	1.2953	0.7185	1.6484	0.2372	0.5439	0.2761	0.3360	0.4578
0.2233	0.1084	0.2725	0.5746	0.3096	1.2195	0.1693	2.8480	1.7535	3.1994
0.5355	0.0281	0.4577	0.9886	0.3570	0.4880	0.7886	0.6464	6.9621	0.4395
0.5545	0.1149	0.0821	1.0438	0.0970	1.1984	0.6901	0.5444	0.0791	1.5138
0.0163	1.6258	0.4038	0.5542	0.2152	0.4569	0.4728	2.1605	1.1753	0.0358
1.1129	0.4219	0.4613	0.7902	1.9451	0.0682	0.0067	0.0892	1.7962	0.1651
0.4483	3.2982	0.6813	0.4375	0.1428	0.6159	0.8452	2.8102	3.2282	0.7926
1.7602	0.3388	1.2218	0.4790	0.8864	2.4507	2.1559	2.3118	0.1544	0.1709
0.6097	1.2873	0.6693	1.6206	0.6881	0.5629	0.2347	0.2352	3.8297	0.3592
1.7148	1.8480	0.2305	1.8525	0.1288	0.3194	0.5693	0.8343	0.2692	2.5546
0.7045	0.0797	0.9792	1.7046	1.4487	0.4401	0.9002	3.3004	0.4273	0.7470
5.1411	1.5252	0.4152	0.3399	1.5862	1.1284	0.2603	0.0219	1.2524	0.8409
0.5345	2.0084	3.7378	0.0091	0.5290	2.5321	0.3915	0.2326	1.1834	2.2947
1.8014	0.6379	0.2535	2.4549	2.1129	0.9765	0.5183	0.3234	0.9804	0.9316
0.0965	0.0863	0.3828	2.2234	0.6006	2.5960	0.5562	0.7388	0.4843	0.3086
0.0696	0.0164	0.8179	1.4533	0.4580	0.0882	0.5530	0.1249	0.9188	1.0442
2.9240	1.6763	1.0844	0.2994	0.0216	0.3540	0.2473	1.9228	0.9097	0.0706
1.0356	0.8787	4.1542	1.4730	0.7638	3.1557	0.0749	0.8124	3.2439	2.1204
1.7099	0.4228	0.3146	0.0427	4.9240	0.4535	0.0519	0.2886	1.0888	2.9278
1.9067	0.2603	1.2458	2.1826	0.0330	0.3630	0.7221	0.4798	0.6748	1.0014
0.7160	0.0519	1.1099	0.1611	1.8574	0.4757	0.3693	0.1072	0.3277	0.0193
1.2309	0.2504	0.0629	1.9590	0.5589	1.2354	0.9823	0.0538	0.6819	0.1222
0.4223	0.5136	0.8320	0.5826	0.5883	0.8406	0.2686	0.1345	1.3802	0.0766
1.4697	1.4338	0.1059	0.2931	1.7057	0.0158	2.6117	1.3460	0.9927	0.9198
1.1331	2.1323	0.0701	0.1078	0.3921	1.7266	0.1386	1.4767	0.7636	0.5444
0.3738	2.3786	0.4930	1.4381	2.6188	0.9457	0.1261	0.8035	1.2941	1.3657
0.8953	0.1155	5.6212	1.2060	0.5895	1.6094	0.3086	0.3622	0.4592	1.0345
0.3922	0.3893	0.7624	0.6278	3.3524	2.5566	0.2322	0.1354	3.1449	1.4659
1.0988	0.4790	0.6383	0.2182	2.2340	0.3634	0.4640	0.0696	1.2084	0.2201
0.2146	0.1600	0.7314	1.0836	1.2925	1.8054	0.3488	0.6463	0.2516	0.9840
0.7489	2.6789	0.6826	0.2122	0.6610	0.0736	2.6766	0.0830	0.2461	0.6852
0.8515	0.3768	0.3656	0.0757	0.0308	0.6919	0.7830	0.3295	3.6192	1.5458
0.1258	1.0159	0.0182	0.1844	0.4960	0.2649	0.4626	0.5512	0.1497	0.0108
2.3288	2.3537	0.0419	1.5865	0.2152	0.2077	2.1020	1.3048	0.3295	1.0944
0.1882	0.3909	1.1431	0.1479	0.0602	1.3079	0.3594	0.7733	0.2706	2.4657

(contd.)

Table A2.3 (contd.)

3.5202	0.2246	0.2192	0.9789	0.6949	0.1418	1.1849	2.1674	0.1957	3.9806
0.0287	2.6889	0.4187	1.8645	0.3404	1.3758	0.8467	1.2666	1.1033	0.2709
1.2280	0.2721	0.4442	0.1851	0.0453	1.9203	0.0124	0.0920	0.3002	0.2983
0.0297	0.7480	0.8055	0.2170	3.1127	0.7236	0.1325	4.1433	0.8884	0.3438
0.0591	0.2924	0.2276	0.3067	1.5974	0.4718	1.7170	0.0248	0.8436	0.9206
1.0641	0.8135	0.6843	0.8648	0.3294	0.0637	0.9801	1.4394	0.8895	0.3319
1.7783	0.1189	0.5233	0.3184	3.1181	1.3062	0.6397	0.2234	4.1162	4.5862
4.5601	0.5416	0.3030	0.3294	0.7134	1.1931	0.3981	1.3931	0.2265	1.0007
0.1466	0.3228	1.4144	0.0524	0.5553	1.2189	0.9171	0.1416	1.4753	0.5104
0.6415	0.7804	1.0163	0.6759	1.5126	0.4745	1.3837	2.1061	0.5932	0.1320
0.5116	0.6517	1.4896	0.0848	1.5269	0.5994	0.5370	0.2635	0.0426	0.4543
0.8782	0.8502	1.5558	0.2353	1.9698	0.5535	0.3781	1.6151	1.8896	0.6320
0.3076	2.5999	2.2703	0.4799	0.8810	0.7712	2.5527	0.0961	0.1181	0.6099
0.5845	2.8184	0.9106	0.6406	0.0958	2.5912	0.2465	1.6289	0.4201	1.1714
0.4383	1.2734	0.2024	0.2033	0.2913	2.1051	1.1084	0.8697	0.0176	0.1087
0.3505	0.6409	2.3410	1.3040	0.5340	0.4504	0.9270	3.5887	0.0176	4.5769
1.2887	1.4947	1.1008	1.2816	0.2288	0.5403	0.9440	1.6718	1.2628	0.2549
2.2799	1.8089	0.7045	0.4570	1.3930	2.3095	1.3981	0.0348	1.9592	5.4993
0.8962	0.1314	0.3085	0.4515	0.2083	0.2435	0.2162	2.0085	1.2198	2.7758
1.5377	0.9400	1.7723	0.6106	0.0424	0.0028	0.2508	0.9324	0.5952	2.8823
0.7063	0.7870	0.5889	0.2872	0.5340	4.6029	3.3972	0.1139	0.5627	0.4867
1.2034	0.9080	0.2172	0.0169	0.0644	1.8354	0.7608	1.8465	1.2756	1.1670
0.4955	0.2511	0.0872	1.5460	1.8302	0.9005	0.1636	0.3825	3.2568	0.3192
0.4295	0.5593	0.4639	2.7039	0.4622	0.6043	2.0626	0.2370	0.8121	0.1569
0.2431	0.1724	0.2445	0.6547	0.1757	0.2650	0.7652	0.3943	0.3859	0.0425
1.0792	1.1186	0.4233	0.9735	0.2673	0.5750	1.5016	2.7067	0.8731	0.3399
0.4208	1.1568	0.9394	0.4712	0.3919	0.4354	0.2549	0.1986	2.6522	0.0408
1.9315	2.2283	0.4783	0.5306	1.5770	0.2343	0.8735	4.0513	0.2107	0.7635
3.6057	4.8781	0.8152	2.5874	0.2012	0.8300	0.2237	2.2275	1.2300	0.1360
8.4972	1.0523	1.6771	3.3247	0.0836	4.2645	0.9052	0.2409	1.8666	0.0302
0.1932	0.3425	0.4364	1.3182	1.8740	0.3963	0.0274	1.4935	0.9083	0.5664
1.0321	0.7576	0.1101	0.7726	1.9979	0.8429	2.7080	0.6552	0.9545	0.3399
0.1924	0.6298	0.2231	0.0313	0.0132	1.2678	0.6509	0.6169	2.6451	1.3429
0.4964	0.1706	0.8262	1.0182	1.1400	0.0837	0.2614	0.0949	0.7188	1.2772
0.5938	2.2960	0.0643	1.6260	1.1426	3.5810	1.2046	0.4621	1.6437	1.7404
0.7921	2.0433	0.8042	2.3699	0.2993	0.9512	0.2510	3.4331	0.0986	2.9672
0.0627	2.7120	0.1561	4.4399	0.2363	0.5129	0.4065	0.0241	0.5858	0.4217
0.1051	0.2881	0.3167	1.2225	0.1422	0.2583	0.0032	1.4266	0.9464	2.2253
1.0219	3.1888	1.3849	0.6060	0.0858	0.7613	1.4090	0.3017	0.5856	0.2055
2.0679	0.1217	0.1310	0.5555	1.4606	0.4404	0.8616	0.5933	0.4423	0.0088
0.8709	0.8439	0.0513	0.8451	0.5618	1.2807	3.4832	0.2536	0.2057	1.0591
0.6493	0.3464	0.4102	0.8092	0.0219	1.0808	0.2915	0.0460	2.8171	3.6311
2.3879	1.1853	1.2833	0.3341	0.8964	1.6769	0.9345	1.6264	0.2299	0.0826
0.0137	2.0478	1.1258	0.1467	0.5040	0.0605	0.0947	0.0074	0.2776	1.3126
1.2043	0.0229	0.8479	0.1928	0.6007	0.6812	0.2069	2.4509	0.0461	0.1587
0.8571	0.0065	0.9830	4.8011	1.3846	0.7025	0.3283	0.7490	0.6792	0.2111
0.1703	0.0516	0.5281	2.2538	1.5120	0.1247	1.1168	0.2596	0.9805	0.3836
1.1553	0.0295	0.5172	0.9408	1.1564	0.3959	3.6302	1.0429	0.0681	0.8513
0.5860	0.1763	0.0754	0.3237	0.6777	4.4005	0.0454	0.2653	1.3221	0.9655
0.4648	0.1211	0.1251	0.3464	2.1544	3.0242	1.2974	0.4552	0.0796	1.8947

0.3225	1.0650	0.0161	1.6670	0.4674	0.3242	0.0734	0.7709	0.2395	0.3337
1.1911	2.7358	0.6790	0.7600	0.5138	0.1291	0.4222	0.6654	0.6381	1.3100
0.3451	0.2419	3.6555	0.0186	0.0236	0.3375	1.8152	0.1544	1.4845	1.2158
3.5440	2.8569	0.5313	0.4481	0.2044	0.3136	1.2577	1.2306	1.9160	0.0678
0.9039	0.2563	2.6067	0.2537	0.0510	0.2955	0.1571	1.6562	0.2477	1.0020
1.6963	0.9860	2.5091	1.5547	0.9164	0.2915	2.7050	0.4800	0.8942	1.4413
0.2417	0.6101	0.7853	1.4039	0.8590	1.1795	1.2945	0.4256	0.5444	0.4838
0.5561	1.2005	0.0198	0.3045	0.4920	0.0642	0.2946	1.1166	2.6119	0.9079
3.0961	0.5537	0.7118	0.6817	1.6405	1.4653	3.7672	1.2467	1.4338	1.4290
1.1939	1.3661	2.8130	2.9333	0.3383	0.2054	0.4964	0.1413	2.1046	1.1361
0.1992	0.7719	0.2674	1.2932	0.4058	1.1339	1.3010	0.3364	0.9962	0.5214
0.3979	0.6721	0.1517	1.3605	0.0244	0.9889	0.9673	0.4509	0.0230	0.0679
0.4236	0.1123	0.0203	0.1660	1.4428	2.5517	1.8990	0.7391	0.3259	0.1074
0.0059	0.1269	0.0016	0.1855	0.2136	0.7185	0.3999	0.1773	1.8161	0.4270
0.8531	0.9818	0.3410	0.2880	0.3711	1.4076	2.5961	1.0038	0.1402	2.0252
0.2479	3.0019	0.2209	0.5164	1.5471	1.0552	0.2849	0.3364	1.1884	1.1964
1.6535	0.1193	1.3134	0.1291	0.3013	0.8232	2.3469	1.4267	0.8353	1.2455
1.3252	2.5751	2.3465	0.1882	1.8173	0.8323	0.1345	2.3552	0.5207	1.4660
1.9167	0.1135	0.1964	2.2752	0.2066	1.1800	1.0843	0.2859	0.9249	1.9323
1.8726	1.1492	0.4916	1.3554	0.0602	0.9880	1.5715	0.1458	1.1776	2.0868
0.1910	0.7606	0.8482	1.2514	4.2987	0.6483	2.3567	1.0607	0.2287	0.0882
0.0400	0.4862	0.2019	0.0050	1.9056	0.1178	0.2582	3.2765	1.5651	4.7954
0.1453	0.3025	2.4653	4.3034	1.8665	0.2716	1.0640	0.6671	0.5612	1.8890
0.3911	0.3436	1.1703	0.3679	0.1875	1.6154	1.7808	0.8341	1.7471	4.9056
0.0852	2.9668	0.4222	0.1814	0.5304	1.1749	1.2840	0.9586	4.5554	1.3274
1.0043	1.1371	0.8428	0.6316	0.5952	0.9318	0.4654	2.1684	0.3810	1.5733
0.3512	1.1913	0.2460	2.3531	1.8260	0.8944	0.3884	0.3803	0.6048	1.6582
0.0116	0.2001	1.2422	1.2906	1.0025	0.3055	1.9163	2.3040	0.2066	1.0513
0.6008	0.3449	0.3110	0.0055	0.6480	1.5093	0.7446	0.1825	0.1556	1.1540
0.7628	2.5791	0.0502	0.3400	1.4389	0.2751	0.0550	0.5778	4.1245	0.7293

SOLUTIONS AND COMMENTS
FOR SELECTED EXERCISES

Chapter 1

1.1 (a) As n increases, the ratio r/n is seen to stabilize. The 'limiting frequency' definition of probability is in terms of the limits of such ratios as $n \to \infty$.

(b) See, e.g., Feller (1957, p. 84). Note the occurrence of 'long leads', i.e. once $(2r - n)$ becomes positive (or negative) it frequently stays so for many consecutive trials.

1.2 Buffon's needle is discussed further in Exercises 7.1–7.4. Estimate Prob(crossing) by:

no. of crossings/no. of trials

and solve to estimate π ($\hat{\pi}$, say).

(i) $\dfrac{2}{\hat{\pi}} = \dfrac{254}{390}$; $\hat{\pi} = 3.071$

(ii) $\dfrac{2}{\hat{\pi}} = \dfrac{638}{960}$; $\hat{\pi} = 3.009$

1.3 Effects of increasing the traffic at a railway station, due to re-routing of trains.

Effects of instituting fast-service tills at a bank/supermarket.

Effects of promoting more lecturers to senior lecturers in universities (see Morgan and Hirsch, 1976).

The improved stability, in high winds, of high-sided vehicles with roofs with rounded edges.

Election voting patterns.

1.5 The story is taken up by McArthur *et al.* (1976) (see Fig. 8.3).

1.6 Appointment systems are now widely used in British surgeries. For a small-scale study resulting from data collected from a provincial French practice, see Example 8.1.

1.7 If a model did not result in a simplification then it would not be a model. Clearly models such as that of Exercise 1.6 ignore features which would make the models more realistic. Thus one might expect more women patients than men (adults) in the morning, as compared with the afternoon. Were this true, and if there was a sex difference regarding consultation times, or lateness factors, the model should be modified accordingly.

1.9 A system with small mean waiting time may frequently give rise to very small waiting times, but occasionally result in very large waiting times. An alternative system with slightly larger mean waiting time, but no very large waiting times, could be preferable. An example of this kind is discussed by Gross and Harris (1974, p. 430). In some cases a multivariate response may be of interest (see Schruben, 1981).

1.10 The following two examples are taken from Shannon and Weaver (1964, pp. 43–44).
  (a) Words chosen independently but with their appropriate frequencies: 'Representing and speedily is an good apt or come can different natural here he the a in came the to of to expert gray came to furnishes the line message had be these.'
  (b) If we simulate to match the first-order transition frequencies, i.e. matching the frequencies of what follows what, we get: 'The head and in frontal attack on an English writer that the character of this point is therefore another method for the letters that the time of who ever told the problem for an unexpected.'

Chapter 2

2.1 $X = -\log_e U; f_X(x) = f_Y(y) \left| \dfrac{dy}{dx} \right|$

in general, and so here,

$$f_X(x) = 1 \left| \frac{du}{dx} \right| = \frac{1}{u^{-1}}$$

but $u = e^{-x}$, and so $f_X(x) = e^{-x}$, for $x \geq 0$.

2.2 Note that for the exponential, gamma and normal cases X remains, respectively, exponential, gamma and normal.

2.3 $f_X(x) = e^{-x} \qquad$ for $x \geq 0$

$W = \gamma X^{1/\beta}$

$(W/\gamma)^\beta = X$

$f_W(w) = f_X(x)|dx/dw|$

$dw/dx = \dfrac{\gamma}{\beta} x^{1/\beta - 1}$

$f_W(w) = \dfrac{\beta e^{-x}}{\gamma} x^{1-1/\beta} = \dfrac{\beta}{\gamma} e^{-(w/\gamma)^\beta} \left(\dfrac{w}{\gamma}\right)^{(\beta-1)}$

i.e. $f_W(w) = \dfrac{\beta}{\gamma^\beta} w^{\beta-1} e^{-(w/\gamma)^\beta}$ for $w \geq 0$.

2.4 $Y = N^2$; $\Pr(0 \leq Y \leq y) = \Pr(-y^{1/2} \leq N \leq y^{1/2})$

$$= \Phi(y^{1/2}) - \Phi(-y^{1/2})$$

and so

$$f_Y(y) = \tfrac{1}{2} y^{-1/2} \phi(y^{1/2}) + \tfrac{1}{2} y^{-1/2} \phi(-y^{1/2}) = y^{-1/2} \phi(y^{1/2})$$
$$= y^{-1/2} e^{-y/2} / \sqrt{(2\pi)} \qquad \text{for } y \geq 0, \text{ i.e. } \chi_1^2.$$

Note that $Y = N^2$ is not 1–1 and so rote application of Equation (2.3) gives the wrong answer.

2.5 $M_Y(\theta) = \mathcal{E}[e^{Y\theta}] = \displaystyle\int_0^\infty \dfrac{y^{-1/2} e^{y(\theta-1/2)}}{\sqrt{(2\pi)}} \, dy$

Suppose that $\theta < \tfrac{1}{2}$. Let $z = -2y(\theta - \tfrac{1}{2})$,
i.e. $z = y(1 - 2\theta)$; $dz = dy(1 - 2\theta)$,

$$M_Y(\theta) = \int_0^\infty \left(\dfrac{z}{1-2\theta}\right)^{-1/2} e^{-z/2} \dfrac{dz}{\sqrt{(2\pi)}(1-2\theta)}$$

$$= \dfrac{(1-2\theta)^{-1/2}}{\sqrt{(2\pi)}} \int_0^\infty z^{-1/2} e^{-z/2} \, dz,$$

i.e. $M_Y(\theta) = (1-2\theta)^{-1/2}$ (see also Table 2.1).
  Hence, in the notation of this question,

$M_{\sum_{i=1}^{n} Y_i^2}(\theta) = (1-2\theta)^{-n/2}$

i.e. (from Table 2.1), the m.g.f. of a χ_n^2 random variable. (See also Exercise 2.21.)

2.7 Solution:

If $S = \dfrac{1}{n} \displaystyle\sum_{i=1}^n Y_i$ then $f_S(s) = \dfrac{1}{\pi(1+s^2)}$ for $-\infty \leq s \leq \infty$, i.e.
S has the same Cauchy p.d.f. as the component $\{Y_i\}$ random variables.

2.8   (a) $N(\mu_1 + \mu_2, \sigma_1^2 + \sigma_2^2)$

(b) Poisson with parameter $(\lambda + \mu)$

(c) If $\lambda = \mu$ the solution is given by Example 2.6, with a generalization given in Exercise 2.6. If $\lambda \neq \mu$,

$$M_{X+Y}(\theta) = \left(\frac{\lambda}{\lambda - \theta}\right)\left(\frac{\mu}{\mu - \theta}\right) = \frac{1}{(\lambda - \theta)}\frac{\lambda\mu}{(\mu - \lambda)} + \frac{1}{(\mu - \theta)}\frac{\lambda\mu}{(\lambda - \mu)}$$

and so,

$$f_{X+Y}(z) = \left(\frac{\lambda\mu}{\mu - \lambda}\right)e^{-\lambda z} + \left(\frac{\lambda\mu}{\lambda - \mu}\right)e^{-\mu z} \qquad \text{for } z \geq 0$$

i.e. a mixture of exponential densities.

While the solutions for (a) and (b) also follow easily from using generating functions, the solution to any one of these parts of this question is also readily obtained from using the convolution integral. For (a), if the convolution integral is used, note that:

$$\int_{-\infty}^{\infty} \exp\{-\tfrac{1}{2}(ax^2 + 2bx + c)\}\,dx = \sqrt{\left(\frac{2\pi}{a}\right)}\exp\left(\frac{b^2 - ac}{2a}\right).$$

2.9 $\Pr(X = r \mid X + Y = n) = \dfrac{\Pr(X = r \text{ and } Y = n - r)}{\Pr(X + Y = n)}$

$$= \frac{e^{-\lambda}\lambda^r}{r!}\frac{e^{-\mu}\mu^{n-r}}{(n-r)!}\frac{n!}{e^{-(\lambda+\mu)}(\lambda+\mu)^n} = \binom{n}{r}\frac{\lambda^r\mu^{n-r}}{(\lambda+\mu)^n}$$

i.e. the conditional distribution of X is binomial:

$$B(n, \lambda/(\lambda + \mu)).$$

2.10 $\Pr(Z \leq z) = \Pr(\max(X, Y) \leq z) = \Pr(X \leq z \text{ and } Y \leq z)$

$$F_Z(z) = F_X(z)F_Y(z).$$

2.11 $\Pr(Y \leq y) = (1 - e^{-y})^n$

Therefore $f_Y(y) = n(1 - e^{-y})^{n-1}e^{-y}$

Now $M_Z(\theta) = \dfrac{1}{(1 - \theta)}\dfrac{2}{(2 - \theta)} \cdots \dfrac{n}{(n - \theta)} = \displaystyle\sum_{i=1}^{n}\left(\frac{i}{i - \theta}\right)\prod_{\substack{j \neq i \\ j = 1}}^{n}\left(\frac{j}{j - i}\right)$

Note that

$$\prod_{\substack{j \neq i \\ j = 1}}^{n}\left(\frac{j}{j - i}\right) = \frac{1 \cdot 2 \cdot \ldots (i - 1)(i + 1)(i + 2) \ldots n}{(1 - i)(2 - i) \ldots (-1)1 \cdot 2 \cdot \ldots (n - i)} = (-1)^{i-1}\binom{n}{i}$$

therefore $M_Z(\theta) = \displaystyle\sum_{i=1}^{n}(-1)^{i-1}\binom{n}{i}\left(\frac{i}{i - \theta}\right)$

and so $f_Z(z) = \sum_{i=1}^{n} \binom{n}{i}(-1)^{i-1} i e^{-iy}$

and as $n\binom{n-1}{i-1} = i\binom{n}{i}$ then it is clear that Y and Z have the same distribution.

Note that

$$M_Y(\theta) = n \int_0^\infty (1 - e^{-y})^{n-1} e^{-y} e^{\theta y} dy$$

If we let $u = e^{-y}$, $du = -e^{-y} dy$, and so

$$M_Y(\theta) = n \int_0^1 (1 - u)^{n-1} u^{-\theta} du,$$

i.e. is proportional to the beta integral, so that

$$M_Y(\theta) = \frac{n\Gamma(n)\Gamma(1-\theta)}{\Gamma(n-\theta+1)}$$

$$= \frac{n!}{(1-\theta)(2-\theta)\dots(n-\theta)}$$

as above. To prove this result by induction, note that if

$$W = X/(n-1), \quad f_W(w) = (n+1)e^{-(n+1)w} \qquad \text{for } w \geq 0.$$

Let $Z_{n+1} = Z_n + W$, in an obvious notation, then using the convolution integral

$$f_{Z_{n+1}}(z) = n \int_0^z (1 - e^{-y})^{n-1} e^{-y} (n+1) e^{-(n+1)(z-y)} dy$$

$$= n(n+1)e^{-(n+1)z} \int_0^z (1 - e^{-y})^{n-1} e^{ny} dy$$

$$= n(n+1)e^{-(n+1)z} \int_0^z (e^y - 1)^{n-1} e^y dy$$

$$= (n+1)e^{-(n+1)z}(e^z - 1)^n = (n+1)(1 - e^{-z})^n e^{-z}.$$

Thus if the result is true for n, it is true for $(n+1)$. But it is clearly true for $n = 0$, and so it is true for $n \geq 0$.

Both Y and Z may be interpreted as the time to extinction for a population of size n, living according to the rules of a linear birth-process (see Cox and Miller, 1965, p. 156, or Bailey, 1964, p. 88, and the solution to Exercise 8.4).

2.12 $\left. \begin{array}{l} X_1 = Y_1 - Y_2 \\[2mm] X_2 = Y_1 + Y_2 \end{array} \right\}$ $\quad \dfrac{\partial x_1}{\partial y_1} = 1; \quad \dfrac{\partial x_1}{\partial y_2} = -1$

$\dfrac{\partial x_2}{\partial y_1} = 1; \quad \dfrac{\partial x_2}{\partial y_2} = 1$

$$f_{X_1X_2}(x_1x_2) = f_{Y_1Y_2}(y_1y_2) \begin{vmatrix} \dfrac{\partial y_1}{\partial x_1} & \dfrac{\partial y_1}{\partial x_2} \\ \dfrac{\partial y_2}{\partial x_1} & \dfrac{\partial y_2}{\partial x_2} \end{vmatrix}$$

$$= \frac{\lambda^2}{2} e^{-\lambda(y_1+y_2)} = \frac{\lambda^2}{2} e^{-\lambda x_2}.$$

It is now that this problem becomes tricky—we must determine the possible region for X_1 and X_2. Clearly, X_1 and X_2 are not independent, and the ranges of each depend upon the value taken by the other.

If $x_1 \geq 0$, then $x_2 \geq x_1$, while if $x_1 \leq 0$, $x_2 \geq -x_1$. So, to obtain the marginal distributions:

$$f_{X_1}(x_1) = \lambda^2 \int_{x_1}^{\infty} \frac{e^{-\lambda x_2}}{2} dx_2 \qquad \text{for } x_1 \geq 0$$

$$= \lambda^2 \int_{-x_1}^{\infty} \frac{e^{-\lambda x_2}}{2} dx_2 \qquad \text{for } x_1 \leq 0$$

$$= \frac{\lambda}{2} e^{-\lambda x_1} \qquad \text{for } x_1 \geq 0$$

$$= \frac{\lambda}{2} e^{\lambda x_1} \qquad \text{for } x_1 \leq 0$$

and $\qquad f_{X_2}(x_2) = \displaystyle\int_{-x_2}^{x_2} \frac{\lambda^2}{2} e^{-\lambda x_2} dx_1 = \lambda^2 x_2 e^{-\lambda x_2} \qquad \text{for } x_2 \geq 0$

i.e. $\Gamma(2, \lambda)$, as anticipated.

2.13 $\Pr(|X_1 - X_2| \leq x, \min(X_1, X_2) \leq y)$

$= \Pr(|X_1 - X_2| \leq x, \min(X_1, X_2) \leq y, X_1 < X_2) + \dots$

$= 2\Pr(X_1 - X_2 \leq x, X_2 \leq y, X_1 > X_2)$

$= 2\displaystyle\int_0^y f(u)\Pr(X_1 - X_2 \leq x, X_2 \leq y, X_1 > X_2 | X_2 = u) du$

$= 2\displaystyle\int_0^y f(u)\Pr(X_1 \leq x + u, X_1 > u) du = 2\displaystyle\int_0^y f(u)[F(x+u) - F(u)] du$

Therefore the joint p.d.f. is

$$\frac{d}{dx}[2f(y)\{F(x+y) - F(y)\}] = 2f(y) f(x+y) = g(x, y), \text{ say.}$$

$$f(x) = \lambda e^{-\lambda x} \Rightarrow g(x, y) = 2\lambda e^{-\lambda y} \lambda e^{-\lambda(x+y)} = 2\lambda^2 e^{-2\lambda y} e^{-\lambda x}$$

$$\Rightarrow \text{independence}$$

Independence $\Rightarrow g(x, y) = 2f(y)f(x + y) = l(x)h(y)$
for some functions l and h.

Put $x = 0$, and let $K = l(0)$, then
$$2[f(y)]^2 = Kh(y) \Rightarrow h(y) = \frac{2}{K}[f(y)]^2$$

Also, $\quad h(y) = \int_0^\infty g(x, y)\,dx = 2f(y)[1 - F(y)]$

Thus, $\quad 2f(y)[1 - F(y)] = \frac{2}{K}[f(y)]^2$

$$1 - F(y) = f(y)/K$$

$$F(y) + \frac{dF}{dy}\bigg/ K = 1$$

$F(y) = 1 + e^{-Ky}$, as required, Finally,

$$\Pr(X_1 + X_2 \le 3\min(X_1, X_2) \le 3b) = \Pr(U + 2V < 3V, V \le b)$$

where $U = |X_1 - X_2|$ and $V = \min(X_1, X_2)$.

U, V are independent, from above, and $X_1 + X_2 = U + 2V$. Hence required probability $= \Pr(U < V \le b)$

$$f_U(u) = \lambda e^{-\lambda u} \text{ (see Exercise 2.12) for some } \lambda > 0$$

(cf. Exercise 2.10), and $f_V(v) = 2\lambda e^{-2\lambda v}$, and using the independence property, we have required probability

$$\int_0^b \int_0^v \lambda e^{-\lambda u} 2\lambda e^{-2\lambda v}\, du\, dv = \frac{1}{3} - e^{-2\lambda b} + \frac{2}{3} e^{-3\lambda b}.$$

2.14 $\quad f_X(x) = \dfrac{e^{-x/2}}{2^a} \dfrac{x^{a-1}}{\Gamma(a)} \qquad$ for $x \ge 0$

$$f_Y(y) = \frac{e^{-y/2} y^{b-1}}{2^b \Gamma(b)} \qquad \text{for } y \ge 0$$

$$\begin{cases} S = X + Y & \dfrac{\partial s}{\partial x} = 1 & \dfrac{\partial s}{\partial y} = 1 \\[2ex] T = X/(X + Y) & \dfrac{\partial t}{\partial x} = \dfrac{y}{(x+y)^2} & \dfrac{\partial t}{\partial y} = -\dfrac{x}{(x+y)^2} \end{cases}$$

therefore $ST = X, Y = S(1 - T)$.

$$f_{ST}(s, t) = f_{XY}(x, y)(x + y) = \frac{e^{-((x+y)/2)} x^{a-1} y^{b-1}(x+y)}{2^{a+b}\Gamma(a)\Gamma(b)}$$

$$= \frac{se^{-s/2}(st)^{a-1}(s(1-t))^{b-1}}{2^{a+b}\Gamma(a)\Gamma(b)}$$

i.e.
$$f_{ST}(s, t) = \frac{s^{a+b-1}e^{-s/2}t^{a-1}(1-t)^{b-1}\Gamma(a+b)}{2^{a+b}\Gamma(a+b)\Gamma(a)\Gamma(b)}.$$

We see that S and T are independent. T is $\chi^2_{2(a+b)}$ and S is $B_e(a, b)$.

2.15 (a) $M_{N_1 N_2}(\theta) = \displaystyle\int_{-\infty}^{\infty}\int_{-\infty}^{\infty} \frac{\exp[\theta x_1 x_2]}{2\pi} \exp[-\tfrac{1}{2}(x_1^2 + x_2^2)]\, dx_1\, dx_2$

$$= \frac{1}{\sqrt{(2\pi)}}\int_{-\infty}^{\infty} \exp[-x_1^2/2]$$

$$\times \left(\frac{1}{\sqrt{(2\pi)}}\int_{-\infty}^{\infty} \exp[\theta x_1 x_2 - x_2^2/2]\, dx_2\right) dx_1$$

$$= \frac{1}{\sqrt{(2\pi)}}\int_{-\infty}^{\infty} \exp[-x_1^2/2]\exp[+\tfrac{1}{2}\theta^2 x_1^2]$$

$$\times \left(\frac{1}{\sqrt{(2\pi)}}\int_{-\infty}^{\infty} \exp[-\tfrac{1}{2}(\theta x_1 - x_2)^2]\, dx_2\right) dx_1$$

therefore $= \dfrac{1}{\sqrt{(2\pi)}} \dfrac{1}{\sqrt{(1-\theta^2)}}\displaystyle\int_{-\infty}^{\infty} \exp[-y^2/2]\, dy$

$$\left(\text{from setting } x_1 = \frac{y}{\sqrt{(1-\theta^2)}}\right)$$

$$= \frac{1}{\sqrt{(1-\theta^2)}}$$

therefore $M_{N_1 N_2 + N_3 N_4}(\theta) = \dfrac{1}{(1-\theta^2)}$

But this is the m.g.f. of a random variable with the Laplace distribution:

$$f_X(x) = \tfrac{1}{2}\exp[-|x|] \qquad \text{for } -\infty \le x \le \infty,$$

To see this:

$$M_X(\theta) = \frac{1}{2}\int_{-\infty}^{0} e^x e^{\theta x}\, dx + \frac{1}{2}\int_{0}^{\infty} e^{\theta x - x}\, dx$$

$$= \frac{1}{2(1+\theta)} - \frac{1}{2(\theta-1)}, \qquad \text{for } \theta < 1$$

$$= \frac{1}{(1-\theta^2)}$$

(This is the distribution of $Y_1 - Y_2$ in Exercise 2.12.)

Hence

$$f_{|N_1 N_2 + N_3 N_4|}(x) = e^{-x} \qquad \text{for } x \geq 0$$

(b) $C = N_1/N_2$. Also, set $D = N_1$, say, to give a 1–1 transformation from (N_1, N_2) to (C, D)

$$\frac{\partial c}{\partial n_1} = \frac{1}{n_2} \qquad \frac{\partial c}{\partial n_2} = -\frac{n_1}{n_2^2} \qquad \frac{\partial d}{\partial n_1} = 1 \qquad \frac{\partial d}{\partial n_2} = 0$$

Hence, $\quad f_{C,D}(c, d) = \dfrac{\exp\left[-\frac{1}{2}(n_1^2 + n_2^2)\right]}{2\pi} \begin{vmatrix} n_2^2 \\ n_1 \end{vmatrix}$

$$(d \equiv y,\, x \equiv c) = \frac{\exp\left[-\frac{1}{2}y^2(1 + 1/x^2)\right]}{2\pi} \begin{vmatrix} y \\ x^2 \end{vmatrix} = f_{X,Y}(x, y),$$

and now we form the marginal distribution of X:

$$f_X(x) = \int_{-\infty}^{\infty} f_{X,Y}(x, y)\,dy = \int_{-\infty}^{0} f_{X,Y}(x, y)\,dy + \int_{0}^{\infty} f_{X,Y}(x, y)\,dy$$

$$= \int_{-\infty}^{0} \frac{1}{2\pi x^2} (-y)\exp\left[-(y^2/2)(1 + 1/x^2)\right] dy$$

$$+ \int_{0}^{\infty} \frac{1}{2\pi x^2} y\exp\left[(-y^2/2)(1 + 1/x^2)\right] dy$$

i.e. $f_X(x) = \dfrac{1}{2\pi x^2} \dfrac{1}{(1 + 1/x^2)} + \dfrac{1}{2\pi x^2} \dfrac{1}{(1 + 1/x^2)} = \dfrac{1}{\pi(1 + x^2)}$,

for $-\infty \leq x \leq \infty$,

i.e. X has the standard Cauchy distribution.

If X, Y are independent, identically distributed *exponential* random variables, with probability density function, e^{-x} for $x \geq 0$, then for $Z = X/Y, f_Z(z) = (1 + z)^{-2}$ for $z \geq 0$, and $\mathscr{E}[Z] = \infty$.

2.16 $\phi(\mathbf{x}) = (2\pi)^{-p/2} |\mathbf{\Sigma}|^{-1/2} \exp\left\{-\frac{1}{2}\mathbf{x}'\mathbf{\Sigma}^{-1}\mathbf{x}\right\} d\mathbf{x}$

Suppose, first of all, that $\boldsymbol{\mu} = \mathbf{0}$:

$$\mathbf{x} = \mathbf{A}^{-1}\mathbf{z} \qquad \text{and the Jacobian is: } \|\mathbf{A}^{-1}\|$$

and so, $\phi(\mathbf{z}) = (2\pi)^{-p/2} |\mathbf{\Sigma}|^{-1/2} \exp\left\{-\frac{1}{2}(\mathbf{A}^{-1}\mathbf{z})'\mathbf{\Sigma}^{-1}\mathbf{A}^{-1}\mathbf{z}\right\} \|\mathbf{A}^{-1}\|$

i.e. $\quad \phi(\mathbf{z}) = (2\pi)^{-p/2} |\mathbf{\Sigma}|^{-1/2} \|\mathbf{A}\|^{-1} \exp\left\{-\frac{1}{2}(\mathbf{A}^{-1}\mathbf{z})'\mathbf{\Sigma}^{-1}\mathbf{A}^{-1}\mathbf{z}\right\}$

$$= (2\pi)^{-p/2} |\mathbf{A}\mathbf{\Sigma}\mathbf{A}'|^{-1/2} \exp\left\{-\frac{1}{2}\mathbf{z}'(\mathbf{A}^{-1})'\mathbf{\Sigma}^{-1}\mathbf{A}^{-1}\mathbf{z}\right\}$$

$$= (2\pi)^{-p/2} |\mathbf{A}\mathbf{\Sigma}\mathbf{A}'|^{-1/2} \exp\left\{-\frac{1}{2}\mathbf{z}'(\mathbf{A}\mathbf{\Sigma}\mathbf{A}')^{-1}\mathbf{z}\right\}$$

i.e. $\mathbf{Z} = \mathbf{A}\mathbf{X}$ has an $N(\mathbf{0}, \mathbf{A}\mathbf{\Sigma}\mathbf{A}')$ distribution.

If $\Sigma = I$, as in the question, Z has an $N(0, AA')$ distribution. The translation: $\tilde{Z} = Z + \mu$ is readily shown to result in $N(\mu, AA')$, as required.

2.17 $\Pr(k \text{ events}) = \int_0^\infty \lambda_1 e^{-\lambda_1 t} e^{-\lambda_2 t} \dfrac{(\lambda_2 t)^k}{k!} \, dt$ for $k \geq 0$

$$= \frac{\lambda_2^k \lambda_1}{k!} \int_0^\infty e^{-(\lambda_1 + \lambda_2)t} t^k \, dt = \frac{\lambda_2^k \lambda_1}{k!(\lambda_1 + \lambda_2)^{k+1}} \int_0^\infty e^{-\theta} \theta^k \, d\theta$$

$$= \frac{\lambda_1 \lambda_2^k}{(\lambda_1 + \lambda_2)^{k+1}}$$ for $k \geq 0$

i.e. a geometric distribution.

2.18 $\Pr(Y \geq n) = \displaystyle\sum_{k=n}^{n+m} \binom{n+m}{k} (1-\theta)^k \theta^{n+m-k}$

$$= \sum_{k=0}^{m} \binom{n+m}{n+k} (1-\theta)^{n+k} \theta^{m-k}$$

while $\Pr(X \leq n) = \displaystyle\sum_{k=0}^{m} \binom{n+k-1}{k} \theta^k (1-\theta)^n$

Therefore we require

$$\sum_{k=0}^{m} \binom{n+k-1}{k} \theta^k = \sum_{k=0}^{m} \binom{n+m}{n+k} (1-\theta)^k \theta^{m-k}$$

i.e. we require

$$\binom{n+i-1}{i} = \sum_{k=m-i}^{m} \binom{n+m}{n+k}\binom{k}{m-i} (-1)^{k-m+i}$$ for $0 \leq i \leq m$,

and this follows simply from considering the coefficient of z^i on both sides of the identity

$$(1+z)^{n+i-1} = (1+z)^{n+m}/(1+z)^{m+1-i}$$

2.19 $e^{-n} \displaystyle\sum_{r=0}^{n} \dfrac{n^r}{r!} = \Pr(X_n \leq n)$, where X_n has a Poisson distribution of parameter n. Now (see Exercise 2.8(b)), if X_n is of this form, we can write

$$X_n = \sum_{i=1}^{n} Z_i$$

where the Z_i are independent, identically distributed Poisson random variables of parameter 1. Hence by a simple central limit theorem,

$$\Pr(X_n \leq n) \to \Phi(0) = \tfrac{1}{2} \qquad \text{as } \mathscr{E}[X_n] = n$$

Therefore $\lim_{n \to \infty} \left(e^{-n} \sum_{r=0}^{n} \frac{n^r}{r!} \right) = \frac{1}{2}$

2.22 See *ABC*, p. 389.

2.25 $y = g(x)$ does not, in this example, have a continuous derivative. Thus Equation (2.3) is not appropriate. However, we always have:

$$\Pr(Y \leq y) = \Pr(X \leq x)$$

and for $y \leq 1$,

$$\Pr(Y \leq y) = 1 - e^{-y}$$

while for $y \geq 1$,

$$\Pr(Y \leq y) = 1 - e^{-(1+y)/2}$$

2.26 See Exercise 7.25.

2.27 For $\lambda < \mu$ the queue settles down to an 'equilibrium' state. For $\lambda \geq \mu$ no such state exists, and the queue size increases ultimately without limit. See Exercise 7.24.

Chapter 3

3.1 Pr(respond 'Yes') = Pr(respond 'Yes' to (i))Pr((i) is the question)

+ Pr(respond 'Yes' to (ii))Pr((ii) is the question).

If one responds 'Yes' to (i) then one responds 'No' to (ii)
 Pr(respond 'Yes' to (ii)) = 1 − Pr(respond 'Yes' to (i)) = 1 − η, say
 Pr(respond 'Yes') = η Pr((i)) + (1 − η) Pr(ii)
Pr(i) and Pr(ii) are determined by a randomization device, and Pr(respond 'Yes') is estimated from the responses. For example, from a class survey of first-year mathematics students, in which group X approved of sit-ins as a form of protest (a question topical at the time), Pr(i) = 3/10; Pr(ii) = 7/10, and 34 responded 'Yes' out of 56, resulting in $\hat{\eta} = 0.38$. The students also wrote their opinion anonymously on slips which were collected, and which gave rise to $\hat{\eta} = 0.33$. See Warner (1965) for further discussion on confidence intervals, etc.

3.2 Conduct two surveys with two different probabilities of answering the innocent question, resulting in two equations in two unknowns. For further discussion, see Moors (1971). Innocent questions with known frequency of response might include month of birth, or whether an identity-number of some kind is even or odd (Campbell and Joiner, 1973).

3.3 Let $\theta = $ Pr(respondent has had an abortion)
Let proportions of balls be: p_r, p_w, p_b.
Then Pr(respond 'Yes') = $\theta p_r + p_w$.
See Greenberg *et al.* (1971).

3.4 With the question as stated in Example 3.1, 'Yes' is potentially incriminating, whereas 'No' is not. See Abdul-Ela *et al.* (1967) and Warner (1971).

3.5 In experiments of this kind, individuals typically overestimate μ, introducing a judgement bias. Here we find:

judgement: $\bar{x} = 45.56; s = 13.69$
random : $\bar{x} = 37.21; s = 10.28$.

3.6 Let X denote the recorded value.

$$\Pr(X = 0) = \sum_{i=4}^{6} \Pr(\text{1st die} = i) \; \Pr(\text{2nd die} = 10 - i)$$

$$= 3 \cdot \tfrac{1}{6} \cdot \tfrac{1}{6} = \tfrac{1}{12} \qquad \text{(assuming independence)}$$

$$\Pr(X = 1) = \sum_{i=5}^{6} \Pr(\text{1st die} = i) \; \Pr(\text{2nd die} = 11 - i) = \tfrac{1}{18}.$$

For $2 \le i \le 9$,

$$\Pr(X = i) = \sum_{j=\max(1, i-6)}^{\min(6, i-1)} \Pr(\text{1st die} = j) \; \Pr(\text{2nd die} = i - j).$$

Thus $\Pr(X = 2) = \tfrac{1}{36}$, $\Pr(X = 3) = \tfrac{1}{18}$, $\Pr(X = 4) = \tfrac{1}{12}$, etc.
For equiprobable random digits, the method suggested is clearly unsuitable.

3.7 $\Pr(\text{2nd coin is } H \mid \text{two tosses differ}) = \dfrac{pq}{pq + qp} = \tfrac{1}{2}$

$(p = \Pr(H) = 1 - q)$.

3.8 There are 6 possibilities for both A and B, and so there are 36 possibilities in all. If we reject (i, i) results, for $1 \le i \le 6$, we get 30 possibilities, which may be used to generate uniform random digits from 0–9 as follows:

Possible outcomes from dice			Digit selected
(1, 2)	(1, 3)	(1, 4)	0
(1, 5)	(1, 6)	(2, 1)	1
(2, 3)	(2, 4)	(2, 5)	2
(2, 6)	(3, 1)	(3, 2)	3
(3, 4)	(3, 5)	(3, 6)	4
(4, 1)	(4, 2)	(4, 3)	5
(4, 5)	(4, 6)	(5, 1)	6
(5, 2)	(5, 3)	(5, 4)	7
(5, 6)	(6, 1)	(6, 2)	8
(6, 3)	(6, 4)	(6, 5)	9

Thus, for example, we choose digit 5 if we get one of (4, 1), (4, 2), (4, 3), and the conditional probability of this is:

$$(3 \times (\tfrac{1}{6}) \times (\tfrac{1}{6}))/(30/36) = \tfrac{1}{10} \qquad \text{as required.}$$

The pairs given thus result in the sequence:

$$0, 3, 1, 5, -, 9, 6, 6, 4, 0.$$

If we take these digits in fours we obtain the required numbers; e.g. 0315, 9664.

3.9 Regard HCY 7F as HCY 007F, etc. One set of 600 digits obtained in this manner is given below:

157	741	602	823	438	455	816	493	681	241
260	765	308	684	564	918	422	772	471	072
217	192	159	274	946	068	017	230	889	812
235	801	517	582	277	573	808	623	641	770
601	319	100	153	976	015	506	460	342	357
485	803	335	844	370	556	724	900	935	195
800	360	263	427	280	419	515	991	296	712
297	122	007	388	186	876	581	793	352	053
285	307	996	988	973	794	981	677	212	464
246	893	373	113	723	725	778	645	028	611
395	288	291	370	744	142	486	374	548	580
591	454	733	986	484	423	594	938	670	323
792	355	642	059	803	356	278	500	840	383
416	453	461	412	851	560	978	483	772	615
885	520	441	909	435	802	055	933	659	554
801	726	501	651	828	941	570	164	104	380
253	882	072	848	909	249	147	309	522	503
015	813	421	805	702	342	920	170	226	312
832	562	730	801	704	965	728	387	761	360
028	331	334	202	479	916	953	930	462	369

3.11 It is easy to demonstrate degeneration of this method:

(i) 55 02 00

(ii) 66 35 22 48 30 90 10 10 10

3.12 If this sequence is operated to small decimal-place accuracy then it can degenerate, as the following example shows:

0.1 0.925 0.309 0.180 0.326 0.349 0.198 0.401

0.964 0.485 0.327 0.073 0.265 0.776 0.776

3.13 We require sample mean and variance for the sample:

$$\{i/m, \, 0 \le i \le m-1\}$$

$$\bar{x} = \frac{1}{m^2} \frac{m(m-1)}{2} = \frac{1}{2}\left(1 - \frac{1}{m}\right)$$

$$(m-1)s^2 = \left(\frac{1}{m^2} \sum_{i=1}^{m-1} i^2 - m\bar{x}^2\right) = \left(\frac{1}{6m^2}(m-1)m(2m-1) - m\bar{x}^2\right)$$

whence $s^2 = \dfrac{1}{12}\left(1 + \dfrac{1}{m}\right).$

3.14 If $A \times U0 + B = (k \times 1000) + r$

where $\qquad 0 \leq r < 1000, \; U1 = k + r/1000 + \varepsilon$

where $\varepsilon \ll 0.001$ is the round-off error, and $\mathrm{INT}(U1) = r/1000 + \varepsilon$,

$$Y = (U1 - \mathrm{INT}(U1)) \times 1000 = r + 1000\varepsilon$$

We require r, therefore set $U1 = \mathrm{INT}(r + 1000\varepsilon + \theta)$
where θ is such that $\theta + 1000\varepsilon > 0$, but $\theta + 1000\varepsilon < 1$.
$\theta = 0.5$ will do.
An example is:

$$1 \quad 168 \quad 595 \quad 82 \quad 429 \quad 435.964 \quad 874.781 \ldots$$

With the additional line we get:

$$1 \quad 168 \quad 595 \quad 82 \quad 429 \quad 436 \quad 903 \ldots$$

3.15 $x_{i+1} = ax_i + b \quad (\mathrm{mod}\ m)$

i.e. $x_{i+1} = ax_i + b - \kappa m \qquad$ for some κ and $0 \leq ax_i + b < m$

and so $\qquad \left(\dfrac{x_{i+1}}{m}\right) = a\left(\dfrac{x_i}{m}\right) + \dfrac{b}{m} - \kappa$

i.e. $u_{i+1} = au_i + \dfrac{b}{m} \quad (\mathrm{mod}\ 1) \qquad$ and $\quad 0 \leq au_i + \dfrac{b}{m} < 1.$

3.16 For any $n \geq 0$, $x_{n+1} = ax_n + b \qquad (\mathrm{mod}\ m)$

Therefore $ax_n + b = \gamma m + x_{n+1} \qquad$ for some integral $\gamma \geq 0$ and
$0 \leq x_{n+1} < m$

and so $x_{n+2} = ax_{n+1} + b \qquad (\mathrm{mod}\ m)$

$$= a^2 x_n + ab - \gamma am + b \qquad (\mathrm{mod}\ m)$$

$$= a^2 x_n + (a+1)b \qquad (\mathrm{mod}\ m)$$

and this is the approach which may be used to prove this result by
induction on k for any n:

$$x_{n+k+1} = \left(a^{k+1} x_n + \left(\frac{a^k - 1}{a-1}\right) ba + b\right) \qquad \mathrm{mod}(m)$$

$$= a^{k+1} x_n + (a^{k+1} - 1)b\,(a-1)^{-1} \qquad \mathrm{mod}(m)$$

Thus every kth term of the original series is another congruential series,
with multiplier a^k and increment $(a^k - 1)(a-1)^{-1}b$, or, equivalently,
$a^k \;(\mathrm{mod}\ m)$ and $(a^k - 1)(a-1)^{-1}b \;(\mathrm{mod}\ m)$, respectively.

3.17 The illustration below is taken from Wichmann and Hill (1982a):

Value of U_2 to 1 d.p.	Value of U_1 to 1 d.p.									
	0.0	0.1	0.2	0.3	0.4	0.5	0.6	0.7	0.8	0.9
0.0	0.0	0.1	0.2	0.3	0.4	0.5	0.6	0.7	0.8	0.9
0.1	0.1	0.2	0.3	0.4	0.5	0.6	0.7	0.8	0.9	0.0
0.2	0.2	0.3	0.4	0.5	0.6	0.7	0.8	0.9	0.0	0.1
0.3	0.3	0.4	0.5	0.6	0.7	0.8	0.9	0.0	0.1	0.2
0.4	0.4	0.5	0.6	0.7	0.8	0.9	0.0	0.1	0.2	0.3
0.5	0.5	0.6	0.7	0.8	0.9	0.0	0.1	0.2	0.3	0.4
0.6	0.6	0.7	0.8	0.9	0.0	0.1	0.2	0.3	0.4	0.5
0.7	0.7	0.8	0.9	0.0	0.1	0.2	0.3	0.4	0.5	0.6
0.8	0.8	0.9	0.0	0.1	0.2	0.3	0.4	0.5	0.6	0.7
0.9	0.9	0.0	0.1	0.2	0.3	0.4	0.5	0.6	0.7	0.8

The values above give the fractional part of $(U_1 + U_2)$. If U_1 and U_2 are independent, then whatever the value of U_2, if U_1 is uniform then so is the fractional part of $(U_1 + U_2)$, and U_2 need not be uniform.

3.20 Note the generalization of this result.

3.21 (a) Numbers in the two half-periods differ by 16:

   0 13 2 31 4 17 6 3 8 21 10 7 12 25 14 11
  16 29 18 15 20 1 22 19 24 5 26 23 28 · 9 30 27

We have:

u_r	$9u_r + 13$	u_{r+1}	Binary form of u_{r+1}	Decimal form of $u_{r+1}/32$
0	13	13	01101	0.40625
13	130	2	00010	0.06250
2	31	31	11111	0.96875
31	292	4	00100	0.12500
4	49	17	10001	0.53125
17	166	6	00110	0.18750
6	67	3	00011	0.09375
3	40	8	01000	0.25000
8	85	21	10101	0.65625
21	202	10	01010	0.31250
10	103	7	00111	0.21875
7	76	12	01100	0.37500
12	121	25	11001	0.78125
25	238	14	01110	0.43750
14	139	11	01011	0.34375
11	112	16	10000	0.50000
16	157	29	11101	0.90625
29	274	18	10010	0.56250
⋮	⋮	⋮	⋮	⋮

revealing further clear patterns.

3.22 (a) Procedures are equivalent if the indicator digit is from a generator of period g.

(b) The sequence is: 1, 8, 11, 10, 5, 12, 15, 14, 9, 0, 3, 2, 13, 4, 7, 6, 1. Suppose $g = 4$, for illustration. We start with (1, 8, 11, 10). Let X denote the next number in the sequence, and suppose that if

$$(*)\begin{cases} 0 \leq X \leq 3 & \text{we replace the 1st stored term} \\ 4 \leq X \leq 7 & \text{we replace the 2nd stored term} \\ 8 \leq X \leq 11 & \text{we replace the 3rd stored term} \\ 12 \leq X \leq 15 & \text{we replace the 4th stored term} \end{cases}$$

This gives:

$$\left.\begin{array}{ll} (1, 8, 11, 10) & \\ (1, 5, 11, 10) & \text{use } 8 \\ (1, 5, 11, 12) & \text{use } 10 \\ (1, 5, 11, 15) & \text{use } 15 \end{array}\right\}$$

etc.

Once the store contains (1, 5, 9, 14) in this example then it enters a (full) cycle. Note that before a cycle can commence, the numbers in the store must correspond, in order, to the four different ranges in $(*)$, and the sequence of numbers for any part of the store must correspond to that in the original sequence for the numbers in that range.

3.23 Omit trailing decimal places, when the x_i are divided by m, to give: $u_i = x_i/m$.

3.25 For suitable $\alpha, \beta, \theta,$

$$\begin{aligned} x_{i+1} &= (2^{16} + 3)x_i + \alpha 2^{31} \\ &= 6x_i + x_i(2^{16} - 3) + \alpha 2^{31} \\ &= 6x_i + (2^{16} + 3)(2^{16} - 3)x_{i-1} + (2^{16} - 3)\beta 2^{31} + \alpha 2^{31} \\ &= 6x_i + (2^{32} - 9)x_{i-1} + \theta 2^{31} \\ &= 6x_i - 9x_{i-1} + \theta 2^{31} \\ &= 6x_i - 9x_{i-1} \quad (\text{mod } 2^{31}). \end{aligned}$$

3.26 (b) One-sixth of the time.

$0 \leq x_i < m$

therefore $x_n + x_{n-1} = x_{n+1} + \kappa m$ where $\kappa = 0$ or $\kappa = 1$.

Now suppose $x_{n-1} < x_{n+1} < x_n$ $(*)$

then $x_{n-1} + \kappa m < x_n + x_{n-1} < x_n + \kappa m$

If $\kappa = 1$, this implies $x_n > m$ $\Big\}$ neither can occur, so $(*)$ is false.

If $\kappa = 0$, this implies $x_{n-1} < 0$

3.27 (b) No zero values with a multiplicative congruential generator.

3.28 FORTRAN programs are provided by Law and Kelton (1982, p. 227); see also Nance and Overstreet (1975). Some versions of FORTRAN allow one to declare extra-large integers.

3.29 Set $y_n = \theta^n$ and solve the resulting quadratic equation in θ (roots θ_1 and θ_2). The general solution is then $y_n = A\theta_1^n + B\theta_2^n$, where A and B are determined by the values of y_0 and y_1.

3.32 From considering the outcome of the first two tosses,

$$p_n = \tfrac{1}{2}p_{n-1} + \tfrac{1}{4}p_{n-2} \qquad \text{for } n \geq 2.$$

$$np_n = \frac{n}{2}p_{n-1} + \frac{n}{4}p_{n-2}$$

$$= \frac{(n-1)}{2}p_{n-1} + \tfrac{1}{2}p_{n-1} + \frac{(n-2)}{4}p_{n-2} + \tfrac{1}{2}p_{n-2} \qquad \text{for } n \geq 2$$

Summing over n: if $\mu = \sum_{n=1}^{\infty} np_n$,

$$\mu - 2p_2 - p_1 = \frac{\mu}{2} + \frac{\mu}{4} + \tfrac{1}{2}(1 - p_1) + \tfrac{1}{2}; \quad p_1 = 0, p_2 = \tfrac{1}{4}, \mu = 6.$$

Chapter 4

4.2

```
10   LET P = .5
20   LET Q = 1-P
30   LET S = 0
40   RANDOMIZE
50   LET U = RND
60   FOR I = 1 TO 3
70   IF U < P THEN 100
80   LET U = U/Q
90   GOTO 120
100  LET U = U/P
110  LET S = S+1
120  NEXT I
130  PRINT S
140  END
```

4.6 $N_1 = (-2 \log_e U_1)^{1/2} \cos 2\pi U_2$

$N_2 = (-2 \log_e U_1)^{1/2} \sin 2\pi U_2$

$\dfrac{n_2}{n_1} = \tan(2\pi u_2); \quad u_1 = \exp[-\tfrac{1}{2}(n_1^2 + n_2^2)]$

$$\begin{vmatrix} \dfrac{\partial u_1}{\partial n_1} & \dfrac{\partial u_1}{\partial n_2} \\[2mm] \dfrac{\partial u_2}{\partial n_1} & \dfrac{\partial u_2}{\partial n_2} \end{vmatrix} = \begin{vmatrix} -n_1 u_1 & -n_2 u_1 \\[2mm] \dfrac{-n_2}{n_1^2 2\pi \sec^2(2\pi u_2)} & \dfrac{1}{2\pi n_1 \sec^2(2\pi u_2)} \end{vmatrix}$$

$$= \frac{u_1(1 + n_2^2/n_1^2)}{2\pi \sec^2(2\pi u_2)} = \frac{u_1}{2\pi} = \frac{\exp[-\tfrac{1}{2}(n_1^2 + n_2^2)]}{2\pi}, \text{ as required.}$$

4.8   The simplest case is when $n = 2$, resulting in the triangular distribution:

$$f(x) = x \qquad \text{for} \qquad 0 \le x \le 1$$
$$f(x) = 2 - x \qquad \text{for} \qquad 1 \le x \le 2$$

For illustrations and applications, see Section 5.5.

4.9   Define $Y_0 = 0$.

$K = i$ if and only if $Y_i \le 1 < Y_{i+1}$ for $i \ge 0$.

Thus $K \le k$ if and only if $Y_{k+1} > 1$. Therefore

$$\Pr(K \le k) = \Pr(Y_{k+1} > 1) = \int_1^\infty \frac{\lambda(\lambda y)^k e^{-\lambda y} dy}{k!}$$

and $\Pr(K = k) = \Pr(K \le k) - \Pr(K \le k - 1)$ for $k \ge 0$

$[\Pr(K \le -1) \equiv \Pr(Y_0 > 1) = 0]$

$$= \Pr(Y_{k+1} > 1) - \Pr(Y_k > 1)$$

$$= \int_1^\infty \left\{ \frac{\lambda^{k+1} y^k e^{-\lambda y}}{k!} - \frac{\lambda^k y^{k-1} e^{-\lambda y}}{(k-1)!} \right\} dy$$

But

$$\int_1^\infty \frac{\lambda^{k+1} y^k e^{-\lambda y}}{k!} dy = -\frac{\lambda^k}{k!} \int_1^\infty y^k d(e^{-\lambda y})$$

$$= -\left[\frac{\lambda^k}{k!} y^k e^{-\lambda y} \right]_1^\infty + \frac{\lambda^k}{(k-1)!} \int_1^\infty e^{-\lambda y} y^{k-1} dy$$

and so (integrals are finite)

$$\Pr(K = k) = \frac{e^{-\lambda} \lambda^k}{k!} \qquad \text{for } k \ge 0.$$

4.10   Lenden–Hitchcock (1980) considered the following four-dimensional method.

Let N_i, $1 \le i \le 4$, be independent random variables, each with the half-normal density:

$$f_{N_i}(x) = \sqrt{\left(\frac{2}{\pi}\right)} e^{-x^2/2} \qquad \text{for } x \ge 0.$$

Changing from Cartesian to polar co-ordinates we get:

$$(*)\begin{cases} N_1 = R \cos \Theta_1 \cos \Theta_2 \cos \Theta_3 \\ N_2 = R \sin \Theta_1 \cos \Theta_2 \cos \Theta_3 \qquad 0 < \Theta_i < \dfrac{\pi}{2}, \text{ for } 1 \le i \le 4 \\ N_3 = R \sin \Theta_2 \cos \Theta_3 \qquad\qquad 0 < R < \infty. \\ N_4 = R \sin \Theta_3 \end{cases}$$

The Jacobian of the transformation is $R^3 \cos \Theta_2 \cos^2 \Theta_3$, and so

$$f_{R, \Theta_1, \Theta_2, \Theta_3}(r, \theta_1, \theta_2, \theta_3) = \frac{4}{\pi^2} r^3 e^{-r^2/2} \cos \theta_1 \cos^2 \theta_2.$$

Thus R, Θ_1, Θ_2 and Θ_3 are independent, and to obtain the $\{N_i\}$ we simply simulate R, Θ_1, Θ_2 and Θ_3 using their marginal distributions and transform back to the $\{N_i\}$ by (*). R^2 has a χ_4^2 distribution, and so is readily simulated, as described in Section 4.3.

(Had a three-dimensional generalization been used, R^2 would have had a χ_3^2 distribution, which is less readily simulated.) Lenden–Hitchcock suggested that the four-dimensional method is more efficient than the standard two-dimensional one.

4.12　Let　$\Psi = \Pr(\text{Accept } (V_2, V_3) \text{ and Accept } (V_1, V_2))$

$$= \Pr(V_2^2 + V_3^2 \le 1 \text{ and } V_2^2 + V_1^2 \le 1)$$

$$f_{V^2}(x) = \frac{1}{2x^{1/2}} \quad \text{for } 0 \le x \le 1; \quad F_{V^2}(x) = x^{1/2}$$

Therefore　$\Psi = \displaystyle\int_0^1 \Pr(V_3^2 \le 1 - x) \Pr(V_1^2 \le 1 - x) \frac{1}{2x^{1/2}} \, dx$

$$= \frac{1}{2} \int_0^1 (x^{-1/2} - x^{1/2}) \, dx = 1 - \frac{1}{3} = \frac{2}{3} \ne \frac{\pi^2}{16}.$$

4.13　Logistic.

4.14　It is more usual to obtain the desired factorization using Choleski's method (see Conte and de Boor, 1972, p. 142), which states that we can obtain the desired factorization with a lower triangular matrix $\mathbf{A}$. The elements of $\mathbf{A}$ can be computed recursively as follows.

Let $\Sigma = \{\sigma_{ij}\}$, $\mathbf{A} = \{a_{ij}\}$. We have $a_{ij} = 0$ for $j > i$, and so $\sigma_{ij} = \sum_{k=1}^{\min(i,j)} a_{ik} a_{jk}; \sigma_{11} = a_{11}^2$, so that $a_{11} = \sigma_{11}^{1/2}$, and then $a_{i1} = (\sigma_{i1}/\sigma_{11}^{1/2})$ for $i > 1$. This gives the first column of $\mathbf{A}$, which may be used to give the second column, and so on. Given the first $(j-1)$ columns of $\mathbf{A}$,

$$a_{jj} = \left(\sigma_{jj} - \sum_{k=1}^{j-1} a_{jk}^2 \right)^{1/2}$$

and　　$a_{ij} = \left(\sigma_{ij} - \sum_{k=1}^{j-1} a_{ik} a_{jk}/a_{jj} \right) \quad \text{for } i > j.$

4.16　The result is easily shown by the transformation-of-variable theory, with the Jacobian of the transformation $= (1 - \rho^2)^{1/2}$. Alternatively, the joint p.d.f. of Y_1 and X_1 can be written as the product of the conditional p.d.f. of $Y_1 | X_1$ and the marginal distribution of X_1, to yield, directly:

$$f_{Y_1 X_1}(y, x) = \frac{\exp[-\frac{1}{2}(y - \rho x)^2 / (1 - \rho^2)]}{(1 - \rho^2)^{1/2} \sqrt{(2\pi)}} \frac{\exp[-\frac{1}{2}x^2]}{\sqrt{(2\pi)}}$$

(See also the solution to Exercise 6.5.)

4.17 Median $= m$ if and only if one value $= m$, $(n - 1)$ of the other values are less than m, and $(n - 1)$ are greater than m. The value to be m can be chosen $(2n - 1)$ ways; the values to be $< m$ can be chosen $\binom{2n-2}{n-1}$ ways, hence,

$$f_M(m)\, dm = (2n - 1) \binom{2n - 2}{n - 1} m^{n-1}(1 - m)^{n-1}\, dm$$

i.e. M has a $B_e(n, n)$ distribution.

4.19 $\displaystyle \Pr(X = k) = \frac{1}{k!} \int_0^\infty \frac{e^{-\lambda} \lambda^k e^{-\theta\lambda} \theta^n \lambda^{n-1}}{\Gamma(n)}\, d\lambda$

$$= \frac{\theta^n}{k!\,\Gamma(n)} \int_0^\infty \lambda^{n+k-1} e^{-\lambda(\theta+1)}\, d\lambda$$

$$= \frac{\theta^n}{k!\,\Gamma(n)} \frac{\Gamma(n + k)}{(\theta + 1)^{n+k}} \qquad \text{as required.}$$

4.21 $y = e^x; \quad \dfrac{dy}{dx} = e^x = y; \quad x = \log_e(y)$

$$f_Y(y) = \frac{1}{y\sigma \sqrt{(2\pi)}} \exp\left\{ -\frac{1}{2} \left(\frac{\log_e y - \mu}{\sigma} \right)^2 \right\} \qquad \text{for } y \ge 0.$$

4.22 $\displaystyle M(\theta) = \sum_{k=1}^\infty e^{\theta k} p_k = -\frac{1}{\log(1 - \alpha)} \sum_{k=1}^\infty \frac{(\alpha e^\theta)^k}{k}$

Note that $\quad dM(\theta)/d\theta = -\dfrac{1}{\log(1 - \alpha)} \displaystyle\sum_{j=0}^\infty (\alpha e^\theta)^j \alpha e^\theta$

i.e. $\quad \dfrac{dM(\theta)}{d\theta} = \dfrac{\alpha e^\theta}{\log(1 - \alpha)(\alpha e^\theta - 1)} \qquad \text{for } \alpha e^\theta < 1$

Therefore $\quad M(\theta) = \kappa + \dfrac{1}{\log(1 - \alpha)} \log(1 - \alpha e^\theta)$

$$M(0) = 1 \text{ so } \kappa = 0.$$

$$\Pr(X = k) = -\frac{1}{\log(1 - \alpha)} \int_0^\alpha y^{k-1}\, dy = -\frac{\alpha^k}{k \log(1 - \alpha)} \qquad \text{for } k \ge 1.$$

This question is continued in Exercise 5.14.

Chapter 5

5.1   Let $W = 1 - U$; $f_W(w) = f_U(u)\left|\dfrac{du}{dw}\right|$

$$\left|\frac{du}{dw}\right| = 1 \qquad \text{and the result is proved.}$$

5.2 $F_{\tilde{X}}(w) = \Pr(\tilde{X} \le w) = \Pr(\tilde{X} \le w \,|\, \tilde{X} = X) \Pr(\tilde{X} = X)$

$$+ \Pr(\tilde{X} \le w \,|\, \tilde{X} = -X) \Pr(\tilde{X} = -X)$$

$$= \tfrac{1}{2}\Pr(X \le w) + \tfrac{1}{2}\Pr(-X \le w)$$

$$= \tfrac{1}{2}\Pr(X \le w) + \tfrac{1}{2}\Pr(X \ge -w)$$

Therefore if $w \ge 0$,

$$F_{\tilde{X}}(w) = \frac{1}{2}\int_0^w f_X(x)\,dx + \frac{1}{2} = \Phi(w)$$

and if $w \le 0$,

$$F_{\tilde{X}}(w) = 0 + \tfrac{1}{2}\Pr(X \ge -w) = \Phi(-w).$$

5.3   Poisson random variables with large mean values, μ, say, can be simulated as the sum of independent Poisson variables with means which sum to μ.

5.4   We can take a Poisson random variable X, with mean 3 as an illustration:

i	0	1	2	3	4	5	6
$\Pr(X = i)$	0.0498	0.1494	0.2240	0.2240	0.1680	0.1008	0.0504

i	7	8	9	10	≥ 11
$\Pr(X = i)$	0.0216	0.008	0.003	0.0008	0.0002

(b) Here we can set $\theta = 2$, say; $\Pr(X \le 2) = 0.4232$.
    As a first stage, check to see if $U > 0.4232$. If so, then it is not necessary to check U against $\Sigma_{i=0}^{j}\Pr(X = i)$, for $j \le \theta = 2$.

(c) Here we could take $\theta_1 = 1$ and $\theta_2 = 4$
    $\Pr(X \le 1) = 0.1992$; $\Pr(X \le 4) = 0.8152$, and as a first stage we check to see where U lies:

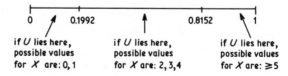

if U lies here,   if U lies here,   if U lies here,
possible values   possible values   possible values
for X are: 0, 1   for X are: 2, 3, 4   for X are: ≥ 5

(d) When the probabilities are ordered we have:

i	2	3	4	1	5	6
$\Pr(X = i)$	0.224	0.224	0.168	0.1494	0.1008	0.0504

i	0	7	8	9	10	≥ 11
$\Pr(X = i)$	0.0498	0.0216	0.008	0.003	0.0008	0.0002

Ordering is a time-consuming operation but once it is done we obtain the obvious benefit of checking the most likely intervals first. As a first step one might check to see if $U \leq 0.616$. If so, X takes one of the values 2, 3 or 4.

5.5 (a) Illustrate $F(x)$ by means of a diagram. Note the symmetry. If $U \leq 0.5$, set $X = \sqrt{(2U)}$; if $U \geq 0.5$, set $X = 2 - \sqrt{(2 - 2U)}$.

(b) We also obtain such an X by setting $X = U_1 + U_2$, where U_1 and U_2 are independent, identically distributed $U(0, 1)$ random variables (see Exercise 4.8).

5.7 $$F_X(x) = 6 \int_0^x y(1 - y)\,dy = 6 \left[\frac{y^2}{2} - \frac{y^3}{3} \right]_0^x = 3x^2 - 2x^3$$

Therefore set $U = 3X^2 - 2X^3$ and solve this cubic equation in X.

5.8 (a) $$F(x) = \frac{1}{\pi} \int_{-\infty}^x \frac{dy}{(1 + y^2)}, \quad -\infty \leq x \leq \infty.$$

Set $y = \tan\theta$; $$F(x) = \frac{1}{\pi} \int_{-\pi/2}^{\tan^{-1}x} d\theta = \frac{1}{\pi}\left(\tan^{-1}x + \frac{\pi}{2}\right)$$

for $-\dfrac{\pi}{2} \leq \theta \leq \dfrac{\pi}{2}$

Therefore set $U\pi = \tan^{-1}x + \dfrac{\pi}{2}$,

i.e. $\tan(U\pi + \pi/2) = X$
or, equivalently,
$X = \tan(\pi U)$

(b) We saw, from Section 4.2.1, that we can write
$$N_1 = (-2\log_e U_1)^{1/2}\sin 2\pi U_2$$
$$N_2 = (-2\log_e U_1)^{1/2}\cos 2\pi U_2$$
where U_1 and U_2 are independent $U(0, 1)$ random variables. Thus $N_1/N_2 = \tan(2\pi U_2)$, which clearly has the same distribution as $X = \tan(\pi U)$ above, i.e. the standard Cauchy.

5.9   Let $u^{-1} = 1 + \exp(-2a_1 \tilde{x}(1 + a_2 \tilde{x}^2))$
Therefore we need to solve

$$\log_e(u^{-1} - 1) = -2a_1 \tilde{x}(1 + a_2 \tilde{x}^2)$$

Use the result: If $ax^3 + x - b = 0$, $x = c - 1/(3ac)$
where

$$2ac^3 = b + \left(b^2 + \frac{4}{27a}\right)^{1/2} \qquad \text{(see Page, 1977).}$$

5.10   See Kemp and Loukas (1978a, b). They suggest truncating the distribution—e.g. to give values of $\{p_{ij}\}$ summing, over i and j, to give unity to, say, 4 decimal places. Then use the approach of Exercise 5.4(d). They found this approach faster than first simulating a marginal variable and then simulating the bivariate distribution via the conditional distribution.

5.11   One approach is via the conditional and marginal distributions, as above, in the comments on Exercise 5.10.

5.12 $\Pr(Y = i - 1) = e^{-\lambda(i-1)} - e^{-\lambda i} = e^{-\lambda i}(e^\lambda - 1)$ for $i \geq 1$

$$= e^{-\lambda(i-1)}(1 - e^{-\lambda}) \quad \text{for } i \geq 1.$$

Hence, from Section 2.6, $(Y + 1)$ has a geometric distribution with $q = e^{-\lambda}$. This relationship is not surprising since both geometric and exponential random variables measure waiting-times, as stated in Section 2.10.

5.13   (a) $F(x) = 1/(1 + e^{-x})$
Therefore set $U = 1/(1 + e^{-X})$

$$U^{-1} = 1 + e^{-X}$$

$$X = -\log_e(U^{-1} - 1).$$

(b) $F(x) = \dfrac{\beta}{\gamma^\beta} \displaystyle\int_0^x w^{\beta - 1} \exp[-(w/\gamma)^\beta]\,dw$

$$= \left[\exp[-(w/\gamma)^\beta]\right]_x^0 = 1 - \exp[-(x/\gamma)^\beta]$$

Therefore set $U = 1 - \exp[-(X/\gamma)^\beta]$

i.e. $\exp[-(X/\gamma)^\beta] = 1 - U$; $-\left(\dfrac{X}{\gamma}\right)^\beta = \log_e(1 - U)$,

$$X = \gamma(-\log_e(1 - U))^{1/\beta}$$

or equivalently, and more simply (see Exercise 5.1)

$$X = \gamma(-\log_e U)^{1/\beta}$$

(c) Set $U = 1 - (k/X)^a$; $\quad 1 - U = (k/X)^a$
$$k(1-U)^{-1/a} = X$$

or equivalently, as in (b)
$$X = kU^{-1/a}$$

(d) Set $U = \exp(-\exp((\xi - X)/\theta))$
$\quad -\log_e U = \exp((\xi - X)/\theta)$
$\quad \theta \log_e(-\log_e U) = \xi - X,$
$\quad X = \xi - \theta \log_e(-\log_e U)$

5.14 To complete the simulation of a random variable with the logarithmic distribution of Exercise 4.22, we need to simulate the random variable Y, for which
$$F_Y(y) = \log(1-y)/\log(1-\alpha) \qquad \text{for } 0 \leq y \leq \alpha.$$

To do this by the inversion method, set $U = \log(1-Y)/\log(1-\alpha)$, i.e.
$\log(1-Y) = U \log(1-\alpha)$; $\quad 1-Y = (1-\alpha)^U$;
$$Y = 1 - (1-\alpha)^U$$

5.15 (a) First we find $f(x, y)$:
$$f(x, y) = f(x)f(y)(1 - \alpha(1 - F(x))(1 - F(y))) + f(x)F(y)\alpha f(y)$$
$$\times (1 - F(x))$$
$$= \alpha F(x)f(y)f(x)(1 - F(y)) - \alpha f(x)f(y)F(x)F(y)$$

Now make the transformation of variable: $U = F(X)$, $V = F(Y)$:
$$f(u, v) = 1 - \alpha(1-u)(1-v) + \alpha v(1-u) + \alpha u(1-v) - \alpha uv$$
$$= 1 - \alpha(1 - 2u)(1 - 2v)$$

(b) $f(u, v) = (1 - \alpha \log u)(1 - \alpha \log v) - \alpha) vu^{-\alpha \log v}$

(c) $f(u, v) = \dfrac{\pi}{2} \dfrac{\{1 + \tan^{-2}(\pi u)\}\{1 + \tan^{-2}(\pi v)\}}{\{1 + \tan^{-2}(\pi u) + \tan^{-2}(\pi v)\}^{3/2}}$

5.16 Rejection, with a high probability of rejection, can be applied simply by generating a uniform distribution of points over an enveloping rectangle, and accepting the abscissae of the points that lie below the curve. High rejection here is not too important as this density is only sampled with probability 0.0228.

5.17 $X = \cos \pi U$; $\quad f_X(x) = \left| \dfrac{du}{dx} \right|$

$\dfrac{dx}{du} = -\pi \sin(\pi u)$; $\quad f_X(x) = \dfrac{1}{\pi \sin \pi u} = \dfrac{1}{\pi \sqrt{(1-x^2)}}$

$$\text{for } -1 \leq x \leq 1.$$

To simulate X without using the 'cos' function we can use the same approach as in the Polar Marsaglia method:

$$\cos \pi U = 2 \cos^2 \left(\frac{\pi U}{2} \right) - 1$$

Thus, if U_1, U_2 are independent $U(0, 1)$ random variables

if $U_1^2 + U_2^2 < 1$, set $X = \dfrac{2U_1^2}{U_1^2 + U_2^2} - 1 = \left(\dfrac{U_1^2 - U_2^2}{U_1^2 + U_2^2} \right)$.

5.18 Using a rectangular envelope gives an acceptance probability of 2/3. Using a symmetric triangular envelope gives an acceptance probability of 2/3 also, but the acceptance probability of 8/9 results from using a symmetric trapezium. Simulation from a trapezium density is easily done using inversion.

5.19 For the exponential envelope we simulate from the half-logistic density:

$$f(x) = 2e^{-x}/(1 + e^{-x})^2 \qquad \text{for } x \geq 0$$

We need to choose $k > 1$ so that $ke^{-x} > 2e^{-x}/(1 + e^{-x})^2$ for all $x \geq 0$, i.e. choose k so that $k > 2(1 + e^{-x})^{-2}$ for all $x \geq 0$, and this is done for smallest k by setting $k = 2$ ($x = \infty$).
If U_1, U_2 are independent $U(0, 1)$ random variables, set $X = -\log_e U_1$ if $U_2 < (1 + U_1)^{-2}$. Finally transform to $(-\infty, \infty)$ range.

5.20
$$\left. \begin{array}{l} f(x) = \dfrac{1}{\sqrt{(2\pi)}} e^{-x^2/2} \\[2mm] h(x) = e^{-x}(1 + e^{-x})^{-2} \end{array} \right\} \qquad -\infty \leq x \leq \infty$$

Set $q(x) = \dfrac{1}{\sqrt{(2\pi)}} e^{-x^2/2} (1 + e^{-x})^2 e^x$.

If $l(x) = \log_e q(x)$

$$l(x) = -\frac{1}{2} \log_e (2\pi) + x - \frac{x^2}{2} + 2 \log_e (1 + e^{-x})$$

$$\frac{dl(x)}{dx} = 1 - x - \frac{2}{(1 + e^x)} = 0 \qquad \text{when } x = 0$$

$$\frac{d^2l(x)}{dx^2} = -1 + \frac{2e^x}{(1 + e^x)^2} < 0 \qquad \text{when } x = 0$$

Therefore $x = 0$ maximizes $q(x)$, and $k = \max_x (q(x)) = \dfrac{4}{\sqrt{(2\pi)}}$
$= 1.596$.

To operate the rejection method here we need to simulate from $h(x)$, and this is readily done by the inversion method of Exercise 5.13(a); probability of rejection $= 1 - 1/k = 0.37$. This may seem surprisingly high, but recall that the two distributions have different variances (1 and $\pi^2/3$) and a better approach would be to use a logistic distribution of unit variance (see Fig. 5.14).

5.21 $k = \max\limits_{x} \left\{ \sqrt{\left(\dfrac{2}{\pi}\right)} \exp\left[-x^2/2\right] / (\lambda \exp\left[-\lambda x\right]) \right\}$

i.e. $k = \max\limits_{x} \left\{ \left(\dfrac{1}{\lambda} \sqrt{\left(\dfrac{2}{\pi}\right)}\right) \exp\left(\lambda x - x^2/2\right) \right\}$

Let $y = \lambda x - x^2/2$; $\dfrac{dy}{dx} = \lambda - x = 0$ when $x = \lambda$, $\dfrac{d^2 y}{dx^2} = -1$.

Hence k is obtained for $x = \lambda$, to give:

$$k = \frac{1}{\lambda} \sqrt{\left(\frac{2}{\pi}\right)} \exp\left(\lambda^2/2\right)$$

The method becomes: accept $X = -\dfrac{1}{\lambda} \log_e U_1$ if

$U_1 U_2 \le \exp\left[-(\lambda^2 + X^2)/2\right]$, for independent $U(0, 1)$ random variables U_1 and U_2.

Probability of rejection $= 1 - \lambda \sqrt{\left(\dfrac{\pi}{2}\right)} \exp\left(-\lambda^2/2\right)$

Let $y = \lambda \exp\left(-\lambda^2/2\right)$

let $z = \log y = \log \lambda - \lambda^2/2$; $\dfrac{dz}{d\lambda} = \dfrac{1}{\lambda} - \lambda = 0$ when $\lambda = 1$.

$\dfrac{d^2 z}{d\lambda^2} = -\dfrac{1}{\lambda^2} - 1, < 0$ when $\lambda = 1$

Thus taking $\lambda = 1$ minimizes the probability of rejection.

5.22 (a) Use the inversion method:

$$F(x) = \lambda \mu \int_0^x y^{\lambda - 1} (\mu + y^\lambda)^{-2}\, dy = \left[\mu(\mu + y^\lambda)^{-1} \right]_x^0$$

$$= 1 - \mu(\mu + x^\lambda)^{-1} \quad \text{for } x \ge 0.$$

Set $U = 1 - \mu(\mu + X^\lambda)^{-1}$; $1 - U = \mu(\mu + X^\lambda)^{-1}$

or equivalently

$$\frac{\mu}{U} = \mu + X^\lambda; \quad X = \left\{ \mu\left(\frac{1}{U} - 1\right) \right\}^{1/\lambda}$$

(b) $k = 4\alpha^{\alpha} e^{-\alpha} (\Gamma(\alpha) \sqrt{(2\alpha - 1)})^{-1}$

resulting in $1 - k^{-1}$ as the probability of rejection.

5.23 The ratio of the k values from Example 5.6, and Exercise 5.22 is:
$4e^{-1} (\sqrt{(2\alpha - 1)})^{-1}$, which is > 1 when $4 > e \sqrt{(2\alpha - 1)}$,
i.e. $\alpha < 8e^{-2} + 0.5 = 1.5827$.

Thus for values of $1 < \alpha < 1.5827$, Cheng's algorithm has the smaller probability of rejection. A full comparison also would include the relative speeds of the algorithms.
Cheng's algorithm becomes:

(i) Set $V = (2\alpha - 1)^{-1/2} \log_e [U_1/(1 - U_1)]$, and $X = \alpha e^V$
(ii) Accept X if $\alpha - \log_e 4 + (\alpha + (2\alpha - 1)^{1/2}) V - X \geq \log_e (U_1^2 U_2)$

Cheng noted that $\log_e x$ is a concave function of $x > 0$, and so for any given $\theta > 0$, $\theta x - \log \theta - 1 \geq \log x$.
Thus at stage (ii) the left-hand-side can first of all be checked against $\theta U_1^2 U_2 - \log \theta - 1$, for some fixed θ (he suggested $\theta = 4.5$, irrespective of α). If the inequality is satisfied for $\theta U_1^2 U_2 - \log \theta - 1$, then *a fortiori* it is satisfied for $\log(U_1^2 U_2)$ and a time-consuming logarithm need not be evaluated. Cheng's algorithm can also be adapted for the case $\alpha < 1$. Cheng and Atkinson and Pearce provide timings of these algorithms run on two different computers (see also Atkinson, 1977a).

5.24 Set $X = 9 - 2 \log_e U$.
If this is not obvious, show that it results from applying the inversion method to the conditional density:

$$f(x) = \tfrac{1}{2} \exp((9 - x)/2) \qquad \text{for } x \geq 9$$

5.25 The objective is, as in the solution to Exercise 5.23, to improve efficiency, by evaluating simpler functions much of the time.

5.26 $f_X(x) = \left| \dfrac{du}{dx} \right|; \quad x^2 = a^2 - 2 \log u; \quad u = \exp[\tfrac{1}{2}(a^2 - x^2)]$

$2x \dfrac{dx}{du} = -\dfrac{2}{u}; \quad f_X(x) = x \exp[\tfrac{1}{2}(a^2 - x^2)]$

$\Pr(U_2 X < a) = a \displaystyle\int_a^{\infty} \dfrac{1}{x} x \exp[\tfrac{1}{2}(a^2 - x^2)] \, dx$

$\qquad = a \exp[a^2/2] \displaystyle\int_a^{\infty} \exp[-x^2/2] \, dx$

$\qquad = \sqrt{(2\pi)} a \exp[a^2/2] (1 - \Phi(a))$

Therefore with the conditioning event: $\{U_2 X < a\}$,

$$f_X(x) = \int_0^{a/x} \frac{x \exp[\frac{1}{2}(a^2 - x^2)]\,du}{\sqrt{(2\pi)}a \exp[a^2/2](1 - \Phi(a))} = \frac{\phi(x)}{(1 - \Phi(a))} \qquad \text{for } x \geq a.$$

5.27 Solution follows from the Polar Marsaglia algorithm, symmetry and Exercise 5.8(b).

5.28 $f_{X_1 X_2}(x_1 x_2) = e^{-x_1 - x_2}$ for $x_1 \geq 0; x_2 \geq 0$.

Probability of conditioning event $= \displaystyle\int_0^\infty e^{-x_1} \int_{\frac{1}{2}(x_1 - 1)^2}^\infty e^{-x_2}\,dx_2\,dx_1$,

$$= \int_0^\infty e^{-x_1} e^{-\frac{1}{2}(x_1 - 1)^2}\,dx_1 = e^{-1/2} \int_0^\infty e^{-\frac{1}{2}x_1^2}\,dx_1 = \frac{\sqrt{(2\pi)}e^{-1/2}}{2}$$

Therefore under the conditioning

$$f_{X_1}(x_1) = \frac{e^{-x_1} \displaystyle\int_{\frac{1}{2}(x_1 - 1)^2}^\infty e^{-x_2}\,dx_2}{\sqrt{(\pi/(2e))}}$$

$$= \sqrt{\left(\frac{2}{\pi}\right)} e^{-x_1^2/2} \qquad \text{for } x_1 \geq 0, \text{ as required.}$$

5.29 This is simply an alternative way of writing the rejection method: in the notation of Section 5.3,

$$k = \max_x \left\{\frac{f(x)}{h(x)}\right\}$$

Therefore $f(x) \leq kh(x)$

$$f(x) = kh(x)\left\{\frac{f(x)}{kh(x)}\right\} = kh(x)s(x) \qquad \text{where } s(x) \leq 1.$$

In this case we have $h(x)$ as a p.d.f. We simulate X from $h(x)$, and then accept it if $Ukh(x) < f(X)$, i.e. $U < f(x)/kh(x)$. We obtain the situation of the question by setting:

$$c = k/m,$$

$$g(x) = h(x)$$

$$\gamma(x) = (mf(x))/(kh(x))$$

Probability of rejection $= 1 - \dfrac{1}{k} = 1 - \dfrac{1}{(mc)}$.

5.31 See the discussion of Example 5.5.

5.32 (a) One possibility is to set $Y = V_1 + V_2 - 2/\alpha$
where V_1 and V_2 are independent $U(0, 2/\alpha)$ random variables.
Set $X = |Y|$.

(b) Set $q(x) = \dfrac{f(x)}{f_1(x)} = \dfrac{\lambda e^{-\lambda x}}{\left(\alpha - \dfrac{\alpha^2 x}{2}\right)}$

Let $w(x) = \log q(x) = \kappa - \lambda x - \log(1 - \alpha x/2)$, for constant κ.

$$\frac{dw(x)}{dx} = -\lambda + \frac{\alpha}{2 - \alpha x} = 0 \text{ when } \lambda = \alpha/(2 - \alpha x),$$

i.e. when $x = (2/\alpha - 1/\lambda)$. At this point,

$$\frac{d^2 w(x)}{dx^2} > 0$$

and so we have a minimum. Note that as $2\lambda > \alpha$, $\dfrac{2}{\alpha} > \dfrac{1}{\lambda}$.

The shrinking factor is $q\left(\dfrac{2}{\alpha} - \dfrac{1}{\lambda}\right) = \dfrac{2\lambda^2}{\alpha^2} e^{1 - 2\lambda/\alpha}$.

Let $u(\alpha) = \log\left(q\left(\dfrac{2}{\alpha} - \dfrac{1}{\lambda}\right)\right) = \kappa - \dfrac{2\lambda}{\alpha} - 2\log\alpha$, for constant κ.

$$\frac{du(\alpha)}{d\alpha} = +\frac{2\lambda}{\alpha^2} - \frac{2}{\alpha} = 0 \text{ when } \lambda = \alpha.$$

This value maximizes $u(\alpha)$, and so the probability of simulating from $f_1(x)$ is less than or equal to $2/e$.

5.33 (b) $f(y) = \dfrac{e^y}{(e-1)} = (e-1)^{-1} \displaystyle\sum_{i=0}^{\infty} \dfrac{y^i}{i!} = \displaystyle\sum_{i=0}^{\infty} \dfrac{(i+1)y^i}{(i+1)!(e-1)}$

5.34 Sample from the density:

$$f_m(x) = \frac{e^{m-x}}{(e-1)} \qquad \text{for } (m-1) < x \le m,$$

and m is an integer ≥ 1, with probability $(e-1)e^{-m}$ for $m \ge 1$.

5.35 For (ii), $\Pr(W \le w) = \displaystyle\prod_{i=1}^{I+1} \Pr(U_i \le w) = w^{I+1}$

Therefore $f_W(w) = (I+1)w^I$ for $1 \ge w \ge 0$

Thus we see, from Exercise 5.33(b) that steps (i) and (ii) return a random variable W with p.d.f.

$$f_W(y) = \frac{e^y}{(e-1)} \qquad \text{for } 0 \le y < 1$$

But we see from Exercise 5.34 that to simulate from e^{-x} we simulate from $e^{m-x}/(e-1)$ with probability $(e-1)e^{-m}$. Step (iii) selects m and then, from Exercise 5.33(a), step (iv) converts from W to X.

This is a way of simulating exponential random variables without evaluating a logarithm. The two discrete distributions involved have the forms:

i	$(e-1)e^{-i}$	$\dfrac{1}{i!(e-1)}$
1	0.63	0.58
2	0.23	0.29
3	0.09	0.10
4	0.03	0.02
$\vdots$	$\vdots$	$\vdots$

On average only 1.582 $U(0, 1)$ random variables need to be considered at the maximization stage.

5.37 This provides a fast algorithm but many constants have to be stored.

5.38 This algorithm utilizes a number of interesting features for improving efficiency. Step 3 uses the algorithm of Marsaglia (1964) as modified by Ahrens and Dieter (1972) for sampling X values in the tail, $|X| > \zeta$. cf. Exercise 5.26. Steps 5, 7 and 8 correspond to the three triangle rejection procedures for the three main triangles illustrated in Fig. 5.16 (see Exercise 5.25). 'Generate' means select an independent $U(0, 1)$ random variable.

5.42 This proof is rather difficult—see the source paper for details.

Chapter 6

6.2 In order, values of $(n-1)s^2/\bar{x}$ are: 514.97, 464.07, 452.09, 472.05, 551.89, 470.56. Using the approximation to χ^2 of Exercise 6.24 gives approximate standard normal values in the range $(-1.51, 1.65)$.

6.3 If dice are fair, and thrown independently,

$\Pr(0 \text{ sixes}) = 25/36$; $\Pr(1 \text{ six}) = 5/18$; $\Pr(2 \text{ sixes}) = 1/36$, resulting in:

No. of sixes	0	1	2	Total
Expected frequency	150	60	6	216

$\chi_2^2 = 9.6$, significant at the 1% level using a 1-tail test.
If p is unknown, the likelihood of the data L is given by:

$$L \propto (1-p)^{n_1 + 2n_0} p^{n_1 + 2n_2}$$

and the maximum-likelihood estimator of p is then $\hat{p} = (n_1 + 2n_2)/\{2(n_0 + n_1 + n_2)\}$ and here $\hat{p} = 0.2222$, resulting in the expected values:

No. of sixes	0	1	2	Total
Expected frequency	130.7	74.7	10.7	216.1

Now $\chi_1^2 = 0.072$, not significant at the 5% level, using a 1-tail test.

6.4 $\Pr(R \le r) = \displaystyle\int_{-\infty}^{\infty} \Pr(n-1 \text{ values lie in } (x, x+r) \mid \text{smallest of } n$
values $= x)m(x)\mathrm{d}x$ where $m(x)$ is the p.d.f. of the smallest of n values.
Thus

$$\Pr(R \le r) = \int_{-\infty}^{\infty} \left(\frac{\Phi(x+r) - \Phi(x)}{1 - \Phi(x)} \right)^{n-1} n(1 - \Phi(x))^{n-1} \phi(x)\, \mathrm{d}x$$

(cf. Exercise 6.13)

$$= n \int_{-\infty}^{\infty} (\Phi(x+r) - \Phi(x))^{n-1} \phi(x)\, \mathrm{d}x$$

6.5   For a p-dimensional proof, see Kendall and Stuart (1969, p. 355). Equation (6.3) states that

$$D^2 = \frac{1}{(1-\rho^2)} (N_1^2 - 2\rho N_1 N_2 + N_2^2)$$

is χ_2^2, where N_1, N_2 are bivariate normal, with zero means, unit variances, and correlation ρ. By Exercise 4.16, we can therefore write

$\left.\begin{array}{l} N_1 = X_1 \\ N_2 = \rho X_1 + (1-\rho^2)^{1/2} X_2 \end{array}\right\}$ where X_1 and X_2 are independent and $N(0, 1)$

and so

$$D^2 = \frac{1}{(1-\rho^2)} \{X_1 + X_2(X_2 - 2\rho X_1)\} = \frac{1}{(1-\rho^2)} \{X_1^2 + (\rho X_1$$
$$+ (1-\rho^2)^{1/2} X_2) \times ((1-\rho^2)^{1/2} X_2 - \rho X_1)\}$$
$$= \frac{1}{(1-\rho^2)} \{(1-\rho^2)(X_1^2 + X_2^2)\},$$

i.e. D^2 has a χ_2^2 distribution.

6.6   (i) Time to first new digit is simply 1. Time to next new digit is X_2, etc.

$$G_{X_i}(z) = (11 - i)z(10 - (i-1)z)^{-1} \qquad \text{for } 2 \leq i \leq 10,$$

and so $\displaystyle G_S(z) = z \prod_{i=1}^{9} \frac{iz}{(10 - (10 - i)z)}$

$$G_S(z) = \sum_{j=10}^{\infty} z^j \frac{9}{10^8} \sum_{i=1}^{9} \binom{8}{i-1} \frac{i^8}{(10-i)} (-1)^{9-i}$$

$$\times \left(1 - \frac{i}{10}\right)\left(\frac{i}{10}\right)^{j-10}$$

$$= \sum_{j=10}^{\infty} z^j \frac{9}{10^9} \sum_{i=1}^{9} \binom{8}{i-1} (-1)^{9-i} \frac{i^{j-2}}{10^{j-10}}$$

$$= \sum_{j=10}^{\infty} z^j 9 \cdot 10^{1-j} \sum_{i=1}^{9} \binom{8}{i-1} (-1)^{9-i} i^{j-2}$$

$$= \sum_{j=10}^{\infty} z^j 10^{1-j} \sum_{v=1}^{9} (-1)^{v+1} \binom{9}{v-1} (10-v)^{j-1}.$$

(ii) Pr(at least one cell is empty) $= S_1 - S_2 + S_3 - S_4 \ldots \pm S_n$,

where $\displaystyle S_1 = \sum_i p_i; \; S_2 = \sum\sum_{i<j} p_{ij}$, etc., and

$$p_i = \left(1 - \frac{1}{n}\right)^r$$

$$p_{ij} = \left(1 - \frac{2}{n}\right)^r$$

$$p_{ijk} = \left(1 - \frac{3}{n}\right)^r, \text{ etc.} \qquad \text{(see Feller, 1957, p. 89).}$$

Therefore for every $v \leq n$, $S_v = \binom{n}{v}\left(1 - \frac{v}{n}\right)^r$

and so $u(r, n) = 1 - S_1 + S_2 \ldots$

$$= \sum_{v=0}^{n} (-1)^v \binom{n}{v}\left(1 - \frac{v}{n}\right)^r$$

and so Pr$(j) = u(j, 10) - u(j-1, 10)$

$$\mathrm{Pr}(j) = 10^{-j} \sum_{v=0}^{10} v(-1)^{v+1} \binom{10}{v} (10-v)^{j-1}$$

$$= 10^{1-j} \sum_{v=1}^{9} (-1)^{v+1} \binom{9}{v-1} (10-v)^{j-1}$$

for $j \geq 10$.

6.7 Failure of the frequency test does not necessarily imply failure of other tests.

6.12 For example, take an acceptable sequence of digits and replace i by ii, for all digits i.

6.13 $\Pr(M \le x) = x^n = F_M(x)$; $f_M(x) = nx^{n-1}$ for $0 \le x \le 1$ (cf. Exercises 2.10 and 2.11).

Let $Y = M^n$; $\dfrac{dy}{dm} = nm^{n-1}$, and Y is $U(0, 1)$, facilitating the application of this test.

6.18 Use $-2\sin^2(2\pi U_2)\log_e U_1$ and $-2(1 - \sin^2 2\pi U_2)\log_e U_1$, thus avoiding the use of cos and square root.

6.22 Variables can be summed in groups and tests applied to the sums, using the results of Exercise 4.17. We saw in Fig. 4.5 that putting exponential variates end-to-end is equivalent to simulating a Poisson process. Conditional upon k events in a Poisson process (in time, say) occurring within some fixed interval, then the times of occurrence are uniformly distributed over that interval, and for a standard-length interval, when an odd number of events occurs the median time may be referred to the appropriate beta p.d.f.

6.24 Due to Fisher, this result is discussed by Kendall and Stuart (1969, p. 371). A more accurate approximation, due to Wilson and Hilferty, is that if X has a χ_v^2 p.d.f.,

$$(X/v)^{1/2} \approx N(1 - 2/9v, \, 2/9v)$$

6.28 As the entire cycle is used, the frequency test would reject the numbers as too uniform. In practice one only uses a fraction of a cycle (cf. comments in Section 3.5).

6.29 See Knuth (1981, p. 517) for an efficient procedure.

Chapter 7

7.1 $2l/\pi d$

7.2 When $l \le d$,

$$\Pr(\text{needle intersects line}) = 2\int_{-\pi/2}^{\pi/2} \frac{l\cos\theta}{2\pi d}\, d\theta = \frac{2l}{\pi d}$$

When $l \ge d$,

$$\Pr(\text{needle intersects line}) = \int_{-\pi/2}^{\pi/2} \frac{\min(l\cos\theta, d)}{2\pi d}\, d\theta$$

$$= \frac{2}{\pi d}\left\{ d\cos^{-1}\left(\frac{d}{l}\right) + l(1 - \sqrt{(1 - d^2/l^2)}) \right\}$$

7.4

	Estimates of π		
Experiment 1:	2.72	3.013	3.125
Experiment 2:	3.212	3.212	3.207

7.5   If Pr (coin lands totally within square) $= 0.5$, then $(a-b)^2/a^2 = \frac{1}{2}$

$$2(a-b)^2 = a^2; \quad 2b^2 - 4b + 1 = 0$$

if we set $a = 1$ and $b = 1 - 1/\sqrt{2}$.

7.6   See Fletcher (1971).

7.7   The variability of the integrand is greater when $x > 0$ and the range of integration is $(-\infty, x)$. For $x > 0$, it is better to set

$$\Phi(x) = \frac{1}{2} + \int_0^x \phi(x)\,dx$$

and then evaluate the integral over the $(0, x)$ range.

7.8   Write

$$\int_0^1 e^{-x^2}\,dx = \int_0^1 (1 + (1 - e^{-1})x)\,dx + \int_0^1 (e^{-x^2} - 1 - (1 - e^{-1})x)\,dx.$$

The first integral on the right-hand side is known $= (3 - e^{-1})/2$, while the variability of $e^{-x^2} - 1 - (1 - e^{-1})x$ is less than e^{-x^2}.

7.10   With just two strata we might have:

$$\tilde{\theta} = \sum_{i=1}^{n_1} \frac{\alpha}{n_1} f(\alpha U_i) + \sum_{j=n_1+1}^{n} \frac{(1-\alpha)}{(n-n_1)} f(\alpha + (1-\alpha)U_j)$$

If we choose $n_1 = n - n_1 = m$ then we can incorporate antithetic variates from

$$\tilde{\theta} = \frac{1}{m} \sum_{i=1}^{m} \{\alpha f(\alpha U_i) + (1 - \alpha)f[\alpha + (1 - \alpha)(1 - U_i)]\}$$

7.11   Let $I = \int_0^1 [x(x-2)(x^2-1)]^{1/2}\,dx$

Set $x = z + \frac{1}{2}$

$$I = \int_{-1/2}^{1/2} [(z^2 - \frac{9}{4})(z^2 - \frac{1}{4})]^{1/2}\,dz$$

Set $2z = \sin\theta$

$$I = \frac{3}{8} \int_{-\pi/2}^{\pi/2} \left(1 - \frac{\sin^2\theta}{9}\right)^{1/2} \cos^2\theta\,d\theta$$

i.e., $I = \dfrac{3}{8} \displaystyle\int_{-\pi/2}^{\pi/2} \cos^2\theta \left\{ 1 - \dfrac{\sin^2\theta}{18} - \sum_{k=2}^{\infty} \dfrac{\sin^{2k}\theta\,(2k-3)!}{k!\,3^{2k}\,2^{2k-2}\,(k-2)!} \right\} d\theta$

Now use:

$$\int_0^{\pi/2} \cos^2\theta\,d\theta = \frac{\pi}{4}$$

$$\int_0^{\pi/2} \sin^{2k}\theta\,d\theta = \pi \binom{2k-1}{k} 2^{-2k}$$

$$\int_0^{\pi} \cos^2\theta \sin^{2k}\theta\,d\theta = \frac{\pi\,2^{-2k-1}\,(2k-1)!}{(k+1)!\,(k-1)!}$$

and sum the infinite series using an appropriate truncation rule.

7.12 For related discussion, see Simulation I (1976, pp. 44–46).

7.13 $\hat{\pi} = \dfrac{4}{n} \displaystyle\sum_{i=1}^{n} \sqrt{(1 - U_i^2)}$

$\mathscr{E}[\hat{\pi}] = \pi$

$\mathrm{Var}\,(\hat{\pi}) = \dfrac{16}{n}\,\mathrm{Var}\,(\sqrt{1 - U^2}) = \dfrac{16}{n}\left(\dfrac{2}{3} - \dfrac{\pi^2}{16}\right) \approx \dfrac{16\,(0.0498)}{n}$

By a central limit theorem, $\hat{\pi}$ is approximately distributed as $N(\pi,\, 16(0.0498)/n)$, and so an approximate 95 % confidence interval for $\hat{\pi}$ has width $\approx 3.4997/\sqrt{n} = 0.01$, for example, when $n \approx 122\,480$.

7.16 Variance reduction y, is given by

$$y = -c^2\,\mathrm{Var}\,(Z) + 2c\,\mathrm{Cov}\,(X, Z)$$

$$\frac{dy}{dc} = 2(-c\,\mathrm{Var}\,(Z) + \mathrm{Cov}\,(X, Z))$$

$$= 0 \qquad \text{when } c = \mathrm{Cov}\,(X, Z)/\mathrm{Var}\,(Z) \text{ and } \frac{d^2 y}{dc^2} < 0, \text{ so we have a}$$

maximum.

7.17 Proof from Kleijnen (1974, p. 254):

$$x_i = ax_{i-1} \quad (\mathrm{mod}\; m) \qquad i \geq 1$$

and so antithetic variates are: $r_i = 1 - x_i/m$.

Consider the sequence: $y_i = ay_{i-1}$ (mod m), where $y_0 = m - x_0$, and let $\tilde{r}_i = y_i/m$. First show, by induction, that $x_i + y_i = m$, for $i \geq 0$. Then deduce that $r_i - \tilde{r}_i$ is an integer, but $0 < r_i,\, \tilde{r}_i < 1$, and so we must have $r_i = \tilde{r}_i$.

7.21 Cf. Example 7.3. See Cox and Smith (1961, p. 135) and Rubinstein (1981, p. 150) for discussion.

Chapter 8

8.3

Type of event	Time from start of simulation	Queue size	Total waiting-times
A1	0 ⎤	1	
D1	0.30 ⎦	0	0.30
A2	1.34	1	
A3	⌐1.56	2	
A4	|2.98	3	
D2	|3.04 ⌐	2	1.70
A5	|3.58	3	
D3	⌊3.68 ⌐	2	2.12
A6	4.92	3	
A7	5.31 ⌐	4	
D4	⌊ 5.56 ↓↓↓	3	2.58

8.4 Let the n individual lifetimes be: $X_1, X_2, \ldots, X_n$. From the specification of the process, these are independent random variables with an exponential, e^{-x} p.d.f. for $x \geq 0$. The time to extinction is therefore $Y = \max(X_1, X_2, \ldots, X_n)$. Also, time to first death has the exponential p.d.f., ne^{-nx} for $x \geq 0$, and the time between the $(i-1)$th and ith deaths has the exponential p.d.f., $(n-i+1)e^{-(n-i+1)x}$, for $1 \leq i \leq n$, which is the p.d.f. of $X/(n-i+1)$, where X has the e^{-x} p.d.f. Thus we see that we can also write the time to extinction as,

$$Z = X_1 + \frac{X_2}{2} + \ldots + \frac{X_n}{n}$$

8.8

$$\mathrm{Var}\,(\bar{X}) = \mathrm{Var}\left(\sum_{i=1}^{n} X_i/n\right)$$

$$= \frac{1}{n^2}\left\{\sum_{i=1}^{n} \mathrm{Var}\,(X_i) + 2\sum\sum_{i<j}\mathrm{Cov}\,(X_i, X_j)\right\}$$

$$= \frac{\sigma^2}{n} + \frac{2}{n^2}\sum_{i=1}^{n-1}\sum_{s=1}^{n-i}\mathrm{Cov}\,(X_i, X_{i+s})$$

$$= \frac{\sigma^2}{n} + \frac{2}{n^2}\sum_{s=1}^{n-1}\sum_{i=1}^{n-s}\mathrm{Cov}\,(X_i, X_{i+s})$$

$$= \frac{\sigma^2}{n} + \frac{2\sigma^2}{n^2}\sum_{s=1}^{n-1}\rho_s(n-s)$$

$$= \frac{\sigma^2}{n}\left(1 + 2\sum_{s=1}^{n-1}\left(1-\frac{s}{n}\right)\rho_s\right)$$

8.10 We can envelop e^{-x} by a suitable multiple, k, of the 'half-logistic' p.d.f.

$$f(x) = \frac{2e^{-x}}{(1 + e^{-x})^2} \qquad \text{for } x \geq 0$$

of Exercise 5.19. Thus here we are reversing the rôles played by these p.d.f.'s in that exercise. Clearly we take $k = 2$, and the probability of rejection is 0.5. Using this method we would expect a greatly reduced variance reduction, as the likelihood of rejection interferes with the antithetic variates.

8.11 The following flow diagram is taken from Gross and Harris (1974, p. 384). We see that traffic selects first of all a size of queue. The test $U \leq 0.4$, checks which drivers have the correct change for the automatic booth. Automatic booths are assumed to have constant service-times,

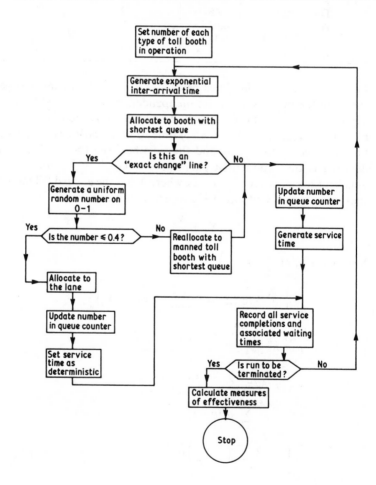

while manual booths have variable service-times from some distribution. Changing booths is a complex procedure, and an alternative possibility would be to select first of all on the availability of the correct change.

8.12 The numbers of individuals turned away per month are all rather large, whatever the distribution of beds.

8.14 Using the formula derived in solution to Exercise 8.8,

$$\text{Var}(\bar{X}) = \frac{\sigma^2}{n}\left(1 + 2\sum_{s=1}^{n-1}\left(1 - \frac{s}{n}\right)\rho_s\right)$$

and the required expression follows from setting $\rho_s = \rho^s$.

8.16 See Donnelly and Shannon (1981) and Schruben (1981) for recent discussion.

8.17 A good discussion is provided by Fishman (1973, p. 233). Empirical distributions are useful in a validation context, but more general results are likely when simulating from a fitted distribution.

Chapter 9

9.1 Ten $N(0, 1)$ variates are put in column vector C1. C2 contains the squares of these, which are summed and put in the constant K1. K1 thus is a realization of a χ^2_{10} random variable, by the result of Exercise 2.5. 100 such values are put into C3 and finally a histogram of these values is produced.

For logistic random variables:

```
NOPRINT
URANDOM 100 OBSERVATIONS, PUT IN C1
LET C2=-LOGE(1/C1-1)
HISTO C2
```

Here the second line simulates 100 $U(0, 1)$ variates and puts them in C1. The third line implements the inversion method, described in the solution to Exercise 5.13, on each element of C1, putting the result in the corresponding element of C2.

For beta random variables: use the MINITAB program of the exercise and the result of Exercise 2.14.

9.3 As stated in Section 1.6, an analytic solution is more elegant and powerful than a solution obtained using simulation. However, as this exercise shows, an analytic solution can be quite complicated, and needs a computer program for its complete enumeration.

9.4   If the second-stage sampling occurs with replacement (i.e. animals are released after capture) then if $p = m/n$, if one samples X animals until $\tilde{m}$ marked animals are recaptured, then

$$\Pr(X = \tilde{m} + k) = \binom{\tilde{m} + k - 1}{\tilde{m} - 1} p^{\tilde{m}}(1 - p)^k, \qquad \text{for} \qquad k \geq 0$$

i.e. X has the negative-binomial distribution of Section 2.6, and

$$\mathscr{E}[X] = \tilde{m}/p = \tilde{m}n/m$$

Thus we can estimate n by:

$$\hat{n} = Xm/\tilde{m}$$

an unbiased estimator of n. For further discussion, see Chapman (1952).

9.5   (i) The percentages of variance explained by the first three principal components are given below:

Principal component	Whole class (%)	13 men (%)	15 women (%)
1	67.6	59.6	53.2
2	11.8	18.0	19.5
3	9.9	9.4	13.8
	89.3	87.0	86.5

The coefficients in the components are given by:

Men

Measurement	Principal component		
	1	2	3
Chest	0.90	0.09	−0.01
Waist	0.94	−0.08	0.07
Wrist	0.60	0.59	−0.34
Hand	0.46	0.70	0.52
Head	0.83	−0.10	−0.41
Height	0.87	−0.23	0.03
Forearm	0.67	−0.59	0.31

Women

	Principal component		
Measurement	1	2	3
Chest	0.86	−0.30	−0.12
Waist	0.91	0.03	0.03
Wrist	0.51	0.73	0.11
Hand	0.63	0.66	0.03
Head	0.53	−0.22	0.80
Height	0.77	−0.51	−0.00
Forearm	0.79	−0.06	−0.54

(ii) The linear discriminant function for all the measurements has the coefficients:

0.78	chest
−1.02	waist
0.25	wrist
−1.47	hand
0.19	head
−0.53	height
0.01	forearm

If we use only chest and waist measurements, the linear discriminant function has the coefficients:

−0.81	chest
2.20	waist

If both of these discriminant functions are applied to the data that gave rise to them, then in each case only one individual is misclassified.

9.6 One would expect a high positive correlation between $\hat{\delta}$ and $\Phi(\hat{\delta}/2)$, and the distribution of $\hat{\delta}^2$ is known, as it is proportional to Hotelling's T^2 statistic—see Mardia *et al.* (1979, p. 77). This therefore suggests using $\hat{\delta}$ as a control variate for $\Phi(\hat{\delta}/2)$, as described in Section 9.4.2.

9.8 For the best subsets, the minimum value R_m of the multiple correlation coefficient (see Mardia *et al.*, 1979, pp. 167–169) between the variables of the set and any of the other variables is maximized, at a value of R_m^*, say. For the 'good' subsets, $0.7 R_m^* \le R_m < R_m^*$, as shown by Jolliffe (1972a).

9.9 The cluster centres for the samples were obtained from sample averages. The two A samples agree well with expectation; however, the C statistic for the second sample from population A does not perform as well as in the other structured examples, possibly due to overlap of the sub-samples. The C statistic provides no clear indication for the unstructured data of population B.

9.10 To be conservative, we can take the first row of Table 9.3 ($n = 8$). The two-dimensional stress value of 6.44% is ≈ 3 standard deviations smaller than the average shown in Table 9.3, adding weight to the 'good' interpretation of this stress value.

9.12 Again we see that the method has worked well in recapturing much of the known structure in the simulated population. Here we would also want to investigate structure using a measure as in Exercise 9.9, and consider 2-, 3-, etc., cluster solutions.

9.13

Winter temperature

October mean pressure (terciles)	Quintiles	1–2	Quintiles	3–5	Total
Upper	(6.4)	11	(9.6)	5	16
Middle	(5.6)	3	(8.4)	11	14
Lower	(6.0)	4	(9)	11	15
Total		18		27	45

Expected values under the hypothesis of no association are given in parentheses. Ignoring the ordering of both the row and column categories, the standard chi-square test of association results in a statistic 8.63, significant at the 2.5% level when referred to χ^2_2 tables. The known (asymptotic) chi-square distribution of the pooled test-statistic may be used to improve precision if that statistic is used as a control variate, as in Section 9.4.2.

9.14 See Tocher (1975, p. 92) for an example with sample median and sample mean, for samples of size 3 from a normal distribution. Here the sample mean is the control variate.

9.16 A BASIC program is given below (cf. Patefield, 1981).

```
10    REM PROGRAM TO PERFORM THE MONTE-CARLO TEST OF
20    REM EXAMPLE 9.3
30    RANDOMIZE
40    DIM A(3,5),B(499)
50    LET N = 499
60    REM 499 RANDOM TABLES WILL BE SIMULATED
70    FOR L = 1 TO N
80      FOR I = 1 TO 3
```

```
90     FOR J = 1 TO 5
100      LET A(I,J) = -3
110      NEXT J
120    NEXT I
130    REM FOR EACH COLUMN WE DISTRIBUTE THE COLUMN TOTAL
140    REM OF 9, UNIFORMLY OVER THE 3 ROWS
150    FOR J = 1 TO 5
160      FOR K = 1 TO 9
170        LET U = RND
180        IF U < .3333333 THEN 220
190        IF U > .666666667 THEN 240
200        LET A(2,J) = A(2,J)+1
210        GOTO 250
220        LET A(1,J) = A(1,J)+1
230        GOTO 250
240        LET A(3,J) = A(3,J)+1
250      NEXT K
260    NEXT J
270    REM S CONTAINS THE RELEVANT PART OF THE CHI-
280    REM SQUARE STATISTIC
290    LET S = 0
300    FOR I = 1 TO 3
310      FOR J = 1 TO 5
320        LET S = S+A(I,J)*A(I,J)/3
330      NEXT J
340    NEXT I
350    LET B(L) = S
360    NEXT L
370    REM THE N VALUES OF THE S STATISTIC ARE NOW
380    REM ORDERED USING A SHELL-SORT ALGORITHM,
390    REM TAKEN FROM COOKE,CRAVEN AND CLARKE, 1982, P18.
400    LET G = N
410    LET G = INT(G/2)
420    IF G = 0 THEN 550
430    FOR I = 1 TO (N-G)
440      FOR J = I TO 1 STEP -G
450        LET J1 = J+G
460        IF B(J) > B(J1) THEN 490
470        LET J = 0
480        GOTO 520
490        LET W = B(J)
500        LET B(J) = B(J1)
510        LET B(J1) = W
520      NEXT J
530    NEXT I
540    GOTO 410
550    FOR I = 1 TO N
560      PRINT B(I)
570    NEXT I
580    END
```

9.17 For small values of the probability, the binomial distribution may be approximated by the Poisson, and $\text{Var}(\hat{p}) \approx p/n$.

9.18 The modern substitute for pieces of cardboard:

```
NOPRINT
  STORE
  NRANDOM 4 OBSERVATIONS, WITH MU = 0.0, SIGMA=1.0, PUT IN C1
  AVERAGE C1, PUT IN K1
  STANDARD DEV. C1, PUT IN K2
  LET K3=2*K1/K2
  JOIN K3 TO C4, PUT IN C4
  END
EXECUTE 750 TIMES
HISTO C4
```

9.19
```
NOPRINT
  STORE
  URANDOM 3 OBSERVATIONS, PUT IN C1
  SUM C1, PUT IN K1
  JOIN K1 TO C2, PUT IN C2
  END
EXECUTE 100 TIMES
HISTO C2
```

This program draws a histogram of 100 sums of k $U(0, 1)$ variates, where $k = 3$. The program may be repeated for different values of k.

BIBLIOGRAPHY

This bibliography contains all the books, papers and reports referred to in the text. In addition, it contains references to other work relevant to simulation, some of which provide further examples of applications. Bibliographies which concentrate on particular areas are to be found in Nance and Overstreet (1972) and Sowey (1972, 1978).

Abdul-Ela, A-L, A., Greenberg, B. G., and Horvitz, D. G. (1967) A multi-proportions randomized response model. *J. Amer. Statist. Assoc.*, **62**, 990–1008.

Abramowitz, M., and Stegun, I. A. (Eds.) (1965) *Handbook of Mathematical Functions*. Dover Publications, New York.

Adam, N. and Dogramaci, A. (Eds.) (1979) *Current Issues in Computer Simulation*. Academic Press, New York.

Ahrens, J. H., and Dieter, U. (1972) Computer methods for sampling from the exponential and normal distributions. *Commun. Ass. Comput. Mach.*, **15**, 873–882.

Ahrens, J. H., and Dieter, U. (1973) Extensions of Forsythe's method for random sampling from the normal distribution. *Math. Comput.*, **27**, 927–937.

Ahrens, J. H., and Dieter, U. (1974a) *Non-uniform Random Numbers*. Technische Hoshschule in Graz, Graz, Austria.

Ahrens, J. H., and Dieter, U. (1974b) Computer methods for sampling from gamma, beta, Poisson and binomial distributions. *Computing (Vienna)*, **12**, 223–246.

Ahrens, J. H., and Dieter, U. (1980) Sampling from binomial and Poisson distributions: a method with bounded computation times. *Computing*, **25**, 193–208.

Ahrens, J. H., and Dieter, U. (1982) Generation of Poisson deviates from modified normal distributions. *Assoc. Comput. Mach. Trans. Math. Soft.*, **8**, 163–179.

Aitchison, J., and Brown, J. A. C. (1966) *The Lognormal Distribution, With Special Reference to its Uses in Economics.* C.U.P.

Akima, H. (1970) A new method of interpolation and smooth curve fitting based on local procedures. *J. Assoc. Comput. Mach.*, **17** (4), 589–602.

Anderson, O. D. (1976) *Time Series Analysis and Forecasting.* Butterworths, London and Boston.

Andrews, D. F. (1976) Discussion of: 'The computer generation of beta, gamma and normal random variables' by A. C. Atkinson and M. C. Pearce. *J. Roy. Statist. Soc.*, **139**, 4, 431–461.

Andrews, D. F., Bickel, P. J., Hampel, F. R., Huber, P. J., Rogers, W. H., and Tukey, J. W. (1972) *Robust Estimates of Location.* Princeton University Press, Princeton.

Apostol, T. M. (1963) *Mathematical Analysis.* Addison-Wesley, Tokyo.

Appleton, D. R. (1976) Discussion of paper by Atkinson and Pearce. *J. Roy. Statist. Soc.*, A, **139**, 4, 449–451.

Arabie, P. (1973) Concerning Monte-Carlo evaluations of nonmetric multi-dimensional scaling algorithms. *Psychometrika*, **38**, 4, 607–608.

Arvidson, N. I., and Johnsson, T. (1982) Variance reduction through negative correlation; a simulation study. *J. Stat. Computation and Simulation*, **15**, 119–128.

Ashcroft, H. (1950) The productivity of several machines under the care of one operator. *J. Roy. Statist. Soc.*, B, **12**, 145–151.

Ashton, W. D. (1971) Gap acceptance problems at a traffic intersection. *Appl. Stats.*, **20** (2), 130–138.

Atkinson, A. C. (1977a) An easily programmed algorithm for generating gamma random variables. *J. Roy. Statist. Soc.*, A, **140**, 232–234.

Atkinson, A. C. (1977b) Discussion of: 'Modelling spatial patterns' by B. D. Ripley. *J. Roy. Statist. Soc.*, B, **39** (2), 172–212.

Atkinson, A. C. (1979a) The computer generation of Poisson random variables. *Appl. Stats.*, **28** (1), 29–35.

Atkinson, A. C. (1979b) A family of switching algorithms for the computer generation of beta random variables. *Biometrika*, **66** (1), 141–146.

Atkinson, A. C. (1979c) Recent developments in the computer generation of Poisson random variables. *Appl. Stats.*, **28** (3), 260–263.

Atkinson, A. C. (1980) Tests of pseudo-random numbers. *Appl. Stats.*, **29** (2), 164–171.

Atkinson, A. C., and Pearce, M. C. (1976) The computer generation of beta, gamma and normal random variables. *J. Roy. Statist. Soc.*, A, **139** (44), 431–461.

Atkinson, A. C., and Whittaker, J. (1976) A switching algorithm for the generation of beta random variables with at least one parameter less than 1. *J. Roy. Statist. Soc.*, A, **139** (4), 462–467.

Bailey, B. J. R. (1981) Alternatives to Hastings' approximation to the inverse of the normal cumulative distribution function. *Appl. Stats.*, **30**, 275–276.

Bailey, N. T. J. (1952) A study of queues and appointment systems in hospital out-patient departments with special reference to waiting times. *J. Roy. Statist. Soc.*, B, **14**, 185–199.

Bailey, N. T. J. (1964) *The Elements of Stochastic Processes with Applications to the Natural Sciences*. Wiley, New York.

Bailey, N. T. J. (1967) The simulation of stochastic epidemics in two dimensions. *Proc. 5th Berkeley Symp. Math. Statist. Prob.*, **4**, 237–257.

Bailey, N. T. J. (1975) *The Mathematical Theory of Infectious Diseases and its Applications* (2nd edn). Griffin, London.

Barnard, G. A. (1963) Discussion of Professor Bartlett's paper. *J. Roy. Statist. Soc.*, B, **25**, 294.

Barnett, V. D. (1962a) The Monte Carlo solution of a competing species problem. *Biometrics*, **18**, 76–103.

Barnett, V. D. (1962b) The behaviour of pseudo-random sequences generated on computers by the multiplicative congruential method. *Math. Comp.*, **16**, 63–69.

Barnett, V. D. (1965) *Random Negative Exponential Deviates*. Tracts for Computers XXVII. C.U.P.

Barnett, V. (1974) *Elements of Sampling Theory*. The English Universities Press Ltd, London.

Barnett, V. (1976) Discussion of: 'The computer generation of beta, gamma and normal random variables' by A. C. Atkinson and M. C. Pearce. *J. Roy. Statist. Soc.*, A, **139** (4), 431–461.

Barnett, V. (1980) Some bivariate uniform distributions. *Commun. Statist. – Theory Meth.*, A, **9** (4), 453–461.

Barnett, V., and Lewis, T. (1978) *Outliers in Statistical Data*. Wiley, Chichester.

Barrodale, I., Roberts, F. D. K., and Ehle, B. L. (1971) *Elementary Computer Applications in Science, Engineering and Business*. Wiley, New York.

Bartholomew, D. J., and Forbes, A. F. (1979) *Statistical Techniques for Manpower Planning*. Wiley, Chichester.

Bartlett, M. S. (1953) Stochastic processes or the statistics of change. *Appl. Stats.*, **2**, 44–64.

Bartlett, M. S., Gower, J. C., and Leslie, P. H. (1960) A comparison of theoretical and empirical results for some stochastic population models. *Biometrika*, **47**, 1–11.

Bays, C., and Durham, S. D. (1976) Improving a poor random number generator. *Assoc. Comput. Math. Trans. Math. Soft.*, **2**, 59–64.

Beale, E. M. L. (1969) Euclidean cluster analysis. *Bull. I.S.I.*, **43**, Book 2, 92–94.

Beasley, J. D., and Springer, S. G. (1977) Algorithm AS111. The percentage points of the normal distribution. *Appl. Stats.*, **26**, 118–121.

Bebbington, A. C. (1975) A simple method of drawing a sample without replacement. *Appl. Stats.*, **24** (1), 136.

Bebbington, A. C. (1978) A method of bivariate trimming for robust estimation of the correlation coefficient. *Appl. Stats.*, **27** (3), 221–226.

Besag, J., and Diggle, P. J. (1977) Simple Monte Carlo tests for spatial pattern. *Appl. Stats.*, **26** (3), 327–333.

Best, D. J. (1979) Some easily programmed pseudo-random normal generators. *Aust. Comp. J.*, **11**, 60–62.

Best, D. J., and Fisher, N. I. (1979) Efficient simulation of the von Mises distribution. *Appl. Stats.*, **28** (2), 152–157.

Best, D. J., and Winstanley, J. (1978) *Random Number Generators*. CSIRO DMS Newsletter 42.

Bevan, J. M., and Draper, G. J. (1967) *Appointment Systems in General Practice*. Oxford Univ. Press for Nuffield Provincial Hospitals Trust, London.

Bishop, J. A., and Bradley, J. S. (1972) Taxi-cabs as subjects for a population study. *J. Biol. Educ.*, **6**, 227–231.

Bishop, J. A., and Sheppard, P. M. (1973) An evaluation of two capture–recapture models using the technique of computer simulations. In, M. S. Bartlett and R. W. Hiorns (eds.). *The Mathematical Theory of the Dynamics of Biological Populations*. Academic Press, London, pp. 235–252.

Blackman, R. G., and Tukey, J. W. (1958) *The Measurement of Power Spectra*. Dover, New York.

Blake, I. F. (1979) *An Introduction to Applied Probability*. Wiley, New York.

Blanco White, M. J., and Pike, M. C. (1964) Appointment systems in out-patients' clinics and the effect of patients' unpunctuality. *Medical Care*, **2**, 133–142.

Bofinger, E., and Bofinger, V. J. (1961) A note on the paper by W. E. Thomson on 'Ernie – a mathematical and statistical analysis'. *J. Roy. Statist. Soc.*, A, **124** (2), 240–243.

Bowden, D. C., and Dick, N. P. (1973) Maximum likelihood estimation for mixtures of 2 normal distributions. *Biometrics*, **29**, 781–790.

Box, G. E. P., Hunter, W. G., and Hunter, J. S. (1978) *Statistics for Experimenters: An introduction to design, data analysis and model building*. Wiley, New York.

Box, G. E. P., and Müller, M. E. (1958) A note on the generation of random normal deviates. *Ann. Math. Stat.*, **29**, 610–611.

Boyett, J. M. (1979) Random $r \times c$ tables with given row and column totals. Algorithm AS144. *Appl. Stats.*, **28**, 329–332.

Bratley, P., Fox, B. L., and Schrage, L. E. (1983) *A Guide to Simulation.* Springer–Verlag, Berlin.

Bremner, J. M. (1981) Calculator warning. *R.S.S. News & Notes,* **8** (1), 10–11.

Brent, R. P. (1974) A Gaussian pseudo random number generator. *Comm. Assoc. Comput. Mach.,* **17**, 704–706.

Brown, D., and Rothery, P. (1978) Randomness and local regularity of points in a plane. *Biometrika,* **65**, 115–122.

Brown, S., and Holgate, P. (1974) The thinned plantation. *Biometrika,* **61**, 2, 253–261.

Buckland, S. T. (1982) A mark–recapture survival analysis. *J. Animal Ecol.,* **51**, 833–847.

Bunday, B. D., and Mack, C. (1973) Efficiency of bi-directionally traversed machines. *Appl. Stats.,* **22**, 74–81.

Burnham, K. P., Anderson, D. R., and Laake, J. L. (1980) Estimation of density from line transect sampling of biological populations. *Wildlife Monographs,* No. 72. Supplement to *The Journal of Wildlife Management,* **44**, No. 2.

Burnham, K. P., and Overton, W. S. (1978) Estimation of the size of a closed population when capture probabilities vary among animals. *Biometrika,* **65** (3), 625–634.

Burr, I. W. (1967) A useful approximation to the normal distribution function, with application to simulation. *Technometrics,* **9**, 647.

Butcher, J. C. (1960) Random sampling from the normal distribution. *Comp. J.,* **3**, 251–253.

Calinski, T., and Harabasz, J. (1974) A dendrite method of cluster analysis. *Comm. in Statis.,* **3**, 1–27.

Cammock, R. M. (1973) *Health Centres, Reception, Waiting and Patient Call.* HMSO, London.

Campbell, C., and Joiner, B. L. (1973) How to get the answer without being sure you've asked the question. *The Amer. Statist.,* **27** (5), 229–230.

Cane, V. R., and Goldblatt, P. O. (1978) The perception of probability. *Manchester/Sheffield res. report. 69/VRC/6.*

Carlisle, P. (1978) Investigation of a doctor's waiting-room using simulation. Unpublished undergraduate dissertation, Univ. of Kent.

Carter, C. O. (1969) *Human Heredity.* Penguin Books Ltd, Harmondsworth.

Chambers, J. M. (1970) Computers in statistical research: Simulation and computer-aided mathematics. *Technometrics,* **12** (1), 1–15.

Chambers, J. M. (1977) *Computational Methods for Data Analysis.* Wiley, New York.

Chambers, J. M. (1980) Statistical computing: History and trends. *The Amer. Statist.,* **34** (4), 238–243.

Chambers, J. M., Mallows, C. L., and Sturk, B. W. (1976) A method for simulating stable random variables. *J. Amer. Statist. Assoc.,* **71**, 340–344.

Chapman, D. G. (1952) Inverse multiple and sequential sample censuses. *Biometrics*, **8**, 286–306.

Chapman, D. G. (1956) Estimating the parameters of a truncated gamma distribution. *Ann. Math. Stat.*, **27**, 498–506.

Chay, S. C., Fardo, R. D., and Mazumdar, M. (1975) On using the Box–Müller transform with congruential pseudo-random generators. *Appl. Stats.*, **24** (1), 132–134.

Chen, H., and Asau, Y. (1974) Generating random variates from an empirical distribution. *Amer. Inst. Indust. Engineers Trans.*, **6**, 163–166.

Cheng, R. C. H. (1977) The generation of gamma variables with non-integral shape parameters. *Appl. Stats.*, **26** (1), 71–74.

Cheng, R. C. H. (1978) Generating beta variates with non-integral shape parameters. *Communications of the ACM*, **21** (4), 317–322.

Cheng, R. C. H. (1982) The use of antithetic variates in computer simulations. *J. Opl. Res. Soc.*, **33**, 229–237.

Cheng, R. C. H., and Feast, G. M. (1979) Some simple gamma variate generators. *Appl. Stats.*, **28** (3), 290–295.

Cheng, R. C. H., and Feast, G. M. (1980) Control variables with known mean and variance. *J. Opl. Res. Soc.*, **31**, 51–56.

Chernoff, H. (1969) Optimal design applied to simulation. *Bull. Int. Statist. Inst.*, **43**, 2264.

Chernoff, H., and Lieberman, G. J. (1956) The use of generalized probability paper for continuous distributions. *Ann. Math. Stat.*, **27**, 806–818.

Clark, C. E. (1964) Sampling efficiency in Monte Carlo analyses. *Ann. Inst. Stat. Math.*, **15**, 197.

Clarke, G. M., and Cooke, D. (1983) *A Basic Course in Statistics*. (2nd Edn) Edward Arnold, London.

Clements, G. (1978) The study of the estimation of animal population sizes. Unpublished undergraduate dissertation, Univ. of Kent, Canterbury.

Cochran, W. G. (1952) The χ^2 test of goodness of fit. *Ann. Math. Stat.*, **23**, 315–346.

Cochran, W. G. (1952) The χ^2 test of goodness of fit. *Ann. Math. Stat.*, **23**, 315–346.

Conte, S. D., and de Boor, C. (1972) *Elementary Numerical Analysis*. McGraw-Hill, Kogakusha, Tokyo.

Conway, R. W. (1963) Some tactical problems in digital simulation. *Management Science*, **10**, 47–61.

Cooke, D., Craven, A. H., and Clarke, G. M. (1982) *Basic Statistical Computing*. Edward Arnold, London.

Cooper, B. E. (1976) Discussion of 'Computer generation of beta, gamma and normal random variables' by A. C. Atkinson and M. C. Pearce. *J. Roy. Statist. Soc.*, A, **139** (4), 431–461.

Cormack, R. M. (1966) A test for equal catchability. *Biometrics*, **22**, 330–342.

Cormack, R. M. (1968) The statistics of capture–recapture methods. *Oceanogr. Mar. Bio. Ann. Rev.*, **6**, 455–506.

Cormack, R. M. (1971) A review of classification. *J. Roy. Statist. Soc.*, A, **134**, 321–367.

Cormack, R. M. (1973) Commonsense estimates from capture-recapture studies. pp. 225–234, In *'The Mathematical Theory of the Dynamics of Biological Populations'* Eds. M. S. Bartlett and R. W. Hiorns. Academic Press, London.

Coveyou, R. R., and MacPherson, R. D. (1967) Fourier analysis of uniform random number generators. *J. Assoc. Comput. Mach.*, **14**, 100–119.

Cox, D. R. (1955) Some statistical methods connected with series of events. *J. Roy. Statist. Soc.*, B, **17**, 129–157.

Cox, D. R. (1958) *Planning of Experiments*. Wiley, New York.

Cox, D. R. (1966) Notes on the analysis of mixed frequency distributions. *Brit. J. Math. & Stat. Psychol.*, **19** (1), 39–47.

Cox, D. R., and Lewis, P. A. W. (1966) *The Statistical Analysis of Series of Events*. Chapman and Hall, London.

Cox, D. R., and Miller, H. D. (1965) *The Theory of Stochastic Processes*. Chapman and Hall, London.

Cox, D. R., and Smith, W. L. (1961) *Queues*. Chapman and Hall, London.

Cox, M. A. A., and Plackett, R. L. (1980) Small samples in contingency tables. *Biometrika*, **67**, 1, 1–14.

Craddock, J. M., and Farmer, S. A. (1971) Two robust methods of random number generation. *The Statistician*, **20** (3), 55–66.

Craddock, J. M., and Flood, C. R. (1970) The distribution of the χ^2 statistic in small contingency tables. *Appl. Stats.*, **19**, 173–181.

Cramér, H. (1954) *Mathematical Methods of Statistics*. Univ. Press, Princeton.

Cugini, J. V., Bowden, J. S., and Skall, M. W. (1980) NBS Minimal BASIC Test programs – version 2, Users Manual. Vol. 2, Source Listings and Sample Output. National Bureau of Standards, Special Publication 500/70/2.

Cunningham, S. W. (1969) From normal integral to deviate. Algorithm AS24. *Appl. Stats.*, **18**, 290–293.

Daley, D. J. (1968) The serial correlation coefficients of waiting times in a stationary single server queue. *J. Austr. Math. Soc.*, **8**, 683–699.

Daley, D. J. (1974) Computation of bi- and tri-variate normal integrals. *Appl. Stats.*, **23** (3), 435–438.

Daniels, H. E. (1982) A tribute to L. H. C. Tippett. *J. Roy. Statist. Soc.*, A, **145** (2), 261–263.

Darby, K. V. (1984) Statistical analysis of British bird observatory data. Unpublished Ph.D. thesis, University of Kent, Canterbury.

Darroch, J. N. (1958) The multiple-recapture census I: Estimation of a closed population. *Biometrika*, **45**, 343–359.

David, F. N. (1973) p. 146 in *Chamber's Encyclopedia*. International Learning Systems Ltd., London.

Davis, P. J., and Rabinowitz, P. (1967) *Numerical Integration*. Blaisdell Pub. Co., Waltham, Mass.

Davis, P. J., and Rabinowitz, P. (1975) *Methods of Numerical Integration*. Academic Press, London.

Day, N. E. (1969) Estimating the components of a mixture of normal distributions. *Biometrika*, **56**, 463–474.

Déak, I. (1981) An economical method for random number generation and a normal generator. *Computing*, **27**, 113–121.

Devroye, L. (1981) The computer generation of Poisson random variables. *Computing*, **26**, 197–207.

Dewey, G. (1923) *Relative Frequency of English Speech Sounds*. Harvard University Press.

de Wit, C. T., and Goudriaan, J. (1974) Simulation of ecological processes. Simulation Monographs, Wageningen; Centre for Agricultural Pub. and Doc.

Diaconis, P., and Efron, B. (1983) Computer-intensive methods in statistics. *Scientific American*, **248** (5), 96–109.

Dieter, U., and Ahrens, J. H. (1974) *Uniform random numbers*. Produced by: Institut für Math. Statistik, Technische Hochschule in Graz. A 8010 Graz, Hamerlinggasse 6, Austria.

Diggle, P. J. (1975) Robust density estimation using distance methods. *Biometrika*, **62**, 39–48.

Diggle, P. J., Besag, J., and Gleaves, J. T. (1976) Statistical analysis of spatial point patterns by means of distance methods. *Biometrics*, **32**, 3, 659–668.

Dixon, W. J. (1971) Notes on available materials to support computer-based statistical teaching. *Rev. Int. Statist. Inst.*, **39**, 257–286.

Donnelly, J. H., and Shannon, R. E. (1981) Minimum mean-squared-error estimators for simulation experiments. *Comm. Assoc. Comp. Mach.*, **24** (4), 253–259.

Donnelly, K. (1978) Simulations to determine the variance and edge effect of total nearest neighbour distance. In: *Simulation Methods in Archaeology*. I. Hodder (Ed.), Cambridge Univ. Press, London.

Douglas, J. B. (1980) *Analysis with Standard Contagious Distributions*. International Co-operative Publishing House, Fairland, Maryland.

Downham, D. Y. (1969) The runs up and down test. *Comp. J.*, **12**, 373–376.

Downham, D. Y. (1970) Algorithm AS29. The runs up and down test. *Appl. Stats.*, **19**, 190–192.

Downham, D. Y., and Roberts, F. D. K. (1967) Multiplicative congruential pseudo-random number generators. *Comp. J.*, **10**, 74–77.

Dudewicz, E. J., and Ralley, T. G. (1981) *The Handbook of Random Number Generation and Testing with TESTRAND Computer Code*. American Sciences Press, Columbus, Ohio.

Duncan, I. B., and Curnow, R. N. (1978) Operational research in the health and social services. *J. Roy. Statist. Soc.*, A, **141**, 153–194.

Dunn, O. J., and Varady, P. D. (1966) Probabilities of correct classification in discriminant analysis. *Biometrics*, **22**, 908–924.

Dwass, M. (1972) Unbiased coin tossing with discrete random variables. *Ann. Math. Stats.*, **43** (3), 860–864.

Efron, B. (1979a) Bootstrap methods: Another look at the jackknife. *Ann. Stat.* **7**, 1–26.

Efron, B. (1979b) Computers and the theory of statistics: Thinking the unthinkable. *SIAM Review*, **21**, 460–480.

Efron, B. (1981) Nonparametric estimates of standard error: The jackknife, the bootstrap and other methods. *Biometrika*, **68** (3), 589–600.

Egger, M. J. (1979) Power transformations to achieve symmetry in quantal bioassay. *Technical Report No. 47*, Division of Biostatistics, Stanford, California.

Ehrenfeld, S., and Ben–Turia, S. (1962) The efficiency of statistical simulation procedures. *Technometrics*, **4**, 257–275.

Ekblom, H. (1972) A Monte Carlo investigation of mode estimators in small samples. *Appl. Stats.*, **21** (2), 177–184.

Ernvall, J., and Nevalainen, O. (1982) An algorithm for unbiased random sampling. *Comp. J.*, **25**, 45–47.

Estes, W. K. (1975) Some targets for mathematical psychology. *J. Math. Psychol.*, **12**, 263–282.

Evans, D. H., Herman, R., and Weiss, G. H. (1964) The highway merging and queueing problem. *Oper. Res.*, **12**, 832–857.

Evans, D. H., Herman, R., and Weiss, G. H. (1965) Queueing at a stop sign. *Proc. 2nd. Internat. Symp. on the Theory of Traffic Flow, London, 1963*. OECD, Paris.

Everitt, B. S. (1977) *The analysis of contingency tables*. Chapman and Hall, London.

Everitt, B. S. (1979) Unresolved problems in cluster analysis. *Biometrics*, **35** (1), 169–182.

Everitt, B. S. (1980) *Cluster Analysis* (2nd edn). Heinemann Educational, London.

Farlie, D. J. and Keen, J. (1967) Quick ways to the top: a team game illustrating steepest ascent techniques. *Appl. Stats.*, **16**, 75–80.

Feller, W. (1957) *An Introduction to Probability Theory and its Applications*, Vol. 1 (2nd edn). Wiley, New York.

Feller, W. (1971) *An Introduction to Probability Theory and its Applications*, Vol. 2 (2nd edn). Wiley, New York.

Felton, G. E. (1957) Electronic computers and mathematicians, pp. 12–21, in: *Oxford Mathematical Conference for Schoolteachers and Industrialists.* Oxford university delegacy for extra-mural studies. Times. Pub. Co. Ltd., London.

Fieller, E. C., and Hartley, H. O. (1954) Sampling with control variables. *Biometrika*, **41**, 494–501.

Fieller, E. C., Lewis, T., and Pearson, E. S. (1955) Correlated random normal deviates. *Tracts for Computers*, No. 26, Cambridge University Press.

Fienberg, S. E. (1980) *The Analysis of Cross-classified Categorical Data.* M.I.T. Press.

Fine, P. E. M. (1977) A commentary on the mechanical analogue to the Reed–Frost epidemic model. *American J. Epidemiology*, **106** (2), 87–100.

Fisher, L., and Kuiper, F. K. (1975) A Monte Carlo comparison of six clustering procedures. *Biometrics*, **31**, 777–784.

Fisher, R. A., and Yates, F. (1948) *Statistical Tables.* Oliver & Boyd, pp. 104–109.

Fishman, G. S. (1973) *Concepts and Methods in Discrete Event Digital Simulation.* Wiley-Interscience, New York.

Fishman, G. S. (1974) Correlated simulation experiments. *Simulation*, **23**, 177–180.

Fishman, G. S. (1976) Sampling from the Poisson distribution on a computer. *Computing*, **17**, 147–156.

Fishman, G. S. (1978) *Principles of Discrete Event Simulation.* Wiley, New York.

Fishman, G. S. (1979) Sampling from the binomial distribution on a computer. *J. Amer. Statist. Assoc.*, **74**, 418–423.

Fletcher, R. (1971) A general quadratic programming algorithm. *J. Inst. Math. Applic.*, **7**, 76–91.

Folks, J. L. (1981) *Ideas of Statistics.* Wiley, New York.

Forsythe, G. E. (1972) Von Neumann's comparison method for random sampling from the normal and other distributions. *Math. Comput.*, **26**, 817–826.

Fox, B. L. (1963) Generation of random samples from the beta *F* distributions. *Technometrics*, **5**, 269–270.

Freeman, P. R. (1979) Exact distribution of the largest multinomial frequency. Algorithm AS145. *Appl. Stats.*, **28**, 333–336.

Fuller, A. T. (1976) The period of pseudo-random numbers generated by Lehmer's congruential method. *Comp. J.*, **19**, 173–177.

Fuller, M. F., and Lury, D. A. (1977) *Statistics Workbook for Social Science Students.* Philip Allan Pub. Ltd., Oxford.

GAMS (1981) *Guide to Available Mathematical Software.* Centre for applied mathematics. National Bureau of Standards, Washington, DC. 20234.

Gani, J. (1980) The problem of Buffon's needle. *Mathematical Spectrum*, **13**, 14–18.

Gardner, G. (1978) A discussion of the techniques of cluster analysis with reference to an application in the field of social services. Unpublished M.Sc. dissertation, Univ. of Kent at Canterbury.

Gardner, M. (1966) *New Mathematical Diversions from Scientific American.* Simon & Schuster.

Gates, C. E. (1969) Simulation study of estimators for the line transect sampling method. *Biometrics*, **25**, 317–328.

Gauss, C. F. (1809) *Theoria Motus Corporum Coelestium.* Perthes & Besser, Hamburg.

Gaver, D. P. and Thompson, G. L. (1973) *Programming and Probability Models in Operations Research.* Brooks/Cole, Monterey, California.

Gebhardt, F. (1967) Generating pseudo-random numbers by shuffling a Fibonacci sequence. *Math. Comp.* **21**, 708–709.

Gerontidis, I., and Smith, R. L. (1982) Monte Carlo generation of order statistics from general distributions. *Appl. Stats.*, **31** (3), 238–243.

Gerson, M. (1975) The technique and uses of probability plotting. *The Statistician*, **24** (4), 235–257.

Gibbs, K. (1979) Appointment systems in general practice. Unpublished undergraduate dissertation, Univ. of Kent at Canterbury.

Gibbs, K. (1980) Mathematical models for population growth with special reference to the African mountain gorilla. Unpublished M.Sc. dissertation, Univ. of Kent at Canterbury.

Gilchrist, R. (1976) The use of verbal call systems to call patients from waiting areas to consulting rooms. *Appl. Stats.*, **25**, 217–227.

Gipps, P. G. (1977) A queueing model for traffic flow. *J. Roy. Statist. Soc.*, B, **39** (2), 276–282.

Gleser, L. J. (1976) A canonical representation for the noncentral Wishart distribution useful for simulation. *J. Amer. Statist. Assoc.*, **71**, 690–695.

Gnanadesikan, R. (1977) *Methods for Statistical Data Analysis of Multivariate Observations.* Wiley, New York.

Gnedenko, B. V. (1976) *The Theory of Probability.* MIR Publishers, Moscow (English translation).

Golder, E. R. (1976a) Algorithm AS98. The spectral test for the evaluation of congruential pseudo-random number generators. *Appl. Stats.*, **25**, 173–180.

Golder, E. R. (1976b) Remark ASR18. The spectral test for the evaluation of congruential pseudo-random generators. *Appl. Stats.*, **25**, 324.

Golder, E. R., and Settle, J. G. (1976) The Box–Müller method for generating pseudo-random normal deviates. *Appl. Stats.*, **25** (1), 12–20.

Good, I. J. (1953) The serial test for sampling numbers and other tests for randomness. *Proc. Camb. Phil. Soc.*, **49**, 276–284.

Good, I. J. (1957) On the serial test for random sequences. *Ann. Math. Statist.*, **28**, 262–264.

Gordon, A. D. (1981) *Classification*. Chapman & Hall, London.

Gordon, R. (1970) On Monte Carlo algebra. *J. Appl. Prob.*, **7**, 373–387.

Gosper, W. G. (1975) Numerical experiments with the spectral test. *Stan-CS-75–490*. Computer Science Department, Stanford University.

Gower, J. C. (1968) Simulating multidimensional arrays in 1D. Algorithm AS1. *Appl. Stats.*, **17**, 180 (see also **18** (1), 116).

Gower, J. C., and Banfield, C. F. (1975) Goodness-of-fit criteria for hierarchical classification and their empirical distributions. In: *Proceedings of the 8th International Biometric Conference*. (Eds L.C.A. Corsten and T. Postelnicu), 347–361.

Grafton, R. G. T. (1981) Algorithm AS157: The runs-up and runs-down tests. *Appl. Stats.*, **30**, 81–85.

Green, P. J. (1978) Small distances and Monte Carlo testing of spatial pattern. *Advances in Applied Prob.*, **10**, 493.

Greenberg, B. G., Kuebler, R. T., Jr., Abernathy, J. R., and Horvitz, D. G. (1971) Application of randomized response technique in obtaining quantitative data. *J. Amer. Statist. Assoc.*, **66**, 243–250.

Greenberger, M. (1961) An a priori determination of serial correlation in computer generated random numbers. *Math. Comp.*, **15**, 383–389.

Greenwood, A. J. (1974) A fast generator for gamma distributed random variables. pp. 19–27 in *Compstat 1974* (G. Bruckman *et al.*, eds.) Physica–Verlag: Vienna.

Greenwood, J. A., and Durand, D. (1960) Aids for fitting the gamma distribution by maximum likelihood. *Technometrics*, **2**, 55–65.

Greenwood, R. E. (1955) Coupon collector's test for random digits. *Mathematical Tables and Other Aids to Computation*, **9**, 1–5.

Griffiths, J. D., and Cresswell, C. (1976) A mathematical model of a Pelican Crossing. *J. Inst. Maths. & Applics.*, **18** (3), 381–394.

Grimmett, G. R., and Stirzaker, D. R. (1982) *Probability and Random Processes*. Oxford University Press.

Grogono, P. (1980) *Programming in PASCAL*. Addison-Wesley Pub. Co. Inc., Reading, Massachusetts.

Gross, A. M. (1973) A Monte Carlo swindle for estimators of location. *Appl. Stats.*, **22** (3), 347–353.

Gross, D., and Harris, C. M. (1974) *Fundamentals of Queueing Theory*. Wiley, Toronto.

Gruenberger, F., and Mark, A. M. (1951) The d^2 test of random digits. *Math. Tables Other Aids Comp.*, **5**, 109–110.

Guerra, V. M., Tapia, R. A., and Thompson, J. R. (1972) A random number generator for continuous random variables. *ICSA Technical Report*, Rice University, Houston, Texas.

Haight, F. A. (1967) *Handbook of the Poisson distribution*. Wiley, New York.

Halton, J. H. (1970) A retrospective and prospective survey of the Monte Carlo method. *SIAM Rev.*, **12** (1), 1–63.

Hamaker, H. C. (1978) Approximating the cumulative normal distribution and its inverse. *Appl. Stats.*, **27** (1), 76–77.

Hammersley, J. M., and Handscomb, D. C. (1964) *Monte Carlo Methods*. Methuen, London.

Hammersley, J. M., and Morton, K. W. (1956) A new Monte Carlo technique: antithetic variates. *Proc. Camb. Phil. Soc.*, **52**, 449–475.

Hannan, E. J. (1957) The variance of the mean of a stationary process. *J. Roy. Statist. Soc.*, B, **19**, 282–285.

Harcourt, A. H., Stewart, K. J., and Fossey, D. (1976) Male emigration and female transfer in wild mountain gorilla. *Nature*, **263**, 226–227.

Harris, R., Norris, M. E., and Quaterman, B. R. (1974) Care of the elderly: structure of the group flow model. *Report 825*. Institute for Operational Research.

Harter, H. L. (1964) *New tables of the incomplete gamma-function ratio and of percentage points of the chi-square and beta distributions*. U.S. Government printing office, Washington, D.C.

Hastings, C., Jr. (1955) *Approximations for Digital Computers*. Princeton University Press, Princeton, N. J.

Hastings, W. K. (1970) Monte Carlo sampling methods using Markov chains and their applications. *Biometrika*, **57**, 97–109.

Hastings, W. K. (1974) Variance reduction and non-normality. *Biometrika*, **61**, 143–149.

Hauptman, H., Vegh, E., and Fisher, J. (1970) Table of all primitive roots for primes less than 5000. *Naval Research Laboratory Report No. 7070*. Washington, D.C.

Hawkes, A. G. (1965) Queueing for gaps in traffic. *Biometrika*, **52**, 1–6.

Hawkes, A. G. (1966) Delay at traffic intersections. *J. Roy. Statist. Soc.*, B, **28** (1), 202–212.

Hay, R. F. M. (1967) The association between autumn and winter circulations near Britain. *Met. Mag.*, **96**, 167–177.

Healy, M. J. R. (1968a) Multivariate normal plotting. *Appl. Stats.*, **17**, 157–161.

Healy, M. J. R. (1968b) Triangular decomposition of a symmetric matrix. Algorithm AS6. *Appl. Stats.*, **17**, 195–197.

Heathcote, C. R., and Winer, P. (1969) An approximation for the moments of waiting times. *Oper. Res.*, **17**, 175–186.

Heidelberger, P., and Welch, P. D. (1981) A spectral method for confidence interval generation and mean length control in simulations. *Comm. Assoc. Comp. Mach.*, **24** (4), 233–245.

Heiberger, R. M. (1978) Generation of random orthogonal matrices. Algorithm. AS127. *Appl. Stats.*, **27**, 199–206.

Hews, R. J. (1981) Stopping rules and goodness of fit criteria for hierarchical clustering methods. Unpublished M.Sc. dissertation. University of Kent.

Hoaglin, D. C., and Andrews, D. F. (1975) The reporting of computation-based results in statistics. *The American Statistician*, **29**, 122–126.

Hoaglin, D. C., and King, M. L. (1978) Remark ASR24. A remark on Algorithm AS98. The spectral test for the evaluation of congruential pseudo-random generators. *Appl. Stats.*, **27**, 375.

Hoeffding, W., and Simons, G. (1970) Unbiased coin tossing with a biased coin. *Ann. Math. Stat.*, **41**, 341–352.

Hoel, P. G. (1954) *Introduction to Mathematical Statistics* (3rd edn). Wiley, New York.

Holgate, P. (1981) Studies in the history of probability and statistics. XXXIX. Buffon's cycloid. *Biometrika*, **68** (3), 712–716.

Hollier, R. H. (1968) A simulation study of sequencing in batch production. *Op. Res. Quart.*, **19**, 389–407.

Hollingdale, S. H. (Ed.) (1967) *Digital Simulation in Operational Research*. English Universities Press, London.

Hope, A. C. A. (1968) A simplified Monte Carlo significance test procedure. *J. Roy. Statist. Soc.*, B, **30**, 582–598.

Hopkins, T. R. (1980) PBASIC—A verifier for BASIC. *Software Practice and Experience*, **10**, 175–181.

Hopkins, T. R. (1983a) A revised algorithm for the spectral test. In Algorithm AS 193, *Applied Statistics*, **32** (3), 328–335.

Hopkins, T. R. (1983b) The collision test—an empirical test for random number generators. Unpublished Computing Lab. Report, University of Kent.

Hordijk, A., Iglehart, D. L., and Schassberger, R. (1976) Discrete time methods for simulating continuous time Markov chains. *Adv. Appl. Prob.*, **8**, 772–788.

Hsu, D. A., and Hunter, J. S. (1977) Analysis of simulation-generated responses using autoregressive models. *Manag. Sci.*, **24**, 181–190.

Hsuan, F. (1979) Generating uniform polygonal random pairs. *Appl. Stats.*, **28**, 170–172.

Hull, T. E., and Dobell, A. R. (1962) Random number generators. *SIAM Rev.*, **4**, 230–254.

Hunter, J. S., and Naylor, T. H. (1970) Experimental designs for computer simulation experiments. *Manag. Sci.*, **16**, 422–434.

Hutchinson, T. P. (1979) The validity of the chi-square test when expected frequences are small: a list of recent research references. *Communications in Statistics*, Part A: *Theory and Methods*, **A8**, 327–335.

Hwang, F. K., and Lin, S. (1971) On generating a random sequence. *J. Appl. Prob.*, **8**, 366–373.

Inoue, H., Kumahora, H., Yoshizawa, Y., Ichimura, M. and Miyatake, O.

(1983) Random numbers generated by a physical device. *Applied Statistics*, **32** (2), 115–120.

Isida, M. (1982) Statistics and micro-computer. *Compstat 1982*, Part II, 141–142. Physica–Verlag, Vienna.

Jackson, R. R. P., Welch, J. D., and Fry, J. (1964) Appointment systems in hospitals and general practice. *Oper. Res. Quart.*, **15**, 219–237.

Jeffers, J. N. R. (1967) Two case studies in the application of principal component analysis. *Appl. Stats.*, **16**, 225–236.

Jeffers, J. N. R. (Ed.) (1972) *Mathematical Models in Ecology*. Blackwell Scientific Publications, Oxford.

Jenkinson, G. (1973) Comparing single-link dendrograms. Unpublished M.Sc. thesis, University of Kent, Canterbury.

Jöhnk, M. D. (1964) Erzeugung von Betarerteilten und Gammaverteilten Zuffallszahlen. *Metrika*, **8**, 5–15.

Johnson, N. L., and Kotz, S. (1969) *Distributions in Statistics: Discrete Distributions*. Houghton Mifflin, Boston.

Johnson, N. L., and Kotz, S. (1970a) *Continuous Univariate Distributions* 1. Houghton Mifflin, Boston.

Johnson, N. L., and Kotz, S. (1970b) *Continuous Univariate Distributions* 2. Houghton Mifflin, Boston.

Johnson, N. L., and Kotz, S. (1972) *Distributions in Statistics: Continuous Multivariate Distributions*. Wiley, New York.

Johnson, S. C. (1967) Hierarchical clustering schemes. *Psychometrika*, **32**, 241–254.

Johnston, W. (1971) The case history of a simulation study. *Appl. Stats.*, **20** (3), 308–312.

Jolliffe, I. T. (1972a) Discarding variables in a principal component analysis, I. Artificial data. *Appl. Stats.*, **21** (2), 160–173.

Jolliffe, I. T. (1972b) Discarding variables in a principal component analysis, II: Real data. *Appl. Stats.*, **22** (1), 21–31.

Jolliffe, I. T., Jones, B., and Morgan, B. J. T. (1982) Utilising clusters: a case-study involving the elderly. *J. Roy. Statist. Soc.*, A, **145** (2), 224–236.

Jolly, G. M. (1965) Explicit estimates from capture–recapture data with both death and immigration—stochastic model. *Biometrika*, **52**, 225–247.

Jones, G. T. (1972) *Simulation and Business Decisions*. Penguin Books, Harmondsworth.

Joseph, A. W. (1968) A criticism of the Monte Carlo method as applied to mathematical computations. *J. Roy. Statist. Soc.*, A, **131**, 226–228.

Jowett, G. H. (1955) The comparison of means of sets of observations from sections of independent stochastic series. *J. Roy. Statist. Soc.*, B, **17**, 208–227.

Kahan, B. C. (1961) A practical demonstration of a needle experiment designed to give a number of concurrent estimates of π. *J. Roy. Statist. Soc.*, A, **124**, 227–239.

Kahn, H. (1956) *Application of Monte Carlo*. Rand Corp., Santa Monica, California.

Kelker, D. (1973) A random walk epidemic simulation. *J. Amer. Statist. Assoc.*, **68**, 821–823.

Kelly, F. P. (1979) *Reversibility and Stochastic Networks*. Wiley, Chichester.

Kemp, A. W. (1981) Efficient generation of logarithmically distributed pseudo-random variables. *Appl. Stats.*, **30** (3), 249–253.

Kemp, C. D. (1982) Low-storage Poisson generators for microcomputers. *Compstat Proceedings*, Part II, 145–146. Physica–Verlag, Vienna.

Kemp, C. D., and Loukas, S. (1978a) The computer generation of bivariate discrete random variables. *J. Roy. Statist. Soc.*, A, **141** (4), 513–517.

Kemp, C. D., and Loukas, S. (1978b) Computer generation of bivariate discrete random variables using ordered probabilities. *Proc. Stat. Comp. Sec. Am. Stat. Assoc.*, San Diego Meeting, 115–116.

Kempton, R. A. (1975) A generalized form of Fisher's logarithmic series. *Biometrika*, **62** (1), 29–38.

Kendall, D. G. (1949) Stochastic processes and population growth. *J. Roy. Statist. Soc.*, **11** (2), 230–264.

Kendall, D. G. (1950) An artificial realisation of a simple birth and death process. *J. Roy. Statist. Soc.*, B, **12**, 116–119.

Kendall, D. G. (1965) Mathematical models of the spread of infection. In: *Mathematics and Computer Science in Biology and Medicine*. HMSO, Leeds.

Kendall, D. G. (1974) Pole-seeking Brownian motion and bird navigation. *J. Roy. Statist. Soc.*, B, **36** (3), 365–417.

Kendall, M. G., and Babbington–Smith, B. (1938) Randomness and random sampling numbers. *J. Roy. Statist. Soc.*, **101**, 147–166.

Kendall, M. G., and Babbington–Smith, B. (1939a) Tables of random sampling numbers. *Tracts for Computers*, XXIV. C.U.P.

Kendall, M. G., and Babbington-Smith, B. (1939b) Second paper on random sampling numbers. *J. Roy. Statist. Soc., Suppl.*, **6**, 51–61.

Kendall, M. G., and Moran, P. A. P. (1963) *Geometrical Probability*. Griffin & Co. Ltd., London.

Kendall, M. G., and Stuart, A. (1961) *The Advanced Theory of Statistics*, Vol. 2. Hafner Pub. Co., New York.

Kendall, M. G. and Stuart, A. (1969) *The Advanced Theory of Statistics*, Volume 1. Hafner Pub. Co., New York.

Kennedy, W. J. and Gentle, J. E. (1980) *Statistical Computing*. Marcel Dekker, New York.

Kenward, A. J. (1982) Computer simulation of two queueing systems. Unpublished undergraduate project, University of Kent at Canterbury.

Kermack, W. O., and McKendrick, A. G. (1937) Tests for randomness in a series of numerical observations. *Proc. Roy. Soc. of Edinburgh*.

Kerrich, J. E. (1946) *An Experimental Introduction to the Theory of Probability*. Einar Munksgaard, Copenhagen.

Kesting, K. W., and Mann, N. R. (1977) A simple scheme for generating multivariate gamma distributions with non-negative covariance matrix. *Technometrics*, **19**, 179–184.

Kimball, B. F. (1960) On the choice of plotting positions on probability paper. *J. Amer. Statist. Assoc.*, **55**, 546–560.

Kinderman, A. J., and Monahan, J. F. (1977) Computer generation of random variables using the ratio of normal deviates. *Assoc. Comput. Mach. Trans. Math. Soft.*, **3**, 257–260.

Kinderman, A. J., and Monahan, J. F. (1980) New methods for generating Student's *t* and gamma variables. *Computing*, **25**, 369–377.

Kinderman, A. J., Monahan, J. F., and Ramage, J. G. (1977) Computer methods for sampling from Student's *t* distribution. *Mathematics of Computation*, **31**, 1009–1017.

Kinderman, A. J., and Ramage, J. G. (1976) Computer generation of normal random variables. *J. Amer. Statist. Assoc.*, **71**, 356, 893–896.

Klahr, D. (1969) A Monte Carlo investigation of the statistical significance of Kruskal's nonmetric scaling procedure. *Psychometrika*, **34** (3), 319–330.

Kleijnen, J. P. C. (1973) Monte Carlo simulation and its statistical design and analysis. *Proceedings of the 29th I.S.I. Conference, Vienna*, pp. 268–279.

Kleijnen, J. P. C. (1974/5) *Statistical Techniques in Simulation* (Parts 1 and 2). Marcel Dekker, Inc., New York.

Kleijnen, J. P. C. (1977) Design and analysis of simulations: Practical Statistical Techniques. *Simulation*, **28**, 81–90.

Knuth, D. E. (1968) *The Art of Computer Programming*, Vol. 1: *Fundamental Algorithms*. Addison-Wesley, Reading, Massachusetts.

Knuth, D. E. (1981) *The Art of Computer Programming*, Vol. 2: *Seminumerical Algorithms*. Addison-Wesley, Reading, Massachusetts.

Kozlov, G. A. (1972) Estimation of the error of the method of statistical tests (Monte-Carlo) due to imperfections in the distribution of random numbers. *Theory of Probability and its Applications*, **17**, 493–509.

Kral, J. (1972) A new additive pseudorandom number generator for extremely short word-length. *Information Processing Letters* **1**, 164–167.

Kronmal, R. (1964) The evaluation of a pseudorandom normal number generator. *J. Assoc. Comp. Mach.*, **11**, 357–363.

Kronmal, R. A., and Peterson, A. V. Jr. (1979) On the alias method for generating random variables from a discrete distribution. *Amer. Statist.*, **33** (4), 214–218.

Kronmal, R. A., and Peterson, A. V. Jr. (1981) A variant of the acceptance–rejection method for computer generation of random variables. *J. Amer. Statist. Assoc.*, **76**, 446–451.

Kruskal, J. B. (1964) Nonmetric multidimensional scaling: a numerical method. *Psychometrika*, **29**, 115–129.

Kruskal, J. B., and Wish, M. (1978) Multidimensional scaling. Vol. 11 of: *Quantitative Applications in the Social Sciences*. Sage, Beverley Hills, California.

Krzanowski, W. J. (1978) Between-group comparison of principal components – some sampling results. *J. Stat. Comput. Simul.*, **15**, 141–154.

Kuehn, H. G. (1961) A 48-bit pseudo-random number generator. *Comm. Ass. Comp. Mach.*, **4**, 350–352.

Kuiker, F. K., and Fisher, L. (1975) A Monte-Carlo comparison of 6 clustering procedures. *Biometrics*, **31**, 777–784.

Lachenbruch, P. A. (1975) *Discriminant Analysis*. Hafner, New York.

Lachenbruch, P. A., and Goldstein, M. (1979) Discriminant analysis. *Biometrics*, **35** (1), 69–86.

Lack, D. (1965) *The Life of the Robin* (4th edn). M.F. & G. Witherby.

Laplace, P. S. (1774) Determiner le milieu que l'on doit prendre entre trois observations données d'un même phénomené. *Mémoires de Mathématique et Physique présentées a l'Académie Royale des Sciences par divers Savans*, **6**, 621–625.

Larntz, K. (1978) Small sample comparisons of exact levels for chi-square goodness-of-fit statistics. *J. Amer. Statist. Assoc.*, **73**, 253–263.

Lavenberg, S. S., and Welch, P. D. (1979) Using conditional expectation to reduce variance in discrete event simulation. *Proc. 1979 Winter Simulation Conference, San Diego, California*, 291–294.

Law, A. M., and Kelton, W. D. (1982) *Simulation Modelling and Analysis*. McGraw-Hill series in industrial engineering and management science, New York.

Learmonth, G. P., and Lewis, P. A. W. (1973) Naval Postgraduate School Random Number Generator Package LLRANDOM, NPS55LW73061A. Naval Postgraduate School, Monterey, California.

Lehman, R. S. (1977) *Computer Simulation and Modelling*. Wiley, Chichester.

Lehmer, D. H. (1951) Mathematical methods in large-scale computing units. *Ann. Comp. Lab. Harvard University*, **26**, 141–146.

Lenden–Hitchcock, K. J. (1980) Aspects of random number generation with particular interest in the normal distribution. Unpublished M.Sc. dissertation, University of Kent.

Leslie, P. H. (1958) A stochastic model for studying the properties of certain biological systems by numerical methods. *Biometrika*, **45**, 16–31.

Leslie, P. H., and Chitty, D. (1951) The estimation of population parameters from data obtained by means of the capture–recapture method, I. The maximum likelihood equations for estimating the death rate. *Biometrika*, **38**, 269–292.

Levene, H., and Wolfowitz, J. (1944) The covariance matrix of runs up and down. *Ann. Math. Statist.*, **15**, 58–69.

Lew, R. A. (1981) An approximation to the cumulative normal distribution with simple coefficients. *Appl. Stats.*, **30** (3), 299–300.

Lewis, J. G., and Payne, W. H. (1973) Generalized feedback shift register pseudorandom number algorithm. *J. Assoc. Comp. Mach.*, **20**, 456–468.

Lewis, P. A. W., Goodman, A. S., and Miller, J. M. (1969) A pseudo-random number generator for the System/360. *IBM Systems J.*, **8**, 136–145.

Lewis, P. A. W., and Shedler, G. S. (1976) Simulation of nonhomogeneous Poisson process with log linear rate function. *Biometrika*, **63**, 501–506.

Lewis, T. (1975) A model for the parasitic disease bilharziasis. *Advances in Applied Probability*, **7**, 673–704.

Lewis, T. G. (1975) *Distribution Sampling for Computer Simulation*. Lexington Books (P. C. Heath).

Lewis, T. G., and Payne, W. H. (1973) Generalized feedback shift register pseudo random number algorithm. *J. Assoc. Comp. Mach.*, **20** (3), 456–468.

Lewis, T. G., and Smith, B. J. (1979) *Computer Principles of Modelling and Simulation*. Houghton Mifflin, Boston.

Lieberman, G. J., and Owen, D. B. (1961) *Tables of the hypergeometric probability distribution*. Stanford University Press, Stanford, California.

Little, R. J. A. (1979) Maximum likelihood inference for multiple regression with missing values—a simulation study. *J. Roy. Statist. Soc.*, B, **41** (1), 76–87.

Lotka, A. J. (1931) The extinction of families. *J. Wash. Acad. Sci.*, **21**, 377–380 and 453–459.

Loukas, S. and Kemp, C. D. (1983) On computer sampling from trivariate and multivariate discrete distributions. *J. Stat. Comput. Simul.*, **17**, 113–123.

Luck, G. M., Luckman, J., Smith, B. W., and Stringer, J. (1971) *Patients, Hospitals and Operational Research*. Tavistock, London.

McArthur, N., Saunders, I. W., and Tweedie, R. L. (1976) Small population isolates: a micro-simulation study. *J. Polynesian Soc.*, **85**, 307–326.

Macdonell, W. R. (1902) On criminal anthropometry and the identification of criminals. *Biometrika*, **1**, 177–227.

MacLaren, M. D., and Marsaglia, G. (1965) Uniform random number generators. *J. Assoc. Comp. Mach.*, **12**, 83–89.

McLeod, A. I. and Bellhouse, D. R. (1983) A convenient algorithm for drawing a simple random sample. *Applied Statistics*, **32** (2), 182–184.

Maher, M. J., and Akçelik, R. (1977) Route control—simulation experiments. *Transportation Research*, **11**, 25–31.

Manly, B. F. J. (1971) A simulation study of Jolly's method for analysing capture–recapture data. *Biometrics*, **27**, 415–424.

Mann, H. B., and Wald, A. (1942) On the choice of the number of class intervals in the application of the chi-squared test. *Ann. Math. Stats.*, **13**, 306–317.

Mantel, N. (1953) An extension of the Buffon needle problem. *Ann. Math. Stats.*, **24**, 674–677.

Mardia, K. V. (1970) *Families of Bivariate Distributions.* Griffin, London.

Mardia, K. V., Kent, J. T., and Bibby, J. M. (1979) *Multivariate Analysis.* Academic Press, London.

Mardia, K. V. (1980) Tests of univariate and multivariate normality. pp. 279–320. In: P. R. Krishnaiah (Ed.) *Handbook of Statistics*, Vol. 1. North Holland, Amsterdam.

Mardia, K. V., and Zemroch, P. J. (1978) *Tables of the F- and Related Distributions with Algorithms.* Academic Press, London.

Marks, S., and Dunn, O. J. (1974) Discriminant functions when covariance matrices are unequal. *J. Amer. Statist. Assoc.*, **69**, 555–559.

Marriott, F. H. C. (1972) Buffon's problems for non-random distributions. *Biometrics*, **28**, 621–624.

Marriott, F. H. C. (1979) Barnard's Monte Carlo tests: How many simulations? *Appl. Stats.*, **28** (1), 75–77.

Marsaglia, G. (1961a) Generating exponential random variables. *Ann. Math. Stats.*, **32**, 899–900.

Marsaglia, G. (1961b) Expressing a random variable in terms of uniform random variables. *Ann. Math. Stats.*, **32**, 894–900.

Marsaglia, G. (1964) Generating a variable from the tail of the normal distribution. *Technometrics*, **6**, 101–102.

Marsaglia, G. (1968) Random numbers fall mainly in the planes. *Proc. Nat. Acad. Sci., USA*, **61**, 25–28.

Marsaglia, G. (1972a) Choosing a point from the surface of a sphere. *Ann. Math. Stats.*, **3** (2), 645–646.

Marsaglia, G. (1972b) The structure of linear congruential sequences. In: *Applications of Number Theory to Numerical Analysis* (Ed. S. K. Zaremba), pp. 249–285. Academic Press, London.

Marsaglia, G., and Bray, T. A. (1964) A convenient method for generating normal variables. *SIAM Rev.*, **6**, 260–264.

Marsaglia, G., MacLaren, M. D., and Bray, T. A. (1964) A fast procedure for generating normal random variables. *Communications of the Ass. Comp. Mach.*, **7** (1), 4–10.

Mawson, J. C. (1968) A Monte Carlo study of distance measures in sampling for spatial distribution in forest stands. *Forestry Science*, **14**, 127–139.

Mead, R., and Freeman, K. H. (1973) An experiment game. *Appl. Stats.*, **22**, 1–6.

Mead, R., and Stern, R. D. (1973) The use of a computer in the teaching of statistics. *J. Roy. Statist. Soc.*, A, **136** (2), 191–225.

Metropolis, N. C., Reitwiesner, G., and von Neumann, J. (1950) Statistical treatment of values of first 2000 decimal digits of e and π calculated on the ENIAC. *Math. Tables & Other Aids to Comp.*, **4**, 109–111.

Meynell, G. G. (1959) Use of superinfecting phage for estimating the division rate of lysogenic bacteria in infected animals. *J. Gen. Microbiol.*, **21**, 421–437.

Michael, J. R., Schucany, W. R., and Haas, R. W. (1976) Generating random variates using transformations with multiple roots. *Amer. Statist.*, **30** (2), 88–90.

Mikes, G. (1946) *How to be an Alien*. Andre Deutch, Tonbridge.

Miller, A. J. (1977a) Random number generation on the SR52. CSIRO, *DMS Newsletter*, No. 28, p. 4.

Miller, A. J. (1977b) On some unrandom numbers. CSIRO, *DMS Newsletter*, No. 29, p. 2.

Miller, A. J. (1980a) Another random number generator. CSIRO, *DMS Newsletter*, No. 65, p. 5.

Miller, A. J. (1980b) On random numbers. CSIRO, *DMS Newsletter*, No. 68, p. 7.

Miller, J. C. P., and Prentice, M. J. (1968) Additive congruential pseudo-random number generators. *Comp. J.*, **11** (3), 341–346.

Mitchell, B. (1971) Variance reduction by antithetic variates in GI/G/1 queueing simulations. *Oper. Res.*, **21**, 988–997.

Mitchell, G. H. (1969) Simulation. *Bull. Inst. Math. Applics.*, **5** (3), 59–62.

Mitchell, G. H. (1972) *Operational Research: Techniques and Examples*. The English Universities Press Ltd., London.

Mitchell, K. J. (1975) Dynamics and simulated yield of Douglas fir. *Forest. Science*, Monograph, 17.

Moder, J. J., and Elmaghraby, S. E. (1978) *Handbook of Operations Research: Foundations and Fundamentals*. Van Nostrand Reinhold, New York.

Mojena, R. (1977) Hierarchical grouping methods and stopping rules: an evaluation. *Computer J.*, **20**, 359–363.

Moore, P. G. (1953) A sequential test for randomness. *Biometrika*, **40**, 111–115.

Moors, J. A. A. (1971) Optimization of the unrelated question randomized response model. *J. Amer. Statist. Assoc.*, **66**, 627–629.

Moran, P. A. P. (1975) The estimation of standard errors in Monte Carlo simulation experiments. *Biometrika*, **62**, 1–4.

Moran, P. A. P., and Fazekas de St Groth, S. (1962) Random circles on a sphere. *Biometrika*, **49**, 384–396.

Morgan, B. J. T. (1974) On the distribution of inanimate marks over a linear birth-and-death process. *J. Appl. Prob.*, **11**, 423–436.

Morgan, B. J. T. (1976) Markov properties of sequences of behaviours. *Appl. Stats.*, **25** (1), 31–36.

Morgan, B. J. T. (1978) Some recent applications of the linear birth-and-death process in biology. *Math. Scientist*, **3**, 103–116.

Morgan, B. J. T. (1979a) A simulation model of the social life of the African Mountain Gorilla. Unpublished manuscript.

Morgan, B. J. T. (1979b) Four approaches to solving the linear birth-and-death (and similar) processes. *Int. J. Math. Educ. Sci. Technol.*, **10** (1), 51–64.

Morgan, B. J. T. (1981) Three applications of methods of cluster-analysis. *The Statistician*, **30** (3), 205–223.

Morgan, B. J. T. (1983) Illustration of three-dimensional surfaces. *BIAS*, **10**, 2.

Morgan, B. J. T., Chambers, S. M., and Morton, J. (1973) Acoustic confusion of digits in memory and recognition. *Perception and Psychophysics*, **14** (2), 375–383.

Morgan, B. J. T., and Leventhal, B. (1977) A model for blue-green algae and gorillas. *J. Appl. Prob.*, **14**, 675–688.

Morgan, B. J. T., and North, P. M. (Eds.) (1984) *Statistics in Ornithology*. Springer-Verlag (to appear).

Morgan, B. J. T., and Robertson, C. (1980) Short-term memory models for choice behaviour. *J. Math. Psychol.*, **21** (1), 30–52.

Morgan, B. J. T. and Watts, S. A. (1980) On modelling microbial infections. *Biometrics*, **36**, 317–321.

Morgan, B. J. T., Woodhead, M. M., and Webster, J. C. (1976) On the recovery of physical dimensions of stimuli, using multidimensional scaling. *J. Acoust. Soc. Am.*, **60** (1), 186–189.

Morgan, R., and Hirsch, W. (1976) Stretch a point and clear the way. *Times Higher Education Supplement*, 23 July.

Morrison, D. F. (1976) *Multivariate Statistical Methods* (2nd edn). McGraw-Hill Kogakusha Ltd., Tokyo.

Morton, K. W. (1957) A generalisation of the antithetic variate technique for evaluating integrals. *J. Math. Phys.*, **36** (3), 289–293.

Moses, L. E. and Oakford, R. F. (1963) *Tables of Random Permutations*. Allen & Unwin, London.

Mosteller, F. (1965) *Fifty Challenging Problems in Probability with Solutions*. Addison-Wesley Publishing Co. Inc., Reading, Massachusetts.

Mountford, M. D. (1982) Estimation of population fluctuations with application to the Common Bird Census. *Appl. Stats.*, **31**, 135–143.

Mudholkar, G. S., and George, E. O. (1978) A remark on the shape of the logistic distribution. *Biometrika*, **65** (3), 667–668.

Murdoch, J., and Barnes, J. A. (1974) *Statistical Tables for Science, Engineering, Management and Business Studies* (2nd edn). Macmillan Press Ltd., London.

Myers, R. H. (1971) *Response Surface Methodology*. Allyn and Bacon, Boston.

Nance, R. E. (1971) On time flow mechanisms for discrete system simulation. *Management Sciences*, **18** (1), 59–73.

Nance, R. E., and Overstreet, C. J., Jr. (1972) A bibliography on random number generation. *Comp. Rev.*, **13**, 495–508.

Nance, R. E., and Overstreet, C., Jr., (1975) Implementation of Fortran random number generators on computers with One's Complement arithmetic. *J. Statist. Comput. Simul.*, **4**, 235–243.

Nance, R. E., and Overstreet, C., Jr. (1978) Some experimental observations on the behaviour of composite random number generators. *Oper. Res.*, **26**, 915–935.

Naylor, T. H. (1971) *Computer Simulation Experiments with Models of Economic Systems*. Wiley, New York.

Naylor, T. H., Burdick, D. S., and Sasser, W. E. (1967) Computer simulation experiments with economic systems—the problem of experimental design. *J. Amer. Statist. Assoc.*, **62**, 1315–1337.

Naylor, T. H. and Finger, J. M. (1967) Verification of computer simulation models. *Management Sciences*, **14**, 92–101.

Naylor, T. H., Wallace, W. H., and Sasser, W. E. (1967) A computer simulation model of the textile industry. *J. Amer. Statist. Assoc.*, **62**, 1338–1364.

Neave, H. (1972) Random number package. Computer applications in the natural and social sciences, No. 14. Department of Geography, University of Nottingham.

Neave, H. R. (1973) On using the Box–Müller transformation with multiplicative congruential pseudo-random number generators. *Appl. Stats.*, **22**, 92–97.

Neave, H. R. (1978) *Statistics Tables*. George Allen & Unwin, London.

Neave, H. R. (1981) *Elementary Statistics Tables*. Alden Press, Oxford.

Nelsen, R. B., and Williams, T. (1968) Randomly delayed appointment streams. *Nature*, **219**, 573–574.

Newman, T. G., and Odell, P. L. (1971) *The Generation of Random Variates*. Griffin, London.

Niederreiter, H. (1978) Quasi-Monte Carlo methods and pseudo-random numbers. *Bull. Amer. Math. Soc.*, **84**, 957–1041.

Norman, J. E., and Cannon, L. E. (1972) A computer programme for the generation of random variables from any discrete distribution. *J. Statist. Comput. Simul.*, **1**, 331–348.

Oakenfull, E. (1979) Uniform random number generators and the spectral test. In: *Interactive Statistics*. Ed. D. McNeil, North-Holland, pp. 17–37.

Odeh, R. E., and Evans, J. O. (1974) Algorithm AS70. The percentage points of the normal distribution. *Appl. Stats.*, **23**, 96–97.

Odeh, R. E., Owen, D. B., Birnbaum, Z. W., and Fisher, L. (1977) *Pocket Book of Statistical tables*. Marcel Dekker, New York & Basel.

Odell, P. L., and Feireson, A. H. (1966) A numerical procedure to generate a sample covariance matrix. *J. Amer. Statist. Assoc.*, **61**, 199–203.

O'Donovan, T. M. (1979) *GPSS Simulation Made Simple*. Wiley, Chichester.

Odoroff, C. L. (1970) A comparison of minimum logit chi-square estimation and maximum likelihood estimation in $2 \times 2 \times 2$ and $3 \times 2 \times 2$ contingency tables: tests for interaction. *J. Amer. Statist. Assoc.*, **65**, 1617–1631.

Ord, J. K. (1972) *Families of Frequency Distributions*. Griffin, London.

Page, E. S. (1959) Pseudo-random elements for computers. *Appl. Stats.*, **8**, 124–131.

Page, E. S. (1965) On Monte Carlo methods in congestion problems: II Simulation of queueing systems. *Oper. Res.*, **13**, 300–305.

Page, E. S. (1967) A note on generating random permutations. *Appl. Stats.*, **16**, 273–274.

Page, E. S. (1977) Approximations to the cumulative normal function and its inverse for use on a pocket calculator. *Appl. Stats.*, **26**, 75–76.

Pangratz, H., and Weinrichter, H. (1979) Pseudo-random number generator based on binary and quinary maximal-length sequences. *IEEE Transactions on Computers*.

Parker, R. A. (1968) Simulation of an aquatic ecosystem. *Biometrics*, **24**, 803–821.

Parzen, E. (1960) *Modern Probability Theory and its Applications*. Wiley, Tokyo.

Patefield, W. M. (1981) An efficient method of generating random $r \times c$ tables with given row and column totals. Algorithm AS159. *Appl. Stats.*, **30** (1), 91–97.

Payne, J. A. (1982) *Introduction to Simulation: Programming Techniques and Methods of Analysis*. McGraw-Hill, New York.

Peach, P. (1961) Bias in pseudo-random numbers. *J. Amer. Statist. Assoc.*, **56**, 610–618.

Pearson, E. S., D'Agostino, R. B., and Bowman, K. O. (1977) Tests for departure from normality: comparison of powers. *Biometrika*, **64** (2), 231–246.

Pearson, E. S. and Hartley, H. O. (1970) *Biometrika Tables for Statisticians*, I. C.U.P.

Pearson, E. S., and Hartley, H. O. (1972) *Biometrika Tables for Statisticians*, II. C.U.P.

Pearson, E. S., and Wishart, J. (Eds.) (1958) '*Student's*' Collected Papers. C.U.P. (2nd reprinting; 1st issued 1942).

Pennycuick, L. (1969) A computer model of the Oxford Great Tit population. *J. Theor. Biol.*, **22**, 381–400.

Pennycuick, C. J., Compton, R. M., and Beckingham, L. (1968) A computer model for simulating the growth of a population, or of two interacting populations. *J. Theor. Biol.* **18**, 316–329.

Perlman, M. D. and Wichura, M. J. (1975) Sharpening Buffon's needle. *Amer. Statist.*, **29**, 157–163.

Peskun, P. H. (1973) Optimal Monte-Carlo sampling using Markov chains. *Biometrika*, **60**, 607–612.

Peskun, P. H. (1980) Theoretical tests for choosing the parameters of the general mixed linear congruential pseudorandom number generator. *J. Stat. Comp. Simul.*, **11**, 281–305.

Petersen, G. G. J. (1896) The yearly immigration of young plaice into the Liemfjord from the German sea, etc. *Rept. Danish. Biol. Sta. for 1895*, **6**, 1–48.

Peterson, A. V., Jr., and Kronmal, R. A. (1980) A representation for discrete distributions by equiprobable mixtures. *J. Appl. Prob.*, **17**, 102–111.

Peterson, A. V., Jr., and Kronmal, R. A. (1982) On mixture methods for the computer generation of random variables. *Amer. Statist.*, **36** (3), 1, 184–191.

Pike, M. C., and Hill, I. D. (1965) Algorithm 266. Pseudo-random numbers (G5). *Comm. A.C.M.*, **8** (10), 605–606.

Plackett, R. L. (1965) A class of bivariate distributions. *J. Amer. Statist. Assoc.*, **60**, 516–522.

Poynter, D. J. (1979) The techniques of randomized response. Unpublished undergraduate dissertation: University of Kent at Canterbury.

Press, S. J. (1972) *Applied Multivariate Analysis*. Holt, Rinehart and Winston, Inc., New York.

Rabinowitz, M., and Berenson, M. L. (1974) A comparison of various methods of obtaining random order statistics for Monte Carlo computation. *Amer. Statist.*, **28** (1), 27–29.

Ralph, C. J. and Scott, J. M. (1981) *Estimating Numbers of Terrestrial Birds*. Allen Press Inc., Kansas.

Ramaley, J. F. (1969) Buffon's needle problem. *American Mathematical Monthly*, **76**, 916–918.

Ramberg, J. S., and Schmeiser, B. W. (1972) An approximate method for generating symmetric random variables. *Comm. Assoc. Comput. Mach.*, **15**, 987–990.

Ramberg, J. S., and Schmeiser, B. W. (1974) An approximate method for generating asymmetric random variables. *Commun. Ass. Comput. Mach.*, **17**, 78–82.

Rao, C. R. (1961) Generation of random permutations of given number of elements using sampling numbers. *Sankhya*, **23**, 305–307.

Read, K. L. Q., and Ashford, J. R. (1968) A system of models for the life cycle of a biological organism. *Biometrika*, **55**, 211–221.

Reid, N. (1981) Estimating the median survival time. *Biometrika*, **68** (3), 601–608.

Relles, D. A. (1970) Variance reduction techniques for Monte Carlo sampling from Student distributions. *Technometrics*, **12**, 499–515.

Relles, D. A. (1972) A simple algorithm for generating binomial random variables when N is large. *J. Amer. Statist. Assoc.*, **67**, 612–613.

Ripley, B. D. (1977) Modelling spatial patterns. *J. Roy. Statist. Soc.*, B, **39** (2), 172–212.

Ripley, B. D. (1979) Simulating spatial patterns: Dependent samples from a multivariate density. *Appl. Stats.*, **28** (1), 109–112.

Ripley, B. D. (1981) *Spatial Statistics*. Wiley, New York.

Ripley, B. D. (1983a) On lattices of pseudo-random numbers. *J. Stat. Comp. Siml.* (Submitted for publication).

Ripley, B. D. (1983b) Computer generation of random variables—a tutorial. *Int. Stat. Rev.*, **51**, 301–319.

Roberts, F. D. K. (1967) A Monte Carlo solution of a two-dimensional unstructured cluster problem. *Biometrika*, **54**, 625–628.

Roberts, C. S. (1982) Implementing and testing new versions of a good, 48-bit, pseudo-random number generator. *The Bell System Technical J.*, **61** (8), 2053–2063.

Ronning, G. (1977) A simple scheme for generating multivariate gamma distributions with non-negative covariance matrix. *Technometrics*, **19**, 179–183.

Rosenhead, J. V. (1968) Experimental simulation of a social system. *Ope. Res. Quart.*, **19**, 289–298.

Rotenberg, A. (1960) A new pseudo-random number generator. *J. Ass. Comp. Mach.*, **7**, 75–77.

Rothery, P. (1982) The use of control variates in Monte Carlo estimation of power. *Appl. Stats.*, **31** (2), 125–129.

Royston, J. P. (1982a) An extension of Shapiro and Wilk's W test for normality to large samples. *Appl. Stats.*, **31** (2), 115–124.

Royston, J. P. (1982b) Algorithm AS177. Expected normal order statistics (exact and approximate). *Appl. Stats.*, **31** (2), 161–165.

Royston, J. P. (1982c) Algorithm AS181: The W test for normality. *Appl. Stats.*, **31** (2), 176–180.

Royston, J. P. (1983) Some techniques for assessing multivariate normality based on the Shapiro-Wilk W. *Applied Statistics*, **32**, 2, 121–133.

Rubinstein, R. Y. (1981) *Simulation and the Monte Carlo Method*. Wiley, New York.

Ryan, T. A., Jr., Joiner, B. L., and Ryan, B. F. (1976) *MINITAB Student Handbook*. Duxbury Press, North Scituate, Massachusetts.

Sahai, H. (1979) A supplement to Sowey's bibliography on random number generation and related topics. *J. Stat. Comp. Siml.*, **10**, 31–52.

Saunders, I. W., and Tweedie, R. L. (1976) The settlement of Polynesia by CYBER 76. *Math. Scientist*, **1**, 15–25.

Scheuer, E. M., and Stoller, D. S. (1962) On the generation of normal random vectors. *Technometrics*, **4**, 278–281.

Schmeiser, B. W. (1979) Approximations to the inverse cumulative normal function for use on hand calculators. *Appl. Stats.*, **28** (2), 175–176.

Schmeiser, B. W. (1980) Generation of variates from distribution tails. *Oper. Res.* **28**, 1012–1017.

Schmeiser, B. W., and Babu, A. J. G. (1980) Beta variate generation via exponential majorizing functions. *Oper. Res.*, 917–926.

Schmeiser, B. W., and Lal, R. (1980) Squeeze methods for generating Gamma variates. *J. Amer. Statist. Assoc.*, **75**, 679–682.

Schruben, L. W. (1981) Control of initialization bias in multivariate simulation response. *Comm. Assoc. Comp. Mach.*, **24**, 4, 246–252.

Schruben, L. W., and Margolin, B. H. (1978) Pseudorandom number assignment in statistically designed simulation and distribution sampling experiments. *J. Amer. Stat. Assoc.*, **73**, 504–525.

Schuh, H.-J., and Tweedie, R. L. (1979) Parameter estimation using transform estimation in time-evolving models. *Math. Biosci.*, **45**, 37–67.

Shanks, D., and Wrench, J. W. (1962) Calculation of π to 100,000 decimals. *Math. Comput.*, **16**, 76–99.

Shannon, C. E., and Weaver, W. (1964) *The Mathematical Theory of Communication*. The University of Illinois Press, Urbana.

Shapiro, S. S., and Wilk, M. B. (1965) An analysis-of-variance test for normality (complete samples). *Biometrika*, **52**, 591–611.

Shreider, Y. A. (1964) *Method of Statistical Testing: Monte Carlo Method.* Elsevier Pub. Co., Amsterdam.

Shubik, M. (1960) Bibliography on simulation, gaming, artificial intelligence and allied topics. *J. Amer. Statist. Assoc.*, **55**, 736–751.

Simon, G. (1976) Computer simulation swindles, with applications to estimates of location and dispersion. *Appl. Stats.*, **25** (3), 266–274.

Simulation I (1976) Unit 13 (Numerical Computation) of M351 (Mathematics). The Open University Press, Milton Keynes.

Simulation II (1976) Unit 14 (Numerical Computation) of M351 (Mathematics) The Open University Press, Milton Keynes.

Smith, J. (1968) *Computer Simulation Models.* Hafner, New York.

Smith, R. H., and Mead, R. (1974) Age structure and stability in models of prey-predator systems. *Theor. Pop. Biol.*, **6**, 308–322.

Smith, W. B., and Hocking, R. R. (1972) Wishart variate generator. Algorithm AS53. *Appl. Stats.*, **21**, 341–345.

Sobol, I. M., and Seckler, B. (1964) On periods of pseudo-random sequences. *Theory of Probability and its Applications*, **9**, 333–338.

Sowey, E. R. (1972) A chronological and classified bibliography on random number generation and testing. *Internat. Stat. Rev.*, **40**, 355–371.

Sowey, E. R. (1978) A second classified bibliography on random number generation and testing. *Internat. Stat. Rev.*, **46** (1), 89–102.

Sparks, D. N. (1973) Euclidean cluster analysis: algorithm AS58. *Appl. Stats.*, **22**, 126–130.

Sparks, D. N. (1975) A remark on algorithm AS58. *Appl. Stats.*, **24**, 160–161.

Spence, I. (1972) A Monte Carlo evaluation of 3 non-metric multidimensional scaling algorithms. *Psychometrika*, **37**, 461–486.

Stenson, H. H., and Knoll, R. L. (1969) Goodness of fit for random rankings in Kruskal's nonmetric scaling procedure. *Psychol. Bull.*, **71**, 122–126.

Stoneham, R. G. (1965) A study of 60 000 digits of the transcendental 'e'. *Amer. Math. Monthly*, **72**, 483–500.

Student (1908a) The probable error of a mean. *Biometrika* VI, p. 1.

Student (1908b) Probable error of a correlation coefficient. *Biometrika* VI, p. 302.

Student (1920) An experimental determination of the probable error of Dr. Spearman's correlation coefficients. *Biometrika* XIII, p. 263.

Sylwestrowicz, J. D. (1982) Parallel processing in statistics. *Compstat Proceedings*, Part I, 131–136. Physica-Verlag, Vienna.

Taussky, O., and Todd, J. (1956) Generation and testing of pseudo-random numbers. In: *Symposium on Monte Carlo Methods*. Ed. H. A. Meyer, pp. 15–28.

Tausworthe, R. C. (1965) Random numbers generated by linear recurrence modulo two. *Math. Comput.*, **19**, 201–209.

Taylor, S. J. L. (1954) *Good General Practice*. Oxford University Press for Nuffield Provincial Hospitals Trust, London.

Teichroew, D. (1965) A history of distribution sampling prior to the era of the computer and its relevance to simulation. *J. Amer. Statist. Assoc.*, **60**, 26–49.

Thompson, W. E. (1959) ERNIE—A mathematical and statistical analysis. *J. Roy. Statist. Soc.*, A, **122**, 301–333.

Tippett, L. H. C. (1925) On the extreme individuals and the range of samples taken from a normal population. *Biometrika*, **17**, 364–387.

Tippett, L. H. C. (1927) *Random Sampling Numbers*. Tracts for Computers, XV, C.U.P.

Titterington, D. M. (1978) Estimation of correlation coefficients by ellipsoidal trimming. *Appl. Stats.*, **27**, (3), 227–234.

Titterington, D. M., Murray, G. D., Murray, L. S., Spiegelhalter, D. J., Skene, A. M., Habbema, J. D. F., and Gelpke, G. J. (1981) Comparison of discrimination techniques applied to a complex data set of head injured patients. *J. Roy. Statist. Soc.*, A, **144**, 145–175.

Tocher, K. D. (1965) A review of simulation languages. *Op. Res. Quart.*, **16** (2), 189–217.

Tocher, K. D. (1975) *The Art of Simulation*. Hodder and Stoughton, London. (First printed 1963).

Tunnicliffe-Wilson, G. (1979) Some efficient computational methods for high order ARMA models. *J. Statist. Comput. and Simulation*, **8**, 301–309.

Uhler, H. S. (1951a) Approximations exceeding 1300 decimals for $\sqrt{3}$, $1/\sqrt{3}$ and $\sin(\pi/3)$ and distribution of digits in them. *Proc. Nat. Acad. Sci. USA*, **37**, 443–447.

Uhler, H. S. (1951b) Many-figure approximations to $\sqrt{2}$, and distribution of digits in $\sqrt{2}$ and $1/\sqrt{2}$. *Proc. Nat. Acad. Sci. USA*, **37**, 63–67.

Upton, G. J. G., and Lampitt, G. A. (1981) A model for interyear change in the size of bird populations. *Biometrics*, **37**, 113–127.

Van Gelder, A. (1967) Some new results in pseudo-random number generation. *J. Assoc. Comput. Mach.*, **14**, 785–792.

von Neumann, J. (1951) Various techniques used in connection with random digits, 'Monte Carlo Method', *US Nat. Bur. Stand. Appl. Math. Ser.*, No. 12, 36–38.

Wagenaar, W. A., and Padmos, P. (1971) Quantitative interpretation of stress in Kruskal's multidimensional scaling technique. *Br. J. Math. & Statist. Psychol.*, **24**, 101–110.

Walker, A. J. (1977) An efficient method for generating discrete random variables with general distributions. *Assoc. Comput. Mach. Trans. Math. Soft.*, **3**, 253–256.

Wall, D. D. (1960) Fibonacci series modulo m. *Amer. Math. Monthly*, **67**, 525–532.

Wallace, N. D. (1974) Computer generation of gamma random variates with non-integral shape parameters. *Commun. Ass. Comput. Mach.*, **17**, 691–695.

Warner, S. L. (1965) Randomized response: a survey technique for eliminating evasive answer bias. *J. Amer. Statist. Assoc.*, **60**, 63–69.

Warner, S. L. (1971) Linear randomized response model. *J. Amer. Statist. Assoc.*, **66**, 884–888.

Wedderburn, R. W. M. (1976) Remark ASR16. A remark on Algorithm AS29. The runs up and down test. *Appl. Stats.*, **25**, 193.

West, J. H. (1955) An analysis of 162332 lottery numbers. *J. Roy. Statist. Soc.*, **118**, 417–426.

Western, A. E. and Miller, J. C. P. (1968) *Tables of Indices and Primitive Roots.* (Royal Society Mathematical Tables, Vol. 9). Cambridge University Press.

Westlake, W. J. (1967) A uniform random number generator based on the combination of two congruential generators. *J. Ass. Comput. Mach.*, **14**, 337–340.

Wetherill, G. B. (1965) An approximation to the inverse normal function suitable for the generation of random normal deviates on electronic computers. *Appl. Stats.*, **14**, 201–205.

Wetherill, G. B. (1982) *Elementary Statistical Methods* (3rd edn). Chapman and Hall, London.

Wetherill, G. B. *et al.* (1984) *Advanced Regression Analysis.* (In preparation).

White, G. C., Anderson, D. R., Burnham, K. P., and Otis, D. L. (1982) Capture–recapture and removal methods for sampling closed populations. Available from: National Technical Information Service, U.S. Department of Commerce, 5285, Port Royal Road, Springfield, VA 22161.

White, G. C. Burnham, K. P., Otis, D. L., and Anderson, D. R. (1978) *User's Manual for Program CAPTURE.* Utah State University Press, 40 pp.

Whittaker, J. (1974) Generating gamma and beta random variables with non-integral shape parameters. *Appl. Stats.*, **23**, 210–214.

Wichern, D. W., Miller, R. B., and Der-Ann Hsu (1976) Changes of variance in first-order autoregressive time series models, with an application. *Appl. Stats.*, **25**, 3, 248–256.

Wichmann, B. A., and Hill, I. D. (1982a) A pseudo-random number generator. *NPL Report DITC 6/82.*

Wichmann, B. A., and Hill, I. D. (1982b) Algorithm AS183: An efficient and portable pseudo-random number generator. *Appl. Stats.*, **31** (2), 188–190.

Wilk, M. B., Gnanadesikan, R., and Huyett, M. J. (1962) Probability plots for the gamma distribution. *Technometrics*, **4**, 1–20.

Williamson, E., and Bretherton, M. H. (1963) *Tables of the Negative Binomial Probability Distribution.* Wiley, Birmingham, England.

Wold, H. (1954) *Random Normal Deviates.* Tracts for computers XXV. C.U.P.

Worsdale, G. J. (1975) Tables of cumulative distribution functions for symmetric stable distributions. *Appl. Stats.*, **24** (1), 123–131.

Wright, K. G. (1978) Output measurement in practice. In: *Economic Aspects of Health Services* (A. J. Culyer and K. G. Wright, Eds). Martin Robertson, pp. 46–64.

Yakowitz, S. J. (1977) *Computational Probability and Simulation.* Addison-Wesley Pub. Co., Reading, Massachusetts.

Yarnold, J. K. (1970) The minimum expectation of χ^2 goodness-of-fit tests and the accuracy of approximation for the null distribution. *J. Amer. Statist. Assoc.*, **65**, 864–886.

Yuen, K. K. (1971) A note on Winsorized *t*. *Appl. Stats.*, **20**, 297–304.

Zeigler, B. P. (1976) *Theory of Modelling and Simulation.* Wiley, New York and London.

Zeigler, B. P. (1979) A cautionary word about antithetic variates. *Simulation Newsletter*, No. 3.

Zelen, M., and Severo, N. C. (1966) Probability functions. In: *Handbook of Mathematical Functions.* (M. Abramowitz and I. A. Stegun, Eds) Department of Commerce of the U.S. Government, Washington, D. C.

AUTHOR INDEX

Abdul-Ela, A-L.A., 275
Abramowitz, M., 107
Ahrens, J. H., 86, 119, 131, 151, 238, 293
Aitchison, J., 90
Akima, H., 239
Anderson, O. D., 6
Andrews, D. F., 61, 63, 160, 216
Apostol, T. M., 46
Appleton, D. R., 119
Arabie, P., 219
Ashcroft, H., 188
Ashford, J. R., 196
Atkinson, A. C., 60, 62, 63, 86, 106, 119,
    130, 135, 138, 140, 151, 204, 237,
    248, 290

Babbington-Smith, B., 55, 137, 140, 141,
    149, 151, 154, 155
Bailey, B. J. R., 121
Bailey, N. T. J., 3, 8, 46, 180, 194, 199,
    211, 268
Baliu, A. J. R., 237
Banfield, C. F., 235
Barnard, G. A., 227
Barnes, J. A., 55
Barnett, V., 5, 24, 51, 119, 126, 139, 150,
    157, 166, 171, 179, 180, 181, 186
Barrodale, I., 8
Bartholomew, D. J., 5
Bays, C., 72
Beale, E. M. L., 220, 221, 233
Bebbington, A. C., 244
Bellhouse, D. R., 244
Besag, J., 227
Best, D. J., 120, 245
Bevan, J. M., 198, 207

Bishop, J. A., 214, 215
Blackman, R. G., 200
Blake, I. F., 46
Blanco White, M. J., 198
Box, G. E. P., 78
Bradley, J. S., 214, 215
Bray, T. A., 81, 113, 129, 132
Bremner, J. M., 68, 157
Brent, R. P., 135, 242
Bretherton, M. H., 120
Brown, J. A. C., 90
Brown, S., 7
Buckland, S. T., 215
Bunday, B. D., 188
Burnham, K. P., 216
Butcher, J. C., 130, 218, 232

Calinski, T., 220
Campbell, C., 52, 67, 274
Carlisle, P., 198
Chambers, J. M., 62, 73, 244
Chambers, S. M., 233
Chapman, D. G., 231, 302
Chay, S. C., 86, 158
Cheng, R. C. H., 86, 129, 171, 181, 198,
    237
Chernoff, H., 146
Clarke, G. M., 2, 60, 71, 235, 305
Clements, G., 214, 215
Cochran, W. G., 46
Conte, S. D., 162, 282
Conway, R. W., 200
Cooke, D., 2, 60, 71, 235, 243, 249, 305
Cooper, B. E., 157
Cormack, R. M., 214, 229
Coveyou, R. S., 60, 138

SUBJECT INDEX